Audio Power Amplifier Design Handbook

Audio Power Amplifier Design Handbook

Fourth edition

Douglas Self MA, MSc

AMSTERDAM • BOSTON • HEIDELBERG • LONDON • NEW YORK • OXFORD
PARIS • SAN DIEGO • SAN FRANCISCO • SINGAPORE • SYDNEY • TOKYO

Newnes is an imprint of Elsevier

Newnes is an imprint of Elsevier
Linacre House, Jordan Hill, Oxford OX2 8DP
30 Corporate Drive, Burlington, MA 01803

First published 2006

British Library Cataloguing in Publication Data
Self, Douglas
Audio power amplifier design handbook. — 4th ed.
1. Audio amplifiers — Design 2. Power amplifiers — Design
I. Title
621.3'81535

Library of Congress Control Number: 2006927666

ISBN-13: 978-0-7506-8072-1
ISBN-10: 0-7506-8072-5

For information on all Newnes publications
visit our web site at www.newnespress.com

Printed and bound in *Great Britain*

06 07 08 09 10 10 9 8 7 6 5 4 3 2 1

Contents

Synopsis

Chapter 1 Introduction and general survey

Chapter 2 History, architecture and negative feedback

Chapter 3 The general principles of power amplifiers

Chapter 4 The small signal stages

Chapter 5 The output stage I

Chapter 6 The output stage II

Chapter 7 Compensation, slew-rate, and stability

Chapter 8 Power supplies and PSRR

Chapter 9 Class-A power amplifiers

Chapter 10 Class-G power amplifiers

Chapter 11 Class-D amplifiers

Chapter 12 FET output stages

Chapter 13 Thermal compensation and thermal dynamics

Chapter 14 The design of DC servos

Chapter 15 Amplifier and loudspeaker protection

Chapter 16 Grounding and practical matters

Chapter 17 Testing and safety

I wish to dedicate this book to my parents Russell and Evelyn, and to all the friends and colleagues who have provided me help, information and encouragement while I was engaged in its writing. In particular I want to acknowledge the active assistance and collaboration of Gareth Connor in the quest for the perfect amplifier, and the fortitude of Peter King in enduring many rambling expositions of my thoughts on the subject.

Preface to Fourth Edition

The Audio Power Amplifier Design Handbook has now reached its Fourth edition, and it is very good to see that it fulfils a real need. It has once again been expanded with a significant amount of new material. This time two whole new chapters have been added, one on the Design of DC Servos for controlling output offsets and another on Class-D Amplifiers, which have grown markedly in importance in recent times.

There have also been major new sections added on the detailed design of DC protection circuitry, and on the Safety regulations. No DC-coupled power amplifier can be regarded as complete until it has proper DC protection – having an amplifier fail is bad enough, but having it fail and take an expensive loudspeaker with it is downright upsetting. If you are designing an amplifier for commercial production, then a knowledge of the various safety requirements is of course vital; such considerations can have a profound effect on how a piece of equipment is constructed.

I would like to thank everyone who has supported me in the production and evolution of The Handbook, not least those who have done so by buying it.

Preface

The design of power amplifiers exerts a deep fascination all of its own in both amateur and professional circles. The job they do is essentially simple, but making a reliable high-performance circuit to do it well is surprisingly difficult, and involves delving into all kinds of byways of electronics. Perhaps this paradox is at the root of the enduring interest they generate. Reliable information on power amplifier design is hard to find, but in this book, I hope to fill at least some of that need.

It is notable how few aspects of amplifier design have received serious scientific investigation. Much of this book is the result of my own research, because the information required simply was not to be found in the published literature.

In the course of my investigations, I was able to determine that power amplifier distortion, traditionally a difficult and mysterious thing to grapple with, was the hydra-headed amalgamation of seven or eight mechanisms, overlaying each other and contributing to a complex result. I have evolved ways of measuring and minimising each distortion mechanism separately, and the result is a design methodology for making Class-B or Class-A amplifiers with distortion performance so good that two or three years ago it would have been regarded as impossible. The methodology gives pleasingly reliable and repeatable results with moderate amounts of negative feedback, and insignificant added cost. It is described and explained in detail here.

This leads to the concept of what I have called a Blameless amplifier, which forms a benchmark for distortion performance that varies surprisingly little, and so forms a well-defined point of departure for more ambitious and radical amplifier designs. The first of these I have undertaken is the Trimodal amplifier (so-called because it can work in any of the modes A, AB and B, as the situation requires) which is fully described in Chapter 9.

Apart from the major issue of distortion and linearity in power amplifier design, I also cover more mundane but important matters such as reliability, power supplies, overload and DC-protection, and so on. In addition there is unique material on reactive loading, unusual forms of compensation, distortion produced by capacitors and fuses, and much more. I have provided a wide and varied selection of references, so that those interested can pursue the issues further.

Sometimes controversies arise in audio; in fact, it would be truer to say that they have become endemic, despite a lack of hard facts on which genuine differences of opinion might be based. Although audio power amplifiers are in many ways straightforward in their doings, they have not escaped the attentions of those who incline more to faith than science. In my writings, I simply go where the facts lead me, and my experiences as an amateur musician, my work designing professional mixing consoles, and my studies in psychology and psychoacoustics have led me to the firm conclusion that inexplicable influences on audio quality simply do not exist, and that any serious book on amplifier design must start from this premise.

I have done my best to make sure that everything in this book is as correct as theory, simulation, practical measurement and late-night worrying can make it. The basic arguments have been validated by the production of more than twenty thousand high-power Blameless amplifiers over the last two years, which is perhaps as solid a confirmation as any methodology can hope to receive. If some minor errors do remain, these are entirely my responsibility, and when alerted I will correct them at the first opportunity.

I hope this book may be interesting and useful to the amplifier designer and constructor, be they an amateur or a professional. However, it is my fondest wish that it may stimulate others to further explore and expand the limits of audio knowledge.

Douglas Self

1

Introduction and general survey

The economic importance of power amplifiers

Audio power amplifiers are of considerable economic importance. They are built in their hundreds of thousands every year, and have a history extending back to the 1920s. It is therefore surprising there have been so few books dealing in any depth with solid-state power amplifier design.

The first aim of this text is to fill that need, by providing a detailed guide to the many design decisions that must be taken when a power amplifier is designed.

The second aim is disseminate the results of the original work done on amplifier design in the last few years. The unexpected result of these investigations was to show that power amplifiers of extraordinarily low distortion could be designed as a matter of routine, without any unwelcome side-effects, so long as a relatively simple design methodology was followed. This methodology will be explained in detail.

Assumptions

To keep its length reasonable, a book such as this must assume a basic knowledge of audio electronics. I do not propose to plough through the definitions of frequency response, THD and signal-to-noise ratio; this can be found anywhere. Commonplace facts have been ruthlessly omitted where their absence makes room for something new or unusual, so this is not the place to start learning electronics from scratch. Mathematics has been confined to a few simple equations determining vital parameters such as open-loop gain; anything more complex is best left to a circuit simulator you trust. Your assumptions, and hence the output, may be wrong, but at least the calculations in-between will be correct . . . .

The principles of negative feedback as applied to power amplifiers are explained in detail, as there is still widespread confusion as to exactly how it works.

Origins and aims

The core of this book is based on a series of eight articles originally published in Electronics World as *'Distortion In Power Amplifiers'*. This series was primarily concerned with distortion as the most variable feature of power amplifier performance. You may have two units placed side by side, one giving 2% THD and the other 0.0005% at full power, and both claiming to provide the ultimate audio experience. The ratio between the two figures is a staggering 4000:1, and this is clearly a remarkable state of affairs. One might be forgiven for concluding that distortion was not a very important parameter. What is even more surprising to those who have not followed the evolution of audio over the last two decades is that the more distortive amplifier will almost certainly be the more expensive. I shall deal in detail with the reasons for this astonishing range of variation.

The original series was inspired by the desire to invent a new output stage that would be as linear as Class-A, without the daunting heat problems. In the course of this work it emerged that output stage distortion was completely obscured by non-linearities in the small-signal stages, and it was clear that these distortions would need to be eliminated before any progress could be made. The small-signal stages were therefore studied in isolation, using *model* amplifiers with low-power and very linear Class-A output stages, until the various overlapping distortion mechanisms had been separated out. It has to be said this was not an easy process. In each case there proved to be a simple, and sometimes well-known cure, and perhaps the most novel part of my approach is that all these mechanisms are dealt with, rather than one or two, and the final result is an amplifier with unusually low distortion, using only modest and safe amounts of global negative feedback.

Much of this book concentrates on the distortion performance of amplifiers. One reason is that this varies more than any other parameter – by up to a factor of a thousand. Amplifier distortion was until recently an enigmatic field – it was clear that there were several overlapping distortion mechanisms in the typical amplifier, but it is the work reported here that shows how to disentangle them, so they may be separately studied and then with the knowledge thus gained, minimised.

I assume here that distortion is a bad thing, and should be minimised; I make no apology for putting it as plainly as that. Alternative philosophies hold that as some forms of non-linearity are considered harmless or even euphonic, they should be encouraged, or at any rate not positively discouraged. I state plainly that I have no sympathy with the latter view; to my mind the goal is to make the audio path as transparent as possible. If some sort of distortion is considered desirable, then surely the logical way to introduce it is by an outboard processor, working at line level. This is not only more cost-effective than generating distortion with directly heated triodes, but has the important attribute that *it can be switched off*. Those who have brought into being our current signal-delivery chain, i.e., mixing consoles, multi-track recorders, CDs, have done us proud in the matter of low distortion, and to wilfully throw away this achievement at the very last stage strikes me as curious at best.

In this book I hope to provide information that is useful to all those interested in power amplifiers. Britain has a long tradition of small and very small audio companies, whose technical and production resources may not differ very greatly from those available to the committed amateur. I hope this volume will be of service to both.

I have endeavoured to address both the quest for technical perfection – which is certainly not over, as far as I am concerned – and also the commercial necessity of achieving good specifications at minimum cost.

The field of audio is full of statements that appear plausible but in fact have never been tested and often turn out to be quite untrue. For this reason, I have confined myself as closely as possible to facts that I have verified myself. This volume may therefore appear somewhat idiosyncratic in places; for example, FET output stages receive much less coverage than bipolar ones because the conclusion appears to be inescapable that FETs are both more expensive and less linear; I have therefore not pursued the FET route very far. Similarly, most of my practical design experience has been on amplifiers of less than 300 W power output, and so heavy-duty designs for large-scale PA work are also under-represented. I think this is preferable to setting down untested speculation.

The study of amplifier design

Although solid-state amplifiers have been around for some 40 years, it would be a great mistake to assume that everything possible is known about them. In the course of my investigations, I discovered several matters which, not appearing in the technical literature, appear to be novel, at least in their combined application:

- The need to precisely balance the input pair to prevent second-harmonic generation.
- The demonstration of how a beta-enhancement transistor increases the linearity and reduces the collector impedance of the Voltage-Amplifier Stage.

- An explanation of why BJT output stages always distort more into 4 Ω than 8 Ω.
- In a conventional BJT output stage, quiescent current as such is of little importance. What is crucial is the voltage between the transistor emitters.
- Power FETs, though for many years touted as superior in linearity, are actually far less linear than bipolar output devices.
- In most amplifiers, the major source of distortion is not inherent in the amplifying stages, but results from avoidable problems such as induction of supply-rail currents and poor power-supply rejection.
- Any number of oscillograms of square-waves with ringing have been published that claim to be the transient response of an amplifier into a capacitive load. In actual fact this ringing is due to the output inductor resonating with the load, and tells you precisely nothing about amplifier stability.

The above list is by no means complete.

As in any developing field, this book cannot claim to be the last word on the subject; rather it hopes to be a snapshot of the state of understanding at this time. Similarly, I certainly do not claim that this book is fully comprehensive; a work that covered every possible aspect of every conceivable power amplifier would run to thousands of pages. On many occasions I have found myself about to write: '*It would take a whole book to deal properly with* . . . ' Within a limited compass I have tried to be innovative as well as comprehensive, but in many cases the best I can do is to give a good selection of references that will enable the interested to pursue matters further. The appearance of a reference means that I consider it worth reading, and not that I think it to be correct in every respect.

Sometimes it is said that discrete power amplifier design is rather unenterprising, given the enormous outpouring of ingenuity in the design of analogue ICs. Advances in op-amp design would appear to be particularly relevant. I have therefore spent some considerable time studying this massive body of material and I have had to regretfully conclude that it is actually a very sparse source of inspiration for new audio power amplifier techniques; there are several reasons for this, and it may spare the time of others if I quickly enumerate them here:

- A large part of the existing data refers only to small-signal MOSFETs, such as those used in CMOS op-amps, and is dominated by the ways in which they differ from BJTs, for example, in their low transconductance. CMOS devices can have their characteristics customised to a certain extent by manipulating the width/length ratio of the channel.
- In general, only the earlier material refers to BJT circuitry, and then it is often mainly concerned with the difficulties of making complementary

circuitry when the only PNP transistors available are the slow lateral kind with limited beta and poor frequency response.

- Many of the CMOS op-amps studied are transconductance amplifiers, i.e., voltage-difference-in, current out. Compensation is usually based on putting a specified load capacitance across the high-impedance output. This does not appear to be a promising approach to making audio power amplifiers.
- Much of the op-amp material is concerned with the common-mode performance of the input stage. This is pretty much irrelevant to power amplifier design.
- Many circuit techniques rely heavily on the matching of device characteristics possible in IC fabrication, and there is also an emphasis on minimising chip area to reduce cost.
- A good many IC techniques are only necessary because it is (or was) difficult to make precise and linear IC resistors. Circuit design is also influenced by the need to keep compensation capacitors as small as possible, as they take up a disproportionately large amount of chip area for their function.

The material here is aimed at all audio power amplifiers that are still primarily built from discrete components, which can include anything from 10 W mid-fi systems to the most rarefied reaches of what is sometimes called the 'high end', though the 'expensive end' might be a more accurate term. There are of course a large number of IC and hybrid amplifiers, but since their design details are fixed and inaccessible they are not dealt with here. Their use is (or at any rate should be) simply a matter of following the relevant application note. The quality and reliability of IC power amps has improved noticeably over the last decade, but low distortion and high power still remain the province of discrete circuitry, and this situation seems likely to persist for the foreseeable future.

Power amplifier design has often been treated as something of a black art, with the implication that the design process is extremely complex and its outcome not very predictable. I hope to show that this need no longer be the case, and that power amplifiers are now designable – in other words it is possible to predict reasonably accurately the practical performance of a purely theoretical design. I have done a considerable amount of research work on amplifier design, much of which appears to have been done for the first time, and it is now possible for me to put forward a design methodology that allows an amplifier to be designed for a specific negative-feedback factor at a given frequency, and to a large extent allows the distortion performance to be predicted. I shall show that this methodology allows amplifiers of extremely low distortion (sub 0.001% at 1 kHz) to be designed and built as a matter of routine, using only modest amounts of global negative feedback.

Misinformation in audio

Few fields of technical endeavour are more plagued with errors, misstatements and confusion than audio. In the last 20 years, the rise of controversial and non-rational audio hypotheses, gathered under the title *Subjectivism* has deepened these difficulties. It is commonplace for hi-fi reviewers to claim that they have perceived subtle audio differences which cannot be related to electrical performance measurements. These claims include the alleged production of a 'three-dimensional sound-stage and protests that the rhythm of the music has been altered'; these statements are typically produced in isolation, with no attempt made to correlate them to objective test results. The latter in particular appears to be a quite impossible claim.

This volume does not address the implementation of Subjectivist notions, but confines itself to the measurable, the rational, and the repeatable. This is not as restrictive as it may appear; there is nothing to prevent you using the methodology presented here to design an amplifier that is technically excellent, and then gilding the lily by using whatever brands of expensive resistor or capacitor are currently fashionable, and doing the internal wiring with cable that costs more per metre than the rest of the unit put together. Such nods to Subjectivist convention are unlikely to damage the real performance; this is however not the case with some of the more damaging hypotheses, such as the claim that negative feedback is inherently harmful. Reduce the feedback factor and you will degrade the real-life operation of almost any design.

Such problems arise because audio electronics is a more technically complex subject than it at first appears. It is easy to cobble together some sort of power amplifier that works, and this can give people an altogether exaggerated view of how deeply they understand what they have created. In contrast, no-one is likely to take a 'subjective' approach to the design of an aeroplane wing or a rocket engine; the margins for error are rather smaller, and the consequences of malfunction somewhat more serious.

The Subjectivist position is of no help to anyone hoping to design a good power amplifier. However, it promises to be with us for some further time yet, and it is appropriate to review it here and show why it need not be considered at the design stage. The marketing stage is of course another matter.

Science and subjectivism

Audio engineering is in a singular position. There can be few branches of engineering science rent from top to bottom by such a basic division as the Subjectivist/rationalist dichotomy. Subjectivism is still a significant issue in the hi-fi section of the industry, but mercifully has made little headway in professional audio, where an intimate acquaintance with the original sound, and the need to earn a living with reliable and affordable equipment, provides an effective barrier against most of the irrational influences. (Note that the opposite of Subjectivist is not 'Objectivist'. This term refers to the followers of the philosophy of Ayn Rand.)

Most fields of technology have defined and accepted measures of excellence; car makers compete to improve MPH and MPG; computer manufacturers boast of MIPs (millions of instructions per second) and so on. Improvement in these real quantities is regarded as unequivocally a step forward. In the field of hi-fi, many people seem to have difficulty in deciding which direction forward is.

Working as a professional audio designer, I often encounter opinions which, while an integral part of the Subjectivist offshoot of hi-fi, are treated with ridicule by practitioners of other branches of electrical engineering. The would-be designer is not likely to be encouraged by being told that audio is not far removed from witchcraft, and that no-one truly knows what they are doing. I have been told by a Subjectivist that the operation of the human ear is so complex that its interaction with measurable parameters lies forever beyond human comprehension. I hope this is an extreme position; it was, I may add, proffered as a flat statement rather than a basis for discussion.

I have studied audio design from the viewpoints of electronic design, psychoacoustics, and my own humble efforts at musical creativity. I have found complete scepticism towards Subjectivism to be the only tenable position. Nonetheless, if hitherto unsuspected dimensions of audio quality are ever shown to exist, then I look forward keenly to exploiting them. At this point I should say that no doubt most of the esoteric opinions are held in complete sincerity.

The Subjectivist position

A short definition of the Subjectivist position on power amplifiers might read as follows:

- Objective measurements of an amplifier's performance are unimportant compared with the subjective impressions received in informal listening tests. Should the two contradict the objective results may be dismissed.
- Degradation effects exist in amplifiers that are unknown to orthodox engineering science, and are not revealed by the usual objective tests.
- Considerable latitude may be employed in suggesting hypothetical mechanisms of audio impairment, such as mysterious capacitor shortcomings and subtle cable defects, without reference to the plausibility of the concept, or the gathering of objective evidence of any kind.

I hope that this is considered a reasonable statement of the situation; meanwhile the great majority of the paying public continue to buy conventional hi-fi systems, ignoring the expensive and esoteric high-end sector where the debate is fiercest.

It may appear unlikely that a sizeable part of an industry could have set off in a direction that is quite counter to the facts; it could be objected that such a loss of direction in a scientific subject would be unprecedented. This is not so.

Parallel events that suggest themselves include the destruction of the study of genetics under Lysenko in the USSR[1]. Another possibility is the study of parapsychology, now in deep trouble because after some 100 years of investigation it has not uncovered the ghost (sorry) of a repeatable phenomenon[2]. This sounds all too familiar. It could be argued that parapsychology is a poor analogy because most people would accept that there was nothing there to study in the first place, whereas nobody would assert that objective measurements and subjective sound quality have no correlation at all; one need only pick up the telephone to remind oneself what a 4 kHz bandwidth and 10% or so THD sounds like.

The most starting parallel I have found in the history of science is the almost-forgotten affair of Blondlot and the N-rays[3]. In 1903, Rene Blondlot, a respected French physicist, claimed to have discovered a new form of radiation he called 'N-rays'. (This was shortly after the discovery of X-rays by Roentgen, so rays were in the air, as it were.) This invisible radiation was apparently mysteriously refracted by aluminium prisms; but the crucial factor was that its presence could only be shown by subjective assessment of the brightness of an electric arc allegedly affected by N-rays. No objective measurement appeared to be possible. To Blondlot, and at least fourteen of his professional colleagues, the subtle changes in brightness were real, and the French Academy published more than a hundred papers on the subject.

Unfortunately N-rays were completely imaginary, a product of the 'experimenter-expectancy' effect. This was demonstrated by American scientist Robert Wood, who quietly pocketed the aluminium prism during a demonstration, without affecting Bondlot's recital of the results. After this the N-ray industry collapsed very quickly, and while it was a major embarrassment at the time, it is now almost forgotten.

The conclusion is inescapable that it is quite possible for large numbers of sincere people to deceive themselves when dealing with subjective assessments of phenomena.

A short history of subjectivism

The early history of sound reproduction is notable for the number of times that observers reported that an acoustic gramophone gave results indistinguishable from reality. The mere existence of such statements throws light on how powerfully mind-set affects subjective impressions. Interest in sound reproduction intensified in the post-war period, and technical standards such as DIN 45–500 were set, though they were soon criticised as too permissive. By the late 1960s it was widely accepted that the requirements for hi-fi would be satisfied by 'THD less than 0.1%, with no significant crossover distortion, frequency response 20–20 kHz, and as little noise as possible, please'. The early 1970s saw this expanded to include slew-rates and properly behaved overload protection, but the approach was always scientific and it was normal to read amplifier reviews in which measurements were dissected but no mention made of listening tests.

Following the growth of subjectivism through the pages of one of the leading Subjectivist magazines (*Hi-Fi News*), the first intimation of what was to come was the commencement of Paul Messenger's column *Subjective Sounds* in September 1976, in which he said: '*The assessment will be (almost) purely subjective, which has both strengths and weaknesses, as the inclusion of laboratory data would involve too much time and space, and although the ear may be the most fallible, it is also the most sensitive evaluation instrument*'. Subjectivism as expedient rather than policy. Significantly, none of the early instalments contained references to amplifier sound. In March 1977, an article by Jean Hiraga was published vilifying high levels of negative feedback and praising the sound of an amplifier with 2% THD. In the same issue, Paul Messenger stated that a Radford valve amplifier sounded better than a transistor one, and by the end of the year the amplifier-sound bandwagon was rolling. Hiraga returned in August 1977 with a highly contentious set of claims about audible speaker cables, and after that no hypothesis was too unlikely to receive attention.

The limits of hearing

In evaluating the Subjectivist position, it is essential to consider the known abilities of the human ear. Contrary to the impression given by some commentators, who call constantly for more psychoacoustical research, a vast amount of hard scientific information already exists on this subject, and some of it may be briefly summarised thus:

- The smallest step-change in amplitude that can be detected is about 0.3 dB for a pure tone. In more realistic situations it is 0.5 to 1.0 dB. This is about a 10% change[4].
- The smallest detectable change in frequency of a tone is about 0.2% in the band 500 Hz–2 kHz. In percentage terms, this is the parameter for which the ear is most sensitive[5].
- The least detectable amount of harmonic distortion is not an easy figure to determine, as there is a multitude of variables involved, and in particular the continuously varying level of programme means that the

level of THD introduced is also dynamically changing. With mostly low-order harmonics present the just-detectable amount is about 1%, though crossover effects can be picked up at 0.3%, and probably lower. There is certainly no evidence that an amplifier producing 0.001% THD sounds any cleaner than one producing 0.005%[6].

It is acknowledged that THD measurements, taken with the usual notch-type analyser, are of limited use in predicting the subjective impairment produced by an imperfect audio path. With music, etc. intermodulation effects are demonstrably more important than harmonics. However, THD tests have the unique advantage that visual inspection of the distortion residual gives an experienced observer a great deal of information about the root cause of the non-linearity. Many other distortion tests exist which, while yielding very little information to the designer, exercise the whole audio bandwidth at once and correlate well with properly-conducted tests for subjective impairment by distortion. The Belcher intermodulation test (the principle is shown in Figure 1.1) deserves more attention than it has received, and may become more popular now that DSP chips are cheaper.

One of the objections often made to THD tests is that their resolution does not allow verification that no non-linearities exist at very low level; a sort of micro-crossover distortion. Hawksford, for example, has stated '*Low-level threshold phenomena . . . set bounds upon the ultimate transparency of an audio system*'[7] and several commentators have stated their belief that some metallic contacts consist of a net of so-called 'micro-diodes'. In fact, this kind of mischievous hypothesis can be disposed of using THD techniques.

I evolved a method of measuring THD down to 0.01% at 200 μV rms, and applied it to large electrolytics, connectors of varying provenance, and lengths of copper cable with and without alleged magic properties. The method required the design of an ultra-low noise (EIN = −150 dBu for a 10 source resistance) and very low THD[8]. The measurement method is shown in Figure 1.2; using an attenuator with a very low value of resistance to reduce the incoming signal keeps the Johnson noise to a minimum. In no case was any unusual distortion detected, and it would be nice to think that this red herring at least has been laid to rest.

- Interchannel crosstalk can obviously degrade stereo separation, but the effect is not detectable until it is worse than 20 dB, which would be a very bad amplifier indeed[9].
- Phase and group delay have been an area of dispute for a long time. As Stanley Lipshitz et al have pointed out, these effects are obviously perceptible if they are gross enough; if an amplifier was so heroically misconceived as to produce the top half of the audio spectrum three hours after the bottom, there would be no room for argument. In more practical terms, concern about phase problems has centred on loudspeakers and

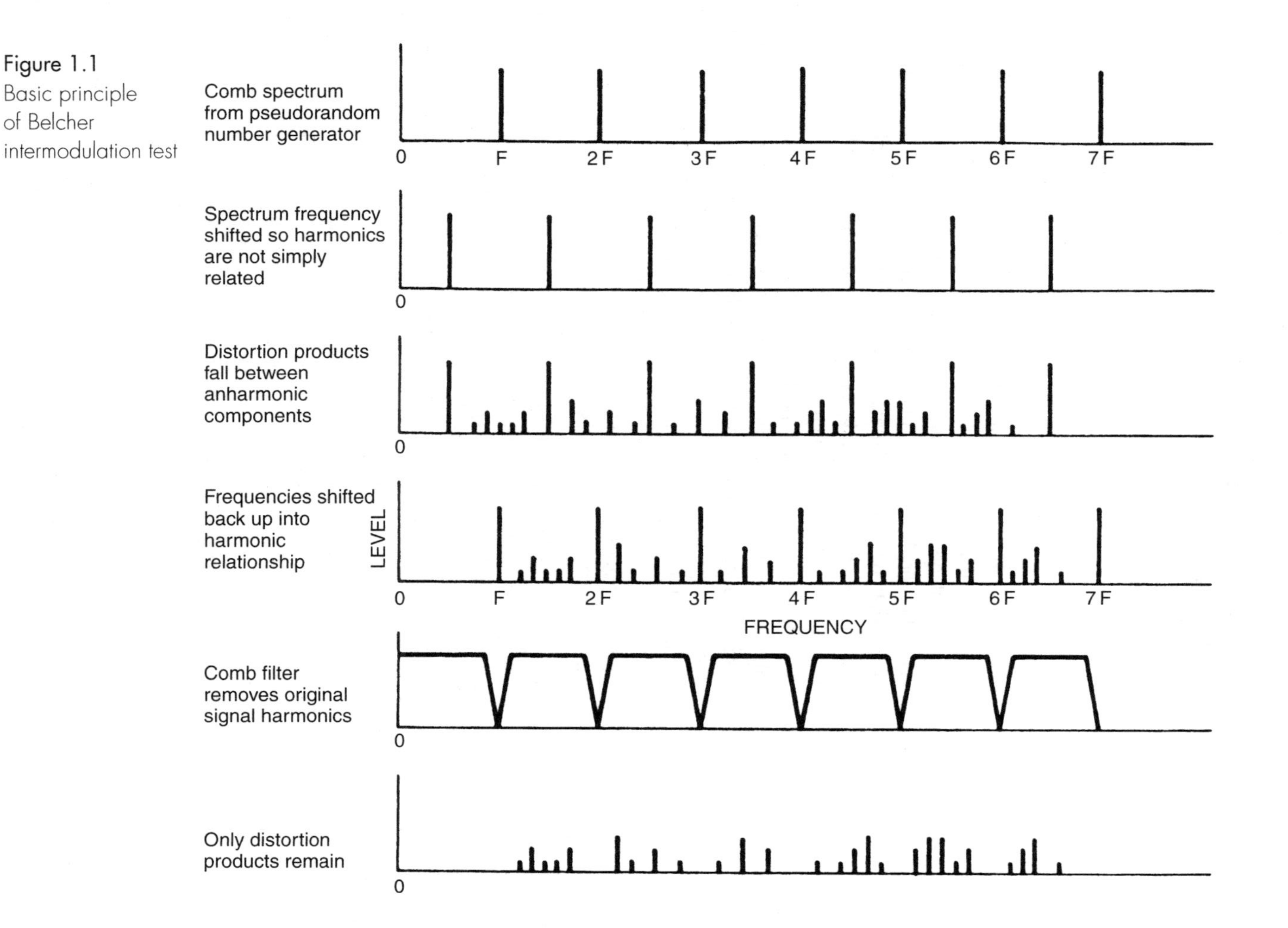

Figure 1.1 Basic principle of Belcher intermodulation test

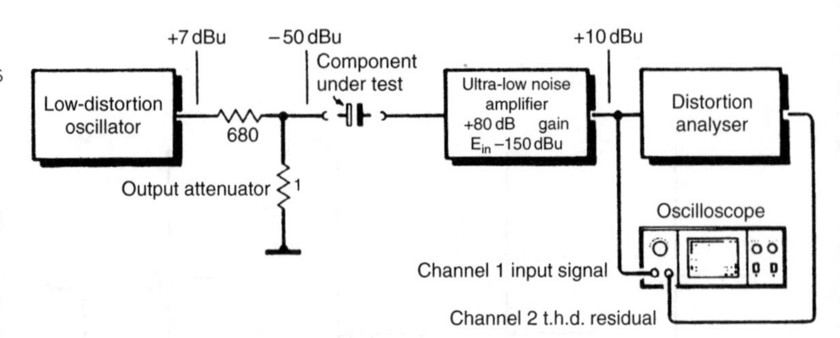

Figure 1.2
THD measurements at very low levels

their crossovers, as this would seem to be the only place where a phase-shift might exist without an accompanying frequency-response change to make it obvious. Lipshitz appears to have demonstrated[10] that a second-order all-pass filter (an all-pass filter gives a frequency-dependant phase-shift without level changes) is audible, whereas BBC findings, reported by Harwood[11] indicate the opposite, and the truth of the matter is still not clear. This controversy is of limited importance to amplifier designers, as it would take spectacular incompetence to produce a circuit that included an accidental all-pass filter. Without such, the phase response of an amplifier is completely defined by its frequency response, and vice-versa; in Control Theory this is Bode's Second Law[12], and it should be much more widely known in the hi-fi world than it is. A properly designed amplifier has its response roll-off points not too far outside the audio band, and these will have accompanying phase-shifts; there is no evidence that these are perceptible[8].

The picture of the ear that emerges from psychoacoustics and related fields is not that of a precision instrument. Its ultimate sensitivity, directional capabilities and dynamic range are far more impressive than its ability to measure small level changes or detect correlated low-level signals like distortion harmonics. This is unsurprising; from an evolutionary viewpoint the functions of the ear are to warn of approaching danger (sensitivity and direction-finding being paramount) and for speech. In speech perception the identification of formants (the bands of harmonics from vocal-chord pulse excitation, selectively emphasised by vocal-tract resonances) and vowel/consonant discriminations, are infinitely more important than any hi-fi parameter. Presumably the whole existence of music as a source of pleasure is an accidental side-effect of our remarkable powers of speech perception: how it acts as a direct route to the emotions remains profoundly mysterious.

Articles of faith: the tenets of subjectivism

All of the alleged effects listed below have received considerable affirmation in the audio press, to the point where some are treated as facts. The reality is that none of them has in the last fifteen years proved susceptible to objective confirmation. This sad record is perhaps equalled only by students of parapsychology. I hope that the brief statements below are considered fair by their proponents. If not I have no doubt I shall soon hear about it:

- *Sinewaves are steady-state signals that represent too easy a test for amplifiers, compared with the complexities of music.*

This is presumably meant to imply that sinewaves are in some way particularly easy for an amplifier to deal with, the implication being that anyone using a THD analyser must be hopelessly naive. Since sines and cosines have an unending series of non-zero differentials, *steady* hardly comes into it. I know of no evidence that sinewaves of randomly varying amplitude (for example) would provide a more searching test of amplifier competence.

I hold this sort of view to be the result of anthropomorphic thinking about amplifiers; treating them as though they think about what they amplify. Twenty sinewaves of different frequencies may be conceptually complex to us, and the output of a symphony orchestra even more so, but to an amplifier both composite signals resolve to a single instantaneous voltage that must be increased in amplitude and presented at low impedance. An amplifier has no perspective on the signal arriving at its input, but must literally take it as it comes.

- *Capacitors affect the signal passing through them in a way invisible to distortion measurements.*

Several writers have praised the technique of subtracting pulse signals passed through two different sorts of capacitor, claiming that the non-zero residue proves that capacitors can introduce audible errors. My view is that these tests expose only well-known capacitor shortcomings such as dielectric absorption and series resistance, plus perhaps the vulnerability of the dielectric film in electrolytics to reverse-biasing. No-one has yet shown how these relate to capacitor audibility in properly designed equipment.

- Passing an audio signal through cables, PCB tracks or switch contacts causes a cumulative deterioration. Precious metal contact surfaces alleviate but do not eliminate the problem. This too is undetectable by tests for non-linearity.

Concern over cables is widespread, but it can be said with confidence that there is as yet not a shred of evidence to support it. Any piece of wire passes a sinewave with unmeasurable distortion, and so simple notions of inter-crystal rectification or 'micro-diodes' can be discounted, quite apart from the fact that such behaviour is absolutely ruled out by established materials science. No plausible means of detecting, let alone measuring, cable degradation has ever been proposed.

The most significant parameter of a loudspeaker cable is probably its lumped inductance. This can cause minor variations in frequency response at the very top of the audio band, given a demanding load impedance. These deviations are unlikely to exceed 0.1 dB for reasonable cable constructions (say inductance less than 4 μH). The resistance of a typical cable (say 0.1 Ω) causes response variations across the band, following the speaker impedance curve, but these are usually even smaller at around 0.05 dB. This is not audible.

Corrosion is often blamed for subtle signal degradation at switch and connector contacts; this is unlikely. By far the most common form of contact degradation is the formation of an insulating sulphide layer on silver contacts, derived from hydrogen sulphide air pollution. This typically cuts the signal altogether, except when signal peaks temporarily punch through the sulphide layer. The effect is gross and seems inapplicable to theories of subtle degradation. Gold-plating is the only certain cure. It costs money.

- *Cables are directional, and pass audio better in one direction than the other.*

Audio signals are AC. Cables cannot be directional any more than 2 + 2 can equal 5. Anyone prepared to believe this nonsense will not be capable of designing amplifiers, so there seems no point in further comment.

- *The sound of valves is inherently superior to that of any kind of semiconductor.*

The 'valve sound' is one phenomenon that may have a real existence; it has been known for a long time that listeners sometimes prefer to have a certain amount of second-harmonic distortion added in[13], and most valve amplifiers provide just that, due to grave difficulties in providing good linearity with modest feedback factors. While this may well sound nice, hi-fi is supposedly about accuracy, and if the sound is to be thus modified it should be controllable from the front panel by a 'niceness' knob.

The use of valves leads to some intractable problems of linearity, reliability and the need for intimidatingly expensive (and once more, non-linear) iron-cored transformers. The current fashion is for exposed valves, and it is not at all clear to me that a fragile glass bottle, containing a red-hot anode with hundreds of volts DC on it, is wholly satisfactory for domestic safety.

A recent development in subjectivism is enthusiasm for single-ended directly-heated triodes, usually in extremely expensive monoblock systems. Such an amplifier generates large amounts of second-harmonic distortion, due to the asymmetry of single-ended operation, and requires a very large output transformer as its primary carries the full DC anode current, and core saturation must be avoided. Power outputs are inevitably very limited at 10 W or less. In a recent review, the Cary CAD-300SEI triode amplifier yielded 3% THD at 9 W, at a cost of £3400[14]. And you still need to buy a preamp.

- *Negative feedback is inherently a bad thing; the less it is used, the better the amplifier sounds, without qualification.*

Negative feedback is not inherently a bad thing; it is an absolutely indispensable principle of electronic design, and if used properly has the remarkable ability to make just about every parameter better. It is usually global feedback that the critic has in mind. Local negative feedback is grudgingly regarded as acceptable, probably because making a circuit with no feedback of any kind is near-impossible. It is often said that high levels of NFB enforce a low slew-rate. This is quite untrue; and this thorny issue is dealt with in detail on page 47. For more on slew-rate see also[15].

- *Tone-controls cause an audible deterioration even when set to the flat position.*

This is usually blamed on *phase-shift*. At the time of writing, tone controls on a pre-amp badly damage its chances of street (or rather sitting-room) credibility, for no good reason. Tone-controls set to *flat* cannot possibly contribute any extra phase-shift and must be inaudible. My view is that they are absolutely indispensable for correcting room acoustics, loudspeaker shortcomings, or tonal balance of the source material, and that a lot of people are suffering sub-optimal sound as a result of this fashion. It is now commonplace for audio critics to suggest that frequency-response inadequacies should be corrected by changing loudspeakers. This is an extraordinarily expensive way of avoiding tone-controls.

- *The design of the power supply has subtle effects on the sound, quite apart from ordinary dangers like ripple injection.*

All good amplifier stages ignore imperfections in their power supplies, op-amps in particular excelling at power-supply rejection-ratio. More nonsense has been written on the subject of subtle PSU failings than on most audio topics; recommendations of hard-wiring the mains or using gold-plated 13 A plugs would seem to hold no residual shred of rationality, in view of the usual processes of rectification and smoothing that the raw AC undergoes. And where do you stop? At the local sub-station? Should we gold-plate the pylons?

- *Monobloc construction (i.e., two separate power amplifier boxes) is always audibly superior, due to the reduction in crosstalk.*

There is no need to go to the expense of monobloc power amplifiers in order to keep crosstalk under control, even when making it substantially better than the −20 dB that is actually necessary. The techniques are conventional; the last stereo power amplifier I designed managed an easy −90 dB at 10 kHz without anything other than the usual precautions. In this area dedicated followers of fashion pay dearly for the privilege, as the cost of the mechanical parts will be nearly doubled.

- *Microphony is an important factor in the sound of an amplifier, so any attempt at vibration-damping is a good idea.*

Microphony is essentially something that happens in sensitive valve preamplifiers, If it happens in solid-state power amplifiers the level is so far below the noise it is effectively non-existent.

Experiments on this sort of thing are rare (if not unheard of) and so I offer the only scrap of evidence I have. Take a microphone pre-amp operating at a gain of +70 dB, and tap the input capacitors (assumed electrolytic) sharply with a screwdriver; the pre-amp output will be a dull thump, at low level. The physical impact on the electrolytics (the only components that show this effect) is hugely greater than that of any acoustic vibration; and I think the effect in power amps, if any, must be so vanishingly small that it could never be found under the inherent circuit noise.

Let us for a moment assume that some or all of the above hypotheses are true, and explore the implications. The effects are not detectable by conventional measurement, but are assumed to be audible. First, it can presumably be taken as axiomatic that for each audible defect some change occurs in the pattern of pressure fluctuations reaching the ears, and therefore a corresponding modification has occurred to the electrical signal passing through the amplifier. Any other starting point supposes that there is some other route conveying information apart from the electrical signals, and we are faced with magic or forces-unknown-to-science. Mercifully no commentator has (so far) suggested this. Hence there must be defects in the audio signals, but they are not revealed by the usual test methods. How could this situation exist? There seem two possible explanations for this failure of detection: one is that the standard measurements are relevant, but of insufficient resolution, and we should be measuring frequency response, etc., to thousandths of a dB. There is no evidence whatsoever that such micro-deviations are audible under any circumstances.

An alternative (and more popular) explanation is that standard sinewave THD measurements miss the point by failing to excite subtle distortion mechanisms that are triggered only by music, the spoken word, or whatever. This assumes that these music-only distortions are also left undisturbed by multi-tone intermodulation tests, and even the complex pseudorandom signals used in the Belcher distortion test[16]. The Belcher method effectively tests the audio path at all frequencies at once, and it is hard to conceive of a real defect that could escape it.

The most positive proof that subjectivism is fallacious is given by subtraction testing. This is the devastatingly simple technique of subtracting before-and-after amplifier signals and demonstrating that nothing audibly detectable remains.

It transpires that these alleged music-only mechanisms are not even revealed by music, or indeed anything else, and it appears the subtraction test has finally shown as non-existent these elusive degradation mechanisms.

The subtraction technique was proposed by Baxandall in 1977[17]. The principle is shown in Figure 1.3; careful adjustment of the roll-off balance network prevents minor bandwidth variations from swamping the true distortion residual. In the intervening years the Subjectivist camp has made no effective reply.

A simplified version of the test was introduced by Hafler[18]. This method is less sensitive, but has the advantage that there is less electronics in the signal path for anyone to argue about. See Figure 1.4. A prominent Subjectivist reviewer, on trying this demonstration, was reduced to claiming that the passive switchbox used to implement the Hafler test was causing so much sonic degradation that all amplifier performance was swamped[19]. I do not feel that this is a tenable position. So far all experiments such as these have been ignored or brushed aside by the Subjectivist camp; no attempt has been made to answer the extremely serious objections that this demonstration raises.

In the twenty or so years that have elapsed since the emergence of the Subjectivist Tendency, no hitherto unsuspected parameters of audio quality have emerged.

The length of the audio chain

An apparently insurmountable objection to the existence of non-measurable amplifier quirks is that recorded sound of almost any pedigree has passed through a complex mixing console at least once; prominent parts like vocals or lead guitar will almost certainly have passed through at least twice, once for recording and once at mix-down. More significantly, it must have passed through the potential quality-bottleneck of an analog tape machine or more likely the A–D converters of digital equipment. In its long path from here to ear the audio passes through at least a hundred op-amps, dozens of connectors and several hundred metres of ordinary screened cable. If mystical degradations can occur, it defies reason to insist that those introduced by the last 1% of the path are the critical ones.

The implications

This confused state of amplifier criticism has negative consequences. First, if equipment is reviewed with results that appear arbitrary, and which are

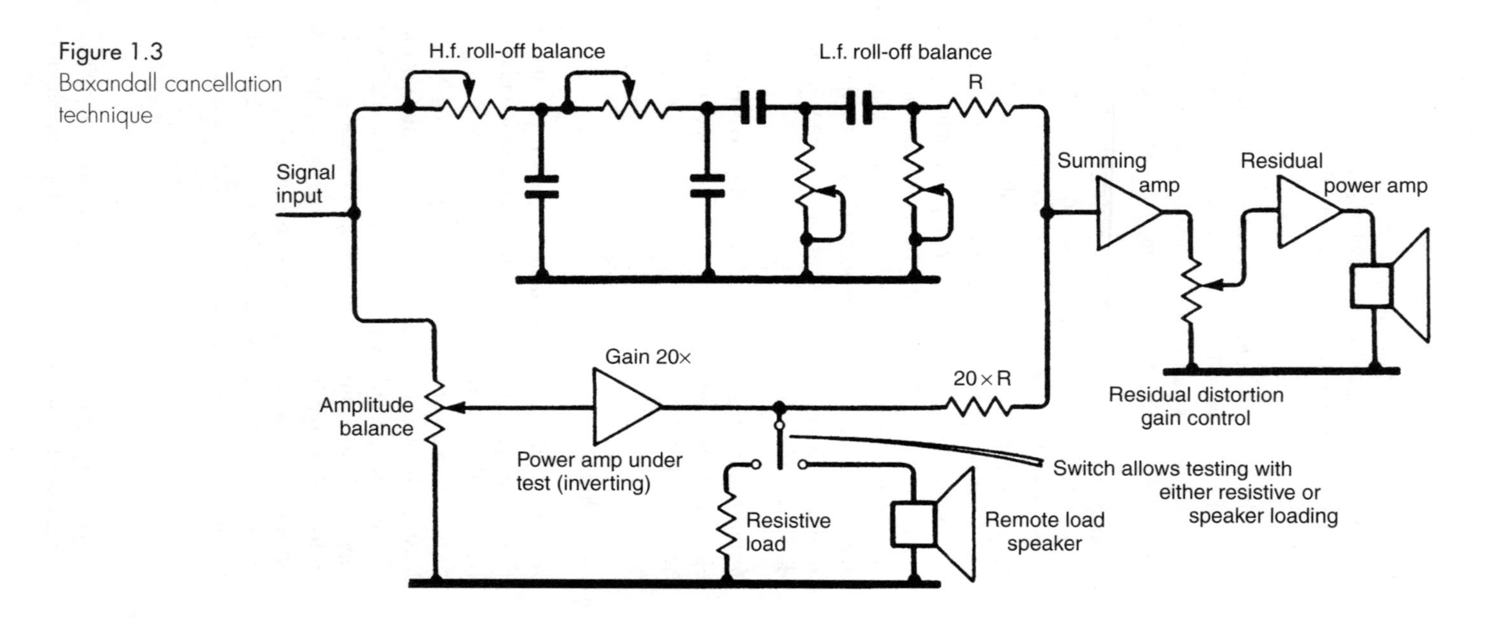

Figure 1.3
Baxandall cancellation technique

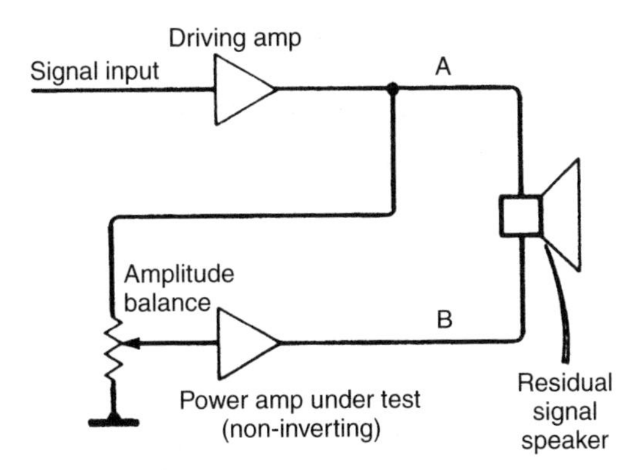

Figure 1.4
Hafler straight-wire differential test

in particular incapable of replication or confirmation, this can be grossly unfair to manufacturers who lose out in the lottery. Since subjective assessments cannot be replicated, the commercial success of a given make can depend entirely on the vagaries of fashion. While this is fine in the realm of clothing or soft furnishings, the hi-fi business is still claiming accuracy of reproduction as its raison d'être, and therefore you would expect the technical element to be dominant.

A second consequence of placing subjectivism above measurements is that it places designers in a most unenviable position. No degree of ingenuity or attention to technical detail can ensure a good review, and the pressure to adopt fashionable and expensive expedients (such as linear-crystal internal wiring) is great, even if the designer is certain that they have no audible effect for good or evil. Designers are faced with a choice between swallowing the Subjectivist credo whole or keeping very quiet and leaving the talking to the marketing department.

If objective measurements are disregarded, it is inevitable that poor amplifiers will be produced, some so bad that their defects are unquestionably audible. In recent reviews[20] it was easy to find a £795 pre-amplifier (Counterpoint SA7) that boasted a feeble 12 dB disc overload margin (another pre-amp costing £2040 struggled up to 15 dB – Burmester 838/846) and another, costing £1550 that could only manage a 1 kHz distortion performance of 1%; a lack of linearity that would have caused consternation ten years ago (Quicksilver). However, by paying £5700 one could inch this down to 0.3% (Audio Research M100–2 monoblocs). This does not of course mean that it is impossible to buy an *audiophile* amplifier that does measure well; another example would be the pre-amplifier/power amplifier combination that provides a very respectable disc overload margin of 31 dB and 1 kHz rated-power distortion below 0.003%; the total cost being £725 (Audiolab 8000C/8000P). I believe this to be a representative sample, and we appear to be in the paradoxical situation that the most expensive equipment provides the worst objective performance. Whatever the rights and wrongs of subjective assessment, I think that most people would agree that this is a strange state of affairs. Finally, it is surely a morally ambiguous position to persuade non-technical people that to get a really good sound they have to buy £2000 pre-amps and so on, when both technical orthodoxy and common sense indicate that this is quite unnecessary.

The reasons why

Some tentative conclusions are possible as to why hi-fi engineering has reached the pass that it has. I believe one basic reason is the difficulty of defining the quality of an audio experience; you cannot draw a diagram to communicate what something sounded like. In the same way, acoustical memory is more evanescent than visual memory. It is far easier to visualise what a London bus looks like than to recall the details of a musical performance. Similarly, it is difficult to 'look more closely'; turning up the volume is more like turning up the brightness of a TV picture; once an optimal level is reached, any further increase becomes annoying, then painful.

It has been universally recognised for many years in experimental psychology, particularly in experiments about perception, that people tend to perceive what they want to perceive. This is often called the *experimenter-expectancy* effect; it is more subtle and insidious than it sounds, and the history of science is littered with the wrecked careers of those who failed to guard against it. Such self-deception has most often occurred in fields like biology, where although the raw data may be numerical, there is no real mathematical theory to check it against. When the only 'results' are vague subjective impressions, the danger is clearly much greater, no matter how absolute the integrity of the experimenter. Thus in psychological work great care is necessary in the use of impartial observers, double-blind techniques, and rigorous statistical tests for significance. The vast majority of Subjectivist writings wholly ignore these precautions, with predictable results. In a few cases properly controlled listening tests have been done, and at the time of writing all have resulted in different amplifiers sounding indistinguishable. I believe the conclusion is inescapable that experimenter expectancy has played a dominant role in the growth of subjectivism.

It is notable that in Subjectivist audio the 'correct' answer is always the more expensive or inconvenient one. Electronics is rarely as simple as that. A major improvement is more likely to be linked with a new circuit topology or new type of semiconductor, than with mindlessly specifying more expensive components of the same type; cars do not go faster with platinum pistons.

It might be difficult to produce a rigorous statistical analysis, but it is my view that the reported subjective quality of a piece of equipment correlates far more with the price than with anything else. There is perhaps here an echo of the Protestant Work Ethic; you must suffer now to enjoy yourself later. Another reason for the relatively effortless rise of subjectivism is the *me-too* effect; many people are reluctant to admit that they cannot detect acoustic subtleties as nobody wants to be labelled as insensitive, outmoded, or just plain deaf. It is also virtually impossible to absolutely disprove any claims, as the claimant can always retreat a fraction and say that there was something special about the combination of hardware in use during the disputed tests, or complain that the phenomena are too delicate for brutal logic to be used on them. In any case, most competent engineers with a taste for rationality probably have better things to do than dispute every controversial report.

Under these conditions, vague claims tend, by a kind of intellectual inflation, to gradually become regarded as facts. Manufacturers have some incentive to support the Subjectivist camp as they can claim that only they understand a particular non-measurable effect, but this is no guarantee that the dice may not fall badly in a subjective review.

The outlook

It seems unlikely that subjectivism will disappear for some time, given the momentum that it has gained, the entrenched positions that some people have taken up, and the sadly uncritical way in which people accept an unsupported assertion as the truth simply because it is asserted with frequency and conviction. In an ideal world every such statement would be greeted by loud demands for evidence. However, the history of the world sometimes leads one to suppose pessimistically that people will believe anything. By analogy, one might suppose that subjectivism would persist for the same reason that parapsychology has; there will always be people who will believe what they want to believe rather than what the hard facts indicate.

Technical errors

Misinformation also arises in the purely technical domain; I have also found that some of the most enduring and widely held technical beliefs to be unfounded. For example, if you take a Class-B amplifier and increase its quiescent current so that it runs in Class-A at low levels, i.e., in Class AB, most people will tell you that the distortion will be reduced as you have moved nearer to the full Class-A condition. This is untrue. A correctly configured amplifier gives more distortion in Class-AB, not less, because of the abrupt gain changes inherent in switching from A to B every cycle.

Discoveries like this can only be made because it is now straightforward to make testbed amplifiers with ultra-low distortion – lower than that which used to be thought possible. The reduction of distortion to the basic or inherent level that a circuit configuration is capable of is a fundamental requirement for serious design work in this field; in Class-B at least this gives a defined and repeatable standard of performance that in later chapters I name a Blameless amplifier, so-called because it avoids error rather than claiming new virtues.

It has proved possible to take the standard Class-B power amplifier configuration, and by minor modifications, reduce the distortion to below the noise floor at low frequencies. This represents approximately 0.0005 to 0.0008% THD, depending on the exact design of the circuitry, and the actual distortion can be shown to be substantially below this if spectrum-analysis techniques are used to separate the harmonics from the noise.

The performance requirements for amplifiers

This section is not a recapitulation of international standards, which are intended to provide a minimum level of quality rather than extend the art. It is rather my own view of what you should be worrying about at the start of the design process, and the first items to consider are the brutally pragmatic ones related to keeping you in business and out of prison.

Safety

In the drive to produce the finest amplifier ever made, do not forget that the Prime Directive of audio design is – Thou Shalt Not Kill. Every other consideration comes a poor second, not only for ethical reasons, but also because one serious lawsuit will close down most audio companies forever.

Reliability

If you are in the business of manufacturing, you had better make sure that your equipment keeps working, so that you too can keep working. It has to be admitted that power amplifiers especially the more powerful ones – have a reputation for reliability that is poor compared with most branches of electronics. The 'high end' in particular has gathered to itself a bad reputation for dependability[21].

Power output

In commercial practice, this is decided for you by the marketing department. Even if you can please yourself, the power output capability needs careful thought as it has a powerful and non-linear effect on the cost.

The last statement requires explanation. As the output power increases, a point is reached when single output devices are incapable of sustaining the thermal dissipation, parallel pairs are required, and the price jumps up. Similarly, transformer laminations come in standard sizes, so the transformer size and cost will also increase in discrete steps.

Domestic hi-fi amplifiers usually range from 20 W to 150 W into 8 Ω though with a scattering of much higher powers. PA units will range from 50 W, for foldback purposes (i.e., the sound the musician actually hears, to monitor his/her playing, as opposed to that thrown out forwards by the main PA stacks; also called stage monitoring) to 1 kW or more. Amplifiers of extreme high power are not popular, partly because the economies of scale are small, but mainly because it means putting all your eggs in one basket, and a failure becomes disastrous. This is accentuated by the statistically unproven but almost universally held opinion that high-power solid-state amplifiers are inherently less reliable than others.

If an amplifier gives a certain output into 8 Ω, it will not give exactly twice as much into 4 Ω loads; in fact it will probably be much less than this, due to the increased resistive losses in 4 Ω operation, and the way that power alters as the square of voltage. Typically, an amplifier giving 180 W into 8 Ω might be expected to yield 260 W into 4 Ω and 350 W into 2 Ω, if it can drive so low a load at all. These figures are approximate, depending very much on power supply design.

Nominally 8 Ω loudspeakers are the most common in hi-fi applications. The *nominal* title accommodates the fact that all loudspeakers, especially multi-element types, have marked changes in input impedance with frequency, and are only resistive at a few spot frequencies. Nominal 8 Ω loudspeakers may be expected to drop to at least 6 Ω in some part of the audio spectrum. To allow for this, almost all amplifiers are rated as capable of 4 Ω as well as 8 Ω loads. This takes care of almost any nominal 8 Ω speaker, but leaves no safety margin for nominal 4 Ω designs, which are likely to dip to 3 Ω or less. Extending amplifier capability to deal with lower load impedances for anything other than very short periods has serious cost implications for the power-supply transformer and heatsinking; these already represent the bulk of the cost.

The most important thing to remember in specifying output power is that you have to increase it by an awful lot to make the amplifier significantly louder. We do not perceive acoustic power as such – there is no way we could possibly integrate the energy liberated in a room, and it would be a singularly useless thing to perceive if we could. It is much nearer the truth to say that we perceive pressure. It is well known that power in watts must be quadrupled to double sound pressure level (SPL) but this is not the same as doubling subjective loudness; this is measured in Sones rather than dB above threshold, and some psychoacousticians have reported that doubling subjective loudness requires a 10 dB rather than 6 dB rise in SPL, implying that amplifier power must be increased tenfold, rather than merely quadrupled[22]. It is any rate clear that changing from a 25 W to a 30 W amplifier will not give an audible increase in level.

This does not mean that fractions of a watt are never of interest. They can matter either in pursuit of maximum efficiency for its own sake, or because a design is only just capable of meeting its output specification.

Some hi-fi reviewers set great value on very high peak current capability for short periods. While it is possible to think up special test waveforms that demand unusually large peak currents, any evidence that this effect is important in use is so far lacking.

Frequency response

This can be dealt with crisply; the minimum is 20 Hz to 20 kHz, ±0.5 dB, though there should never be any *plus* about it when solid-state amplifiers are concerned. Any hint of a peak before the roll-off should be looked at with extreme suspicion, as it probably means doubtful HF stability. This is less true of valve amplifiers, where the bandwidth limits of the output transformer mean that even modest NFB factors tend to cause peaking at both high and low ends of the spectrum.

Having dealt with the issue crisply, there is no hope that everyone will agree that this is adequate. CDs do not have the built-in LF limitations of vinyl and could presumably encode the barometric pressure in the recording studio if this was felt to be desirable, and so an extension to −0.5 dB at 5 or 10 Hz is perfectly feasible. However, if infrabass information does exist down at these frequencies, no domestic loudspeaker will reproduce them.

Noise

There should be as little as possible without compromising other parameters. The noise performance of a power amplifier is not an irrelevance[23], especially in a domestic setting.

Distortion

Once more, a sensible target might be: *As little as possible without messing up something else*. This ignores the views of those who feel a power amplifier is an appropriate device for adding distortion to a musical performance. Such views are not considered in the body of this book; it is, after all, not a treatise on fuzz-boxes or other guitar effects.

I hope that the techniques explained in this book have a relevance beyond power amplifiers. Applications obviously include discrete op-amp-based pre-amplifiers[24], and extend to any amplifier aiming at static or dynamic precision.

My philosophy is the simple one that distortion is bad, and high-order distortion is worse. The first part of this statement, is, I suggest, beyond argument, and the second part has a good deal of evidence to back it. The distortion of the *n*th harmonic should be weighted by $n^2/4$ worse, according to many authorities[25]. This leaves the second harmonic unchanged, but scales up the third by 9/4, i.e., 2.25 times, the fourth by 16/4, i.e., 4 times, and so on. It is clear that even small amounts of high-order harmonics could be unpleasant, and this is one reason why even modest crossover distortion is of such concern.

Digital audio now routinely delivers the signal with less than 0.002% THD, and I can earnestly vouch for the fact that analogue console designers work furiously to keep the distortion in long complex signal paths down to similar levels. I think it an insult to allow the very last piece of electronics in the chain to make nonsense of these efforts.

I would like to make it clear that I do not believe that an amplifier yielding 0.001% THD is going to sound much better than its fellow giving 0.002%. However, if there is ever a scintilla of doubt as to what level of distortion is perceptible, then using the techniques I have presented it should be possible to routinely reduce the THD below the level at which there can be any rational argument.

I am painfully aware that there is a school of thought that regards low THD as inherently immoral, but this is to confuse electronics with religion. The implication is that very low THD can only be obtained by huge global NFB factors that require heavy dominant-pole compensation that severely degrades slew-rate; the obvious flaw in this argument is that once the compensation is applied the amplifier no longer has a large global NFB factor, and so its distortion performance presumably reverts to mediocrity, further burdened with a slew-rate of 4 V per fortnight.

To me low distortion has its own aesthetic and philosophical appeal; it is satisfying to know that the amplifier you have just designed and built is so linear that there simply is no realistic possibility of it distorting your favourite material. Most of the linearity-enhancing strategies examined in this book are of minimal cost (the notable exception being resort to Class-A) compared with the essential heatsinks, transformer, etc., and so why not have ultra-low distortion? Why put up with more than you must?

Damping factor

Audio amplifiers, with a few very special exceptions[26], approximate to perfect voltage sources; i.e., they aspire to a zero output impedance across the audio band. The result is that amplifier output is unaffected by loading, so that the frequency-variable impedance of loudspeakers does not give an equally variable frequency response, and there is some control of speaker cone resonances.

While an actual zero impedance is impossible, a very close approximation is possible if large negative-feedback factors are used. (Actually, a judicious mixture of voltage and current feedback will make the output impedance zero, or even negative – i.e., increasing the loading makes the output voltage increase. This is clever, but usually pointless, as will be seen.) Solid-state amplifiers are quite happy with lots of feedback, but it is usually impractical in valve designs.

Damping factor is defined as the ratio of the load impedance *R*load to the amplifier output resistance Rout:

$$\text{Damping factor} = \frac{R\text{load}}{\text{Rout}} \qquad \text{Equation 1.1}$$

A solid-state amplifier typically has output resistance of the order of 0.05 Ω, so if it drives an 8 Ω speaker we get a damping factor of 160 times. This simple definition ignores the fact that amplifier output impedance usually varies considerably across the audio band, increasing with frequency as the negative feedback factor falls; this indicates that the output *resistance* is actually more like an inductive reactance. The presence of an output inductor to give stability with capacitative loads further complicates the issue.

Mercifully, damping factor as such has very little effect on loudspeaker performance. A damping factor of 160 times, as derived above, seems to imply a truly radical effect on cone response – it implies that resonances and such have been reduced by 160 times as the amplifier output takes an iron grip on the cone movement. Nothing could be further from the truth.

The resonance of a loudspeaker unit depends on the total resistance in the circuit. Ignoring the complexities of crossover circuitry in multi-element speakers, the total series resistance is the sum of the speaker coil resistance, the speaker cabling, and, last of all, the amplifier output impedance. The values will be typically 7, 0.5 and 0.05 Ω, so the amplifier only contributes 0.67% to the total, and its contribution to speaker dynamics must be negligible.

The highest output impedances are usually found in valve equipment, where global feedback including the output transformer is low or non-existent; values around 0.5 Ω are usual. However, idiosyncratic semiconductor designs sometimes also have high output resistances; see Olsher[27] for a design with Rout = 0.6 Ω, which I feel is far too high.

This view of the matter was practically investigated and fully confirmed by James Moir as far back as 1950[28], though this has not prevented periodic resurgences of controversy.

The only reason to strive for a high damping factor – which can, after all, do no harm – is the usual numbers game of impressing potential customers with specification figures. It is as certain as anything can be that the subjective difference between two amplifiers, one with a DF of 100, and the other boasting 2000, is undetectable by human perception. Nonetheless, the specifications look very different in the brochure, so means of maximising the DF may be of some interest. This is examined further in Chapter 7.

Absolute phase

Concern for absolute phase has for a long time hovered ambiguously between real audio concerns like noise and distortion, and the Subjective realm where solid copper is allegedly audible. Absolute phase means the preservation of signal phase all the way from microphone to loudspeaker, so that a drum impact that sends an initial wave of positive pressure towards the live audience is reproduced as a similar positive pressure wave from the loudspeaker. Since it is known that the neural impulses from the ear retain the periodicity of the waveform at low frequencies, and distinguish between compression and rarefaction, there is a prima facie case for the audibility of absolute phase.

It is unclear how this applies to instruments less physical than a kickdrum. For the drum the situation is simple – you kick it, the diaphragm moves outwards and the start of the transient must be a wave of compression in the air. (Followed almost at once by a wave of rarefaction.) But what about an electric guitar? A similar line of reasoning – plucking the string moves it in a given direction, which gives such-and-such a signal polarity, which leads to whatever movement of the cone in the guitar amp speaker cabinet – breaks down at every point in the chain. There is no way to know how the pickups are wound, and indeed the guitar will almost certainly have a switch for reversing the phase of one of them. I also suggest that the preservation of absolute phase is not the prime concern of those who design and build guitar amplifiers.

The situation is even less clear if more than one instrument is concerned, which is of course almost all the time. It is very difficult to see how two electric guitars played together could have a *correct* phase in which to listen to them.

Recent work on the audibility of absolute phase[29], [30] shows it is sometimes detectable. A single tone flipped back and forth in phase, providing it has a spiky asymmetrical waveform and an associated harsh sound, will show a change in perceived timbre and, according to some experimenters, a perceived change in pitch. A monaural presentation has to be used to yield a clear effect. A complex sound, however, such as that produced by a musical ensemble, does not in general show a detectable difference.

Proposed standards for the maintenance of absolute phase have just begun to appear[31], and the implication for amplifier designers is clear; whether absolute phase really matters or not, it is simple to maintain phase in a power amplifier (compare a complex mixing console, where correct phase is vital, and there are hundreds of inputs and outputs, all of which must be in phase in every possible configuration of every control) and so it should be done. In fact, it probably already has been done, even if the designer has not given absolute phase a thought, because almost all amplifiers use series negative feedback, and this must be non-inverting. Care is however required if there are stages such as balanced line input amplifiers before the power amplifier itself.

Acronyms

I have kept the number of acronyms used to a minimum. However, those few are used extensively, so a list is given in case they are not all blindingly obvious:

BJT	Bipolar junction transistor
CFP	Complementary-Feedback-Pair
C/L	Closed-loop
CM	Common-mode
EF	Emitter-follower
EIN	Equivalent input noise
FET	Field-effect transistor
HF	Amplifier behaviour above the dominant pole frequency, where the open-loop gain is usually falling at 6 dB/octave
I/P	Input
LF	Relating to amplifier action below the dominant pole, where the open-loop gain is assumed to be essentially flat with frequency
NFB	Negative feedback
O/L	Open loop
P1	The first o/l response pole, and its frequency in Hz (i.e., the −3 dB point of a 6 dB/oct rolloff)
P2	The second response pole, at a higher frequency
PSRR	Power supply rejection ratio
THD	Total harmonic distortion
VAS	Voltage-amplifier stage

References

1. Martin Gardner *Fads & Fallacies in the Name of Science*, Ch. 12, Pub. Dover, pp. 140–151.
2. David, F Mark *Investigating the Paranormal Nature*, Vol 320, 13 March 1986.

3. Randi, J *Flim-Flam! Psychics, ESP Unicorns and Other Delusions* Prometheus Books, 1982, pp. 196–198.
4. Harris, J D *Loudness discrimination* J. Speech Hear. Dis. Monogr. Suppl. 11, pp. 1–63.
5. Moore, B C J *Relation between the critical bandwidth k the frequency-difference limen* J. Acoust. Soc. Am. 55, p. 359.
6. Moir, J *Just Detectable Distortion Levels* Wireless World, Feb 1981, pp. 32–34.
7. Hawksford, M *The Essex Echo* Hi-fi News & RR, May 1986, p. 53.
8. Self, D *Ultra-Low-Noise Amplifiers & Granularity Distortion* JAES, Nov 1987, pp. 907–915.
9. Harwood and Shorter *Stereophony and the effect of crosstalk between left and right channels* BBC Engineering Monograph No 52.
10. Lipshitz et al, *On the audibility of midrange phase distortion in audio systems* JAES, Sept 1982, pp. 580–595.
11. Harwood, H *Audibility of phase effects in loudspeakers* Wireless World, Jan 1976, pp. 30–32.
12. Shinners, S *Modern control system theory and application* pub. Addison-Wesley, p. 310.
13. King, G *Hi-fi reviewing* Hi-fi News & RR, May 1978, p. 77.
14. Harley, R *Review of Cary CAD-300SEI Single-Ended Triode Amplifier* Stereophile, Sept 1995, p. 141.
15. Baxandall, P *Audio power amplifier design* Wireless World, Jan 1978, p. 56.
16. Belcher, R A *A new distortion measurement* Wireless World, May 1978, pp. 36–41.
17. Baxandall, P *Audible amplifier distortion is not a mystery* Wireless World, Nov 1977, pp. 63–66.
18. Hafler, D *A Listening Test for Amplifier Distortion* Hi-fi News & RR, Nov 1986, pp. 25–29.
19. Colloms, M *Hafler XL-280 Test* Hi-Fi News & RR, June 1987, pp. 65–67.
20. *Hi-fi Choice, The Selection* Pub. Sportscene, 1986.
21. Lawry, R H *High End Difficulties* Stereophile, May 1995, p. 23.
22. Moore, B J *An Introduction to the Psychology of Hearing* Academic Press, 1982, pp. 48–50.
23. Fielder, L *Dynamic range issues in the Modern Digital Audio Environment* JAES Vol 43.
24. Self, D *Advanced Preamplifier Design* Wireless World, Nov 1976, p. 41.
25. Moir, J *Just Detectable Distortion Levels* Wireless World, Feb 1981, p. 34.
26. Mills and Hawksford *Transconductance Power Amplifier Systems for Current-Driven Loudspeakers* JAES Vol 37.
27. Olsher, D *Times One RFS400 Power Amplifier Review* Stereophile, Aug 1995, p. 187.

28. Moir, J *Transients and Loudspeaker Damping* Wireless World, May 1950, p. 166.
29. Greiner and Melton *A Quest for the Audibility of Polarity* Audio, Dec 1993, p. 40.
30. Greiner and Melton *Observations on the Audibility of Acoustic Polarity* JAES Vol 42.
31. AES *Draft AES recommended practice Standard for professional audio – Conservation of the Polarity of Audio Signals* Inserted in: JAES Vol 42.

2

History, architecture and negative feedback

A brief history of amplifiers

A full and detailed account of semiconductor amplifier design since its beginnings would be a book in itself – and a most fascinating volume it would be. This is not that book, but I still feel obliged to give a very brief account of how amplifier design has evolved in the last three or four decades.

Valve amplifiers, working in push-pull Class-A or AB1, and perforce transformer-coupled to the load, were dominant until the early 1960s, when truly dependable transistors could be made at a reasonable price. Designs using germanium devices appeared first, but suffered severely from the vulnerability of germanium to even moderately high temperatures; the term *thermal runaway* was born. At first all silicon power transistors were NPN, and for a time most transistor amplifiers relied on input and output transformers for push-pull operation of the power output stage. These transformers were as always heavy, bulky, expensive, and non-linear, and added insult to injury as their LF and HF phase-shifts severely limited the amount of negative feedback that could be safely applied.

The advent of the transformerless Lin configuration[1], with what became known as a quasi-complementary output stage, disposed of a good many problems. Since modestly capable PNP driver transistors were available, the power output devices could both be NPN, and still work in push-pull. It was realised that a transformer was not required for impedance matching between power transistors and 8 Ω loudspeakers.

Proper complementary power devices appeared in the late 1960s, and full complementary output stages soon proved to give less distortion than their quasi-complementary predecessors. At about the same time DC-coupled amplifiers began to take over from capacitor-coupled designs, as the transistor differential pair became a more familiar circuit element.

A much fuller and generally excellent history of power amplifier technology is given in Sweeney and Mantz[2].

Amplifier architectures

This grandiose title simply refers to the large-scale structure of the amplifier; i.e., the block diagram of the circuit one level below that representing it as a single white block labelled *Power Amplifier*. Almost all solid-state amplifiers have a three-stage architecture as described below, though they vary in the detail of each stage.

The three-stage architecture

The vast majority of audio amplifiers use the conventional architecture, shown in Figure 2.1. There are three stages, the first being a transconductance stage (differential voltage in, current out) the second a transimpedance stage (current in, voltage out) and the third a unity-voltage-gain output stage. The second stage clearly has to provide all the voltage gain and I have therefore called it the voltage-amplifier stage or VAS. Other authors have called it the *pre-driver stage* but I prefer to reserve this term for the first transistors in output triples. This three-stage architecture has several advantages, not least being that it is easy to arrange things so that interaction between stages is negligible. For example, there is very little signal voltage at the input to the second stage, due to its current-input

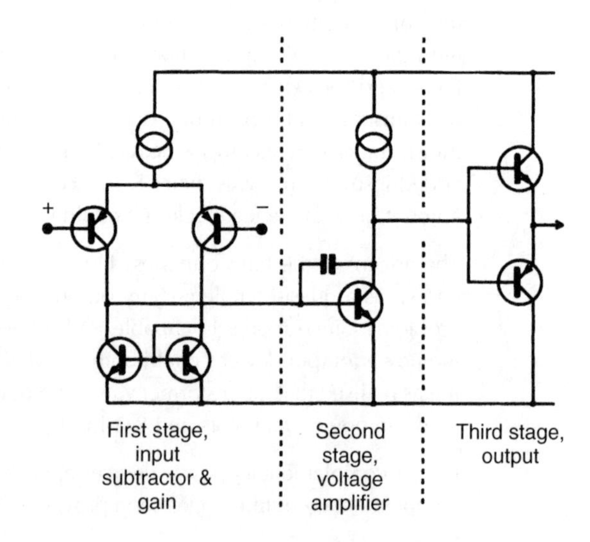

Figure 2.1 The three-stage amplifier structure. There is a transconductance stage, a transadmittance stage (the VAS) and a unity-gain buffer output stage

(virtual-earth) nature, and therefore very little on the first stage output; this minimises Miller phaseshift and possible Early effect in the input devices.

Similarly, the compensation capacitor reduces the second stage output impedance, so that the non-linear loading on it due to the input impedance of the third stage generates less distortion than might be expected. The conventional three-stage structure, familiar though it may be, holds several elegant mechanisms such as this. They will be fully revealed in later chapters. Since the amount of linearising global NFB available depends upon amplifier open-loop gain, how the stages contribute to this is of great interest. The three-stage architecture always has a unity-gain output stage – unless you really want to make life difficult for yourself – and so the total forward gain is simply the product of the transconductance of the input stage and the transimpedance of the VAS, the latter being determined solely by the Miller capacitor Cdom, except at very low frequencies. Typically, the closed-loop gain will be between +20 and +30 dB. The NFB factor at 20 kHz will be 25 to 40 dB, increasing at 6 dB per octave with falling frequency until it reaches the dominant pole frequency P1, when it flattens out. What matters for the control of distortion is the amount of negative feedback (NFB) available, rather than the open-loop bandwidth, to which it has no direct relationship. In my Electronics World Class-B design, the input stage gm is about 9 ma/V, and Cdom is 100 pF, giving an NFB factor of 31 dB at 20 kHz. In other designs I have used as little as 26 dB (at 20 kHz) with good results.

Compensating a three-stage amplifier is relatively simple; since the pole at the VAS is already dominant, it can be easily increased to lower the HF negative-feedback factor to a safe level. The local NFB working on the VAS through Cdom has an extremely valuable linearising effect.

The conventional three-stage structure represents at least 99% of the solid-state amplifiers built, and I make no apology for devoting much of this book to its behaviour. I doubt if I have exhausted its subtleties.

The two-stage amplifier architecture

In contrast, the architecture in Figure 2.2 is a two-stage amplifier, the first stage being once more a transconductance stage, though now without a guaranteed low impedance to accept its output current. The second stage combines VAS and output stage in one block; it is inherent in this scheme that the VAS must double as a phase splitter as well as a generator of raw gain. There are then two quite dissimilar signal paths to the output, and it is not at all clear that trying to break this block down further will assist a linearity analysis. The use of a phase-splitting stage harks back to valve amplifiers, where it was inescapable as a complementary valve technology has so far eluded us.

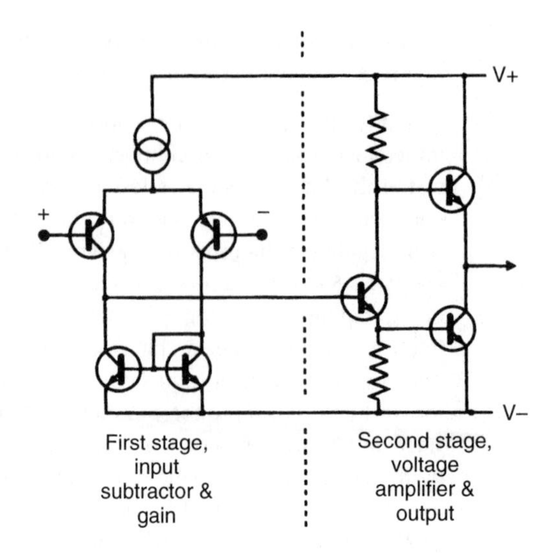

Figure 2.2
The two-stage amplifier structure. A voltage-amplifier output follows the same transconductance input stage

Paradoxically, a two-stage amplifier is likely to be more complex in its gain structure than a three-stage. The forward gain depends on the input stage gm, the input stage collector load (because the input stage can no longer be assumed to be feeding a virtual earth) and the gain of the output stage, which will be found to vary in a most unsettling manner with bias and loading. Choosing the compensation is also more complex for a two-stage amplifier, as the VAS/phase-splitter has a significant signal voltage on its input and so the usual pole-splitting mechanism that enhances Nyquist stability by increasing the pole frequency associated with the input stage collector will no longer work so well. (I have used the term Nyquist stability, or Nyquist oscillation throughout this book to denote oscillation due to the accumulation of phase-shift in a global NFB loop, as opposed to local parasitics, etc.)

The LF feedback factor is likely to be about 6 dB less with a 4Ω load, due to lower gain in the output stage. However, this variation is much reduced above the dominant pole frequency, as there is then increasing local NFB acting in the output stage.

Two-stage amplifiers are not popular; I can quote only two examples, Randi[3] and Harris[4]. The two-stage amplifier offers little or no reduction in parts cost, is harder to design and in my experience invariably gives a poor distortion performance.

Power amplifier classes

For a long time the only amplifier classes relevant to high-quality audio were Class-A and Class-AB. This is because valves were the only active devices, and Class-B valve amplifiers generated so much distortion that they were barely acceptable even for Public Address purposes. All amplifiers with pretensions to high fidelity operated in push-pull Class-A.

Solid-state gives much more freedom of design; all of the amplifier classes below have been commercially exploited. Unfortunately, there will only be space to deal in detail in this book with A, AB, and B, though this certainly covers the vast majority of solid-state amplifiers. Plentiful references are given so that the intrigued can pursue matters further.

Class-A

In a Class-A amplifier current flows continuously in all the output devices, which enables the non-linearities of turning them on and off to be avoided. They come in two rather different kinds, although this is rarely explicitly stated, which work in very different ways. The first kind is simply a Class-B stage (i.e., two emitter-followers working back-to-back) with the bias voltage increased so that sufficient current flows for neither device to cut off under normal loading. The great advantage of this approach is that it cannot abruptly run out of output current; if the load impedance becomes lower than specified then the amplifier simply takes brief excursions into Class AB, hopefully with a modest increase in distortion and no seriously audible distress.

The other kind could be called the controlled-current-source (VCIS) type, which is in essence a single emitter-follower with an active emitter load for adequate current-sinking. If this latter element runs out of current capability it makes the output stage clip much as if it had run out of output voltage. This kind of output stage demands a very clear idea of how low an impedance it will be asked to drive before design begins.

Valve textbooks will be found to contain enigmatic references to classes of operation called AB1 and AB2; in the former grid current did not flow for any part of the cycle, but in the latter it did. This distinction was important because the flow of output-valve grid current in AB2 made the design of the previous stage much more difficult.

AB1 or AB2 has no relevance to semiconductors, for in BJT's base current always flows when a device is conducting, while in power FET's gate current never does, apart from charging and discharging internal capacitances.

Class-AB

This is not really a separate class of its own, but a combination of A and B. If an amplifier is biased into Class-B, and then the bias further increased, it will enter AB. For outputs below a certain level both output devices conduct, and operation is Class-A. At higher levels, one device will be turned completely off as the other provides more current, and the distortion jumps upward at this point as AB action begins. Each device will conduct between 50% and 100% of the time, depending on the degree of excess bias and the output level.

Class AB is less linear than either A or B, and in my view its only legitimate use is as a fallback mode to allow Class-A amplifiers to continue working reasonably when faced with a low-load impedance.

Class-B

Class-B is by far the most popular mode of operation, and probably more than 99% of the amplifiers currently made are of this type. Most of this book is devoted to it, so no more is said here.

Class-C

Class-C implies device conduction for significantly less than 50% of the time, and is normally only usable in radio work, where an LC circuit can smooth out the current pulses and filters harmonics. Current-dumping amplifiers can be regarded as combining Class-A (the correcting amplifier) with Class-C (the current-dumping devices); however it is hard to visualise how an audio amplifier using devices in Class-C only could be built.

Class-D

These amplifiers continuously switch the output from one rail to the other at a supersonic frequency, controlling the mark/space ratio to give an average representing the instantaneous level of the audio signal; this is alternatively called Pulse Width Modulation (PWM). Great effort and ingenuity has been devoted to this approach, for the efficiency is in theory very high, but the practical difficulties are severe, especially so in a world of tightening EMC legislation, where it is not at all clear that a 200 kHz high-power square wave is a good place to start. Distortion is not inherently low[5], and the amount of global negative feedback that can be applied is severely limited by the pole due to the effective sampling frequency in the forward path. A sharp cut-off low-pass filter is needed between amplifier and speaker, to remove most of the RF; this will require at least four inductors (for stereo) and will cost money, but its worst feature is that it will only give a flat frequency response into one specific load impedance. The technique now has a whole chapter of this book to itself. Other references to consult for further information are Goldberg and Sandler[6] and Hancock[7].

Class-E

An extremely ingenious way to operating a transistor so that it has either a small voltage across it or a small current through it almost all the time; in other words the power dissipation is kept very low[8]. Regrettably this is an RF technique that seems to have no sane application to audio.

Class-F

There is no Class-F, as far as I know. This seems like a gap that needs filling . . .

Class-G

This concept was introduced by Hitachi in 1976 with the aim of reducing amplifier power dissipation. Musical signals have a high peak/mean ratio, spending most of the this at low levels, so internal dissipation is much reduced by running from low-voltage rails for small outputs, switching to higher rails current for larger excursions.

The basic series Class-G with two rail voltages (i.e., four supply rails, as both voltage are ±) is shown in Figure 2.3[9],[11]. Current is drawn from the lower ±V1 supply rails whenever possible; should the signal exceed ±V1, TR6 conducts and D3 turns off, so the output current is now drawn entirely from the higher ±V2 rails, with power dissipation shared between TR3 and TR6. The inner stage TR3, 4 is usually operated in Class-B, although AB or A are equally feasible if the output stage bias is suitably increased. The outer devices are effectively in Class-C as they conduct for significantly less than 50% of the time.

In principle movements of the collector voltage on the inner device collectors should not significantly affect the output voltage, but in practice Class-G is often considered to have poorer linearity than Class-B because of glitching due to charge storage in commutation diodes D3, D4. However, if glitches occur they do so at moderate power, well displaced from the crossover region, and so appear relatively infrequently with real signals.

An obvious extension of the Class-G principle is to increase the number of supply voltages. Typically the limit is three. Power dissipation is further reduced and efficiency increased as the average voltage from which the output current is drawn is kept closer to the minimum. The inner devices operate in Class-B/AB as before, and the middle devices are in Class-C. The outer devices are also in Class-C, but conduct for even less of the time.

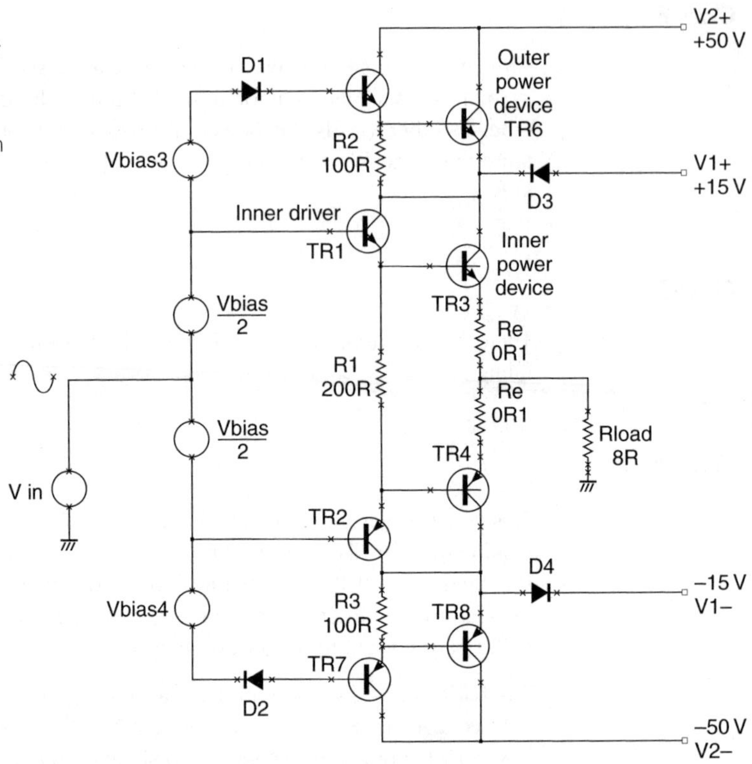

Figure 2.3
Class-G-Series output stage. When the output voltage exceeds the transition level, D3 or D4 turn off and power is drawn from the higher rails through the outer power devices

To the best of my knowledge three-level Class-G amplifiers have only been made in Shunt mode, as described below, probably because in Series mode the cumulative voltage drops become too great and compromise the efficiency gains. The extra complexity is significant, as there are now six supply rails and at least six power devices all of which must carry the full output current. It seems most unlikely that this further reduction in power consumption could ever be worthwhile for domestic hi-fi.

A closely related type of amplifier is Class-G-Shunt[10]. Figure 2.4 shows the principle; at low outputs only Q3, Q4 conduct, delivering power from the low-voltage rails. Above a threshold set by Vbias3 and Vbias4, D1 or D2 conduct and Q6, Q8 turn on, drawing current from the high-voltage rails, with D3, 4 protecting Q3, 4 against reverse bias. The conduction periods of the Q6, Q8 Class-C devices are variable, but inherently less than 50%. Normally the low-voltage section runs in Class-B to minimise dissipation. Such shunt Class-G arrangements are often called 'commutating amplifiers'.

Some of the more powerful Class-G-Shunt PA amplifiers have three sets of supply rails to further reduce the average voltage drop between rail and output. This is very useful in large PA amplifiers.

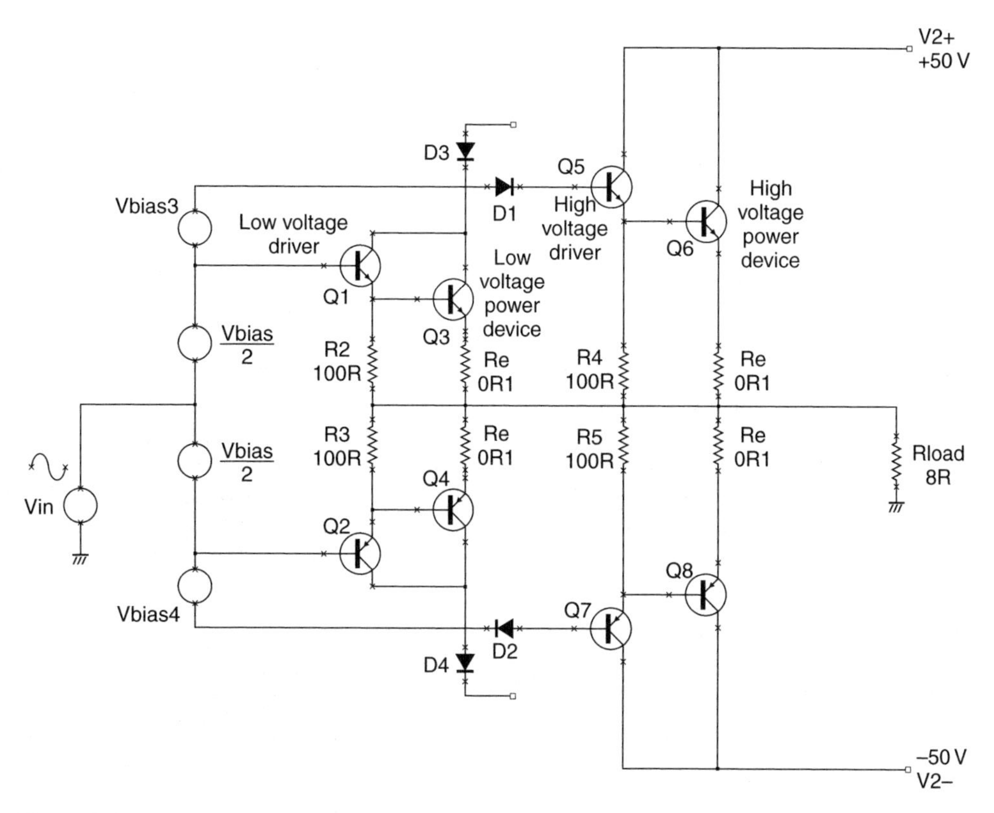

Figure 2.4
A Class-G-Shunt output stage, composed of two EF output stages with the usual drivers. Vbias3, 4 set the output level at which power is drawn from the higher rails

Class-H

Class-H is once more basically Class-B, but with a method of dynamically boosting the single supply rail (as opposed to switching to another one) in order to increase efficiency[12]. The usual mechanism is a form of bootstrapping. Class-H is occasionally used to describe Class-G as above; this sort of confusion we can do without.

Class-S

Class-S, so named by Doctor Sandman[13], uses a Class-A stage with very limited current capability, backed up by a Class-B stage connected so as to make the load appear as a higher resistance that is within the first amplifier's capability.

The method used by the Technics SE-A100 amplifier is extremely similar[14].

I hope that this necessarily brief catalogue is comprehensive; if anyone knows of other bona fide classes I would be glad to add them to the collection. This classification does not allow a completely consistent nomenclature; for example, Quad-style Current-Dumping can only be specified as a mixture of Class A and C, which says nothing about the basic principle of operation, which is error-correction.

Variations on Class-B

The solid-state Class-B three-stage amplifier has proved both successful and flexible, so many attempts have been made to improve it further, usually by trying to combine the efficiency of Class-B with the linearity of Class-A. It would be impossible to give a comprehensive list of the changes and improvements attempted, so I give only those that have been either commercially successful or particularly thought-provoking to the amplifier-design community.

Error-correcting amplifiers

This refers to error-cancellation strategies rather than the conventional use of negative feedback. This is a complex field, for there are at least three different forms of error-correction, of which the best known is error-feedforward as exemplified by the ground-breaking Quad 405[15]. Other versions include error feedback and other even more confusingly named techniques, some at least of which turn out on analysis to be conventional NFB in disguise. For a highly ingenious treatment of the feedforward method by Giovanni Stochino[16].

Non-switching amplifiers

Most of the distortion in Class-B is crossover distortion, and results from gain changes in the output stage as the power devices turn on and off. Several researchers have attempted to avoid this by ensuring that each device is clamped to pass a certain minimum current at all times[17]. This approach has certainly been exploited commercially, but few technical details have been published. It is not intuitively obvious (to me, anyway) that stopping the diminishing device current in its tracks will give less crossover distortion. See also Chapter 9.

Current-drive amplifiers

Almost all power amplifiers aspire to be voltage sources of zero output impedance. This minimises frequency response variations caused by the peaks and dips of the impedance curve, and gives a *universal* amplifier that can drive any loudspeaker directly.

The opposite approach is an amplifier with a sufficiently high output impedance to act as a constant-current source. This eliminates some problems – such as rising voice-coil resistance with heat dissipation – but introduces others such as control of the cone resonance. Current amplifiers therefore appear to be only of use with active crossovers and velocity feedback from the cone[18].

It is relatively simple to design an amplifier with any desired output impedance (even a negative one) and so any compromise between voltage and current drive is attainable. The snag is that loudspeakers are universally designed to be driven by voltage sources, and higher amplifier impedances demand tailoring to specific speaker types[19].

The Blomley principle

The goal of preventing output transistors from turning off completely was introduced by Peter Blomley in 1971[20]; here the positive/negative splitting is done by circuitry ahead of the output stage, which can then be designed so that a minimum idling current can be separately set up in each output device. However, to the best of my knowledge this approach has not yet achieved commercial exploitation.

Geometric mean Class-AB

The classical explanations of Class-B operation assume that there is a fairly sharp transfer of control of the output voltage between the two output devices, stemming from an equally abrupt switch in conduction from one to the other. In practical audio amplifier stages this is indeed the case, but it is not an inescapable result of the basic principle. Figure 2.5 shows a conventional output stage, with emitter resistors Re1, Re2 included to increase quiescent-current stability and allow current-sensing for overload protection; it is these emitter resistances that to a large extent make classical Class-B what it is.

However, if the emitter resistors are omitted, and the stage biased with two matched diode junctions, then the diode and transistor junctions form a *translinear loop*[21] around which the junction voltages sum to zero. This links the two output transistor currents I_p, I_n in the relationship $I_n{*}I_p$ = constant, which in op-amp practice is known as Geometric-Mean Class AB operation. This gives smoother changes in device current at the crossover point, but this does not necessarily mean lower THD. Such techniques are

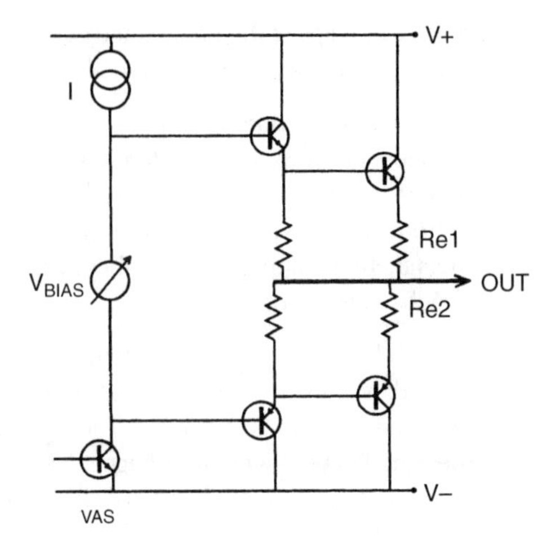

Figure 2.5
A conventional double emitter-follower output stage with emitter resistors Re shown

not very practical for discrete power amplifiers; first, in the absence of the very tight thermal coupling between the four junctions that exists in an IC, the quiescent-current stability will be atrocious, with thermal runaway and spontaneous combustion a near-certainty. Second, the output device bulk emitter resistance will probably give enough voltage drop to turn the other device off anyway, when current flows. The need for drivers, with their extra junction-drops, also complicates things.

A new extension of this technique is to redesign the translinear loop so that $1/I_n + 1/I_p = \text{constant}$, this being known as Harmonic-Mean AB operation[22]. It is too early to say whether this technique (assuming it can be made to work outside an IC) will be of use in reducing crossover distortion and thus improving amplifier performance.

Nested differentiating feedback loops

This is a most ingenious, but conceptually complex technique for significantly increasing the amount of NFB that can be applied to an amplifier. See Cherry[23].

AC- and DC-coupled amplifiers

All power amplifiers are either AC-coupled or DC-coupled. The first kind have a single supply rail, with the output biased to be halfway between this rail and ground to give the maximum symmetrical voltage swing; a large DC-blocking capacitor is therefore used in series with the output. The second kind have positive and negative supply rails, and the output is biased to be at 0 V, so no output DC-blocking is required in normal operation.

The advantages of AC-coupling

1. The output DC offset is always zero (unless the output capacitor is leaky).
2. It is very simple to prevent turn-on thump by purely electronic means. The amplifier output must rise up to half the supply voltage at turn-on, but providing this occurs slowly there is no audible transient. Note that in many designs, this is not simply a matter of making the input bias voltage rise slowly, as it also takes time for the DC feedback to establish itself, and it tends to do this with a snap-action when a threshold is reached.
3. No protection against DC faults is required, providing the output capacitor is voltage-rated to withstand the full supply rail. A DC-coupled amplifier requires an expensive and possibly unreliable output relay for dependable speaker protection.
4. The amplifier should be more easy to make short-circuit proof, as the output capacitor limits the amount of electric charge that can be transferred each cycle, no matter how low the load impedance. This is speculative; I have no data as to how much it really helps in practice.
5. AC-coupled amplifiers do not in general appear to require output inductors for stability. Large electrolytics have significant equivalent series resistance (ESR) and a little series inductance. For typical amplifier output sizes the ESR will be of the order of 100 mΩ; this resistance is probably the reason why AC-coupled amplifiers rarely had output inductors, as it is enough resistance to provide isolation from capacitative loading and so gives stability. Capacitor series inductance is very low and probably irrelevant, being quoted by one manufacturer as 'A few tens of nanoHenrys'. The output capacitor was often condemned in the past for reducing the low-frequency damping factor (DF), for its ESR alone is usually enough to limit the DF to 80 or so. As explained above, this is not a technical problem because 'damping factor' means virtually nothing.

The advantages of DC-coupling

1. No large and expensive DC-blocking capacitor is required. On the other hand the dual supply will need at least one more equally expensive reservoir capacitor, and a few extra components such as fuses.
2. In principle there should be no turn-on thump, as the symmetrical supply rails mean the output voltage does not have to move through half the supply voltage to reach its bias point – it can just stay where it is. In practice the various filtering time-constants used to keep the bias voltages free from ripple are likely to make various sections of the amplifier turn on at different times, and the resulting thump can be substantial. This can be dealt with almost for free, when a protection relay is fitted, by delaying the relay pull-in until any transients are over. The delay required is usually less than a second.

3 Audio is a field where almost any technical eccentricity is permissible, so it is remarkable that AC-coupling appears to be the one technique that is widely regarded as unfashionable and unacceptable. DC-coupling avoids any marketing difficulties.
4 Some potential customers will be convinced that DC-coupled amplifiers give better speaker damping due to the absence of the output capacitor impedance. They will be wrong, as explained on page 25, but this misconception has lasted at least 40 years and shows no sign of fading away.
5 Distortion generated by an output capacitor is avoided. This is a serious problem, as it is not confined to low frequencies, as is the case in small-signal circuitry. See page 176. For a 6800 µF output capacitor driving 40 W into an 8 Ω load, there is significant mid-band third harmonic distortion at 0.0025%, as shown in Figure 2.6. This is at least five times more than the amplifier generates in this part of the frequency range. In addition, the THD rise at the LF end is much steeper than in the small-signal case, for reasons that are not yet clear. There are two cures for output capacitor distortion. The straightforward approach uses a huge output capacitor, far larger in value than required for a good low-frequency response. A 100, 000 µF/40 V Aerovox from BHC eliminated all distortion, as shown in Figure 2.7. An allegedly 'audiophile' capacitor gives some interesting results; a Cerafine Supercap of only moderate size (4700 µF/63 V) gave Figure 2.8, where the mid-band distortion is gone, but the LF distortion rise remains. What special audio properties this component is supposed to have are unknown; as far as I know electrolytics are never advertised as 'low mid-band THD', but that seems to be the case here. The volume of the capacitor case is about twice as great as conventional electrolytics of the same value, so it is possible the

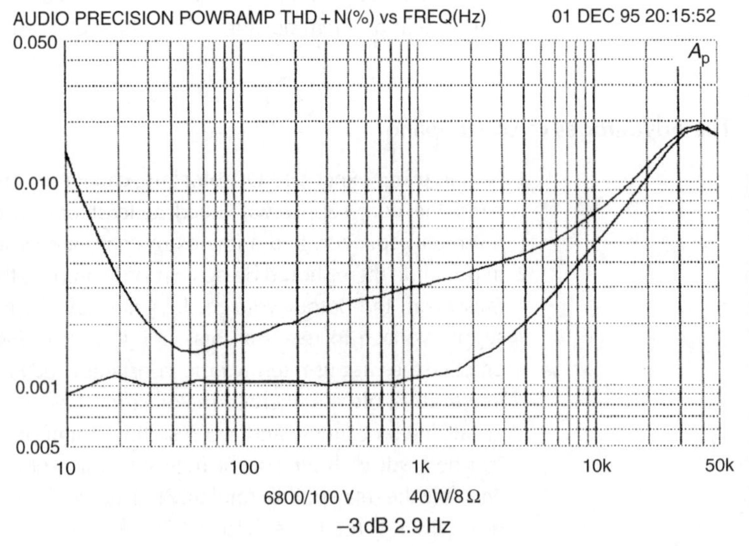

Figure 2.6
The extra distortion generated by an 6800 µF electrolytic delivering 40 W into 8 Ω. Distortion rises as frequency falls, as for the small-signal case, but at this current level there is also added distortion in the mid-band

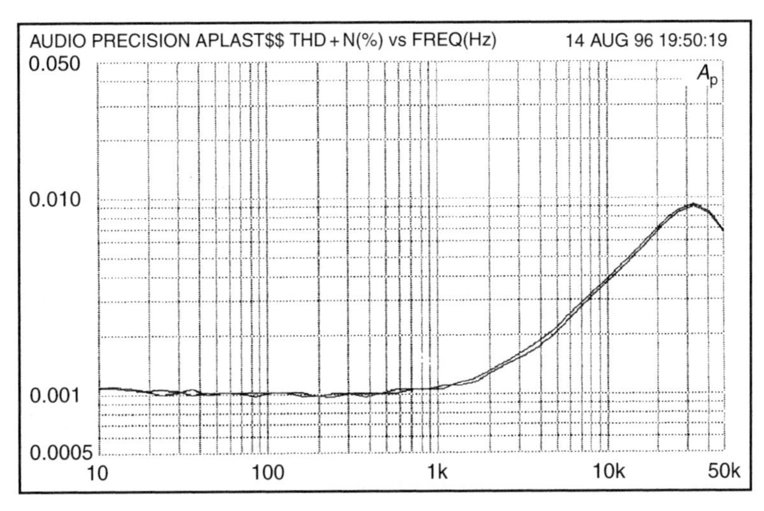

Figure 2.7
Distortion with and without a very large output capacitor, the BHC Aerovox 100,000 μF/40 V (40 watts/8 Ω). Capacitor distortion is eliminated

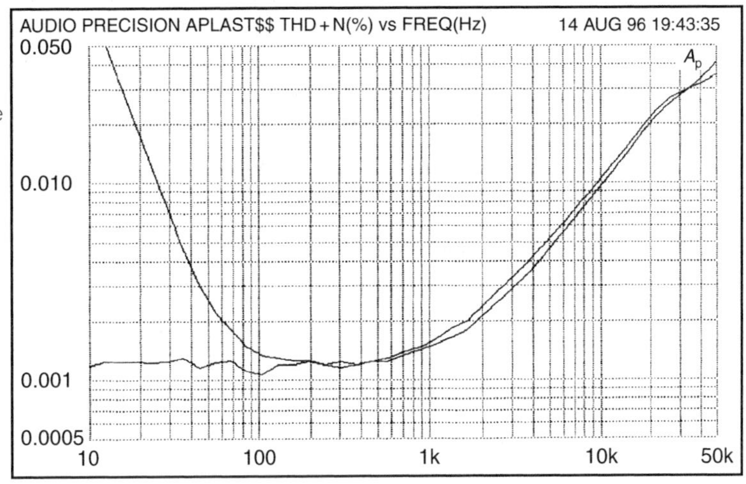

Figure 2.8
Distortion with and without an 'audiophile' Cerafine 4700 μF/63 V capacitor. Mid-band distortion is eliminated but LF rise is much the same as the standard electrolytic

crucial difference may be a thicker dielectric film than is usual for this voltage rating.

Either of these special capacitors costs more than the rest of the amplifier electronics put together. Their physical size is large. A DC-coupled amplifier with protective output relay will be a more economical option.

A little-known complication with output capacitors is that their series reactance increases the power dissipation in the output stage at low frequencies. This is counter-intuitive as it would seem that any impedance added in series must reduce the current drawn and hence the power dissipation. In fact it is the load phase-shift that increases the amplifier dissipation.

6 The supply currents can be kept out of the ground system. A single-rail AC amplifier has half-wave Class-B currents flowing in the 0 V rail, and these can have a serious effect on distortion and crosstalk performance.

Negative feedback in power amplifiers

It is not the role of this book to step through elementary theory which can be easily found in any number of textbooks. However, correspondence in audio and technical journals shows that considerable confusion exists on negative feedback as applied to power amplifiers; perhaps there is something inherently mysterious in a process that improves almost all performance parameters simply by feeding part of the output back to the input, but inflicts dire instability problems if used to excess. I therefore deal with a few of the less obvious points here; much more information is provided in Chapter 7.

The main use of NFB in amplifiers is the reduction of harmonic distortion, the reduction of output impedance, and the enhancement of supply-rail rejection. There are analogous improvements in frequency response and gain stability, and reductions in DC drift, but these are usually less important in audio applications.

By elementary feedback theory, the factor of improvement for all these quantities is:

$$\text{Improvement ratio} = A.\beta \qquad \text{Equation 2.1}$$

where A is the open-loop gain, and β the attenuation in the feedback network, i.e., the reciprocal of the closed-loop gain. In most audio applications the improvement factor can be regarded as simply open-loop gain divided by closed-loop gain.

In simple circuits you just apply negative feedback and that is the end of the matter. In a typical power amplifier, which cannot be operated without NFB, if only because it would be saturated by its own DC offset voltages, there are several stages which may accumulate phase-shift, and simply closing the loop usually brings on severe Nyquist oscillation at HF. This is a serious matter, as it will not only burn out any tweeters that are unlucky enough to be connected, but can also destroy the output devices by overheating, as they may be unable to turn off fast enough at ultrasonic frequencies. (See page 160.)

The standard cure for this instability is compensation. A capacitor is added, usually in Miller-Integrator format, to roll-off the open-loop gain at 6 dB per octave, so it reaches unity loop-gain before enough phase-shift can build up to allow oscillation. This means the NFB factor varies strongly with frequency, an inconvenient fact that many audio commentators seem to forget.

It is crucial to remember that a distortion harmonic, subjected to a frequency-dependent NFB factor as above, will be reduced by the NFB factor corresponding to its own frequency, not that of its fundamental. If you have a choice, generate low-order rather than high-order distortion harmonics, as the NFB deals with them much more effectively.

Negative-feedback can be applied either locally (i.e., to each stage, or each active device) or globally, in other words right around the whole amplifier. Global NFB is more efficient at distortion reduction than the same amount distributed as local NFB, but places much stricter limits on the amount of phase-shift that may be allowed to accumulate in the forward path.

Above the dominant pole frequency, the VAS acts as a Miller integrator, and introduces a constant 90° phase lag into the forward path. In other words, the output from the input stage must be in quadrature if the final amplifier output is to be in phase with the input, which to a close approximation it is. This raises the question of how the ninety-degree phase shift is accommodated by the negative-feedback loop; the answer is that the input and feedback signals applied to the input stage are there subtracted, and the small difference between two relatively large signals with a small phase shift between them has a much larger phase shift. This is the signal that drives the VAS input of the amplifier.

Solid-state power amplifiers, unlike many valve designs, are almost invariably designed to work at a fixed closed-loop gain. If the circuit is compensated by the usual dominant-pole method, the HF open-loop gain is also fixed, and therefore so is the important negative feedback factor. This is in contrast to valve amplifiers, where the amount of negative feedback applied was regarded as a variable, and often user-selectable parameter; it was presumably accepted that varying the negative feedback factor caused significant changes in input sensitivity. A further complication was serious peaking of the closed-loop frequency response at both LF and HF ends of the spectrum as negative feedback was increased, due to the inevitable bandwidth limitations in a transformer-coupled forward path. Solid-state amplifier designers go cold at the thought of the customer tampering with something as vital as the NFB factor, and such an approach is only acceptable in cases like valve amplification where global NFB plays a minor role.

Some common misconceptions about negative feedback

All of the comments quoted below have appeared many times in the hi-fi literature. All are wrong.

Negative feedback is a bad thing. Some audio commentators hold that, without qualification, negative feedback is a bad thing. This is of course completely untrue and based on no objective reality. Negative feedback is one of the fundamental concepts of electronics, and to avoid its use altogether is virtually impossible; apart from anything else, a small amount of local NFB exists in every common-emitter transistor because of the internal emitter resistance. I detect here distrust of good fortune; the uneasy feeling that if something apparently works brilliantly then there must be something wrong with it.

A low negative-feedback factor is desirable. Untrue; global NFB makes just about everything better, and the sole effect of too much is HF oscillation, or poor transient behaviour on the brink of instability. These effects are painfully obvious on testing and not hard to avoid unless there is something badly wrong with the basic design.

In any case, just what does *low* mean? One indicator of imperfect knowledge of negative feedback is that the amount enjoyed by an amplifier is almost always baldly specified as *so many dB* on the very few occasions it is specified at all – despite the fact that most amplifiers have a feedback factor that varies considerably with frequency. A dB figure quoted alone is meaningless, as it cannot be assumed that this is the figure at 1 kHz or any other standard frequency.

My practice is to quote the NFB factor at 20 kHz, as this can normally be assumed to be above the dominant pole frequency, and so in the region where open-loop gain is set by only two or three components. Normally the open-loop gain is falling at a constant 6 dB/octave at this frequency on its way down to intersect the unity-loop-gain line and so its magnitude allows some judgement as to Nyquist stability. Open-loop gain at LF depends on many more variables such as transistor beta, and consequently has wide tolerances and is a much less useful quantity to know. This is dealt with in more detail on page 104.

Negative feedback is a powerful technique, and therefore dangerous when misused. This bland truism usually implies an audio Rakes's Progress that goes something like this: an amplifier has too much distortion, and so the open-loop gain is increased to augment the NFB factor. This causes HF instability, which has to be cured by increasing the compensation capacitance. This is turn reduces the slew-rate capability, and results in a sluggish, indolent, and generally bad amplifier.

The obvious flaw in this argument is that the amplifier so condemned no longer has a high NFB factor, because the increased compensation capacitor has reduced the open-loop gain at HF; therefore feedback itself can hardly be blamed. The real problem in this situation is probably unduly low standing current in the input stage; this is the other parameter determining slew-rate.

NFB may reduce low-order harmonics but increases the energy in the discordant higher harmonics. A less common but recurring complaint is that the application of global NFB is a shady business because it transfers energy from low-order distortion harmonics – considered musically consonant – to higher-order ones that are anything but. This objection contains a grain of truth, but appears to be based on a misunderstanding of one article in an important series by Peter Baxandall[24] in which he showed that if you took an amplifier with only second-harmonic distortion, and then introduced NFB around it, higher-order harmonics were indeed generated as the second harmonic was fed back round the loop. For example, the fundamental and the second-harmonic intermodulate to give a component at third-harmonic frequency. Likewise, the second and third intermodulate to give the fifth harmonic. If we accept that high-order harmonics should be numerically weighted to reflect their greater unpleasantness, there could conceivably be a rise rather than a fall in the weighted THD when negative feedback is applied.

All active devices, in Class A or B (including FETs, which are often erroneously thought to be purely square-law), generate small amounts of high-order harmonics. Feedback could and would generate these from nothing, but in practice they are already there.

The vital point is that if enough NFB is applied, all the harmonics can be reduced to a lower level than without it. The extra harmonics generated, effectively by the distortion of a distortion, are at an extremely low level providing a reasonable NFB factor is used. This is a powerful argument against low feedback factors like 6 dB, which are most likely to increase the weighted THD. For a full understanding of this topic, a careful reading of the Baxandall series is absolutely indispensable.

A low open-loop bandwidth means a sluggish amplifier with a low slew-rate. Great confusion exists in some quarters between open-loop bandwidth and slew-rate. In truth open-loop bandwidth and slew-rate are nothing to do with each other, and may be altered independently. Open-loop bandwidth is determined by compensation *Cdom*, VAS β, and the resistance at the VAS collector, while slew-rate is set by the input stage standing current and *Cdom* · *Cdom* affects both, but all the other parameters are independent. (See Chapter 3 for more details.)

In an amplifier, there is a maximum amount of NFB you can safely apply at 20 kHz; this does not mean that you are restricted to applying the same amount at 1 kHz, or indeed 10 Hz. The obvious thing to do is to allow the NFB to continue increasing at 6 dB/octave – or faster if possible – as frequency falls, so that the amount of NFB applied doubles with each octave as we move down in frequency, and we derive as much benefit as we can. This obviously cannot continue indefinitely, for eventually open-loop gain runs out, being limited by transistor beta and other factors. Hence the NFB factor levels out at a relatively low and ill-defined frequency; this frequency is the open-loop bandwidth, and for an amplifier that can never be used open-loop, has very little importance.

It is difficult to convince people that this frequency is of no relevance whatever to the speed of amplifiers, and that it does not affect the slew-rate. Nonetheless, it is so, and any First-year electronics textbook will confirm this. High-gain op-amps with sub-1 Hz bandwidths and blindingly fast slewing are as common as the grass (if somewhat less cheap) and if that does not demonstrate the point beyond doubt then I really do not know what will.

Limited open-loop bandwidth prevents the feedback signal from immediately following the system input, so the utility of this delayed feedback is limited. No linear circuit can introduce a pure time-delay; the output must begin to respond at once, even if it takes a long time to complete its response. In the typical amplifier the dominant-pole capacitor introduces a 90° phase-shift between input-pair and output at all but the lowest audio frequencies, but this is not a true time-delay. The phrase *delayed feedback* is often used to describe this situation, and it is a wretchedly inaccurate term; if you really delay the feedback to a power amplifier (which can only be done by adding a time-constant to the feedback network rather than the forward path) it will quickly turn into the proverbial power oscillator as sure as night follows day.

Amplifier stability and NFB

In controlling amplifier distortion, there are two main weapons. The first is to make the linearity of the circuitry as good as possible before closing the feedback loop. This is unquestionably important, but it could be argued it can only be taken so far before the complexity of the various amplifier stages involved becomes awkward. The second is to apply as much negative feedback as possible while maintaining amplifier stability. It is well known that an amplifier with a single time-constant is always stable, no matter how high the feedback factor. The linearisation of the VAS by local Miller feedback is a good example. However, more complex circuitry, such as the generic three-stage power amplifier, has more than one time-constant, and these extra poles will cause poor transient response or instability if a high feedback factor is maintained up to the higher frequencies where they start to take effect. It is therefore clear that if these higher poles can be eliminated or moved upward in frequency, more feedback can be applied and distortion will be less for the same stability margins. Before they can be altered – if indeed this is practical at all – they must be found and their impact assessed.

The dominant pole frequency of an amplifier is, in principle, easy to calculate; the mathematics is very simple (see page 64). In practice, two of the most important factors, the effective beta of the VAS and the VAS collector impedance, are only known approximately, so the dominant pole frequency is a rather uncertain thing. Fortunately this parameter in itself has no effect on amplifier stability. What matters is the amount of feedback at high frequencies.

Things are different with the higher poles. To begin with, where are they? They are caused by internal transistor capacitances and so on, so there is no physical component to show where the roll-off is. It is generally regarded as fact that the next poles occur in the output stage, which will use power devices that are slow compared with small-signal transistors. Taking the Class-B design on page 179, the TO-92 MPSA06 devices have an Ft of 100 MHz, the MJE340 drivers about 15 MHz (for some reason this parameter is missing from the data sheet) and the MJ802 output devices an Ft of 2.0 MHz. Clearly the output stage is the prime suspect. The next question is at what frequencies these poles exist. There is no reason to suspect that each transistor can be modelled by one simple pole.

There is a huge body of knowledge devoted to the art of keeping feedback loops stable while optimising their accuracy; this is called Control Theory, and any technical bookshop will yield some intimidatingly fat volumes called things like 'Control System Design'. Inside, system stability is tackled by Laplace-domain analysis, eigenmatrix methods, and joys like the Lyapunov stability criterion. I think that makes it clear that you need to be pretty good at mathematics to appreciate this kind of approach.

Even so, it is puzzling that there seems to have been so little application of Control Theory to audio amplifier design. The reason may be that so much Control Theory assumes that you know fairly accurately the characteristics of what you are trying to control, especially in terms of poles and zeros.

One approach to appreciating negative feedback and its stability problems is SPICE simulation. Some SPICE simulators have the ability to work in the Laplace or s-domain, but my own experiences with this have been deeply unhappy. Otherwise respectable simulator packages output complete rubbish in this mode. Quite what the issues are here I do not know, but it does seem that s-domain methods are best avoided. The approach suggested here instead models poles directly as poles, using RC networks to generate the time-constants. This requires minimal mathematics and is far more robust. Almost any SPICE simulator – evaluation versions included – should be able to handle the simple circuit used here.

Figure 2.9 shows the basic model, with SPICE node numbers. The scheme is to idealise the situation enough to highlight the basic issues and exclude distractions like non-linearities or clipping. The forward gain is simply the transconductance of the input stage multiplied by the transadmittance of the VAS integrator. An important point is that with correct parameter values, the current from the input stage is realistic, and so are all the voltages.

The input differential amplifier is represented by G. This is a standard SPICE element – the VCIS, or voltage-controlled current source. It is inherently

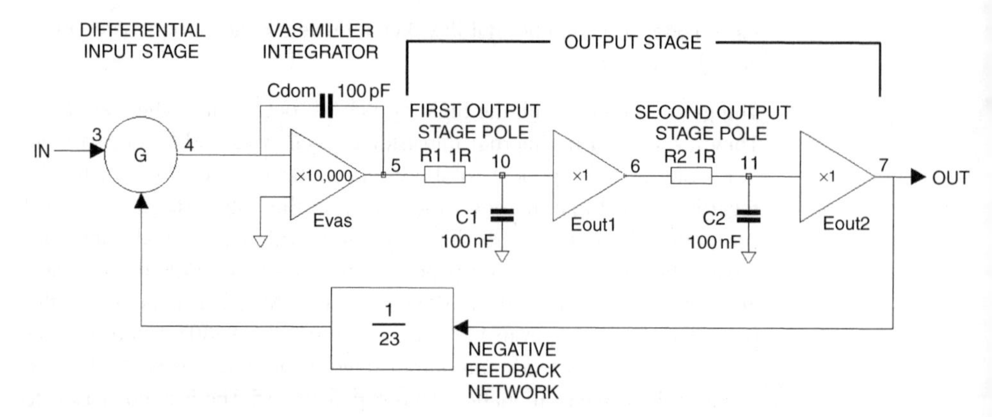

Figure 2.9
Block diagram of system for SPICE stability testing

differential, as the output current from Node 4 is the scaled difference between the voltages at Nodes 3 and 7. The scaling factor of 0.009 sets the input stage transconductance (*gm*) to 9 mA/V, a typical figure for a bipolar input with some local feedback. Stability in an amplifier depends on the amount of negative feedback available at 20 kHz. This is set at the design stage by choosing the input *gm* and *Cdom*, which are the only two factors affecting the open-loop gain (see page 63). In simulation it would be equally valid to change *gm* instead; however, in real life it is easier to alter *Cdom* as the only other parameter this affects is slew-rate. Changing input stage transconductance is likely to mean altering the standing current and the amount of local feedback, which will in turn impact input stage linearity.

The VAS with its dominant pole is modelled by the integrator Evas, which is given a high but finite open-loop gain, so there really is a dominant pole P1 created when the gain demanded becomes equal to that available. With *Cdom* = 100 pF this is below 1 Hz. With infinite (or as near-infinite as SPICE allows) open-loop gain the stage would be a perfect integrator. A explained elsewhere, the amount of open-loop gain available in real versions of this stage is not a well-controlled quantity, and P1 is liable to wander about in the 1–100 Hz region; fortunately this has no effect at all on HF stability. *Cdom* is the Miller capacitor that defines the transadmittance, and since the input stage has a realistic transconductance *Cdom* can be set to 100 pF, its usual real-life value. Even with this simple model we have a nested feedback loop. This apparent complication here has little effect, so long as the open-loop gain of the VAS is kept high.

The output stage is modelled as a unity-gain buffer, to which we add extra poles modelled by R1, C1 and R2, C2. Eout1 is a unity-gain buffer internal to the output stage model, added so the second pole does not load the first.

The second buffer Eout2 is not strictly necessary as no real loads are being driven, but it is convenient if extra complications are introduced later. Both are shown here as a part of the output stage but the first pole could equally well be due to input stage limitations instead; the order in which the poles are connected makes no difference to the final output. Strictly speaking, it would be more accurate to give the output stage a gain of 0.95, but this is so small a factor that it can be ignored.

The component values here are of course completely unrealistic, and chosen purely to make the maths simple. It is easy to appreciate that 1 Ω and 1 μF make up a 1 μsec time-constant. This is a pole at 159 kHz. Remember that the voltages in the latter half of the circuit are realistic, but the currents most certainly are not.

The feedback network is represented simply by scaling the output as it is fed back to the input stage. The closed-loop gain is set to 23 times, which is representative of most power amplifiers.

Note that this is strictly a linear model, so the slew-rate limiting which is associated with Miller compensation is not modelled here. It would be done by placing limits on the amount of current that can flow in and out of the input stage.

Figure 2.10 shows the response to a 1 V step input, with the dominant pole the only time element in the circuit. (The other poles are disabled by making C1, C2 0.00001 pF, because this is quicker than changing the actual circuit.) The output is an exponential rise to an asymptote of 23 V, which is exactly what elementary theory predicts. The exponential shape comes from the way that the error signal which drives the integrator becomes less as the output approaches the desired level. The error, in the shape

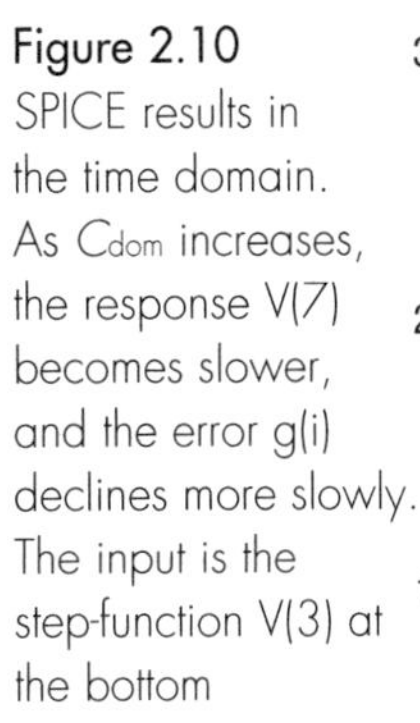

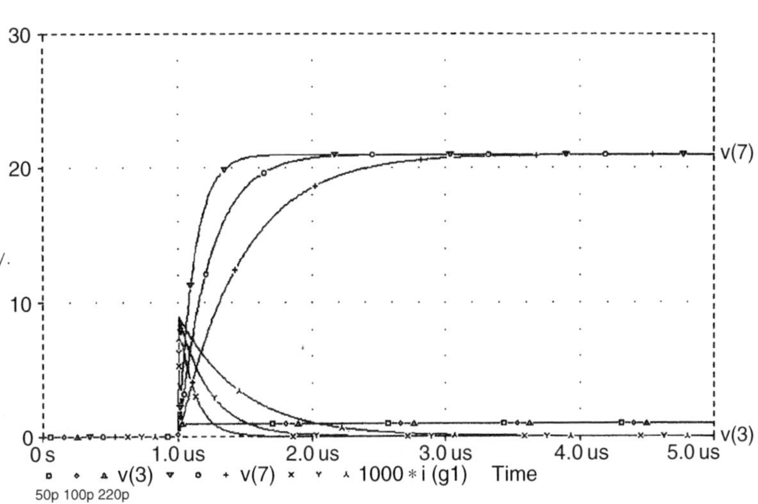

Figure 2.10 SPICE results in the time domain. As C_{dom} increases, the response V(7) becomes slower, and the error g(i) declines more slowly. The input is the step-function V(3) at the bottom

of the output current from G, is the smaller signal shown; it has been multiplied by 1000 to get mA onto the same scale as volts. The speed of response is inversely proportional to the size of Cdom, and is shown here for values of 50 pF and 220 pF as well as the standard 100 pF. This simulation technique works well in the frequency domain, as well as the time domain. Simply tell SPICE to run an AC simulation instead of a TRANS (transient) simulation. The frequency response in Figure 2.11 exploits this to show how the closed-loop gain in a NFB amplifier depends on the open-loop gain available. Once more elementary feedback theory is brought to life. The value of Cdom controls the bandwidth, and it can be seen that the values used in the simulation do not give a very extended response compared with a 20 kHz audio bandwidth.

In Figure 2.12, one extra pole P2 at 1.59 MHz (a time-constant of only 100 ns) is added to the output stage, and Cdom stepped through 50, 100 and 200 pF as before. 100 pF shows a slight overshoot that was not there before; with 50 pF there is a serious overshoot that does not bode well for the frequency response. Actually, it's not that bad; Figure 2.13 returns to the frequency-response domain to show that an apparently vicious overshoot is actually associated with a very mild peaking in the frequency domain.

From here on Cdom is left set to 100 pF, its real value in most cases. In Figure 2.14 P2 is stepped instead, increasing from 100 ns to 5 μs, and while the response gets slower and shows more overshoot, the system does not become unstable. The reason is simple: sustained oscillation (as opposed to transient ringing) in a feedback loop requires positive feedback, which means that a total phase shift of 180° must have accumulated in the forward

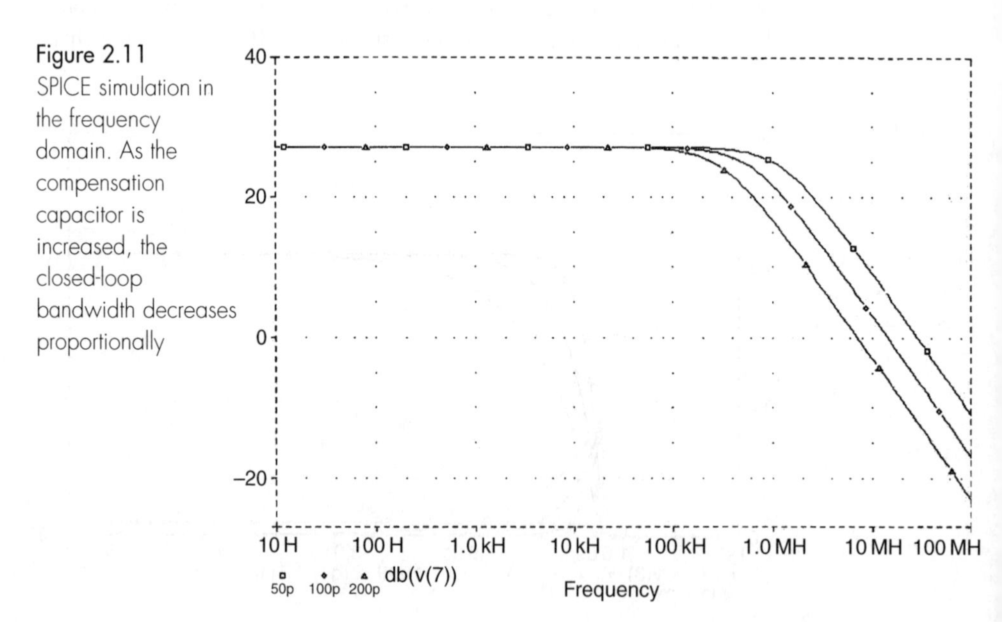

Figure 2.11 SPICE simulation in the frequency domain. As the compensation capacitor is increased, the closed-loop bandwidth decreases proportionally

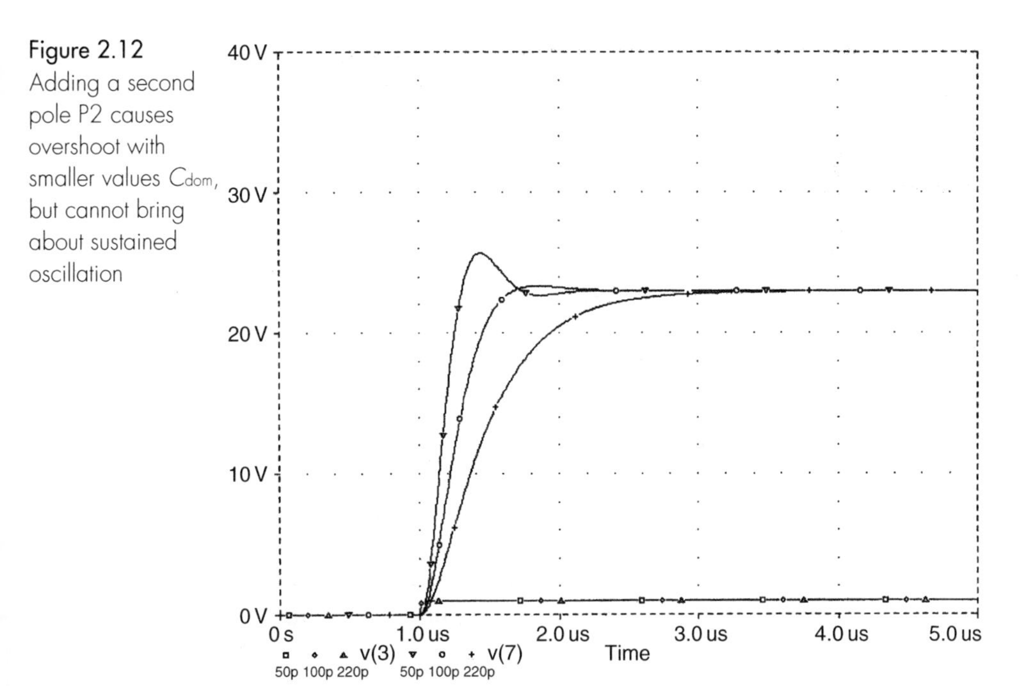

Figure 2.12 Adding a second pole P2 causes overshoot with smaller values C_{dom}, but cannot bring about sustained oscillation

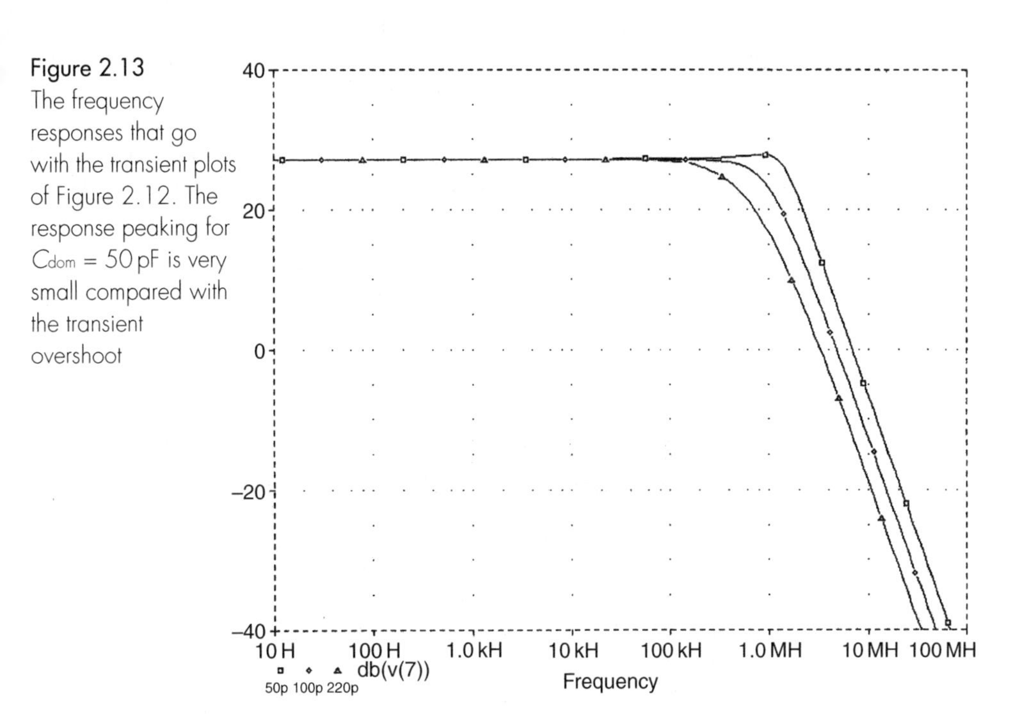

Figure 2.13 The frequency responses that go with the transient plots of Figure 2.12. The response peaking for C_{dom} = 50 pF is very small compared with the transient overshoot

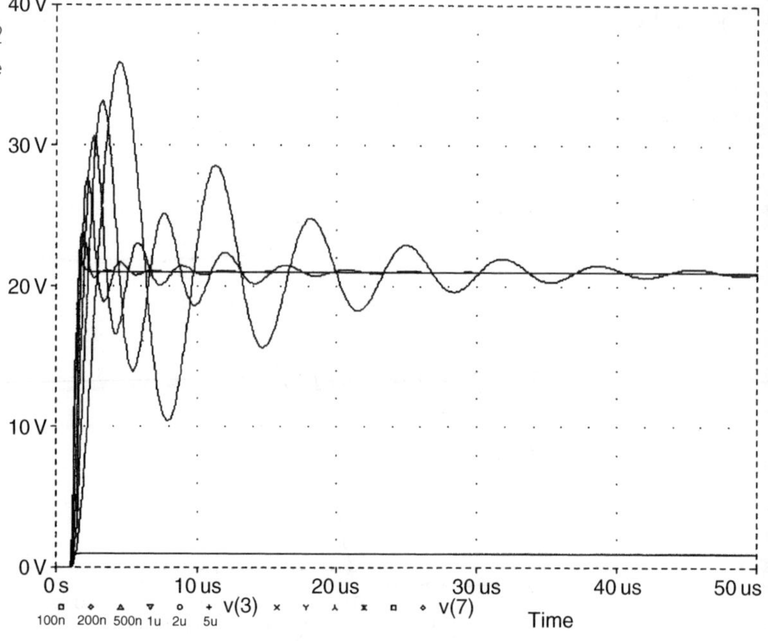

Figure 2.14 Manipulating the P2 frequency can make ringing more prolonged but it is still not possible to provoke sustained oscillation

path, and reversed the phase of the feedback connection. With only two poles in a system the phase shift cannot reach 180°. The VAS integrator gives a dependable 90° phase shift above P1, being an integrator, but P2 is instead a simple lag and can only give 90° phase lag at infinite frequency. So, even this very simple model gives some insight. Real amplifiers do oscillate if Cdom is too small, so we know that the frequency response of the output stage cannot be meaningfully modelled with one simple lag.

A certain president of the United States is alleged to have said: 'Two wrongs don't make a right – so let's see if three will do it'. Adding in a third pole *P*3 in the shape of another simple lag gives the possibility of sustained oscillation.

Stepping the value of *P*2 from 0.1 to 5 µsec with *P*3 = 500 nsec shows sustained oscillation starting to occur at *P*2 = 0.45 µsec. For values such as *P*2 = 0.2 µsec the system is stable and shows only damped oscillation. Figure 2.15 shows over 50 µsec what happens when the amplifier is made very unstable (there are degrees of this) by setting *P*2 = 5 µsec and *P*3 = 500 nsec. It still takes time for the oscillation to develop, but exponentially diverging oscillation like this is a sure sign of disaster. Even in the short time examined here the amplitude has exceeded a rather theoretical half a kilovolt. In reality oscillation cannot increase indefinitely, if only because the supply rail voltages would limit the amplitude. In practice slew-rate limiting is probably the major controlling factor in the amplitude of high-frequency oscillation.

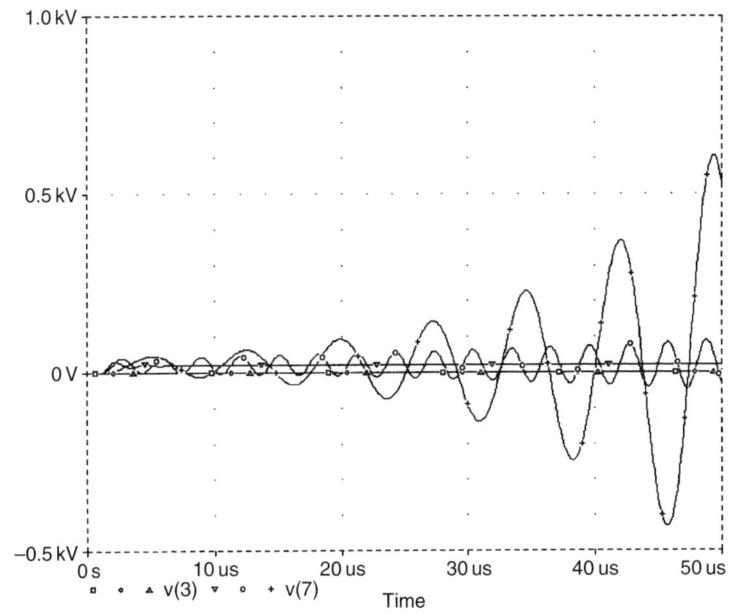

Figure 2.15 Adding a third pole makes possible true instability with exponentially increasing amplitude of oscillation. Note the unrealistic voltage scale on this plot

We have now modelled a system that will show instability. But does it do it right? Sadly, no. The oscillation is about 200 kHz, which is a rather lower frequency than is usually seen when a amplifier misbehaves. This low frequency stems from the low *P*2 frequency we have to use to provoke oscillation; apart from anything else this seems out of line with the known Ft of power transistors. Practical amplifiers are likely to take off at around 500 kHz to 1 MHz when Cdom is reduced, and this seems to suggest that phase shift is accumulating quickly at this sort of frequency. One possible explanation is that there are a large number of poles close together at a relatively high frequency.

A fourth pole can be simply added to Figure 2.9 by inserting another RC–buffer combination into the system. With $P2 = 0.5\,\mu\text{sec}$ and $P3 = P4 = 0.2\,\mu\text{sec}$, instability occurs at 345 kHz, which is a step towards a realistic frequency of oscillation. This is case B in Table 2.1.

Table 2.1 Instability onset. P2 is increased until sustained oscillation occurs

Case	*Cdom*	*P2*	*P3*	*P4*	*P5*	*P6*	
A	100p	0.45	0.5	–	–		200 kHz
B	100p	0.5	0.2	0.2	–		345 kHz
C	100p	0.2	0.2	0.2	0.01		500 kHz
D	100p	0.3	0.2	0.1	0.05		400 kHz
E	100p	0.4	0.2	0.1	0.01		370 kHz
F	100p	0.2	0.2	0.1	0.05	0.02	475 kHz

When a fifth output stage pole is grafted on, so that $P3 = P4 = P5 = 0.2$ μsec the system just oscillates at 500 kHz with $P2$ set to 0.01 μsec. This takes us close to a realistic frequency of oscillation. Rearranging the order of poles so $P2 = P3 = P4 = 0.2$ μsec, while $P5 = 0.01$ μsec, is tidier, and the stability results are of course the same; this is a linear system so the order does not matter. This is case C in Table 2.1.

Having $P2$, $P3$ and $P4$ all at the same frequency does not seem very plausible in physical terms, so case D shows what happens when the five poles are staggered in frequency. $P2$ needs to be increased to 0.3 μsec to start the oscillation, which is now at 400 kHz. Case E is another version with five poles, showing that if $P5$ is reduced $P2$ needs to be doubled to 0.4 μsec for instability to begin.

In the final case F, a sixth pole is added to see if this permitted sustained oscillation is above 500 kHz. This seems not to be the case; the highest frequency that could be obtained after a lot of pole-twiddling was 475 kHz. This makes it clear that this model is of limited accuracy (as indeed are all models – it is a matter of degree) at high frequencies, and that further refinement is required to gain further insight.

Maximising the NFB

Having freed ourselves from Fear of Feedback, and appreciating the dangers of using only a little of it, the next step is to see how much can be used. It is my view that the amount of negative feedback applied should be maximised at all audio frequencies to maximise linearity, and the only limit is the requirement for reliable HF stability. In fact, global or Nyquist oscillation is not normally a difficult design problem in power amplifiers; the HF feedback factor can be calculated simply and accurately, and set to whatever figure is considered safe. (Local oscillations and parasitics are beyond the reach of design calculations and simulations, and cause much more trouble in practice.)

In classical Control Theory, the stability of a servomechanism is specified by its Phase Margin, the amount of extra phase-shift that would be required to induce sustained oscillation, and its Gain Margin, the amount by which the open-loop gain would need to be increased for the same result. These concepts are not very useful in amplifier work, where many of the significant time-constants are only vaguely known. However it is worth remembering that the phase margin will never be better than 90°, because of the phase-lag caused by the VAS Miller capacitor; fortunately this is more than adequate.

In practice the designer must use his judgement and experience to determine an NFB factor that will give reliable stability in production. My own experience leads me to believe that when the conventional three-stage architecture is used, 30 dB of global feedback at 20 kHz is safe, providing an output inductor is used to prevent capacitive loads from eroding the stability margins. I would say that 40 dB was distinctly risky, and I would not care to pin it down any more closely than that.

The 30 dB figure assumes simple dominant-pole compensation with a 6 dB/octave roll-off for the open-loop gain. The phase and gain margins are determined by the angle at which this slope cuts the horizontal unity-loop-gain line. (I am deliberately terse here; almost all textbooks give a very full treatment of this stability criterion.) An intersection of 12 dB/octave is definitely unstable. Working within this, there are two basic ways in which to maximise the NFB factor:

1 while a 12 dB/octave gain slope is unstable, intermediate slopes greater than 6 dB/octave can be made to work. The maximum usable is normally considered to be 10 dB/octave, which gives a phase margin of 30°. This may be acceptable in some cases, but I think it cuts it a little fine. The steeper fall in gain means that more NFB is applied at lower frequencies, and so less distortion is produced. Electronic circuitry only provides slopes in multiples of 6 dB/octave, so 10 dB/octave requires multiple overlapping time-constants to approximate a straight line at an intermediate slope. This gets complicated, and this method of maximising NFB is not popular,
2 the gain slope varies with frequency, so that maximum open-loop gain and hence NFB factor is sustained as long as possible as frequency increases; the gain then drops quickly, at 12 dB/octave or more, but flattens out to 6 dB/octave before it reaches the critical unity loop-gain intersection. In this case the stability margins should be relatively unchanged compared with the conventional situation. This approach is dealt with in Chapter 7.

Maximising linearity before feedback

Make your amplifier as linear as possible before applying NFB has long been a cliché. It blithely ignores the difficulty of running a typical solid-state amplifier without any feedback, to determine its basic linearity.

Virtually no dependable advice on how to perform this desirable linearisation has been published. The two factors are the basic linearity of the forward path, and the amount of negative feedback applied to further straighten it out. The latter cannot be increased beyond certain limits or high-frequency stability is put in peril, whereas there seems no reason why open-loop linearity could not be improved without limit, leading us to what in some senses must be the ultimate goal – a distortionless amplifier. This book therefore takes as one of its main aims the understanding and improvement of open-loop linearity; as it proceeds we will develop circuit blocks culminating in some practical amplifier designs that exploit the techniques presented here.

References

1. Lin, H C *Transistor Audio Amplifier* Electronics, Sept 1956, p. 173.
2. Sweeney and Mantz *An Informal History of Amplifiers* Audio, June 1988, p. 46.
3. Linsley-Hood *Simple Class-A Amplifier* Wireless World, April 1969, p. 148.
4. Olsson, B *Better Audio from Non-Complements?* Electronics World, Dec 1994, p. 988.
5. Attwood, B *Design Parameters Important for the Optimisation of PWM (Class-D) Amplifiers* JAES Vol 31, Nov 1983, p. 842.
6. Goldberg and Sandler *Noise Shaping and Pulse-Width Modulation for All-Digital Audio Power Amplifier* JAES Vol 39, Feb 1991, p. 449.
7. Hancock, J *A Class-D Amplifier Using MOSFETS with Reduced Minority Carrier Lifetime* JAES Vol 39, Sept 1991, p. 650.
8. Peters, A *Class E RF Amplifiers IEEE* J. Solid-State Circuits, June 1975, p. 168.
9. Feldman, L *Class-G High-Efficiency Hi-Fi Amplifier* Radio-Electronics, Aug 1976, p. 47.
10. Raab, F *Average Efficiency of Class-G Power Amplifiers* IEEE Transactions on Consumer Electronics, Vol CE-22, May 1986, p. 145.
11. Sampei et al *Highest Efficiency & Super Quality Audio Amplifier Using MOS-Power FETs in Class-G* IEEE Transactions on Consumer Electronics, Vol CE-24 Aug 1978, p. 300.
12. Buitendijk, P *A 40 W Integrated Car Radio Audio Amplifier* IEEE Conf on Consumer Electronics, 1991 Session THAM 12.4, p. 174. (Class-H)
13. Sandman, A *Class S: A Novel Approach to Amplifier Distortion* Wireless World, Sept 1982, p. 38.
14. Sinclair (ed) *Audio and Hi-Fi Handbook* pub. Newnes 1993, p. 541.
15. Walker, P J *Current Dumping Audio Amplifier* Wireless World, Dec 1975, p. 560.
16. Stochino, G *Audio Design Leaps Forward?* Electronics World, Oct 1994, p. 818.
17. Tanaka, S *A New Biasing Circuit for Class-B Operation* JAES, Jan/Feb 1981, p. 27.
18. Mills and Hawksford *Transconductance Power Amplifier Systems for Current-Driven Loudspeakers* JAES Vol 37, March 1989, p. 809.
19. Evenson, R *Audio Amplifiers with Tailored Output Impedances* Preprint for Nov 1988 AES convention (Los Angeles).
20. Blomley, P *A New Approach to Class-B* Wireless World, Feb 1971, p. 57.
21. Gilbert, B *Current Mode Circuits from a Translinear Viewpoint Ch 2, Analogue IC Design: The Current-Mode Approach* Ed Toumazou, Lidgey & Haigh, IEE 1990.

22. Thus *Compact Bipolar Class AB Output Stage* IEEE Journal of Solid-State Circuits, Dec 1992, p. 1718.
23. Cherry, E *Nested Differentiating Feedback Loops in Simple Audio Power Amplifiers* JAES Vol 30 #5, May 1982, p. 295.
24. Baxandall, P *Audio Power Amplifier Design: Part 5* Wireless World, Dec 1978, p. 53. (This superb series of articles had 6 parts and ran on roughly alternate months, starting in Jan 1978.)

3

The general principles of power amplifiers

How a generic amplifier works

Figure 3.1 shows a very conventional power amplifier circuit; it is as *standard* as possible. A great deal has been written about this configuration, though the subtlety and quiet effectiveness of the topology are usually overlooked, and the explanation below therefore touches on several aspects that seem to be almost unknown. The circuit has the merit of being docile enough to be made into a functioning amplifier by someone who has only the sketchiest of notions as to how it works.

The input differential pair implements one of the few forms of distortion cancellation that can be relied upon to work reliably without adjustment – this is because the transconductance of the input pair is determined by the physics of transistor action rather than matching of ill-defined parameters such as beta; the logarithmic relation between Ic and Vbe is proverbially accurate over some eight or nine decades of current variation.

The voltage signal at the Voltage Amplifier Stage (hereafter VAS) transistor base is typically a couple of millivolts, looking rather like a distorted triangle wave. Fortunately the voltage here is of little more than academic interest, as the circuit topology essentially consists of a transconductance amp (voltage-difference input to current output) driving into a transresistance (current-to-voltage converter) stage. In the first case the exponential Vbe/Ic law is straightened out by the differential-pair action, and in the second the global (overall) feedback factor at LF is sufficient to linearise the VAS, while at HF shunt Negative Feedback (hereafter NFB) through *C*dom conveniently takes over VAS-linearisation while the overall feedback factor is falling.

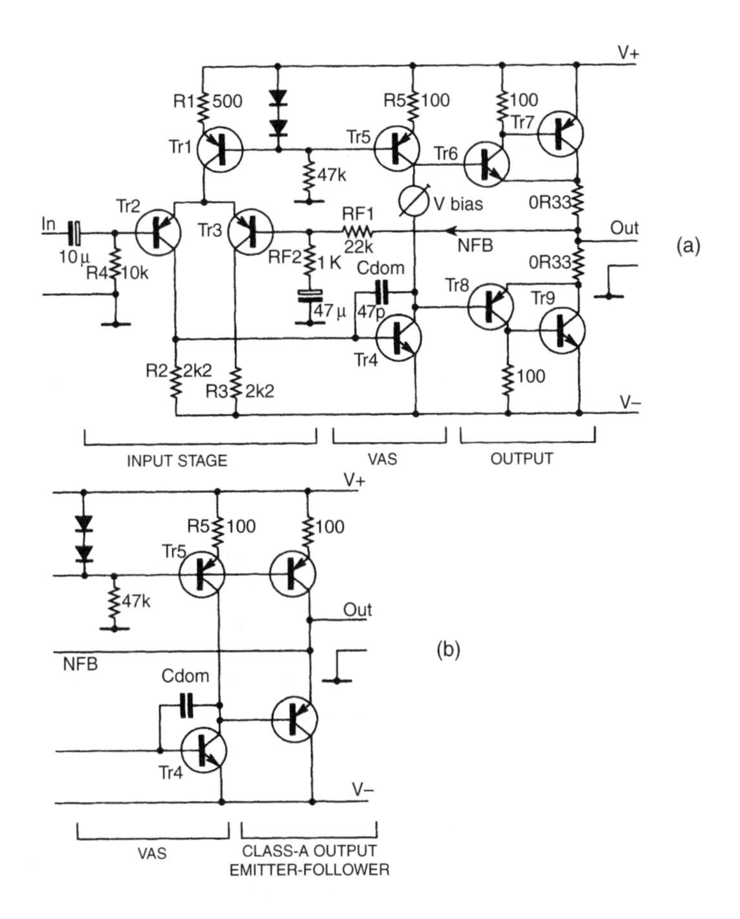

Figure 3.1
(a) A conventional Class-B power amp circuit. (b) With small-signal Class-A output emitter-follower replacing Class-B output to make a model amplifier

The behaviour of Miller dominant-pole compensation in this stage is actually exceedingly elegant, and not at all a case of finding the most vulnerable transistor and slugging it. As frequency rises and Cdom begins to take effect, negative feedback is no longer applied globally around the whole amplifier, which would include the higher poles, but instead is seamlessly transferred to a purely local role in linearising the VAS. Since this stage effectively contains a single gain transistor, any amount of NFB can be applied to it without stability problems.

The amplifier operates in two regions; the LF, where open-loop (o/l) gain is substantially constant, and HF, above the dominant-pole breakpoint, where the gain is decreasing steadily at 6 dB/octave. Assuming the output stage is unity-gain, three simple relationships define the gain in these two regions:

$$\text{LF gain} = gm \times \beta \times Rc \qquad \text{Equation 3.1}$$

At least one of the factors that set this (beta) is not well-controlled and so the LF gain of the amplifier is to a certain extent a matter of pot-luck; fortunately this does not matter, so long as it is high enough to give a suitable level of NFB to eliminate LF distortion. The use of the word *eliminate* is deliberate, as will be seen later. Usually the LF gain, or HF local feedback-factor, is made high by increasing the effective value of the VAS collector impedance *Rc*, either by the use of current-source collector-load, or by some form of bootstrapping.

The other important relations are:

$$\text{HF gain} = gm/(w \times C\text{dom}) \qquad \text{Equation 3.2}$$

$$\text{Dominant pole freq P1} = 1/(w \times C\text{dom} \times \beta \times Rc) \qquad \text{Equation 3.3}$$

(where w = 2 × pi × freq).

In the HF region, things are distinctly more difficult as regards distortion, for while the VAS is locally linearised, the global feedback-factor available to linearise the input and output stages is falling steadily at 6 dB/octave. For the time being we will assume that it is possible to define an HF gain (say N dB at 20 kHz) which will assure stability with practical loads and component variations. Note that the HF gain, and therefore both HF distortion and stability margin, are set by the simple combination of the input stage transconductance and one capacitor, and most components have no effect on it at all.

It is often said that the use of a high VAS collector impedance provides a current drive to the output devices, often with the implication that this somehow allows the stage to skip quickly and lightly over the dreaded crossover region. This is a misconception – the collector impedance falls to a few kilohms at HF, due to increasing local feedback through Cdom, and in any case it is very doubtful if true current drive would be a good thing – calculation shows that a low-impedance voltage drive minimises distortion due to beta-unmatched output halves[1], and it certainly eliminates the effect of Distortion 4, described below.

The advantages of convention

It is probably not an accident that the generic configuration is by a long way the most popular, though in the uncertain world of audio technology it is unwise to be too dogmatic about this sort of thing. The generic configuration has several advantages over other approaches:

- The input pair not only provides the simplest way of making a DC-coupled amplifier with a dependably small output offset voltage, but can also (given half a chance) completely cancel the second-harmonic distortion which would be generated by a single-transistor input stage. One vital condition for this must be met; the pair must be accurately balanced by choosing the associated components so that the two collector currents are equal. (The *typical* component values shown in Figure 3.1 do *not* bring about this most desirable state of affairs.)

- The input devices work at a constant and near-equal *V*ce, giving good thermal balance.
- The input pair has virtually no voltage gain so no low-frequency pole can be generated by Miller effect in the TR2 collector-base capacitance. All the voltage gain is provided by the VAS stage, which makes for easy compensation. Feedback through *C*dom lowers VAS input and output impedances, minimising the effect of input-stage capacitance, and the output-stage capacitance. This is often known as pole-splitting[2]; the pole of the VAS is moved downwards in frequency to become the dominant pole, while the input-stage pole is pushed up in frequency.
- The VAS Miller compensation capacitance smoothly transfers NFB from a global loop that may be unstable, to the VAS local loop that cannot be. It is quite wrong to state that *all* the benefits of feedback are lost as the frequency increases above the dominant pole, as the VAS is still being linearised. This position of *C*dom also swamps the rather variable *C*cb of the VAS transistor.

The eight distortions

My original series of articles on amplifier distortion listed seven important distortion mechanisms, all of which are applicable to any Class-B amplifier, and do not depend on particular circuit arrangements. As a result of further experimentation, I have now increased this to eight.

In the typical amplifier THD is often thought to be simply due to the Class-B nature of the output stage, which is linearised less effectively as the feedback factor falls with increasing frequency. This is, however, only true when all the removable sources of distortion have been eliminated. In the vast majority of amplifiers in production, the true situation is more complex, as the small-signal stages can generate significant distortion of their own, in at least two different ways; this distortion can easily exceed output stage distortion at high frequencies. It is particularly inelegant to allow this to occur given the freedom of design possible in the small-signal section.

If the ills that a class-B stage is heir to are included then there are eight major distortion mechanisms. Note that this assumes that the amplifier is not overloaded, and has proper global or Nyquist stability and does not suffer from any parasitic oscillations; the latter, if of high enough frequency, tend to manifest themselves only as unexpected increases in distortion, sometimes at very specific power outputs and frequencies.

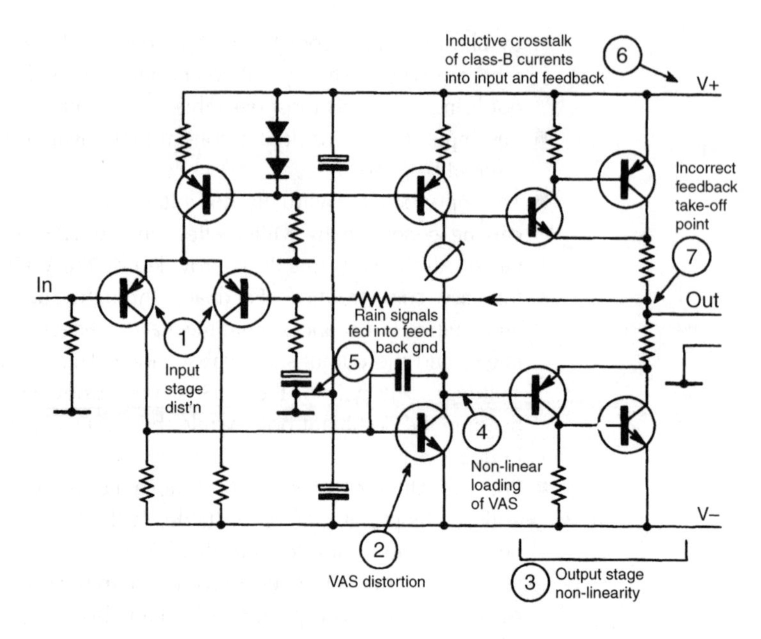

Figure 3.2
The location of the first seven major distortion mechanisms. The eighth (capacitor distortion) is omitted for clarity

In Figure 3.2 an attempt has been made to show the distortion situation diagrammatically, indicating the location of each mechanism within the amplifier. Distortion 8 is not shown.

Distortion one: input stage distortion

Non-linearity in the input stage. If this is a carefully balanced differential pair then the distortion is typically only measurable at HF, rises at 18 dB/octave, and is almost pure third harmonic. If the input pair is unbalanced (which from published circuitry it usually is) then the HF distortion emerges from the noise floor earlier, as frequency increases, and rises at 12 dB/octave as it is mostly second harmonic.

Distortion two: VAS distortion

Non-linearity in the voltage-amplifier stage (which I call the VAS for concision) surprisingly does not always figure in the total distortion. If it does, it remains constant until the dominant-pole freq P1 is reached, and then rises at 6 dB/octave. With the configurations discussed here it is always second harmonic.

Usually the level is very low due to linearising negative feedback through the dominant-pole capacitor. Hence if you crank up the local VAS open-loop gain, for example by cascoding or putting more current-gain in the local VAS-Cdom loop, and attend to Distortion 4) below, you can usually ignore VAS distortion.

Distortion three: output stage distortion

Non-linearity in the output stage, which is naturally the obvious source. This in a Class-B amplifier will be a complex mix of large-signal distortion and crossover effects, the latter generating a spray of high-order harmonics, and in general rising at 6 dB/octave as the amount of negative feedback decreases. Large-signal THD worsens with 4 Ω loads and worsens again at 2 Ω. The picture is complicated by dilatory switch-off in the relatively slow output devices, ominously signalled by supply current increasing in the top audio octaves.

Distortion four: VAS loading distortion

Loading of the VAS by the non-linear input impedance of the output stage. When all other distortion sources have been attended to, this is the limiting distortion factor at LF (say below 2 kHz); it is simply cured by buffering the VAS from the output stage. Magnitude is essentially constant with frequency, though overall effect in a complete amplifier becomes less as frequency rises and feedback through Cdom starts to linearise the VAS.

Distortion five: rail decoupling distortion

Non-linearity caused by large rail-decoupling capacitors feeding the distorted signals on the supply lines into the signal ground. This seems to be the reason that many amplifiers have rising THD at low frequencies. Examining one commercial amplifier kit, I found that rerouting the decoupler ground-return reduced the THD at 20 Hz by a factor of three.

Distortion six: induction distortion

Non-linearity caused by induction of Class-B supply currents into the output, ground, or negative-feedback lines. This was highlighted by Cherry[3] but seems to remain largely unknown; it is an insidious distortion that is hard to remove, though when you know what to look for on the THD residual it is fairly easy to identify. I suspect that a large number of commercial amplifiers suffer from this to some extent.

Distortion seven: NFB takeoff distortion

Non-linearity resulting from taking the NFB feed from slightly the wrong place near where the power-transistor Class-B currents sum to form the output. This may well be another very prevalent defect.

Distortion eight: capacitor distortion

Distortion, rising as frequency falls, caused by non-linearity in the input DC-blocking capacitor or the feedback network capacitor. The latter is more likely.

Non-existent distortions

Having set down what might be called The Eight Great Distortions, we must pause to put to flight a few Paper Tigers .... The first is common-mode distortion in the input stage, a spectre that haunts the correspondence columns. Since it is fairly easy to make an amplifier with less than <0.00065% THD (1 kHz) without paying any special attention to this it cannot be too serious a problem.

Giovani Stochino and I have investigated this a little, and we have independently found that if the common-mode voltage on the input pair is greatly increased, then a previously negligible distortion mechanism is indeed provoked. This CM increase is achieved by reducing the C/L gain to between 1 and 2x; the input signal is much larger for the same output, and the feedback signal must match it, so the input stage experiences a proportional increase in CM voltage.

At present it appears that the distortion produced by this mechanism increases as the square of the CM voltage. It therefore appears that the only precautions required against common-mode distortion are to ensure that the closed-loop gain is at least five times (which is no hardship, as it almost certainly is anyway) and to use a tail current-source for the input pair.

The second distortion conspicuous by its absence in the list is the injection of distorted supply-rail signals directly into the amplifier circuitry. Although this putative mechanism has received a lot of attention[4], dealing with Distortion 5 above by proper grounding seems to be all that is required; once more, if triple-zero THD can be attained using simple unregulated supplies and without paying any attention to the Power Supply Rejection Ratio beyond keeping the amplifier free from hum (which it reliably can be) then there seems to be no problem. There is certainly no need for regulated supply rails to get a good performance. PSRR does need careful attention if the hum/noise performance is to be of the first order, but a little RC filtering is usually all that is needed. This is dealt with in Chapter 8.

A third mechanism of very doubtful validity is thermal distortion, allegedly induced by parameter changes in semiconductor devices whose instantaneous power dissipation varies over a cycle. This would surely manifest itself as a distortion rise at very low frequencies, but it simply does not happen. There are several distortion mechanisms that can give a THD rise at LF, but when these are eliminated the typical distortion trace remains flat down to at least 10 Hz. The worst thermal effects would be expected in Class-B output stages where dissipation varies wildly over a cycle; however drivers and output devices have relatively large junctions with high thermal inertia. Low frequencies are of course also where the NFB factor is at its maximum. This contentious issue is dealt with at greater length in Chapter 5.

To return to our list of the unmagnificent eight, note that only Distortion 3 is directly due to O/P stage non-linearity, though Distortion 4–7 all result from the Class-B nature of the typical output stage. Distortion 8 can happen in any amplifier stage.

The performance of a standard amplifier

The THD curve for the standard amplifier is shown in Figure 3.3. As usual, distortion increases with frequency, and as we shall see later, would give grounds for suspicion if it did not. The flat part of the curve below 500 Hz represents non-frequency-sensitive distortion rather than the noise floor, which for this case is at the 0.0005% level. Above 500 Hz the distortion rises at an increasing rate, rather than a constant number of dB/octave, due to the combination of Distortions 1, 2, 3 and 4. (In this case, Distortions 5, 6 and 7 have been carefully eliminated to keep things simple; this is why the distortion performance looks good already, and the significance of this should not be overlooked.) It is often written that having distortion constant across the audio band is a Good Thing; a most unhappy conclusion, as the only practical way to achieve this with a Class-B amplifier is

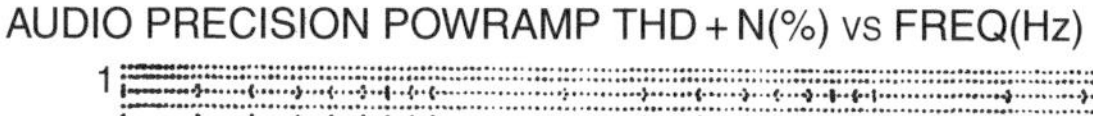

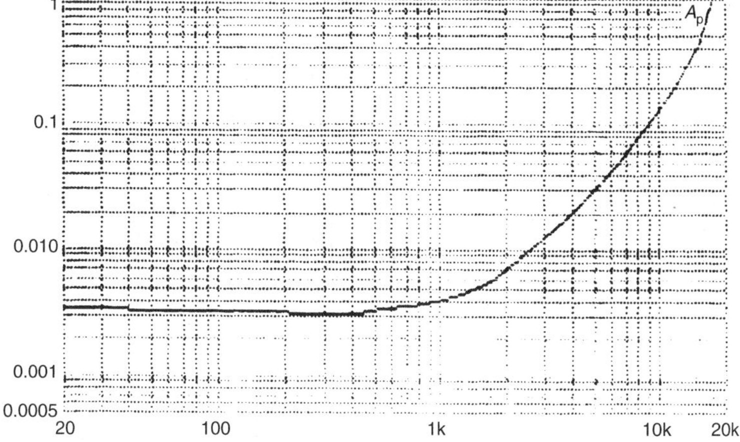

Figure 3.3 The distortion performance of the Class-B amplifier in Figure 3.1

to *increase* the distortion at LF, for example by allowing the VAS to distort significantly.

It should now be clear why it is hard to wring linearity out of such a snake-pit of contending distortions. A circuit-value change is likely to alter at least 2 of the distortion mechanisms, and probably change the o/l gain as well; in the coming chapters I shall demonstrate how each distortion mechanism can be measured and manipulated in isolation.

Open-loop linearity and how to determine it

Improving something demands measuring it, and thus it is essential to examine the open-loop linearity of power-amp circuitry. This cannot be done directly, so it is necessary to measure the NFB factor and calculate open-loop distortion from closed-loop measurements. The closed-loop gain is normally set by input sensitivity requirements.

Measuring the feedback-factor is at first sight difficult, as it means determining the open-loop gain. Standard methods for measuring op-amp open-loop gain involve breaking feedback-loops and manipulating closed-loop (c/l) gains, procedures that are likely to send the average power-amplifier into fits. Nonetheless the need to measure this parameter is inescapable, as a typical circuit modification – e.g. changing the value of R2 changes the open-loop gain as well as the linearity, and to prevent total confusion it is essential to keep a very clear idea of whether an observed change is due to an improvement in o/l linearity or merely because the o/l gain has risen. It is wise to keep a running check on this as work proceeds, so the direct method of open-loop gain measurement shown in Figure 3.4 was evolved.

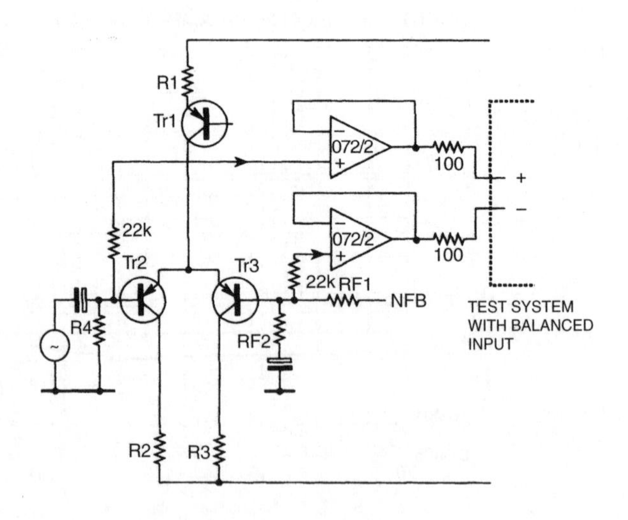

Figure 3.4
Test circuit for measuring open-loop gain directly. The accuracy with which high o/l gains can be measured depends on the testgear CMRR

Direct open-loop gain measurement

The amplifier shown in Figure 3.1 is a differential amplifier, so its open-loop gain is simply the output divided by the voltage difference between the inputs. If output voltage is kept constant by providing a constant swept-frequency voltage at the +ve input, then a plot of open-loop gain versus frequency is obtained by measuring the error-voltage between the inputs, and referring this to the output level. This gives an upside-down plot that rises at HF rather than falling, as the differential amplifier requires more input for the same output as frequency increases, but the method is so quick and convenient that this can be lived with. Gain is plotted in dB with respect to the chosen output level (+16 dBu in this case) and the actual gain at any frequency can be read off simply by dropping the minus sign. Figure 3.5 shows the plot for the amplifier in Figure 3.1.

The HF-region gain slope is always 6 dB/octave unless you are using something special in the way of compensation, and by the Nyquist rules must continue at this slope until it intersects the horizontal line representing the feedback factor, if the amplifier is stable. In other words, the slope is not being accelerated by other poles until the loop gain has fallen to unity, and this provides a simple way of putting a lower bound on the next pole P2; the important P2 frequency (which is usually somewhat mysterious) must be above the intersection frequency if the amplifier is seen to be stable.

Given test-gear with a sufficiently high Common-Mode-Rejection-Ratio balanced input, the method of Figure 3.4 is simple; just buffer the differential inputs from the cable capacitance with TL072 buffers, which place negligible loading on the circuit if normal component values are used. In particular be wary of adding stray capacitance to ground to the −ve input, as this directly imperils amplifier stability by adding an extra

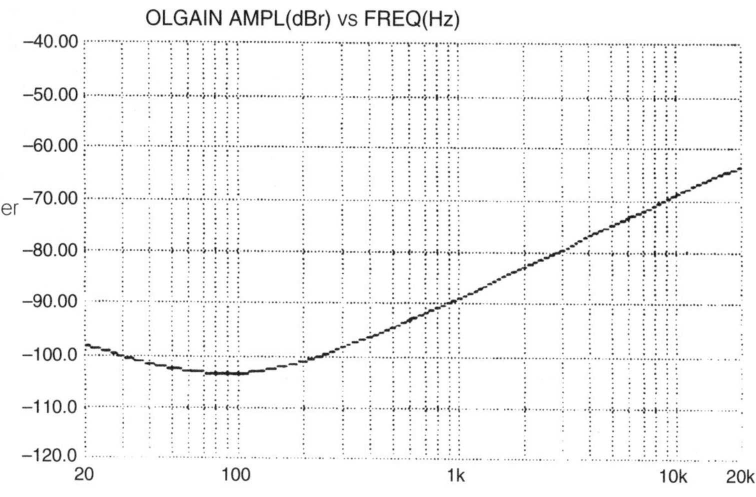

Figure 3.5
Open-loop gain versus freq plot for Figure 3.1. Note that the curve rises as gain falls, because the amplifier error is the actual quantity measured

feedback pole. Short wires from power amplifier to buffer IC can usually be unscreened as they are driven from low impedances.

The testgear input CMRR defines the maximum open-loop gain measurable; I used an Audio Precision System-1 without any special alignment of CMRR. A calibration plot can be produced by feeding the two buffer inputs from the same signal; this will probably be found to rise at 6 dB/octave, being set by the inevitable input asymmetries. This must be low enough for amplifier error signals to be above it by at least 10 dB for reasonable accuracy. The calibration plot will flatten out at low frequencies, and may even show an LF rise due to imbalance of the test-gear input-blocking capacitors; this can make determination of the lowest pole P1 difficult, but this is not usually a vital parameter in itself.

Using model amplifiers

Distortions 1 and 2 can dominate amplifier performance and need to be studied without the manifold complications introduced by a Class-B output stage. This can be done by reducing the circuit to a *model* amplifier that consists of the small-signal stages alone, with a very linear Class-A emitter-follower attached to the output to allow driving the feedback network; here *small-signal* refers to current rather than voltage, as the model amplifier should be capable of giving a full power-amp voltage swing, given sufficiently high rail voltages. From Figure 3.2 it is clear that this will allow study of Distortions 1 and 2 in isolation, and using this approach it will prove relatively easy to design a small-signal amplifier with negligible distortion across the audio band, and this is the only sure foundation on which to build a good power amplifier.

A typical plot combining Distortions 1 and 2 from a model amp is shown in Figure 3.6, where it can be seen that the distortion rises with an accelerating slope, as the initial rise at 6 dB/octave from the VAS is contributed to and then dominated by the 12 dB/octave rise in distortion from an unbalanced input stage.

The model can be powered from a regulated current-limited PSU to cut down the number of variables, and a standard output level chosen for comparison of different amplifier configurations; the rails and output level used for the results in this work were ±15 V and +16 dBu. The rail voltages can be made comfortably lower than the average amplifier HT rail, so that radical bits of circuitry can be tried out without the creation of a silicon cemetery around your feet. It must be remembered that some phenomena such as input-pair distortion depend on absolute output level, rather than the proportion of the rail voltage used in the output swing, and will be increased by a mathematically predictable amount when the real voltage swings are used.

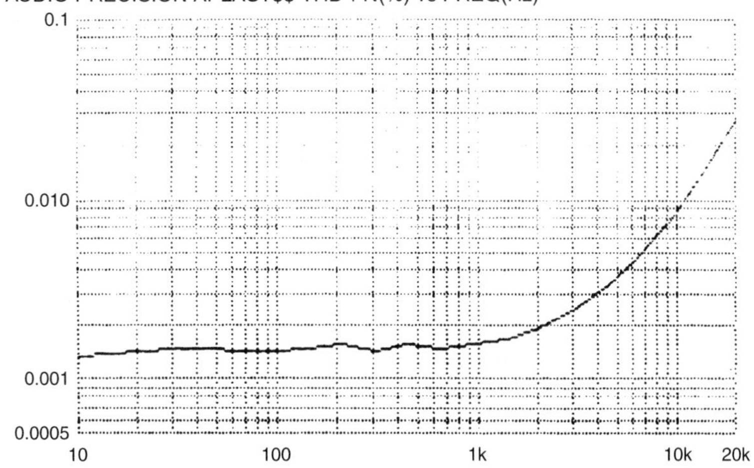

Figure 3.6
The distortion from a model amplifier, produced by the input pair and the Voltage-Amplifier Stage – note increasing slope as input pair distortion begins to add to VAS distortion

The use of such model amplifiers requires some caution, and gives no insight into BJT output stages, whose behaviour is heavily influenced by the sloth and low current gain of the power devices. As a general rule, it should be possible to replace the small-signal output with a real output stage and get a stable and workable power amplifier; if not, then the model is probably dangerously unrealistic.

The concept of the blameless amplifier

Here I introduce the concept of what I have chosen to call a *Blameless* audio power amplifier. This is an amplifier designed so that all the easily-defeated distortion mechanisms have been rendered negligible. (Note that the word *Blameless* has been carefully chosen to *not* imply Perfection, but merely the avoidance of known errors.) Such an amplifier gives about 0.0005% THD at 1 kHz and approximately 0.003% at 10 kHz when driving 8 Ω. This is much less THD than a Class-B amplifier is normally expected to produce, but the performance is repeatable, predictable, and definitely does not require large global feedback factors.

Distortion 1 cannot be totally eradicated, but its onset can be pushed well above 20 kHz by the use of local feedback. Distortion 2 (VAS distortion) can be similarly suppressed by cascoding or beta-enhancement, and Distortions 4 to 7 can be made negligible by simple topological methods. All these measures will be detailed later. This leaves Distortion 3, which includes the intractable Class-B problems, i.e., crossover distortion (Distortion 3b) and HF switch-off difficulties (Distortion 3c). Minimising 3b requires a Blameless amplifier to use a BJT output rather than FETs.

A Blameless Class-B amplifier essentially shows crossover distortion only, so long as the load is no heavier than 8 Ω; this distortion increases with frequency as the amount of global NFB falls. At 4 Ω loading an extra distortion mechanism (3a) generates significant third harmonic.

The importance of the Blameless concept is that it represents the best distortion performance obtainable from straightforward Class-B. This performance is stable and repeatable, and varies little with transistor type as it is not sensitive to variable quantities such as beta.

Blamelessness is a condition that can be defined with precision, and is therefore a standard other amplifiers can be judged against. A Blameless design represents a stable point of departure for more radical designs, such as the Trimodal concept in Chapter 9. This may be the most important use of the idea.

References

1. Oliver *Distortion In Complementary-Pair Class-B Amplifiers* Hewlett-Packard Journal, Feb 1971, p. 11.
2. Feucht *Handbook of Analog Circuit Design* Academic Press 1990, p. 256 (Pole-splitting).
3. Cherry, E *A New Distortion Mechanism in Class-B Amplifiers* JAES, May 1981, p. 327.
4. Ball, G *Distorting Power Supplies* Electronics World+WW, Dec 1990, p. 1084.

4

The small signal stages

'A beginning is the time for taking the most delicate care that the balances are correct.' Frank Herbert, *Dune*.

The role of the input stage

The input stage of an amplifier performs the critical duty of subtracting the feedback signal from the input, to generate the error signal that drives the output. It is almost invariably a differential transconductance stage; a voltage-difference input results in a current output that is essentially insensitive to the voltage at the output port. Its design is also frequently neglected, as it is assumed that the signals involved must be small, and that its linearity can therefore be taken lightly compared with that of the VAS or the output stage. This is quite wrong, for a misconceived or even mildly wayward input stage can easily dominate the HF distortion performance.

The input transconductance is one of the two parameters setting HF open-loop (o/l) gain, and therefore has a powerful influence on stability and transient behaviour as well as distortion. Ideally the designer should set out with some notion of how much o/l gain at 20 kHz will be safe when driving worst-case reactive loads (this information should be easier to gather now there is a way to measure o/l gain directly) and from this a suitable combination of input transconductance and dominant-pole Miller capacitance can be chosen.

Many of the performance graphs shown here are taken from a *model* (small-signal stages only) amplifier with a Class-A emitter-follower output, at +16 dBu on ±15 V rails; however, since the output from the input pair is in current form, the rail voltage in itself has no significant effect on the linearity of the input stage; it is the current swing at its output that is the crucial factor.

Distortion from the input stage

The motivation for using a differential pair as the input stage of an amplifier is usually its low DC offset. Apart from its inherently lower offset due to the cancellation of the *V*be voltages, it has the important added advantage that its standing current does not have to flow through the feedback network. However a second powerful reason, which seems less well-known, is that linearity is far superior to single-transistor input stages. Figure 4.1 shows three versions, in increasing order of sophistication. The resistor-tail version at 1a has poor CMRR and PSRR and is generally a false economy of the shabbiest kind; it will not be further considered here. The mirrored version at 1c has the best balance, as well as twice the transconductance of 1b.

At first sight, the input stage should generate a minimal proportion of the overall distortion because the voltage signals it handles are very small, appearing as they do upstream of the VAS that provides almost all the voltage gain. However, above the first pole frequency P1, the current required to drive *C*dom dominates the proceedings, and this remorselessly doubles with each octave, thus:

$$i_{pk} = w \times C\text{dom} \times V\text{pk} \quad \text{Equation 4.1}$$

where $w = 2 \times \pi \times \text{freq}$

For example, the current required at 100 W (8 Ω) and 20 kHz, with a 100 pF *C*dom is 0.5 mA peak, which may be a large proportion of the input standing current, and so the linearity of transconductance for large current excursions will be of the first importance if we want low distortion at high frequencies.

Curve A in Figure 4.2 shows the distortion plot for a model amplifier (at +16 dBu output) designed so the distortion from all other sources is negligible compared with that from the carefully balanced input stage; with a small-signal class A stage this reduces to making sure that the VAS is properly linearised. Plots are shown for both 80 kHz and 500 kHz

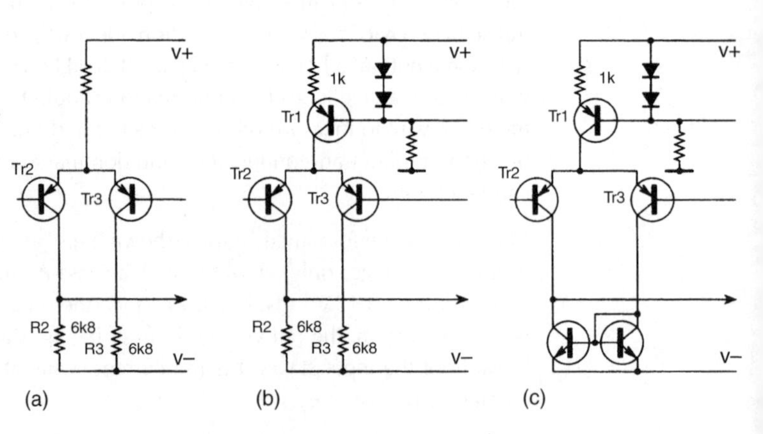

Figure 4.1
Three versions of an input pair. (a) Simple tail resistor. (b) Tail current-source. (c) With collector current-mirror to give inherently good Ic balance

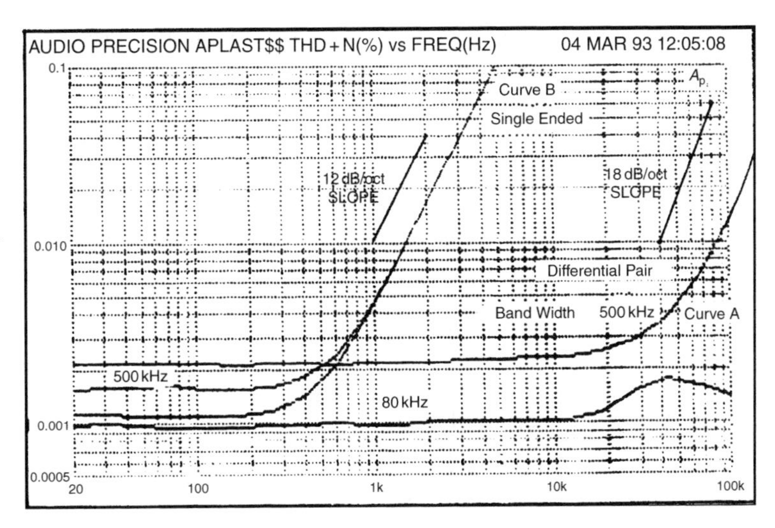

Figure 4.2 Distortion performance of model amplifier-differential pair at A compared with singleton input at B. The singleton generates copious second-harmonic distortion

measurement bandwidths, in an attempt to show both HF behaviour and the vanishingly low LF distortion. It can be seen that the distortion is below the noise floor until 10 kHz, when it emerges and heaves upwards at a precipitous 18 dB/octave. This rapid increase is due to the input stage signal current doubling with every octave, to feed Cdom; this means that the associated third harmonic distortion will quadruple with every octave increase. Simultaneously the overall NFB available to linearise this distortion is falling at 6 dB/octave since we are almost certainly above the dominant-pole frequency P1, and so the combined effect is an octuple or 18 dB/octave rise. If the VAS or the output stage were generating distortion this would be rising at only 6 dB/octave, and so would look quite different on the plot.

This non-linearity, which depends on the rate-of-change of the output voltage, is the nearest thing that exists to the late unlamented TID (Transient Intermodulation Distortion), an acronym that now seems to be falling out of fashion. SID (Slew-Induced Distortion) is a better description of the effect, but implies that slew-limiting is responsible, which is not the case.

If the input pair is *not* accurately balanced, then the situation is more complex. Second as well as third harmonic distortion is now generated, and by the same reasoning this has a slope nearer to 12 dB/octave; this vital point is examined more closely below.

BJTs vs FETs for the input stage

At every stage in the design of an amplifier, it is perhaps wise to consider whether BJTs or FETs are the best devices for the job. I may as well say at once that the predictable Vbe/Ic relationship and much higher transconductance of the bipolar transistor make it, in my opinion, the best choice for all three stages of a generic power amplifier. To quickly summarise the position:

Advantages of the FET input stage

There is no base current with FETs, so this is eliminated as a source of DC offset errors. However, it is wise to bear in mind that FET gate leakage currents increase very rapidly with temperature, and under some circumstances may need to be allowed for.

Disadvantages of FET input stage

1 The undegenerated transconductance is low compared with BJTs. There is much less scope for linearising the input stage by adding degeneration in the form of source resistors, and so an FET input stage will be very non-linear compared with a BJT version degenerated to give the same low transconductance.
2 The Vgs offset spreads will be high. Having examined many different amplifier designs, it seems that in practice it is essential to use dual FETs, which are relatively very expensive and not always easy to obtain. Even then, the Vgs mismatch will probably be greater than Vbe mismatch in a pair of cheap discrete BJTs; for example the 2N5912 N-channel dual FET has a specified maximum Vgs mismatch of 15 mV. In contrast the Vbe mismatches of BJTs, especially those taken from the same batch (which is the norm in production) will be much lower, at about 2–3 mV, and usually negligible compared with DC offset caused by unbalanced base currents.
3 The noise performance will be inferior if the amplifier is being driven from a low-impedance source, say 5 kΩ or less. This is almost always the case.

Singleton input stage versus differential pair

Using a single input transistor (Figure 4.3a) may seem attractive, where the amplifier is capacitor-coupled or has a separate DC servo; it at least promises strict economy. However, the snag is that this singleton configuration has no way to cancel the second-harmonics generated in copious quantities by its strongly-curved exponential Vin/Iout characteristic[1]. The result is shown in Figure 4.2 curve-B, where the distortion is much higher, though rising at the slower rate of 12 dB/octave.

The input stage distortion in isolation

Examining the slope of the distortion plot for the whole amplifier is instructive, but for serious research we need to measure input-stage

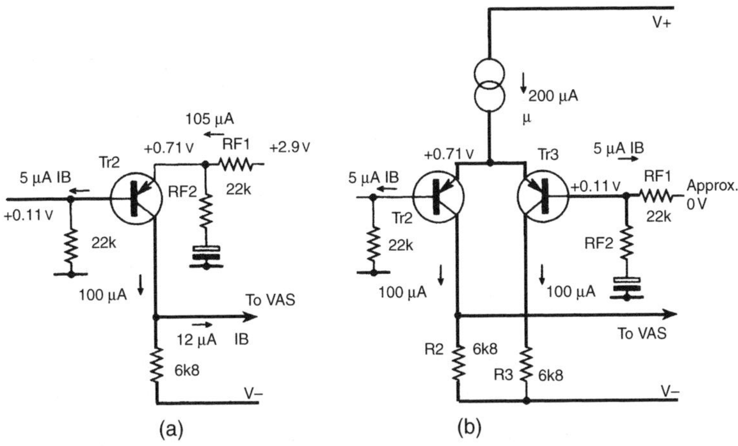

Figure 4.3 Singleton and differential pair input stages, showing typical DC conditions. The large DC offset of the singleton is mainly due to all the stage current flowing through the feedback resistor RF1

non-linearity in isolation. This can be done with the test circuit of Figure 4.4. The op-amp uses shunt feedback to generate an appropriate AC virtual-earth at the input-pair output. Note that this current-to-voltage conversion op-amp requires a third −30V rail to allow the i/p pair collectors to work at a realistic DC voltage, i.e., about one diode-worth above the −15V rail. Rf can be scaled as convenient, to stop op-amp clipping, without the input stage knowing anything has changed. The DC balance of the pair can be manipulated by VR1, and it is instructive to see the THD residual diminish as balance is approached, until at its minimum amplitude it is almost pure third harmonic.

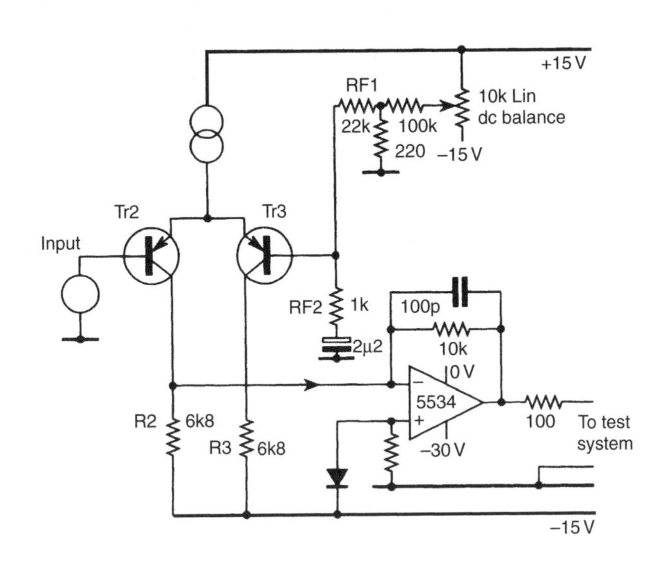

Figure 4.4 Test circuit for examining input stage distortion in isolation. The shunt-feedback op-amp is biased to provide the right DC conditions for TR2

The differential pair has the great advantage that its transfer characteristic is mathematically highly predictable[2]. The output current is related to the differential input voltage *V*in by:

$$I\text{out} = I\text{e}.\tanh(-V\text{in}/2V\text{t}) \qquad \text{Equation 4.2}$$

(where *V*t is the usual *thermal voltage* of about 26 mV at 25°C, and *I*e the tail current).

Two vital facts derived from this equation are that the transconductance (*g*m) is maximal at *V*in = 0, when the two collector currents are equal, and that the value of this maximum is proportional to the tail current *I*e. Device beta does not figure in the equation, and the performance of the input pair is not significantly affected by transistor type.

Figure 4.5a shows the linearising effect of local feedback or degeneration on the voltage-in/current-out law; Figure 4.5b plots transconductance against input voltage and shows clearly how the peak transconductance value

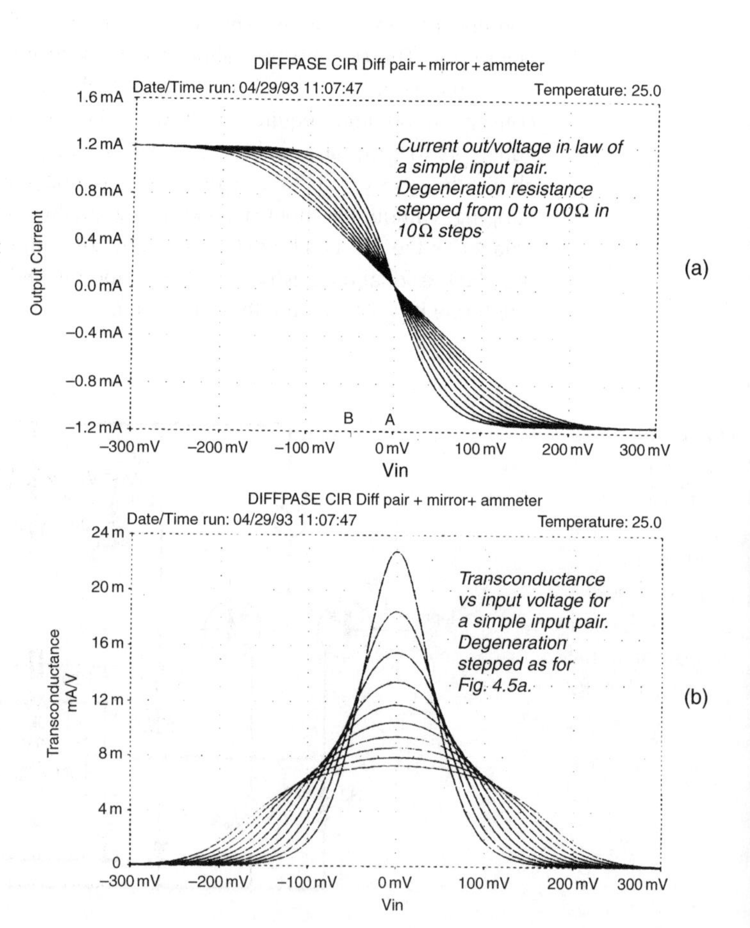

Figure 4.5 Effect of degeneration on input pair V/I law, showing how transconductance is sacrificed in favour of linearity. (SPICE simulation)

is reduced, but the curve made flatter and linear over a wider operating range. Simply adding emitter degeneration markedly improves the linearity of the input stage, but the noise performance is slightly worsened, and of course the overall amplifier feedback factor has been reduced, for as previously shown, the vitally-important HF closed-loop gain is determined solely by the input transconductance and the value of the dominant-pole capacitor.

Input stage balance

Exact DC balance of the input differential pair is essential in power amplifiers. It still seems almost unknown that minor deviations from equal I_c in the pair seriously upset the second-harmonic cancellation, by moving the operating point from A to B in Figure 4.5a. The average slope of the characteristic is greatest at A, so imbalance also reduces the open-loop gain if serious enough. The effect of small amounts of imbalance is shown in Figure 4.6 and Table 4.1; for an input of –45 dBu a collector-current imbalance of only 2% gives a startling worsening of linearity, with THD increasing from 0.10% to 0.16%; for 10% imbalance this deteriorates badly to 0.55%. Unsurprisingly, imbalance in the other direction (Ic1 > Ic2) gives similar results.

Imbalance defined as deviation of Ic (per device) from that value which gives equal currents in the pair.

This explains the complex distortion changes that accompany the apparently simple experiment of altering the value of R2[3]. We might design an input stage like Figure 4.7a, where R1 has been selected as 1k by uninspired guesswork and R2 made highish at 10k in a plausible but misguided

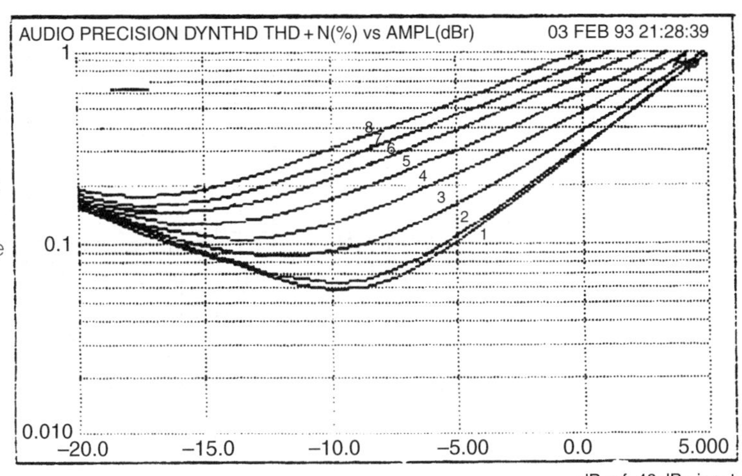

Figure 4.6 Effect of collector-current imbalance on an isolated input pair; the second harmonic rises well above the level of the third if the pair moves away from balance by as little as 2%

Table 4.1
(Key to Figure 4.6)

Curve No.	*Ic* Imbalance (%)	Curve No.	*Ic* Imbalance (%)
1	0	5	5.4
2	0.5	6	6.9
3	2.2	7	8.5
4	3.6	8	10

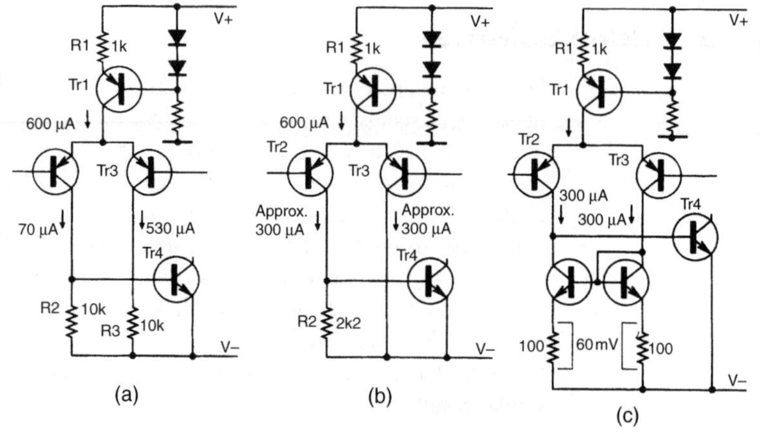

Figure 4.7
Improvements to the input pair. (a) Poorly designed version. (b) Better; partial balance by correct choice of R2. (c) Best; near-perfect Ic balance enforced by mirror

attempt to maximise o/l gain by minimising loading on Q1 collector. R3 is also 10k to give the stage a notional *balance*, though unhappily this is a visual rather than electrical balance. The asymmetry is shown in the resulting collector currents; the design generates a lot of avoidable second harmonic distortion, displayed in the 10k curve of Figure 4.8.

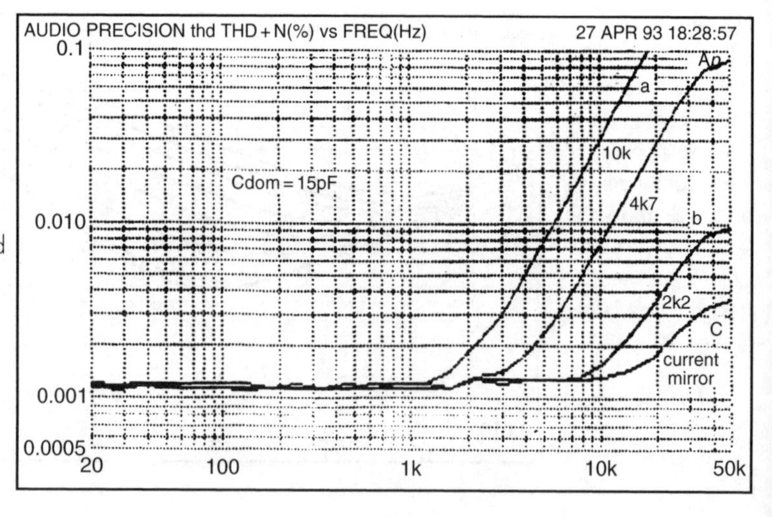

Figure 4.8
Distortion of model amplifier: (a) Unbalanced with R2 = 10k. (b) Partially balanced with R = 2k2. (c) Accurately balanced by current-mirror

Recognising the crucial importance of DC balance, the circuit can be rethought as Figure 4.7b. If the collector currents are to be roughly equal, then R2 must be about 2 × R1, as both have about 0.6 V across them. The dramatic effect of this simple change is shown in the 2k2 curve of Figure 4.8; the improvement is accentuated as the o/l gain has also increased by some 7 dB, though this has only a minor effect on the closed-loop linearity compared with the improved balance of the input pair. R3 has been excised as it contributes very little to input stage balance.

The joy of current-mirrors

Although the input pair can be approximately balanced by the correct values for R1 and R2, we remain at the mercy of several circuit tolerances. Figure 4.6 shows that balance is critical, needing an accuracy of 1% or better for optimal linearity and hence low distortion at HF, where the input pair works hardest. The standard current-mirror configuration in Figure 4.7c forces the two collector currents very close to equality, giving correct cancellation of the second harmonic; the great improvement that results is seen in the current-mirror curve of Figure 4.8. There is also less DC offset due to unequal base-currents flowing through input and feedback resistances; I often find that a power-amplifier improvement gives at least two separate benefits. This simple mirror has well-known residual base-current errors but they are not large enough to affect the distortion performance.

The hyperbolic-tangent law also holds for the mirrored pair[4], though the output current swing is twice as great for the same input voltage as the resistor-loaded version. This doubled output is given at the same distortion as for the unmirrored version, as linearity depends on the input voltage, which has not changed. Alternatively, we can halve the input and get the same output, which with a properly balanced pair generating third harmonic only will give one-quarter the distortion. A pleasing result.

The input mirror is made from discrete transistors, regretfully foregoing the Vbe-matching available to IC designers, and it needs its own emitter-degeneration for good current-matching. A voltage-drop across the current-mirror emitter-resistors in the range 30–60 mV will be enough to make the effect of Vbe tolerances on distortion negligible; if degeneration is omitted then there is significant variation in HF distortion performance with different specimens of the same transistor type.

Putting a current-mirror in a well-balanced input stage increases the total o/l gain by at least 6 dB, and by up to 15 dB if the stage was previously poorly balanced; this needs to be taken into account in setting the compensation. Another happy consequence is that the slew-rate is roughly doubled, as the input stage can now source and sink current into Cdom without wasting it in a collector load. If Cdom is 100 pF, the slew-rate of Figure 4.7b is about

2.8 V/μsec up and down, while 4.7c gives 5.6 V/μsec. The unbalanced pair at 4.7a displays further vices by giving 0.7 V/μsec positive-going and 5 V/μsec negative-going.

Improving input-stage linearity

Even if the input pair has a current-mirror, we may still feel that the HF distortion needs further reduction; after all, once it emerges from the noise floor it octuples with each doubling of frequency, and so it is well worth postponing the evil day until as far as possible up the frequency range. The input pair shown has a conventional value of tail-current. We have seen that the stage transconductance increases with Ic, and so it is possible to increase the *gm* by increasing the tail-current, and then return it to its previous value (otherwise *C*dom would have to be increased proportionately to maintain stability margins) by applying local NFB in the form of emitter-degeneration resistors. This ruse powerfully improves input linearity, despite its rather unsettling flavor of something-for-nothing. The transistor non-linearity can here be regarded as an internal non-linear emitter resistance *r*e, and what we have done is to reduce the value of this (by increasing Ic) and replaced the missing part of it with a linear external resistor Re.

For a single device, the value of *r*e can be approximated by:

$$re = 25/Ic\,\Omega \text{ (for } Ic \text{ in mA)} \qquad \text{Equation 4.3}$$

Our original stage at Figure 4.9a has a per-device Ic of 600 μA, giving a differential (i.e., mirrored) *gm* of 23 mA/V and $re = 41.6\,\Omega$. The improved version at Figure 4.9b has $Ic = 1.35$ mA and so $re = 18.6\,\Omega$; therefore

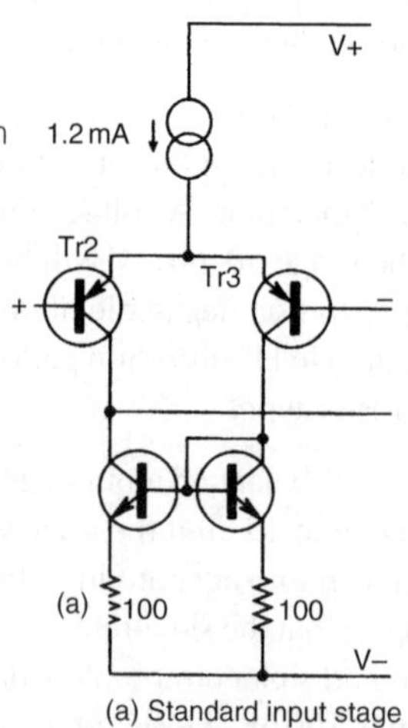

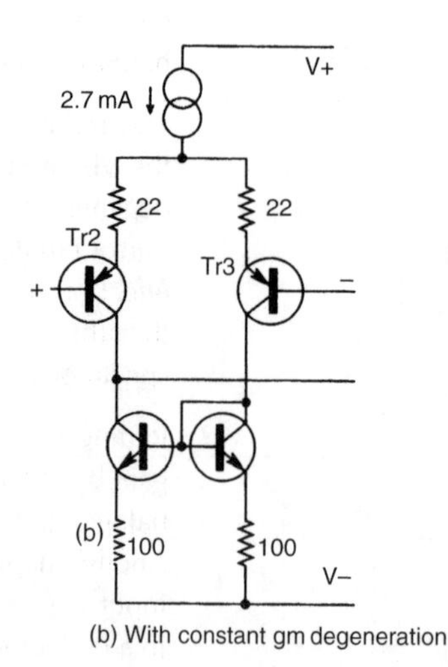

Figure 4.9
Input pairs before and after constant-*gm* degeneration, showing how to double stage current while keeping transconductance constant; distortion is reduced by about ten times

emitter degeneration resistors of 22 Ω are required to reduce the *gm* back to its original value, as 18.6 + 22 = 41.6 Ω. The distortion measured by the circuit of Figure 4.4 for a −40 dBu input voltage is reduced from 0.32% to 0.032%, which is an extremely valuable linearisation, and will translate into a distortion reduction at HF of about five times for a complete amplifier; for reasons that will emerge later the full advantage is rarely gained. The distortion remains a visually pure third-harmonic, so long as the input pair remains balanced. Clearly this sort of thing can only be pushed so far, as the reciprocal-law reduction of *re* is limited by practical values of tail current. A name for this technique seems to be lacking; *constant-*gm *degeneration* is descriptive but rather a mouthful.

The standing current is roughly doubled so we have also gained a higher slew-rate; it has theoretically increased from 10 V/μsec to 20 V/μsec, and once again we get two benefits for the price of one inexpensive modification.

Radical methods of improving input linearity

If we are seeking still better linearity, various techniques exist. Whenever it is needful to increase the linearity of a circuit, it is often a good approach to increase the *local* feedback factor, because if this operates in a tight local NFB loop there is often little effect on the overall global-loop stability. A reliable method is to replace the input transistors with complementary-feedback (CFP or Sziklai) pairs, as shown in the stage of Figure 4.10a. If an isolated input stage is measured using the test circuit of Figure 4.4, the constant-gm degenerated version shown in Figure 4.9b yields 0.35% third-harmonic distortion for a −30 dBu input voltage, while the CFP version

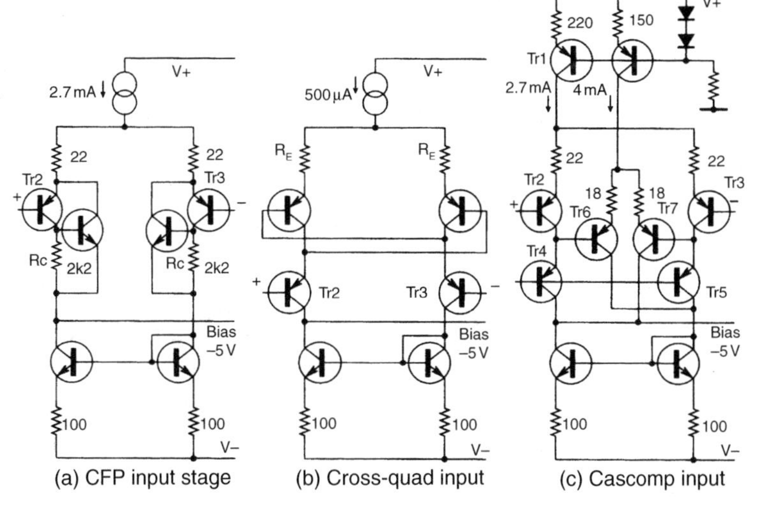

Figure 4.10
Some enhanced differential pairs: (a) The Complementary-Feedback Pair. (b) The Cross-quad. (c) The Cascomp

gives 0.045%. (Note that the input level here is 10 dB up on the previous example, to get well clear of the noise floor.) When this stage is put to work in a model amplifier, the third-harmonic distortion at a given frequency is roughly halved, assuming all other distortion sources have been appropriately minimised. However, given the high-slope of input-stage distortion, this only extends the low-distortion regime up in frequency by less than an octave. See Figure 4.11.

A compromise is required in the CFP circuit on the value of Rc, which sets the proportion of the standing current that goes through the NPN and PNP devices on each side of the stage. A higher value of Rc gives better linearity, but more noise, due to the lower Ic in the NPN devices that are the inputs of the input stage, as it were, causing them to match less well the relatively low source resistances. 2k2 is a good compromise.

Several other elaborations of the basic input pair are possible, although almost unknown in the audio community. We are lucky in power-amp design as we can tolerate a restricted input common-mode range that would be unusable in an op-amp, giving the designer great scope. Complexity in itself is not a serious disadvantage as the small-signal stages of the typical amplifier are of almost negligible cost compared with mains transformers, heatsinks, etc.

Two established methods to produce a linear input transconductance stage (referred to in op-amp literature simply as a transconductor) are the cross-quad[5] and the cascomp[6] configurations. The cross-quad (Figure 4.10b) gives a useful reduction in input distortion when operated in isolation but is hard to incorporate in a practical amplifier because it relies on very low source-resistances to tame the negative conductances inherent in its operation. The cross-quad works by imposing the input voltage to each

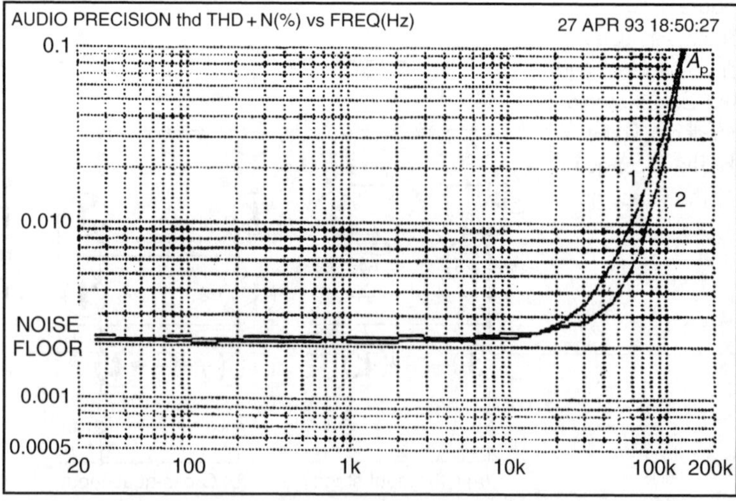

Figure 4.11 Whole-amplifier THD with normal and CFP input stages; input stage distortion only shows above noise floor at 20 kHz, so improvement occurs above this frequency. The noise floor appears high as the measurement bandwidth is 500 kHz

half across two base-emitter junctions in series, one in each arm of the circuit. In theory the errors due to non-linear *re* of the transistors is divided by beta, but in practice the reduction in distortion is modest.

The cascomp (Figure 4.10c) does not have problems with negative impedances, but it is significantly more complex to design. Q2, Q3 are the main input pair as before, delivering current through cascode transistors Q4, Q5 (this does not in itself affect linearity), which, since they carry almost the same current as Q2, Q3 duplicate the input Vbe errors at their emitters. This is sensed by error diff-amp Q6, Q7, whose output currents are summed with the main output in the correct phase for error-correction. By careful optimisation of the (many) circuit variables, distortion at −30 dBu input can be reduced to about 0.016% with the circuit values shown. Sadly, this effort provides very little further improvement in whole-amplifier HF distortion over the simpler CFP input, as other distortion mechanisms are coming into play, one of which is the finite ability of the VAS to source current into the other end of *C*dom.

Input stage cascode configurations

Power amplifiers with pretensions to sophistication sometimes add cascoding to the standard input differential amplifier. This does nothing whatever to improve input-stage linearity, as there is no appreciable voltage swing on the input collectors; its main advantage is reduction of the high *V*ce that the input devices work at. This allows cooler running, and therefore possibly improved thermal balance; a *V*ce of 5 V usually works well. Isolating the input collector capacitance from the VAS input sometimes allows *C*dom to be slightly reduced for the same stability margins, but the improvement is marginal.

Input noise and how to reduce it

The noise performance of a power amplifier is defined by its input stage, and so the issue is examined here. Power-amp noise is not an irrelevance; a powerful amplifier will have a high voltage gain, and this can easily result in a faint but irritating hiss from efficient loudspeakers even when all volume controls are fully retarded[3]. In the design considered here the EIN has been measured at −120 dBu, which is only 7 or 8 dB worse than a first-class microphone preamplifier; the inferiority is largely due to the source resistances seen by the input devices being higher than the usual 150 Ω microphone impedance. By way of demonstration, halving the impedance of the usual feedback network (22k and 1k) reduces the EIN further by about 2 dB.

Amplifier noise is defined by a combination of the active devices at the input and the surrounding resistances. The operating conditions of the input transistors themselves are set by the demands of linearity and slew-rate, so there is little freedom of design here; however the collector currents are already high enough to give near-optimal noise figures with the low source impedances (a few hundred ohms) that we have here, so this is not too great a problem. Noise figure is a weak function of *Ic*, so minor tweakings of the tail-current make no detectable difference. We certainly have the choice of input device type; there are many more possibles if we have relatively low rail voltages. Noise performance is, however, closely bound up with source impedance, and we need to define this before device selection.

Looking therefore to the passives, there are several resistances generating Johnson Noise in the input, and the only way to reduce this noise is to reduce them in value. The obvious candidates are R2, R3 see Figure 4.12 (input stage degeneration resistors) and R9, which determines the output impedance of the negative-feedback network. There is also another

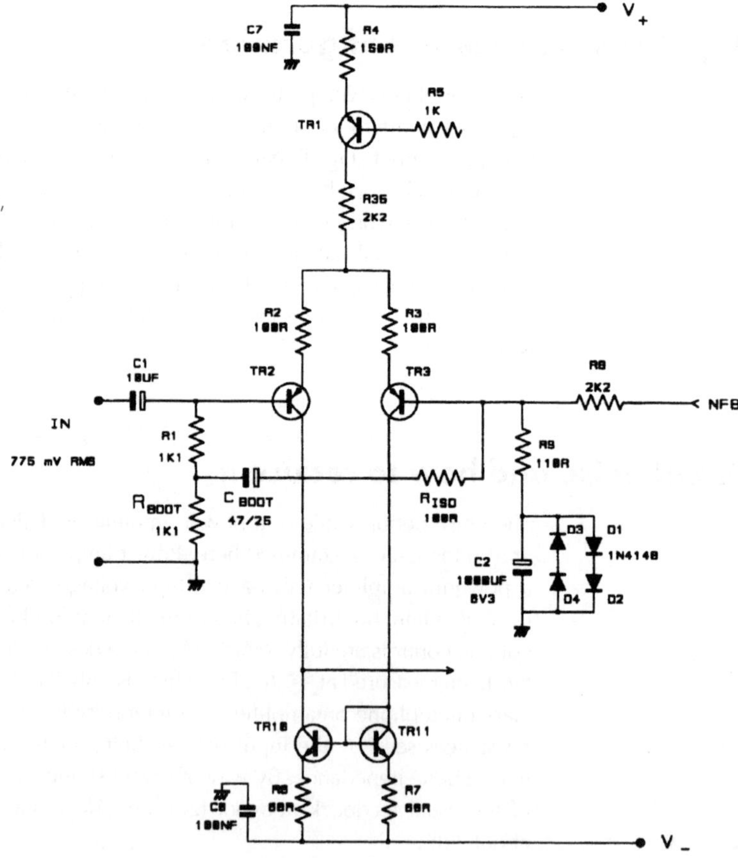

Figure 4.12
Stable input bootstrapping from the feedback point. Riso is essential for HF stability; with 100 Ω, as shown, the input impedance is 13 kΩ

unseen component; the source resistance of the preamplifier or whatever upstream. Even if this equipment were miraculously noise-free, its output resistance would still generate Johnson noise. If the preamplifier had, say, a 20k volume pot at its output (not a good idea, as this gives a poor gain structure and cable dependent HF losses, but that is another story[5]) then the source resistance could be a maximum of 5k, which would almost certainly generate enough Johnson Noise to dominate the power-amplifier's noise behaviour. However, there is nothing that power-amp designers can do about this, so we must content ourselves with minimising the noise-generating resistances we do have control over.

Noise from the input degeneration resistors R2, R3 is the price we pay for linearising the input stage by running it at a high current, and then bringing its transconductance down to a useable value by adding linearising local negative feedback. These resistors cannot be reduced if the HF NFB factor is then to remain constant, for Cdom will have to be proportionally increased, reducing slew-rate. With the original 22k–1k NFB network, these resistors degrade the noise performance by 1.7 dB. (This figure, like all other noise measurements given here, assumes a 50 Ω external source resistance.)

If we cannot alter the input degeneration resistors, then the only course left is the reduction of the NFB network impedance, and this sets off a whole train of consequences. If R8 is reduced to 2k2, then R9 becomes 110 Ω, and this reduces noise output from −93.5 to −95.4 dBu. (Note that if R2, R3 were not present, the respective figures would be −95.2 and −98.2 dBu.) However, R1 must also be reduced to 2k2 to maintain DC balance, and this is too low an input impedance for direct connection to the outside world. If we accept that the basic amplifier will have a low input impedance, there are two ways to deal with it. The simplest is to decide that a balanced line input is essential; this puts an op-amp stage before the amplifier proper, buffers the low input impedance, and can provide a fixed source impedance to allow the HF and LF bandwidths to be properly defined by an RC network using non-electrolytic capacitors. The usual practice of slapping an RC network on an unbuffered amplifier input must be roundly condemned as the source impedance is unknown, and so therefore is the roll-off point. A major stumbling block for subjectivist reviewing, one would have thought.

Another approach is to have a low resistance DC path at the input but a high AC impedance; in other words to use the fine old practice of input bootstrapping. Now this requires a low-impedance unity-gain-with-respect-to-input point to drive the bootstrap capacitor, and the only one available is at the amplifier inverting input, i.e., the base of TR3. While this node has historically been used for the purpose of input bootstrapping[6], it has only been done with simple circuitry employing very low feedback factors.

There is very real reason to fear that any monkey business with the feedback point (TR3 base) will add shunt capacitance, creating a feedback pole that will degrade HF stability. There is also the awkward question of what will happen if the input is left open-circuit . . . .

The input can be safely bootstrapped; Figure 4.12 shows how. The total DC resistance of R1 and *R*boot equals R8, and their central point is driven by *C*boot. Connecting *C*boot directly to the feedback point did not produce gross instability, but it did seem to increase susceptibility to odd bits of parasitic oscillation. *R*iso was then added to isolate the feedback point from stray capacitance, and this seemed to effect a complete cure. The input could be left open-circuit without any apparent ill-effects, though this is not good practice if loudspeakers are connected. A value for *R*iso of 220 Ω increases the input impedance to 7.5k, and 100 Ω raises it to 13.3k, safely above the 10k standard value for a bridging impedance. Despite successful tests, I must admit to a few lingering doubts about the HF stability of this approach, and it might be as well to consider it as experimental until more experience is gained.

One more consequence of a low-impedance NFB network is the need for feedback capacitor C2 to be proportionally increased to maintain LF response, and prevent capacitor distortion from causing a rise in THD at low frequencies; it is the latter requirement that determines the value. (This is a separate distortion mechanism from the seven originally identified, and is given the title Distortion 8.) This demands a value of 1000 μF, necessitating a low-rated voltage such as 6V3 if the component is to be of reasonable size. This means that C2 needs protective shunt diodes in both directions, because if the amplifier fails it may saturate in either direction. Examination of the distortion residual shows that the onset of conduction of back-to-back diodes will cause a minor increase in THD at 10 Hz, from less than 0.001% to 0.002%, even at the low power of 20 W/8 Ω. It is not my practice to tolerate such gross non-linearity, and therefore four diodes are used in the final circuit, and this eliminates the distortion effect. It could be argued that a possible reverse-bias of 1.2 V does not protect C2 very well, but at least there will be no explosion.

We can now consider alternative input devices to the MPSA56, which was never intended as a low-noise device. Several high-beta low-noise types such as 2SA970 give an improvement of about 1.8 dB with the low-impedance NFB network. Specialised low-Rb devices like 2SB737 give little further advantage (possibly 0.1 dB) and it is probably better to go for one of the high-beta types; the reason why will soon emerge.

It could be argued that the above complications are a high price to pay for a noise reduction of some 2 dB; however, with the problems comes a definite advantage, for the above NFB network modification also significantly improves the output DC offset performance.

Offset and match: the DC precision issue

The same components that dominate amplifier noise performance also determine the output DC offset; if R9 is reduced to minimise the source resistance seen by TR3, then the value of R8 is scaled to preserve the same closed-loop gain, and this reduces the voltage drops caused by input transistor base currents.

Most of my amplifier designs have assumed that a ±50 mV output DC offset is acceptable. This allows DC trimpots, offset servos, etc. to be gratefully dispensed with. However, it is not in my nature to leave well enough alone, and it could be argued that ±50 mV is on the high side for a top-flight amplifier. I have therefore reduced this range as much as possible without resorting to a servo; the required changes have already been made when the NFB network was reduced in impedance to minimise Johnson noise. (See page 87.)

With the usual range of component values, the DC offset is determined not so much by input transistor Vbe mismatch, which tends to be only 5 mV or so, but more by a second mechanism – imbalance in beta. This causes imbalance of the base currents (Ib) drawn thorough input bias resistor R1 and feedback resistor R8, and the cancellation of the voltage-drops across these components is therefore compromised.

A third source of DC offset is non-ideal matching of input degeneration resistors R2, R3. Here they are 100 Ω, with 300 mV dropped across each, so two 1% components at opposite ends of their tolerance bands could give a maximum offset of 6 mV. In practice this is most unlikely, and the error from this source will probably not exceed 2 mV.

There are several ways to reduce DC offset. First, low-power amplifiers with a single output pair must be run from modest HT rails and so the requirement for high-Vce input transistors can be relaxed. This allows higher beta devices to be used, directly reducing Ib. The 2SA970 devices used in this design have a beta range of 350–700, compared with 100 or less for MPSA06/56. Note the pinout is not the same.

On page 87, we reduced the impedance of the feedback network by a factor of 4.5, and the offset component due to Ib imbalance is reduced by the same ratio. We might therefore hope to keep the DC output offset for the improved amplifier to within ±15 mV without trimming or servos. Using high-beta input devices, the Ib errors did not exceed ±15 mV for 10 sample pairs (*not* all from the same batch) and only three pairs exceeded ±10 mV. The Ib errors are now reduced to the same order of magnitude as Vbe mismatches, and so no great improvement can be expected from further reduction of circuit resistances. Drift over time was measured at less than 1 mV, and this seems to be entirely a function of temperature equality in the input pair.

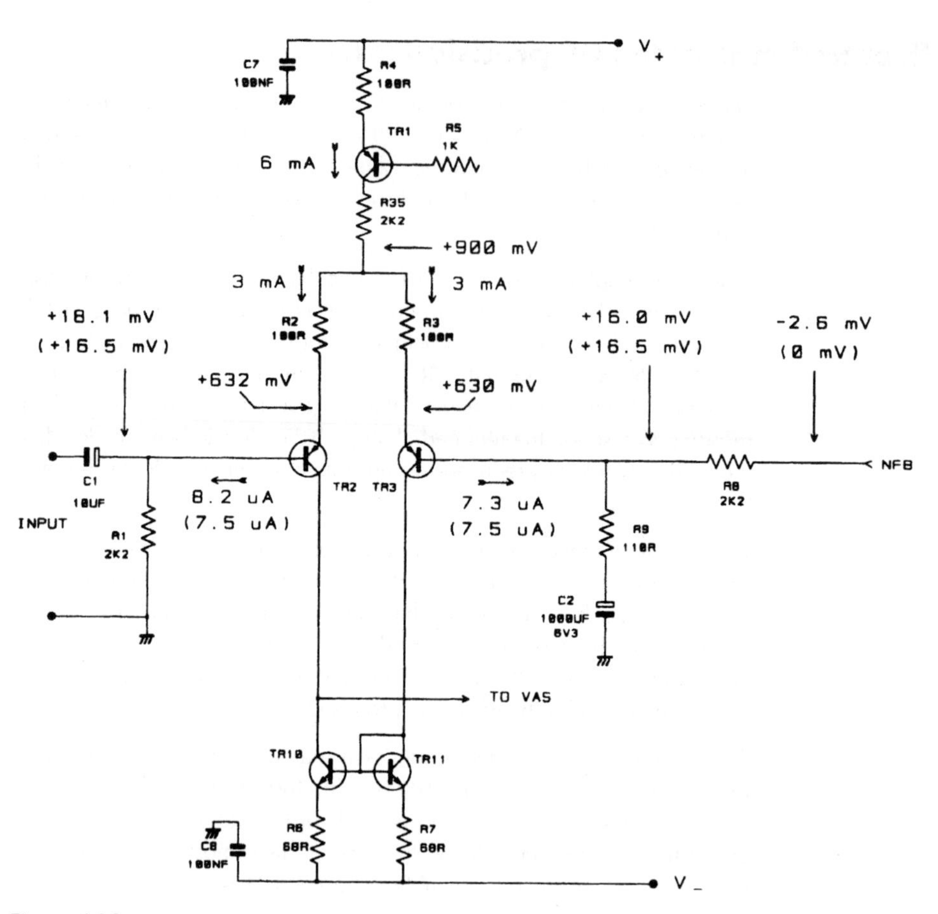

Figure 4.13
The measured DC conditions in a real input stage. Ideal voltages and currents for perfectly matched components are shown in brackets

Figure 4.13 shows the ideal DC conditions in a perfectly balanced input stage, assuming $\beta = 400$, compared with a set of real voltages and currents from the prototype amplifier. In the latter case, there is a typical partial cancellation of offsets from the three different mechanisms, resulting in a creditable output offset of $-2.6\,\text{mV}$.

The input stage and the slew-rate

This is another parameter which is usually assumed to be set by the input stage, and has a close association with HF distortion. A brief summary is therefore given here, but the subject is dealt with in much greater depth in Chapter 7.

An amplifier's slew-rate is proportional to the input stage's maximum-current capability, most circuit configurations being limited to switching the whole of the tail current to one side or the other. The usual differential pair can only manage half of this, as with the output slewing negatively half the tail-current is wasted in the input collector load R2. The addition of an input current-mirror, as advocated above, will double the slew rate in both directions as this inefficiency is abolished. With a tail current of 1.2 mA a mirror improves the slew-rate from about 5 V/μsec to 10 V/μsec (for *Cdom* = 100 pF). The constant-gm degeneration method of linearity enhancement in Figure 4.9 further increases it to 20 V/μsec.

In practice slew rates are not exactly identical for positive and negative-going directions, especially in the conventional amplifier architecture which is the main focus of this book.

The voltage-amplifier stage

The Voltage-Amplifier Stage (or VAS) has often been regarded as the most critical part of a power-amplifier, since it not only provides all the voltage gain but also must give the full output voltage swing. (The input stage may give substantial transconductance gain, but the output is in the form of a current.) However, as is not uncommon in audio, all is not quite as it appears. A well-designed VAS stage will contribute relatively little to the overall distortion total of an amplifier, and if even the simplest steps are taken to linearise it further, its contribution sinks out of sight.

As a starting point, Figure 4.14 shows the distortion plot of a model amplifier with a Class-A output (±15 V rails, +16 dBu out) as per Chapter 3,

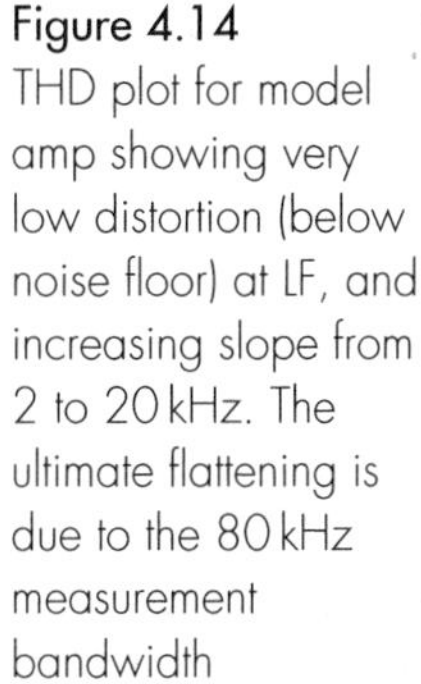

Figure 4.14 THD plot for model amp showing very low distortion (below noise floor) at LF, and increasing slope from 2 to 20 kHz. The ultimate flattening is due to the 80 kHz measurement bandwidth

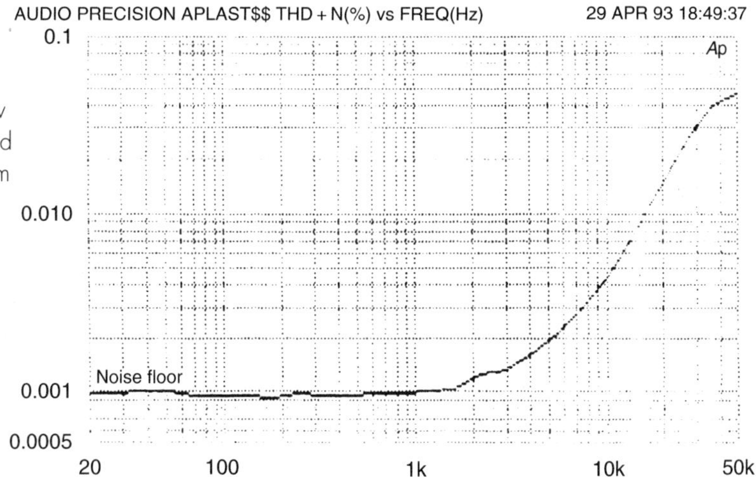

where no special precautions have been taken to linearise the input stage or the VAS; output stage distortion is negligible. It can be seen that the distortion is below the noise floor at LF; however, the distortion slowly rising from about 1 kHz is coming from the VAS. At higher frequencies, where the VAS 6 dB/octave rise becomes combined with the 12 or 18 dB/octave rise of input-stage distortion, we can see the distortion slope of accelerating steepness that is typical of many amplifier designs.

As previously explained, the main reason why the VAS generates relatively little distortion is because at LF, global feedback linearises the whole amplifier, while at HF the VAS is linearised by local NFB through *Cdom*.

Measuring VAS distortion in isolation

Isolating the VAS distortion for study requires the input pair to be specially linearised, or else its steeply rising distortion characteristic will swamp the VAS contribution. This is most easily done by degenerating the input stage; this also reduces the open-loop gain, and the reduced feedback factor mercilessly exposes the non-linearity of the VAS. This is shown in Figure 4.15, where the 6 dB/octave slope suggests that this must originate in the VAS, and increases with frequency solely because the compensation is rolling-off the global feedback factor. To confirm that this distortion is due solely to the VAS, it is necessary to find a method for experimentally varying VAS linearity while leaving all other circuit parameters unchanged. Figure 4.16 shows my arrangement for doing this by varying the VAS V–voltage; this varies the proportion of its characteristic over which the VAS swings, and thus only alters the effective VAS linearity, as the important input stage conditions remain unchanged. The current-mirror must go up

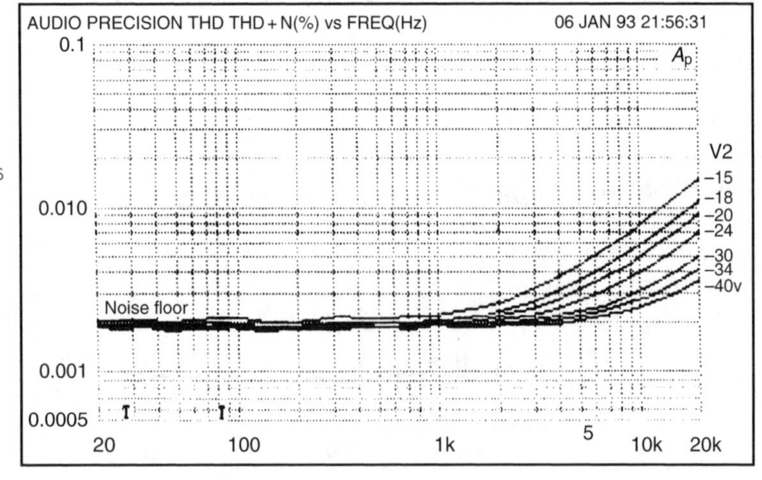

Figure 4.15 The change in HF distortion resulting from varying V– in the VAS test circuit. The VAS distortion is only revealed by degenerating the input stage with 100 Ω resistors

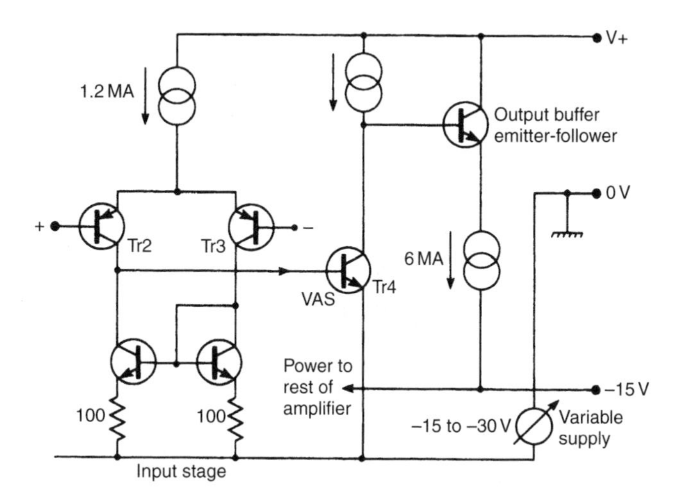

Figure 4.16
VAS distortion test circuit. Although the input pair mirror moves up and down with the VAS emitter, the only significant parameter being varied is the available voltage-swing at the VAS collector

and down with the VAS emitter for correct operation, and so the Vce of the input devices also varies, but this has no significant effect, as can be proved by the unchanged behaviour on inserting cascode stages in the input transistor collectors.

VAS operation

The typical VAS topology as shown in Figure 4.17a is a classical common-emitter voltage-amplifier stage, with a current-drive input into the base. The small-signal characteristics, which set open-loop gain and so on, can be usefully simulated by the *Spice* model shown in Figure 4.18, of a VAS reduced to its conceptual essentials. G is a current-source whose value is controlled by the voltage-difference between Rin and Rf2, and represents the differential transconductance input stage. F represents the VAS transistor, and is a current-source yielding a current of beta times that sensed flowing through *ammeter* V which by *Spice* convention is a voltage-source set to 0 V; the value of beta, representing current-gain as usual, models the relationship between VAS collector current and base current. Rc represents the total VAS collector impedance, a typical real value being 22 k. With suitable parameter values, this simple model provides a good demonstration of the relationships between gain, dominant-pole frequency, and input stage current that were introduced in Chapter 3. Injecting a small signal current into the output node from an extra current-source also allows the fall of impedance with frequency to be examined.

The overall voltage-gain clearly depends linearly on beta, which in real transistors may vary widely. Working on the trusty engineering principle that what cannot be controlled must be made irrelevant, local shunt NFB through Cdom sets the crucial HF gain that controls Nyquist stability. The LF

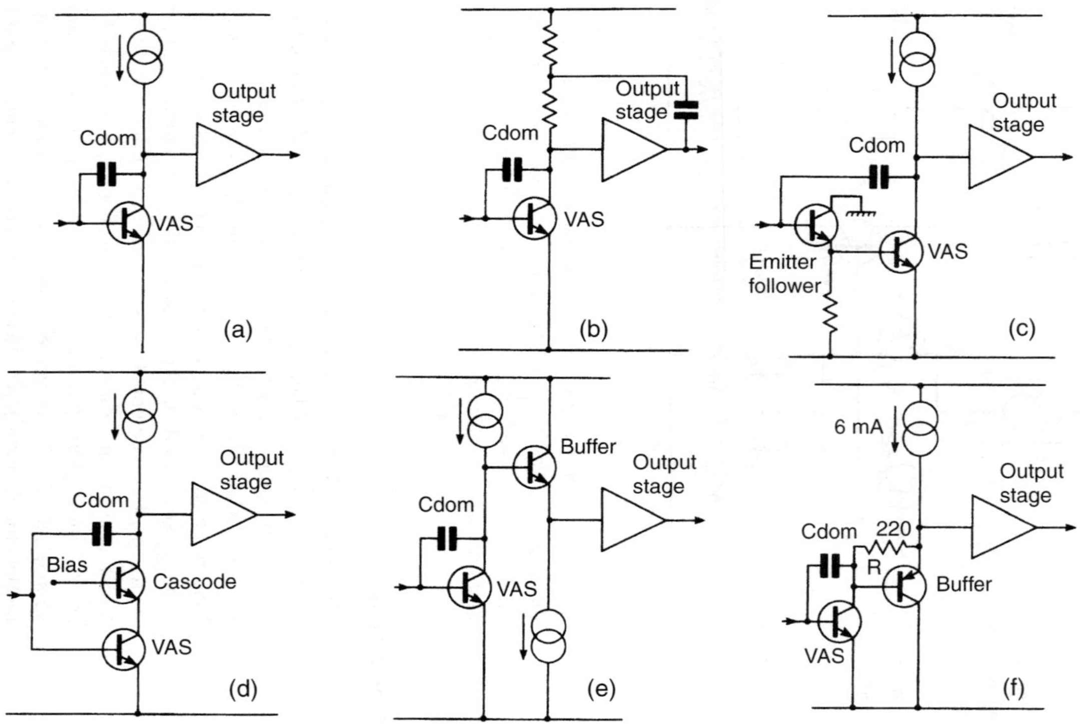

Figure 4.17
Six variations on a VAS:

(a) Conventional VAS with current-source load
(b) Conventional VAS with bootstrapped load
(c) Increase in local NFB by adding beta-enhancing emitter-follower
(d) Increase in local NFB by cascoding VAS
(e) Buffering the VAS collector from the output stage
(f) Alternative buffering, bootstrapping VAS load R

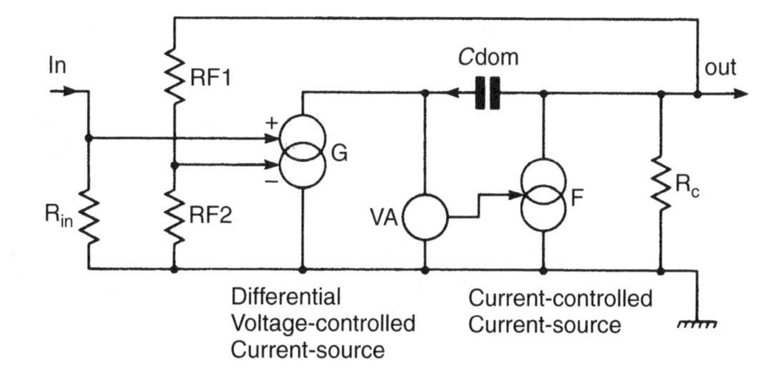

Figure 4.18 Conceptual SPICE model of differential input stage (G) and VAS (F). The current in F is Beta times the current in VA

gain below the dominant-pole frequency P1 remains variable (and therefore so does P1) but is ultimately of little importance; if there is an adequate NFB factor for overall linearisation at HF then there are unlikely to be problems at LF, where the gain is highest. As for the input stage, the linearity of the VAS is not greatly affected by transistor type, given a reasonably high beta.

VAS distortion

VAS distortion arises from the fact that the transfer characteristic of a common-emitter amplifier is curved, being a small portion of an exponential[7]. This characteristic generates predominantly second-harmonic distortion, which in a closed-loop amplifier will increase at 6 dB/ octave with frequency.

VAS distortion does not get worse for more powerful amplifiers as the stage traverses a constant proportion of its characteristic as the supply-rails are increased. This is not true of the input stage; increasing output swing increases the demands on the transconductance amp as the current to drive Cdom increases. The increased Vce of the input devices does not measurably affect their linearity.

It is ironic that VAS distortion only becomes clearly visible when the input pair is excessively degenerated – a pious intention to *linearise before applying feedback* can in fact make the closed-loop distortion worse by reducing the open-loop gain and hence the NFB factor available to linearise the VAS. In a real (non-model) amplifier with a distortive output stage the deterioration will be worse.

Linearising the VAS: active load techniques

As described in Chapter 3, it is important that the local open-loop gain of the VAS (that existing inside the local feedback loop closed by Cdom) be high, so that the VAS can be linearised, and therefore a simple resistive load is unusable.

Increasing the value of Rc will decrease the collector current of the VAS transistor, reducing its transconductance and getting you back where you started in terms of voltage gain.

One way to ensure enough local loop gain is to use an active load to increase the effective collector impedance at TR4 and thus increase the raw voltage gain; either bootstrapping or a current-source will do this effectively, though the current source is perhaps more dependable, and is the usual choice for hi-fi or professional amplifiers. The Bootstrap promises more o/p swing, as the collector of TR4 can in theory soar like a lark above the V+ rail; under some circumstances this can be the overriding concern, and bootstrapping is alive and well in applications such as automotive power-amps that must make the best possible use of a restricted supply voltage[8].

Both active-load techniques have another important role; ensuring that the VAS stage can source enough current to properly drive the upper half of the output stage in a positive direction, right up to the rail. If the VAS collector load was a simple resistor to +V, then this capability would certainly be lacking.

It may not be immediately obvious how to check that impedance-enhancing measures are working properly, but it is actually fairly simple. The VAS collector impedance can be determined by the simple expedient of shunting the VAS collector to ground with decreasing resistance until the open-loop gain reading falls by 6 dB, indicating that the collector impedance is equal to the current value of the test resistor.

The popular current source version is shown in Figure 4.17a. This works well, though the collector impedance is limited by the effective output resistance Ro of the VAS and the current source transistors[9], which is another way of saying that the improvement is limited by Early effect.

It is often stated that this topology provides current-drive to the output stage; this is only partly true. It is important to realise that once the local NFB loop has been closed by adding Cdom the impedance at the VAS output falls at 6 dB/octave for frequencies above P1. With typical values the impedance is only a few kΩ – at 10 kHz, and this hardly qualifies as current-drive at all.

Collector-load bootstrapping (Figure 4.17b) works in most respects as well as a current source load, for all its old-fashioned look. Conventional capacitor bootstrapping has been criticised for prolonging recovery from clipping; I have no evidence to offer on this myself, but one subtle drawback definitely does exist – with bootstrapping the LF open-loop gain is dependent on amplifier output loading. The effectiveness of boot-strapping depends crucially on the output stage gain being unity or very close to it; however the presence of the output-transistor emitter resistors means that there will be a load-dependant gain loss in the output stage, which in turn significantly alters the amount by which the VAS collector impedance is increased; hence the LF feedback factor is dynamically altered by the impedance characteristics of the loudspeaker load and the spectral distribution of the source material. This has a special significance if the load is an *audiophile* speaker that may have impedance dips down to 2 Ω, in which case the gain loss is serious. If anyone needs a new audio-impairment mechanism to fret about, then I humbly offer this one in the confident belief that its effects, while measurable, are not of audible significance. Possibly this is a more convincing reason for avoiding bootstrapping than alleged difficulties with recovery from clipping.

Another drawback of bootstrapping is that the standing DC current through the VAS, and hence the bias generator, varies with rail voltage. Setting and maintaining the quiescent conditions is quite difficult enough already, so an extra source of possible variation is decidedly unwelcome.

A less well-known but more dependable form of bootstrapping is available if the amplifier incorporates a unity-gain buffer between the VAS collector and the output stage; this is shown in Figure 4.17f, where *Rc* is the collector load, defining the VAS collector current by establishing the Vbe of the buffer transistor across itself. This is constant, and *Rc* is therefore bootstrapped and appears to the VAS collector as a constant-current source. In this sort of topology a VAS current of 3 mA is quite sufficient, compared with the 6 mA standing current in the buffer stage. The VAS would in fact work well with lower collector currents down to 1 mA, but this tends to compromise linearity at the high-frequency, high-voltage corner of the operating envelope, as the VAS collector current is the only source for driving current into Cdom.

VAS enhancements

Figure 4.15 shows VAS distortion only, clearly indicating the need for further improvement over that given inherently by Cdom if our amplifier is to be as good as possible. The virtuous approach might be to try to straighten out the curved VAS characteristic, but in practice the simplest method is to increase the amount of *local* negative feedback through Cdom. Equation 4.1 in Chapter 3 shows that the LF gain (i.e., the gain before Cdom is connected) is the product of input stage transconductance, TR4 beta and the collector impedance Rc. The last two factors represent the VAS gain and therefore the amount of local NFB can be augmented by increasing either. Note that so long as the value of Cdom remains the same, the global feedback factor at HF is unchanged and so stability is not affected.

The effective beta of the VAS can be substantially increased by replacing the VAS transistor with a Darlington, or in other words putting an emitter-follower before it (Figure 4.17c). Adding an extra stage to a feedback amplifier always requires thought, because if significant additional phase-shift is introduced, the global loop stability can suffer. In this case the new stage is inside the Cdom Miller-loop and so there is little likelihood of trouble from this. The function of such an emitter-follower is sometimes described as *buffering the input stage from the VAS* but its true function is linearisation by enhancement of local NFB through Cdom.

Alternatively the VAS collector impedance can be increased to get more local gain. This is straightforwardly done with a cascode configuration – (see Figure 4.17d) but it should be said at once that the technique is only really useful when the VAS is not directly driving a markedly non-linear impedance . . . such as that at the input of a Class-B output stage. Otherwise this non-linear loading renders it largely a cosmetic feature. Assuming for the moment that this problem is dealt with, either by use of a Class-A output or by VAS-buffering, the drop in distortion is dramatic, as for the beta-enhancement method. The gain increase is ultimately limited by Early effect in the cascode and current-source transistors, and more seriously by the loading effect of the next stage, but it is of the order of 10 times and gives a useful effect. This is shown by curves A, B in Figure 4.19, where once more the input stage of a model amplifier has been over-degenerated with 100 Ω emitter resistors to bring out the VAS distortion more clearly. Note that in both cases the slope of the distortion increase is 6 dB/octave. Curve C shows the result when a standard undegenerated input pair is combined with the cascoded VAS; the distortion is submerged in the noise floor for most of the audio band, being well below 0.001%. I think this justifies my contention that input-stage and VAS distortions need not be problems; we have all but eliminated Distortions 1 and 2 from the list of eight in Chapter 3.

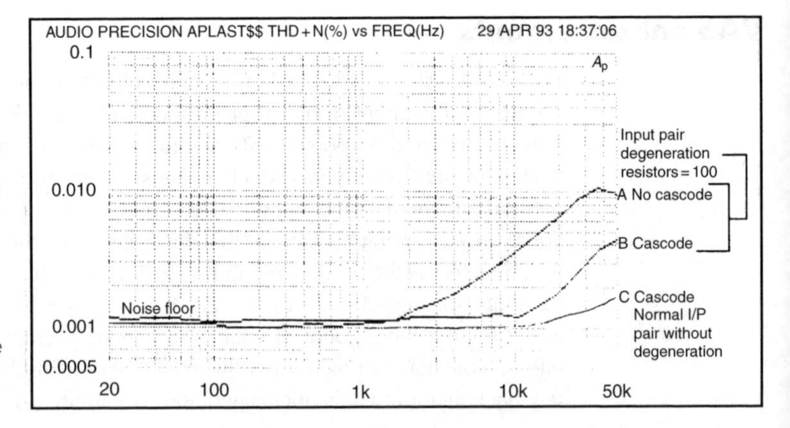

Figure 4.19 Showing the reduction of VAS distortion possible by cascoding. The results from adding an emitter-follower to the VAS, as an alternative method of increasing local VAS feedback, are very similar

Using a cascode transistor also allows the use of a high-beta transistor for the VAS; these typically have a limited Vceo that cannot withstand the high rail voltages of a high-power amplifier. There is a small loss of available voltage swing, but only about 300 mV, which is usually tolerable. Experiment shows that there is nothing to be gained by cascoding the current-source collector load.

A cascode topology is often used to improve frequency response, by isolating the upper collector from the Cbc of the lower transistor. In this case the frequency response is deliberately defined by Cdom, so this appears irrelevant, but in fact it is advantageous that Cbc – which carries the double demerit of being unpredictable and signal-dependent – is rendered harmless. Thus compensation is determined only by a well-defined passive component.

It is hard to say which technique is preferable; the beta-enhancing emitter-follower circuit is slightly simpler than the cascode version, which requires extra bias components, but the cost difference is tiny. When wrestling with these kind of financial decisions it is as well to remember that the cost of a small-signal transistor is often less than a fiftieth of that of an output device, and the entire small-signal section of an amplifier usually represents less than 1% of the total cost, when heavy metal such as the mains transformer and heatsinks are included.

Note that although the two VAS-linearising approaches look very different, the basic strategy of increased *local* feedback is the same. Either method, properly applied, will linearise a VAS into invisibility.

The importance of voltage drive

As explained above, it is fundamental to linear VAS operation that the collector impedance is high, and not subject to external perturbations. Thus a Class-B output stage, with large input impedance variations around the crossover point, is about the worst thing you could connect to it, and it is a tribute to the general robustness of the *standard* amplifier configuration that it can handle this internal unpleasantness gracefully, 100 W/8 Ω distortion typically degrading only from 0.0008% to 0.0017% at 1 kHz, assuming that the avoidable distortions have been eliminated. Note however that the effect becomes greater as the global feedback-factor is reduced. There is little deterioration at HF, where other distortions dominate. To the best of my knowledge I first demonstrated this in reference 10; if I am wrong then I have no doubt I shall soon hear about it.

The VAS buffer is most useful when LF distortion is already low, as it removes Distortion 4, which is (or should be) only visible when grosser non-linearities have been seen to. Two equally effective ways of buffering are shown in Figure 4.17e and f.

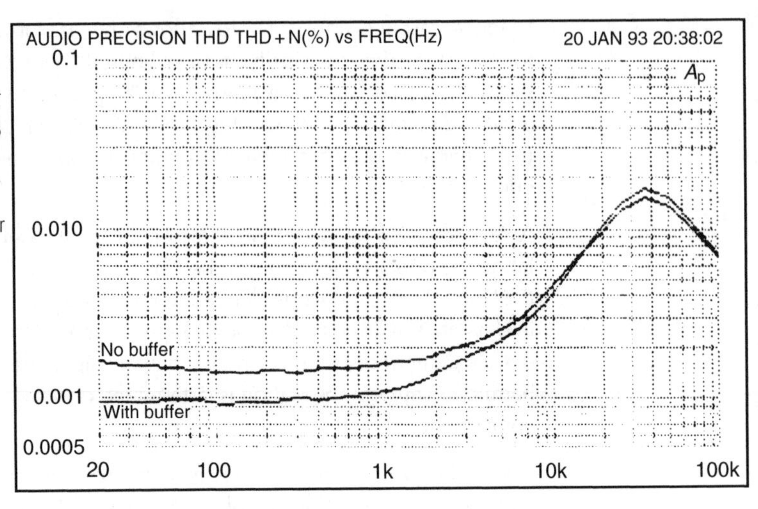

Figure 4.20 The beneficial effect of using a VAS-buffer in a full-scale Class-B amplifier. Note that the distortion needs to be low already for the benefit to be significant

There are other potential benefits to VAS buffering. The effect of beta mismatches in the output stage halves is minimised[11]. Voltage drive also promises the highest fT from the output devices, and therefore potentially greater stability, though I have no data of my own to offer on this point. It is right and proper to feel trepidation about inserting another stage in an amplifier with global feedback, but since this is an emitter-follower its phase-shift is minimal and it works well in practice.

If we have a VAS buffer then, providing we put it the right way up we can implement a form of DC-coupled bootstrapping that is electrically very similar to providing the VAS with a separate current-source. (See Figure 4.17f.)

The use of a buffer is essential if a VAS cascode is to do some good. Figure 4.20 shows before/after distortion for a full-scale power amplifier with cascode VAS driving 100 W into 8 Ω.

The balanced VAS

When we are exhorted to *make the amplifier linear before adding negative feedback* one of the few specific recommendations made is usually the use of a balanced VAS – sometimes combined with a double input stage consisting of two differential amplifiers, one complementary to the other. The latter seems to have little to recommend it, as you cannot balance a stage that is already balanced, but a balanced (and, by implication, more linear) VAS has its attractions. However, as explained above, the distortion contribution from a properly-designed VAS is negligible under most circumstances, and therefore there seems to be little to be gained.

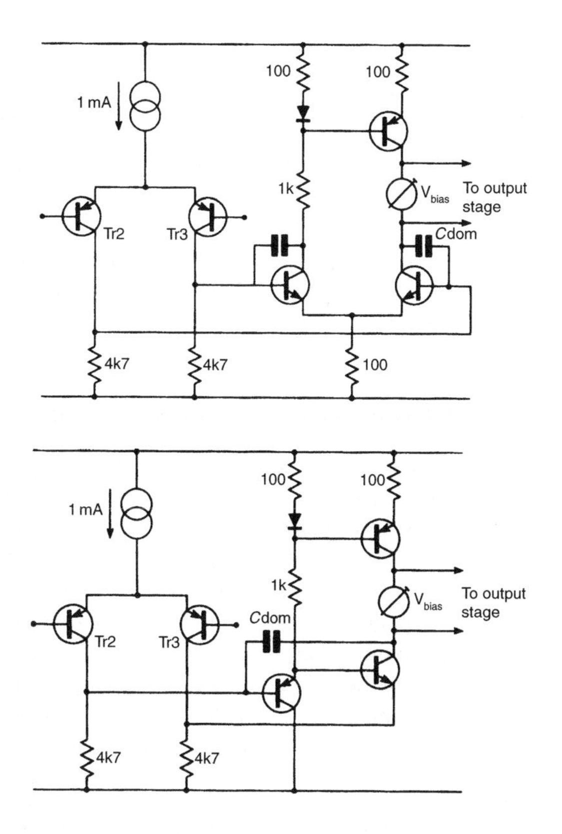

Figure 4.21 Two kinds of balanced VAS: Type 1 gives more open-loop gain, but no better open-loop linearity. Type 2 – the circuit originated by Lender

Two possible versions are shown in Figure 4.21; Type 1 gives approximately 10 dB more o/l gain than the standard, but this naturally requires an increase in Cdom if the same stability margins are to be maintained. In a model amplifier, any improvement in linearity can be wholly explained by this o/l gain increase, so this seems (not unexpectedly) an unpromising approach. Also, as Linsley-Hood has pointed out[12], the standing current through the bias generator is ill-defined compared with the usual current-source VAS; similarly the balance of the input pair is likely to be poor compared with the current-mirror version. A further difficulty is that there are now two signal paths from the input stage to the VAS output, and it is difficult to ensure that these have the same bandwidth; if they do not then a pole-zero doublet is generated in the open-loop gain characteristic that will markedly increase settling-time after a transient. This seems likely to apply to all balanced VAS configurations.

Type 2 is attributed by Borbely to Lender[13]. Figure 4.21 shows one version, with a quasi-balanced drive to the VAS transistor, via both base and emitter. This configuration does not give good balance of the input pair, as this is at the mercy of the tolerances of R2, R3, the Vbe of the VAS, and so on. Borbely has advocated using two complementary versions of this, giving

Type 3, but it is not clear that this in any way overcomes the objections above, and the increase in complexity is significant.

This can be only a brief examination of balanced VAS stages; many configurations are possible, and a comprehensive study of them all would be a major undertaking. All seem to be open to the objection that the vital balance of the input pair is not guaranteed, and that the current through the bias generator is not well-defined. However one advantage would seem to be the potential for sourcing and sinking large currents into Cdom, which might improve the ultimate slew-rate and HF linearity of a very fast amplifier.

The VAS and manipulating open-loop bandwidth

Acute marketing men will by now have realised that reducing the LF o/l gain, leaving HF gain unchanged, must move the P1 frequency upwards, as shown in Figure 4.22 *Open-loop gain is held constant up to 2 kHz* sounds so much better than *the open-loop bandwidth is restricted to 20 Hz* although these two statements could describe near-identical amplifiers, except that the first has plenty of open-loop gain at LF while the second has even more than that. Both amplifiers have the same feedback factor at HF, where the amount available has a direct effect on distortion performance, and could easily have the same slew-rate. Nonetheless the second amplifier somehow reads as sluggish and indolent, even when the truth of the matter is known.

It therefore follows that reducing the LF o/l gain may be of interest to commercial practitioners. Low values of open-loop gain also have their place in the dogma of the subjectivist, and the best way to bring about this state of affairs is worth examining, always bearing in mind that:

1 there is no engineering justification for it,
2 reducing the NFB factor will reveal more of the output stage distortion; since in general NFB is the only weapon we have to deal with this, blunting its edge seems ill-advised.

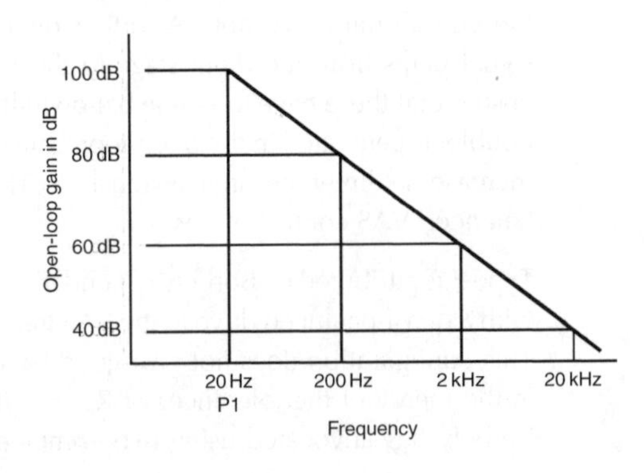

Figure 4.22
Showing how dominant-pole frequency P1 can be altered by changing the LF open-loop gain; the gain at HF, which determines Nyquist stability and HF distortion, is unaffected

It is of course simple to reduce o/l gain by degenerating the input pair, but this diminishes it at HF as well as LF. To alter it at LF only it is necessary to tackle the VAS instead, and Figure 4.23 shows two ways to reduce its gain. Figure 4.23a reduces gain by reducing the value of the collector impedance, having previously raised it with the use of a current-source collector load. This is no way to treat a gain stage; loading resistors low enough to have a significant effect cause unwanted current variations in the VAS as well as shunting its high collector impedance, and serious LF distortion appears. While this sort of practice has been advocated in the past[14], it seems to have nothing to recommend it as it degrades VAS linearity at the same time as syphoning off the feedback that would try to minimise the harm. Figure 4.23b also reduces overall o/l gain, but by adding a frequency-insensitive component to the local shunt feedback around the VAS. The value of RNFB is too high to load the collector significantly and therefore the full gain is available for local feedback at LF, even before *C*dom comes into action.

Figure 4.24 shows the effect on the open-loop gain of a model amplifier for several values of RNFB; this plot is in the format described in Chapter 3, where error-voltage is plotted rather than gain directly, and so the curve once more appears upside down compared with the usual presentation. Note that the dominant-pole frequency is increased from 800 Hz to above

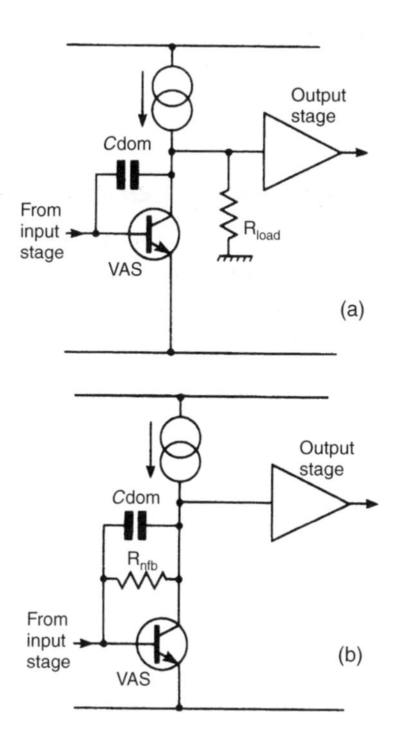

Figure 4.23
Two ways to reduce o/l gain:
(a) by simply loading down the collector. This is a cruel way to treat a VAS; current variations cause extra distortion
(b) local NFB with a resistor in parallel with C_{dom}. This looks crude, but actually works very well

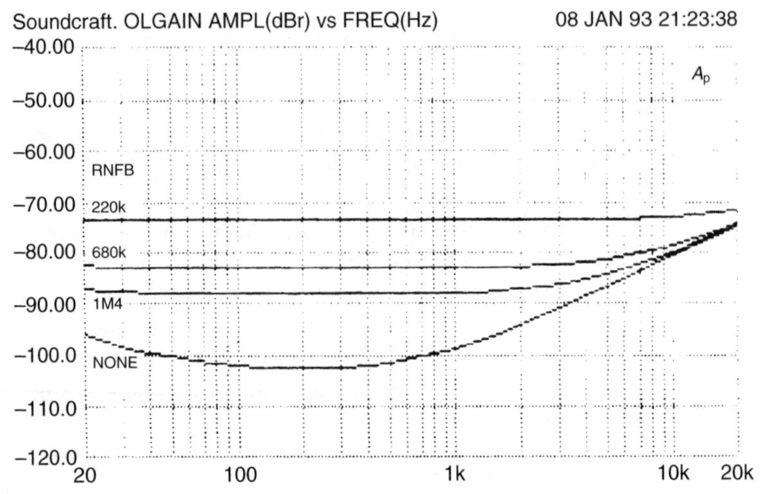

Figure 4.24
The result of VAS gain-reduction by local feedback; the dominant pole frequency is increased from about 800 Hz to about 20 kHz, with high-frequency gain hardly affected

20 kHz by using a 220 k value for RNFB; however the gain at higher frequencies is unaffected and so is the stability. Although the amount of feedback available at 1 kHz has been decreased by nearly 20 dB, the distortion at +16 dBu output is only increased from less than 0.001 to 0.0013%; most of this reading is due to noise.

In contrast, reducing the open-loop gain even by 10 dB by loading the VAS collector to ground requires a load of 4k7, which under the same conditions yields distortion of more than 0.01%.

Manipulating open-loop bandwidth

If the value of RNFB required falls below about 100 K, then the standing current flowing through it can become large enough to upset the amplifier operating conditions (Figure 4.23b). This is revealed by a rise in distortion above that expected from reducing the feedback factor, as the input stage becomes unbalanced as a result of the global feedback straightening things up. This effect can be simply prevented by putting a suitably large capacitor in series with RNFB. A 2μ2 non-electrolytic works well, and does not cause any strange response effects at low frequencies.

An unwelcome consequence of reducing the global negative feedback is that power-supply rejection is impaired (see page 257). To prevent negative supply-rail ripple reaching the output it is necessary to increase the filtering of the V-rail that powers the input stage and the VAS. Since the voltage drop in an RC filter so used detracts directly from the output voltage swing, there are severe restrictions on the highest resistor value that can be tolerated. The only direction left to go is increasing C, but this is also subject to limitations as it must withstand the full supply voltage and rapidly becomes a bulky and expensive item.

That describes the 'brawn' approach to improving PSRR. The 'brains' method is to use the input cascode compensation scheme described on page 253. This solves the problem by eliminating the change of reference at the VAS, and works extremely well with no compromise on HF stability. No filtering at all is now required for the V-supply rail – it can feed the input stage and VAS directly.

Conclusions

Hopefully the first half of this chapter has shown that input stage design is not something to be taken lightly if low noise, low distortion, and low offset are desired. A good design choice even for very high quality requirements is a constant-*gm* degenerated input pair with a degenerated current-mirror; the extra cost of the mirror will be trivial.

The latter half of this chapter showed how the strenuous efforts of the input circuitry can be best exploited by the voltage-amplifier stage following it. At first it appears axiomatic that the stage providing all the voltage gain of an amplifier, at the full voltage swing, is the prime suspect for generating a major part of its non-linearity. In actual fact, this is unlikely to be true, and if we select for an amplifier a cascode VAS with current-source collector-load and buffer it from the output stage, or use a beta-enhancer in the VAS, the second of our eight distortions is usually negligible.

References

1. Gray and Meyer *Analysis and Design of Analog Integrated Circuits* Wiley 1984, p. 172 (exponential law of singleton).
2. Gray and Meyer *Analysis and Design of Analog Integrated Circuits* Wiley 1984, p. 194 (tanh law of simple pair).
3. Self *Sound Mosfet Design* Electronics and Wireless World, Sept 1990, p. 760 (varying input balance with R2).
4. Gray and Meyer *Analysis and Design of Analog Integrated Circuits* Wiley 1984, p. 256 (tanh law of current-mirror pair).
5. Feucht *Handbook of Analog Circuit Design* Academic Press 1990, p. 432 (Cross-quad).
6. Quinn IEEE International Solid-State Circuits Conference, THPM 14.5, p. 188 (Cascomp).
7. Gray and Meyer *Analysis and Design of Analog Integrated Circuits* Wiley 1984, p. 251 (VAS transfer characteristic).
8. Antognetti (Ed) *Power Integrated Circuits* McGraw-Hill 1986 (see page 201).
9. Gray and Meyer *Analysis and Design of Analog Integrated Circuits* Wiley 1984, p. 252 (Rco limit on VAS gain).

10. Self *Sound Mosfet Design* Electronics and Wireless World, Sept 1990, p. 760.
11. Oliver *Distortion In Complementary-Pair Class-B Amplifiers* Hewlett-Packard Journal, Feb 1971, p. 11.
12. Linsley-Hood, J *Solid State Audio Power – 3* Electronics and Wireless World, Jan 1990, p. 16.
13. Borbely *A 60 W MOSFET Power Amplifier* The Audio Amateur, Issue 2, 1982, p. 9.
14. Hefley *High Fidelity, Low Feedback, 200 W* Electronics and Wireless World, June 1992, p. 454.

5

The output stage I

Classes and devices

The almost universal choice in semiconductor power amplifiers is for a unity-gain output stage, and specifically a voltage-follower. Output stages with gain are not unknown – see Mann[1] for a design with ten times gain in the output section – but they have significantly failed to win popularity. Most people feel that controlling distortion while handling large currents is quite hard enough without trying to generate gain at the same time.

In examining the small-signal stages, we have so far only needed to deal with one kind of distortion at a time, due to the monotonic transfer characteristics of such stages, which usually (but not invariably[2]) work in Class A. Economic and thermal realities mean that most output stages are Class B, and so we must now also consider crossover distortion (which remains the thorniest problem in power amplifier design) and HF switchoff effects.

We must also decide what *kind* of active device is to be used; JFETs offer few if any advantages in the small-current stages, but power FETS in the output appear to be a real possibility, providing that the extra cost proves to bring with it some tangible benefits.

The most fundamental factor in determining output-stage distortion is the Class of operation. Apart from its inherent inefficiency, Class-A is the ideal operating mode, because there can be no crossover or switchoff distortion. However, of those designs which have been published or reviewed, it is notable that the large-signal distortion produced is still significant. This looks like an opportunity lost, as of the distortions enumerated in Chapter 3, we now only have to deal with Distortion 1 (input-stage), Distortion 2 (VAS), and distortion 3 (output-stage large-signal non-linearity). Distortions 4, 5, 6 and 7, as mentioned earlier, are direct results of Class-B operation and therefore can be thankfully disregarded in a Class-A design. However, Class-B is overwhelmingly of the greater importance, and is therefore dealt with in detail below.

Class B is subject to much misunderstanding. It is often said that a pair of output transistors operated without any bias are *working in Class-B*, and therefore *generate severe crossover distortion*. In fact, with no bias each output device is operating for slightly less than half the time, and the question arises as to whether it would not be more accurate to call this Class-C and reserve Class-B for that condition of quiescent current which eliminates, or rather minimises, the crossover artefacts.

There is a further complication; it is not generally appreciated that moving into what is usually called Class-AB, by increasing the quiescent current, does *not* make things better. In fact, if the output power is above the level at which Class-A operation can be sustained, the THD reading will certainly increase as the bias control is advanced. This is due to what is usually called *gm-doubling* (i.e., the voltage-gain increase caused by both devices conducting simultaneously in the centre of the output-voltage range, that is, in the Class-A region) putting edges into the distortion residual that generate high-order harmonics much as under-biasing does. This vital fact seems almost unknown, presumably because the gm-doubling distortion is at a relatively low level and is completely obscured in most amplifiers by other distortions.

This phenomenon is demonstrated in Figure 5.1a, b and c, which shows spectrum analysis of the distortion residuals for under-biasing, optimal, and over-biasing of a 150 W/8 Ω amplifier at 1 kHz. As before, all non-linearities except the unavoidable Distortion 3 (output stage) have been effectively eliminated. The over-biased case had its quiescent current increased until the gm-doubling edges in the residual had an approximately 50:50 mark/space ratio, and so was in Class-A about half the time, which represents a rather generous amount of quiescent current for Class-AB. Nonetheless, the higher-order odd harmonics in Figure 5.1c are at least 10 dB greater in amplitude than those for the optimal Class-B case, and the third harmonic is actually higher than for the under-biased case as well. However the under-biased amplifier, generating the familiar sharp spikes on the residual, has a generally greater level of high-order odd harmonics above the fifth; about 8 dB higher than the AB case.

Since high-order odd harmonics are generally considered to be the most unpleasant, there seems to be a clear case for avoiding Class-AB altogether, as it will always be less efficient and generate more high-order distortion than the equivalent Class-B circuit as soon as it leaves Class-A. Class distinction seems to resolve itself into a binary choice between A or B.

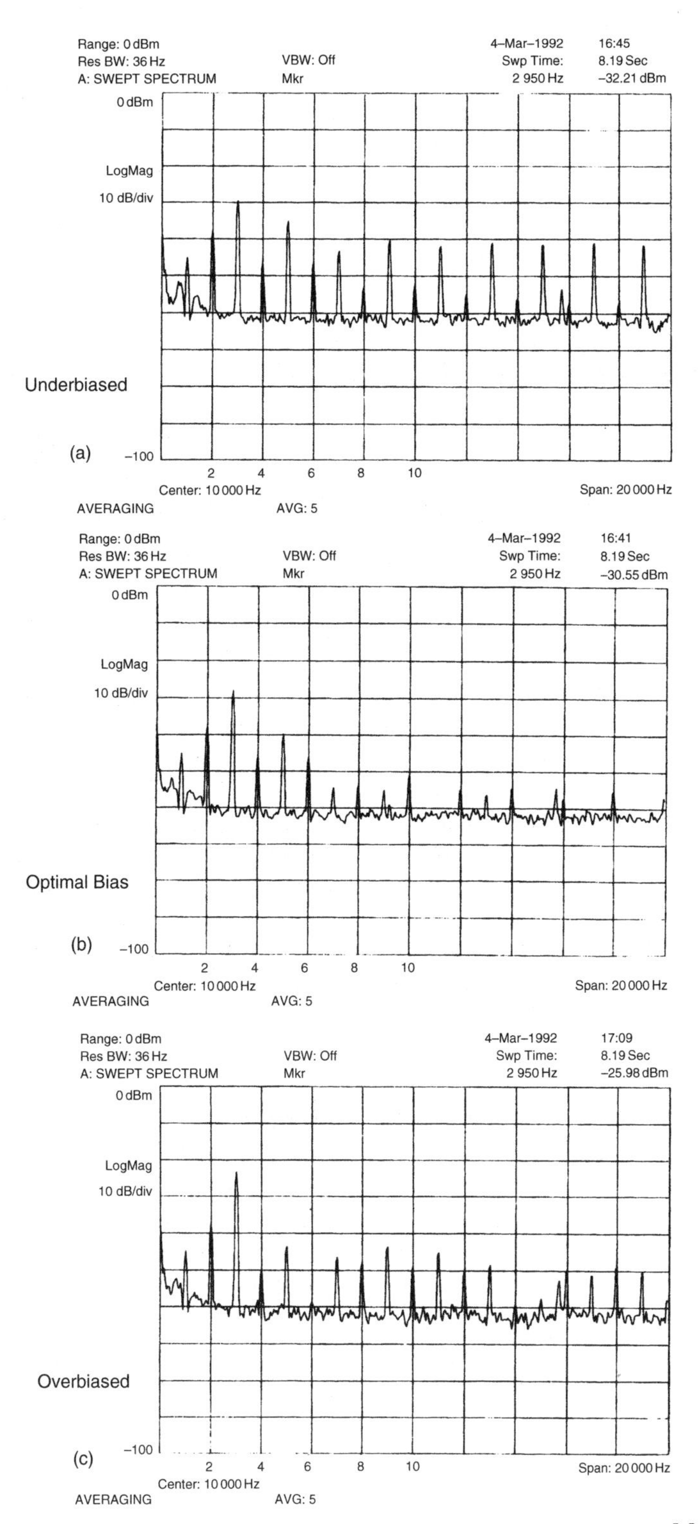

Figure 5.1 Spectrum analysis of Class-B and AB distortion residual

It must be emphasised that these effects are only visible in an amplifier where the other forms of distortion have been properly minimised. The RMS THD reading for Figure 5.1a was 0.00151%, for Figure 5.1b 0.00103%, and for Figure 5.1c 0.00153%. The tests were repeated at the 40 W power level with very similar results. The spike just below 16 kHz is interference from the testgear VDU.

This is complex enough, but there are other and deeper subtleties in Class-B, which are dealt with below.

The distortions of the output

I have called the distortion produced directly by the output stage Distortion 3 (see page 67) and this can now be subdivided into three categories. Distortion 3a describes the large-signal distortion that is produced by both Class-A and B, ultimately because of the large current swings in the active devices; in bipolars, but not FETs, large collector currents reduce the beta, leading to drooping gain at large output excursions. I shall use the term 'LSN' for Large-Signal Non-linearity, as opposed to crossover and switchoff phenomena that cause trouble at all output levels.

These other two contributions to Distortion 3 are associated with Class-B and AB only; Distortion 3b is classic crossover distortion, resulting from the non-conjugate nature of the output characteristics, and is essentially non-frequency dependent. In contrast, Distortion 3c is switchoff distortion, generated by the output devices failing to turn off quickly and cleanly at high frequencies, and is very strongly frequency-dependent. It is sometimes called *switching distortion,* but this allows room for confusion, as some writers use *switching distortion* to cover crossover distortion as well; hence I have used the term *switchoff distortion* to refer specifically to charge-storage turn-off troubles. Since Class-B is almost universal, and introduces all three kinds of non-linearity, we will concentrate on this.

Harmonic generation by crossover distortion

The usual non-linear distortions generate most of their unwanted energy in low-order harmonics that NFB can deal with effectively. However, crossover and switching distortions that warp only a small part of the output swing tend to push energy into high-order harmonics, and this important process is demonstrated here, by Fourier analysis of a SPICE waveform.

Taking a sinewave fundamental, and treating the distortion as an added error signal E, let the ratio WR describe the proportion of the cycle where E is non-zero. If this error is a triangle-wave extending over the whole cycle (WR = 1) this would represent large-signal non-linearity, and Figure 5.2

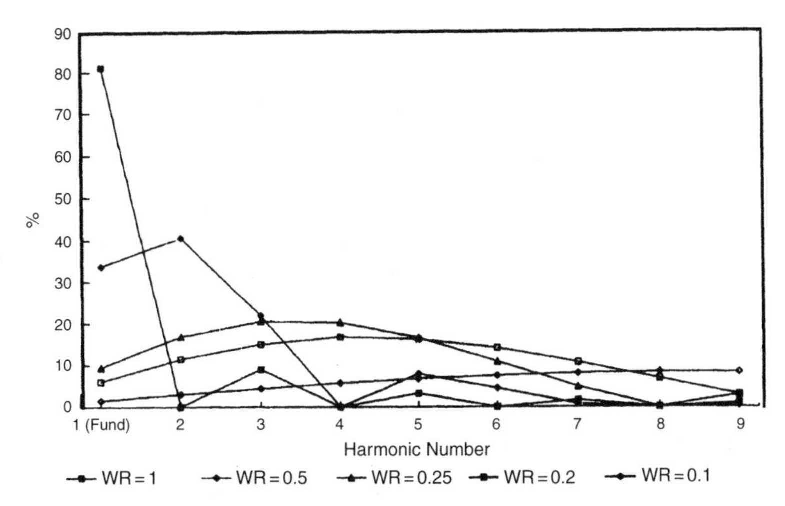

Figure 5.2
The amplitude of each harmonic changes with WR; as the error waveform gets narrower, energy is transferred to the higher harmonics

shows that most of the harmonic energy goes into the third and fifth harmonics; the even harmonics are all zero due to the symmetry of the waveform.

Figure 5.3 shows how the situation is made more like crossover or switching distortion by squeezing the triangular error into the centre of the cycle so that its value is zero elsewhere; now E is non-zero for only half the cycle (denoted by WR = 0.5) and Figure 5.2 shows that the even harmonics are no longer absent. As WR is further decreased, the energy is pushed into higher-order harmonics, the amplitude of the lower falling.

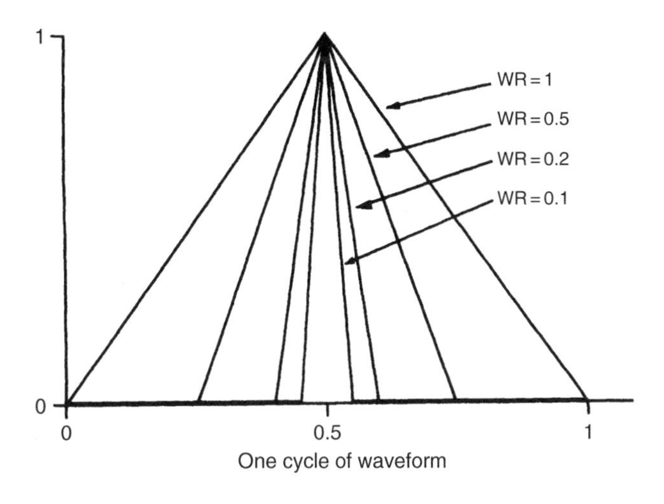

Figure 5.3
Diagram of the error waveform E for some values of WR

The high harmonics have roughly equal amplitude, spectrum analysis (see Figure 5.1 page 111) confirming that even in a Blameless amplifier driven at 1 kHz, harmonics are freely generated from the seventh to the 19th at an equal level to a dB or so. The 19th harmonic is only 10 dB below the third.

Thus, in an amplifier with crossover distortion, the order of the harmonics will decrease as signal amplitude reduces, and WR increases; their lower frequencies allow them to be better corrected by the frequency-dependant NFB. This effect seems to work *against* the commonly assumed rise of percentage crossover distortion as level is reduced.

Comparing output stages

One of my aims in this book is to show how to isolate each source of distortion so that it can be studied (and hopefully reduced) with a minimum of confusion and perplexity. When investigating output behaviour, it is perfectly practical to drive output stages open-loop, providing the driving source-impedance is properly specified; this is difficult with a conventional amplifier, as it means the output must be driven from a frequency-dependant impedance simulating that at the VAS collector, with some sort of feedback mechanism incorporated to keep the drive voltage constant.

However, if the VAS is buffered from the output stage by some form of emitter-follower, as advocated on page 101, it makes things much simpler, a straightforward low-impedance source (e.g., 50 Ω) providing a good approximation of conditions in a VAS-buffered closed-loop amplifier. The VAS-buffer makes the system more designable by eliminating two variables – the VAS collector impedance at LF, and the frequency at which it starts to decrease due to local feedback through *C*dom. This markedly simplifies the study of output stage behaviour.

The large-signal linearity of various kinds of open-loop output stage with typical values are shown in Figures 5.6–5.16. These diagrams were all generated by SPICE simulation, and are plotted as incremental output gain against output voltage, with the load resistance stepped from 16 to 2 Ω, which I hope is the lowest impedance that feckless loudspeaker designers will throw at us. They have come to be known as *wingspread* diagrams, from their vaguely bird-like appearance. The power devices are MJ802 and MJ4502, which are more complementary than many so-called pairs, and minimise distracting large-signal asymmetry. The quiescent conditions are in each case set to minimise the peak deviations of gain around the crossover point for 8 Ω loading; for the moment it is assumed that you can set this accurately and keep it where you want it. The difficulties in actually doing this will be examined later.

If we confine ourselves to the most straightforward output stages, there are at least 16 distinct configurations, without including error-correcting[3], current-dumping[4], or Blomley[5] types. These are:

Emitter-Follower	3 types	Figure 5.4
Complementary-Feedback Pair	1 type	Figure 5.5
Quasi-Complementary	2 types	Figure 5.5
Output Triples	At least 7 types	Figure 5.6
Power FET	3 types	Chapter 11

The emitter-follower output

Three versions of the most common type of output stage are shown in Figure 5.4; this is the double-emitter-follower, where the first follower acts as driver to the second (output) device. I have deliberately called this an Emitter-Follower (EF) rather than a Darlington configuration, as this latter implies an integrated device that includes driver, output, and assorted emitter resistors in one ill-conceived package. As for all the circuitry here, the component values are representative of real practice. Important attributes of this topology are:

1 the input is transferred to the output via two base-emitter junctions in series, with no local feedback around the stage (apart from the very local 100% voltage feedback that makes an EF what it is),
2 there are two dissimilar base-emitter junctions between the bias voltage and the emitter resistor Re, carrying different currents and at different temperatures. The bias generator must attempt to compensate for both at once, though it can only be thermally coupled to one. The output devices have substantial thermal inertia, and so any thermal compensation can only be a time-average of the preceding conditions. Figure 5.4a shows the most prevalent version (Type I) which has its driver emitter resistors connected to the output rail.

The Type II configuration in Figure 5.4b is at first sight merely a pointless variation on Type I, but in fact it has a valuable extra property. The shared driver emitter-resistor Rd, with no output-rail connection, allows the drivers to reverse-bias the base-emitter junction of the output device being turned off. Assume that the output voltage is heading downwards through the crossover region; the current through Re1 has dropped to zero, but that through Re2 is increasing, giving a voltage-drop across it, so TR4 base is caused to go more negative to get the output to the right voltage. This negative excursion is coupled to TR3 base through Rd, and with the values shown can reverse bias it by up to $-0.5\,V$, increasing to $-1.6\,V$ with a $4\,\Omega$ load. The speed-up capacitor Cs markedly improves this action, preventing the charge-suckout rate being limited by the resistance of Rd. While the Type I circuit has a similar voltage drop across Re2, the connection of the

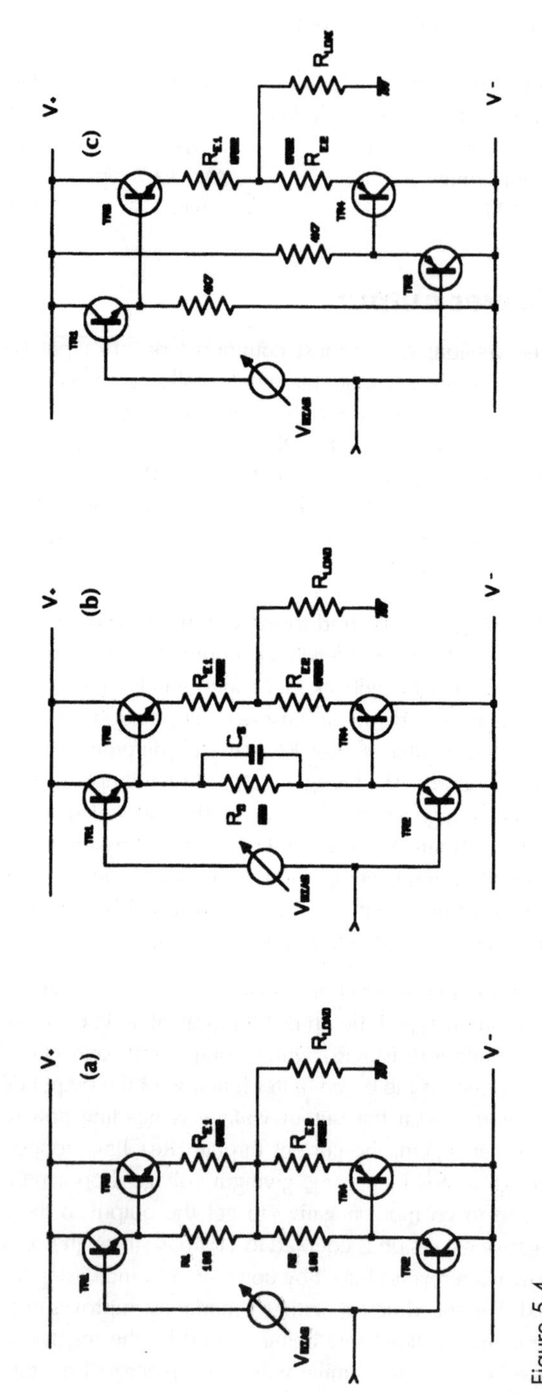

Figure 5.4
Three types of Emitter-Follower output stages

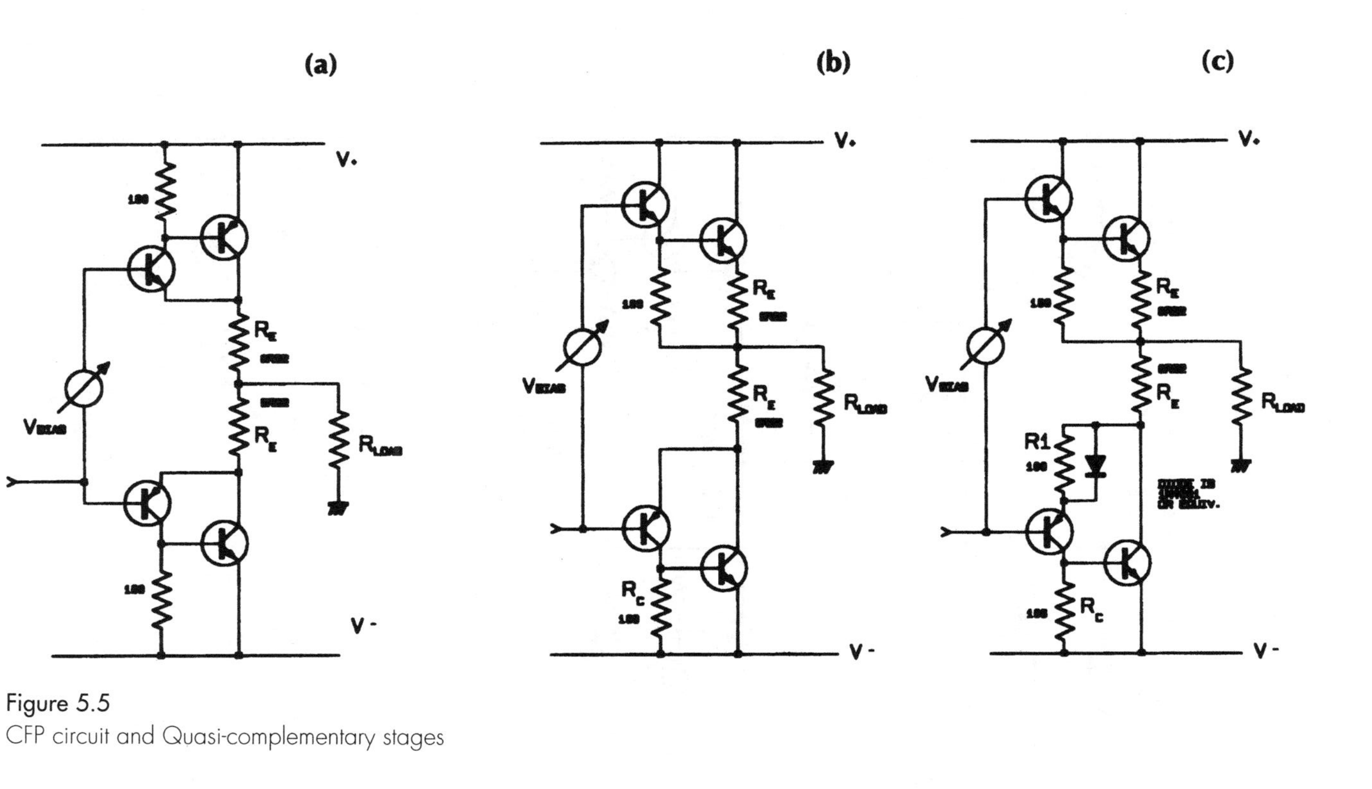

Figure 5.5
CFP circuit and Quasi-complementary stages

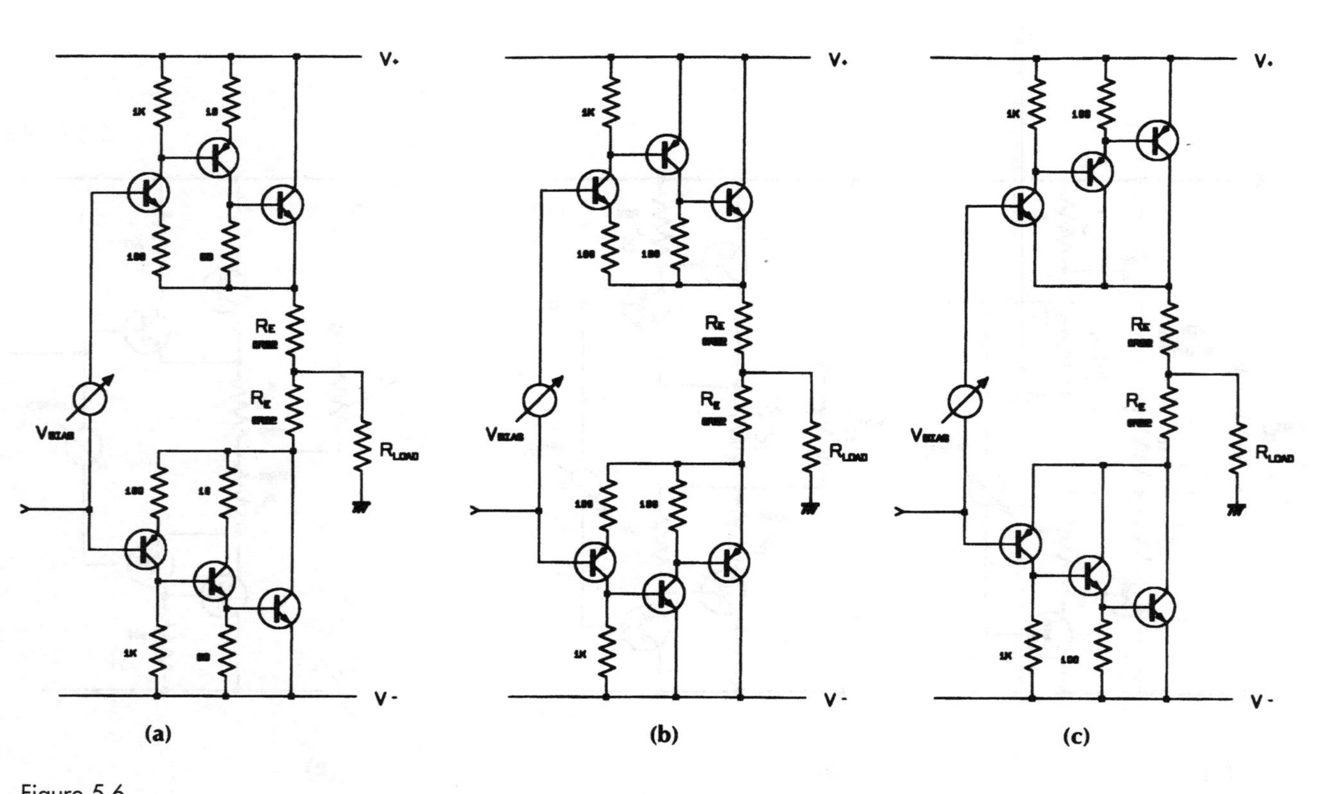

Figure 5.6
Three of the possible Output-Triple configurations

mid-point of R1, R2 to the output rail prevents this from reaching TR3 base; instead TR1 base is reverse-biased as the output moves negative, and since charge-storage in the drivers is usually not a problem, this does little good. In Type II, the drivers are never reverse-biased, though they do turn off. The important issue of output turn-off and switching distortion is further examined on page 156.

The Type III topology shown in Figure 5.4c maintains the drivers in Class-A by connecting the driver Re's to the opposite supply rail, rather than the output rail. It is a common misconception[6] that Class-A drivers somehow maintain better low-frequency control over the output devices, but I have yet to locate any advantage myself. The driver dissipation is of course substantially increased, and nothing seems to be gained at LF as far as the output transistors are concerned, for in both Type I and Type II the drivers are still conducting at the moment the outputs turn off, and are back in conduction before the outputs turn on, which would seem to be all that matters. Type III is equally good as Type II at reverse-biasing the output bases, and may give even cleaner HF turn-off as the carriers are being swept from the bases by a higher resistance terminated in a higher voltage, approximating constant-current drive; this remains to be determined by experiment.

The large-signal linearity of these three versions is virtually identical all have the same feature of two base-emitter junctions in series between input and load. The gain/output voltage plot is shown at Figure 5.7; with BJTs

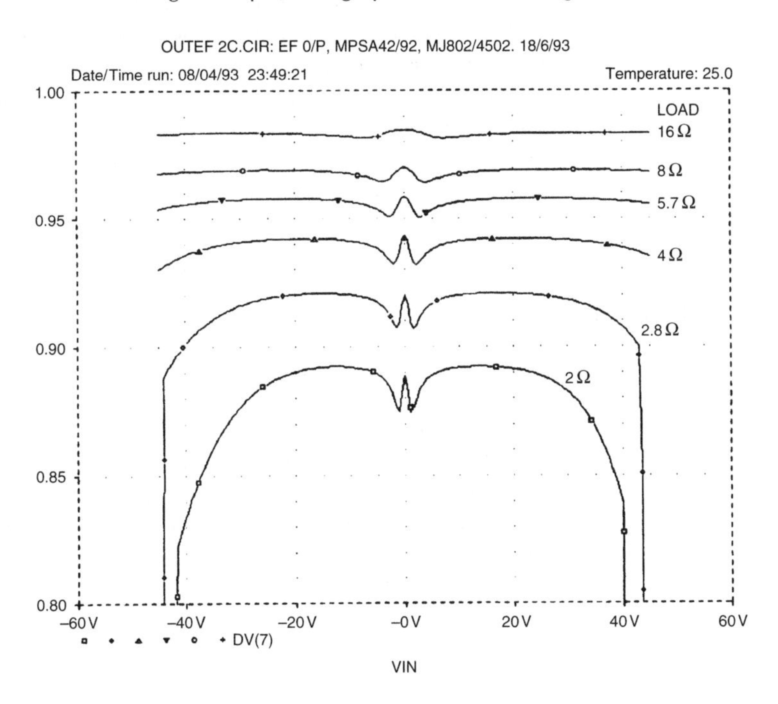

Figure 5.7 Emitter-Follower large-signal gain versus output

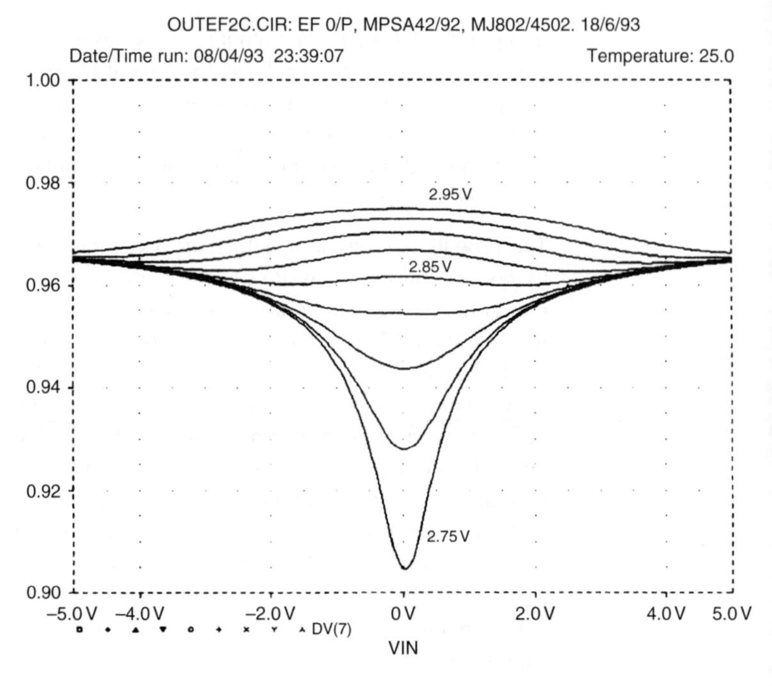

Figure 5.8
EF crossover region gain deviations, ±5 V range

the gain reduction with increasing loading is largely due to the Re's. Note that the crossover region appears as a relatively smooth wobble rather than a jagged shape. Another major feature is the gain-droop at high output voltages and low loads, and this gives us a clue that high collector currents are the fundamental cause of this. A close-up of the crossover region gain for 8 Ω loading only is shown in Figure 5.8; note that no Vbias setting can be found to give a constant or even monotonic gain; the double-dip and central gain peak are characteristic of optimal adjustment. The region extends over an output range of about ±5 V, independent of load resistance.

The CFP output

The other major type of bipolar complementary output is the Complementary-Feedback Pair (hereinafter CFP) sometimes called the Sziklai-Pair, seen in Figure 5.5a. There seems to be only one popular configuration, though versions with gain are possible. The drivers are now placed so that they compare the output voltage with that at the input. Thus wrapping the outputs in a local NFB loop promises better linearity than emitter-follower versions with 100% feedback applied separately to driver and output transistors.

The CFP topology is generally considered to show better thermal stability than the EF, because the Vbe of the output devices is inside the local NFB loop, and only the driver Vbe affects the quiescent conditions. The true situation is rather more complex, and is explored in Chapter 12.

In the CFP output, like the EF, the drivers are conducting whenever the outputs are, so special arrangements to keep them in Class-A seem pointless. The CFP stage, like EF Type I, can only reverse-bias the driver bases, and not the output bases, unless extra voltage rails outside the main ones are provided.

The output gain plot is shown in Figure 5.9; Fourier analysis of this shows that the CFP generates less than half the LSN of an emitter-follower stage. (See Table 5.1.) Given also the greater quiescent stability, it is hard to see why this topology is not more popular.

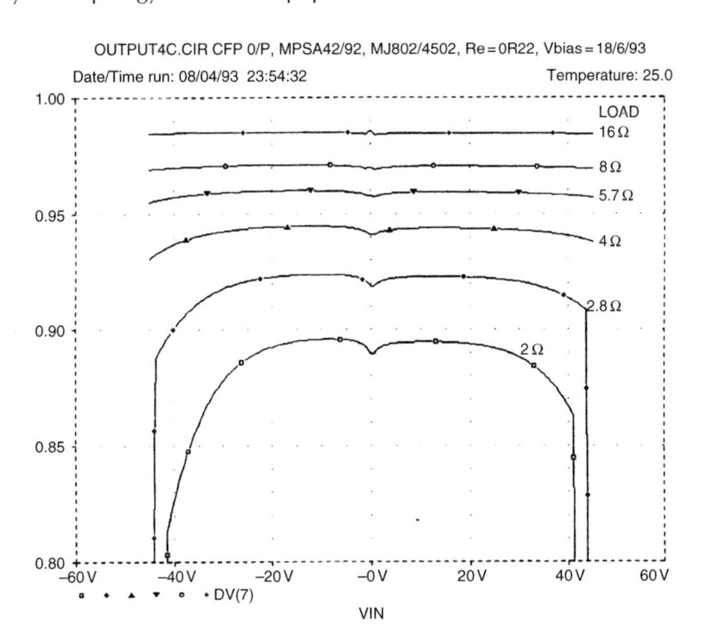

Figure 5.9 Complementary-Feedback-Pair gain versus output

Table 5.1 Summary of output distortion

	Emitter Follower	CFP	Quasi Simple	Quasi Bax	Triple Type 1	Simple MOSFET	Quasi MOSFET	Hybrid MOSFET
8 Ω	0.031%	0.014%	0.069%	0.050%	0.13%	0.47%	0.44%	0.052%
Gain:	0.97	0.97	0.97	0.96	0.97	0.83	0.84	0.97
4 Ω	0.042%	0.030%	0.079%	0.083%	0.60%	0.84%	0.072%	0.072%
Gain:	0.94	0.94	0.94	0.94	0.92	0.72	0.73	0.94

Table 5.1 summarises the SPICE curves for 4 and 8 Ω loadings; FET results from Chapter 11 are included for comparison. Each gain plot was subjected to Fourier analysis to calculate THD % results for a ± 40 V input.

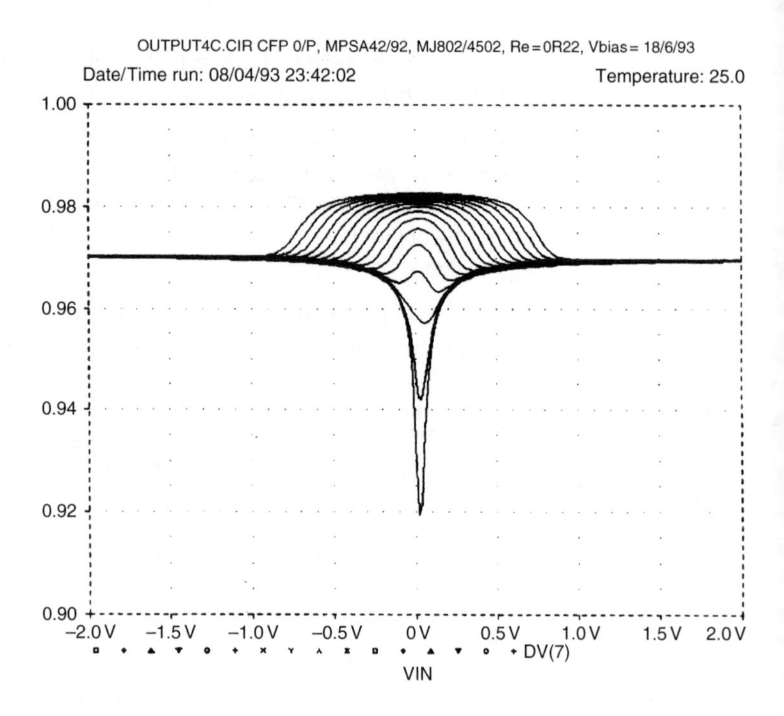

Figure 5.10 CFP crossover region ±2 V, Vbias as a parameter

The crossover region is much narrower, at about ±0.3 V (Figure 5.10). When under-biased, this shows up on the distortion residual as narrower spikes than an emitter-follower output gives. The bad effects of gm-doubling as Vbias increases above optimal (here 1.296 V) can be seen in the slopes moving outwards from the centre.

Quasi-complementary outputs

Originally, the quasi-complementary configuration[7] was virtually mandatory, as it was a long time before PNP silicon power transistors were available in anything approaching complements of the NPN versions. The standard version shown at Figure 5.5b is well known for poor symmetry around the crossover region, as shown in Figure 5.11. Figure 5.12 zooms in to show that the crossover region is a kind of unhappy hybrid of the EF and CFP, as might be expected, and that no setting of Vbias can remove the sharp edge in the gain plot.

A major improvement to symmetry can be made by using a Baxandall diode[8], as shown in Figure 5.5c. This stratagem yields gain plots very similar to those for the true complementary EF at Figures 5.7, 5.8, though in practice the crossover distortion seems rather higher. When this Quasi-Baxandall stage is used closed-loop in an amplifier in which Distortions 1 and 2, and 4 to 7 have been properly eliminated, it is capable of much better performance than is commonly believed; for example,

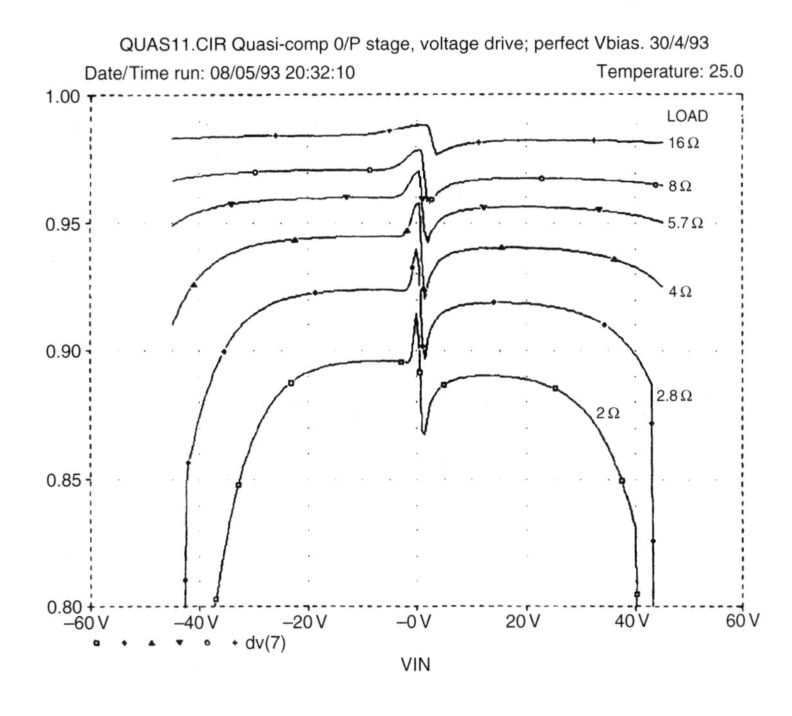

Figure 5.11 Quasi-complementary large-signal gain versus output

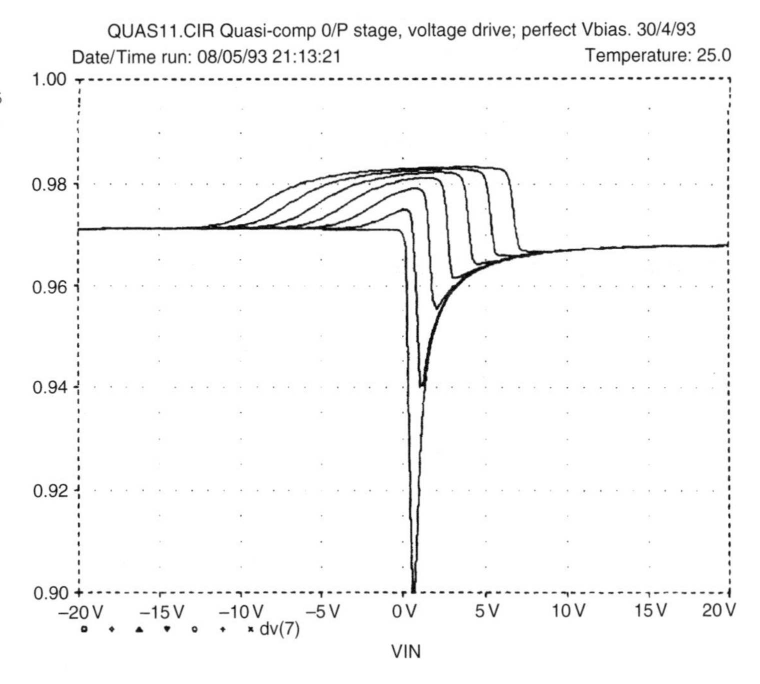

Figure 5.12 Quasi crossover region ±20 V, Vbias as parameter

0.0015% (1 kHz) and 0.015% (10 kHz) at 100 W is straightforward to obtain from an amplifier with a moderate NFB factor of about 34 dB at 20 kHz.

The best reason to use the quasi-Baxandall approach today is to save money on output devices, as PNP power BJTs remain somewhat pricier than NPNs. Given the tiny cost of a Baxandall diode, and the absolutely dependable improvement it gives, there seems no reason why anyone should ever use the *standard* quasi circuit. My experiments show that the value of R1 in Figure 5.5c is not critical; making it about the same as Rc seems to work well.

Triple-based output configurations

If we allow the use of three rather than two bipolar transistors in each half of an output stage, the number of circuit permutations possible leaps upwards, and I cannot provide even a rapid overview in the space available. There are two possible advantages if output triples are used correctly:

1. better linearity at high output voltages and currents,
2. more stable quiescent setting as the pre-drivers can be arranged to handle very little power indeed, and to remain almost cold in use.

However, triples do not abolish crossover distortion, and they are, as usually configured, incapable of reverse-biasing the output bases to improve switchoff. Figure 5.6 shows three of the more useful ways to make a triple output stage – all of those shown (with the possible exception of Figure 5.6c, which I have just made up) have been used in commercial designs, and Figure 5.6a will be recognised as the Quad-303 quasi-complementary triple. The design of triples demands care, as the possibility of local HF instability in each output half is very real.

Triple EF output stages

Sometimes it is necessary to use a triple output stage simply because the currents flowing in the output stage are too big to be handled by two transistors in cascade. If you are driving 2 Ω or 1 Ω loads, then typically there will be multiple output devices in parallel. Providing the base current for five or more output transistors, with their relatively low beta, will usually be beyond the normal driver types, and it is common to use another output device as the driver. This will hopefully have the power-handling capability, but with this comes low beta once again. This means that the driver base currents in turn become too large for a normal VAS stage to source. There are two solutions – make the VAS capable of sourcing hundreds of mA, or insert another stage of current – gain between VAS and drivers. The latter is much easier, and the usual choice. These extra transistors are usually called the pre-drivers (see Figure 5.13).

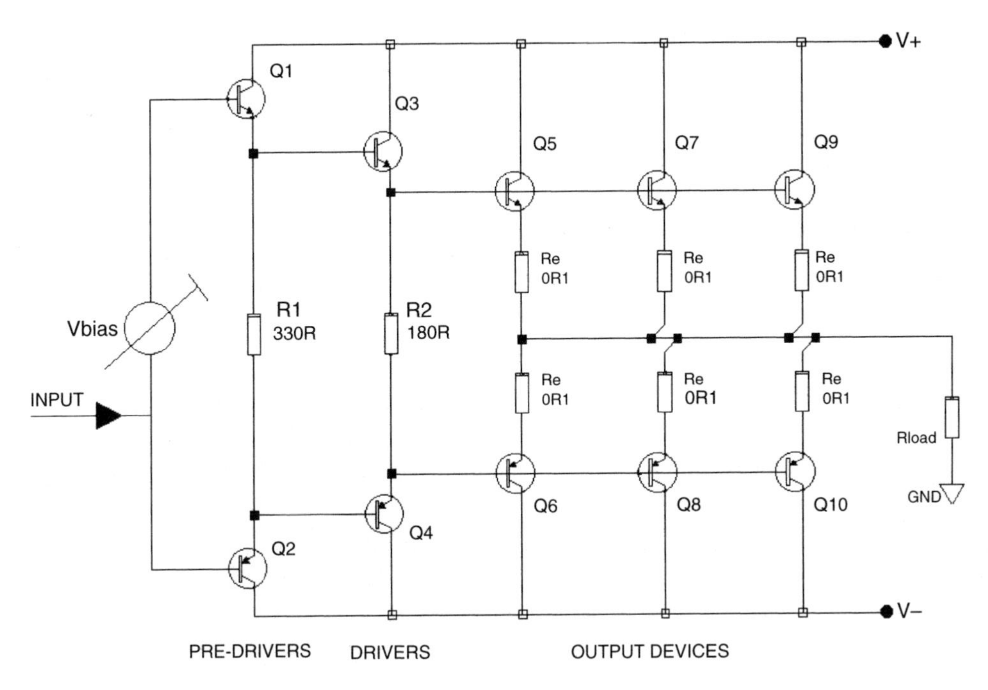

Figure 5.13
A triple-EF output stage. Both pre-drivers and drivers have emitter-resistors

In this circuit the pre-drivers dissipate relatively little power, and providing they are medium-power devices such as those in a TO220 package it is unlikely that they will need heatsinking to cope with the demands made on them. There is, however, another reason to fit pre-drive heatsinks – or at least make room at the layout stage so you have the option.

In Figure 5.13 there is about 1.2 V across R2, so Q3, 4 have to supply a standing current of about 7 mA. This has no effect on the drivers as they are likely to be well cooled to deal with normal load demands. However, the voltage across R1 is two Vbe's higher, at 2.4 V, so the standing current through it is actually higher at 7.3 mA. (The exact figures naturally depend on the values for R1, R2 that are chosen, but it is difficult to make them much higher than shown here without compromising the speed of high-frequency turn-off.) The pre-drivers are usually small devices, and so they are likely to get warm, and this leads to drift in the bias conditions after switch-on. Adding heatsinks cannot eliminate this effect, but does usefully reduce it.

In a triple-EF output stage like this the Vbias generator must produce enough voltage to turn on six base-emitter junctions, plus the small standing voltage Vq across the emitter resistors, totalling about 3.9 V in practice. The Vbe of the bias transistor is therefore being multiplied by a larger factor, and

Vbias will drop more for the same temperature rise. This should be taken into account, as it is easy with this kind of output stage to come up with a bias generator that is overcompensated for temperature.

Distortion and its mechanisms

Subdividing Distortion 3 into Large-Signal Non-linearity, crossover, and switchoff distortion provides a basis for judging which output stage is best. The LSN is determined by both circuit topology and device characteristics, crossover distortion is critically related to quiescent-conditions stability, and switchoff distortion depends strongly on the output stage's ability to remove carriers from power BJT bases. I now look at how these shortcomings can be improved, and the effect they have when an output stage is used closed-loop.

In Chapter 4 it was demonstrated that the distortion from the small-signal stages can be kept to very low levels that will prove to be negligible compared with closed-loop output-stage distortion, by the adroit use of relatively conventional circuitry. Likewise, Chapter 6 will reveal that Distortions 4 to 8 can be effectively eliminated by lesser-known but straightforward methods. This leaves Distortion 3, in its three components, as the only distortion that is in any sense *unavoidable*, as Class-B stages completely free from crossover artefacts are so far beyond us.

This is therefore a good place to review the concept of a *Blameless* amplifier, introduced in Chapter 3; one designed so that all the easily defeated distortion mechanisms have been rendered negligible. (Note that the word *Blameless* has been carefully chosen to not imply Perfection.) Distortion 1 cannot be totally eradicated, but its onset can be pushed well above 20 kHz. Distortion 2 can be effectively eliminated by cascoding, and Distortion 4–Distortion 7 can be made negligible by simple measures to be described later. This leaves Distortion 3, which includes the knottiest Class-B problems, i.e., crossover distortion (Distortion 3b) and HF switchoff difficulties (Distortion 3c).

The design rules presented here will allow the routine design of Blameless Amplifiers. However, this still leaves the most difficult problem of Class-B unsolved, so it is too early to conclude that as far as amplifier linearity is concerned, history is over . . . .

Large-signal distortion (Distortion 3a)

Amplifiers always distort more with heavier loading. This is true without exception so far as I am aware. Why? Is there anything we can do about it?

A Blameless Class-B amplifier typically gives an 8 Ω distortion performance that depends very little on variable transistor characteristics such as beta. At this load impedance output stage non-linearity is almost entirely crossover distortion, which is a voltage-domain effect.

As the load impedance of the amplifier is decreased from infinity to 4 Ω, distortion increases in an intriguing manner. The unloaded THD is not much greater than that from the AP System-1 test oscillator, but as loading increases crossover distortion rises steadily: see Figure 7.25. When the load impedance falls below about 8 Ω, a new distortion begins to appear, overlaying the existing crossover non-linearities. It is essentially third harmonic. In Figure 5.14 the upper trace shows the 4 Ω THD is consistently twice that for 8 Ω, once it appears above the noise floor.

I label this Distortion 3a, or Large Signal Non-linearity (LSN), where 'Large' refers to currents rather than voltages. Unlike crossover Distortion 3b, the amount of LSN generated is highly dependent on device characteristics. The distortion residual is basically third order because of the symmetric and compressive nature of the output stage gain characteristic, with some second harmonic because the beta loss is component dependent and not perfectly symmetrical in the upper and lower output stage halves. Figure 5.15 shows a typical THD residual for Large Signal Nonlinearity, driving 50 W into 4 Ω. The residual is averaged 64 times to reduce noise.

LSN occurs in both emitter-follower (EF) and Complementary-Feedback Pair (CFP) output configurations; this section concentrates on the CFP version, as shown in Figure 5.5a. Figure 5.16 shows the incremental gain of a simulated CFP output stage for 8 and 4 Ω; the lower 4 Ω trace has greater downward curvature, i.e., a greater falloff of gain with increasing current. Note that this falloff is steeper in the negative half, so the THD generated will contain even as well as odd harmonics. The simulated EF behaviour is very similar.

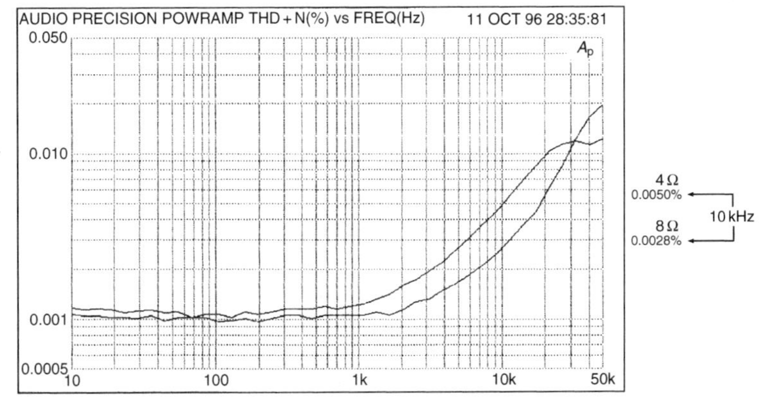

Figure 5.14 Upper trace shows distortion increase due to LSN as load goes from 8 to 4 Ω. Blameless amplifier at 25 W/8 Ω

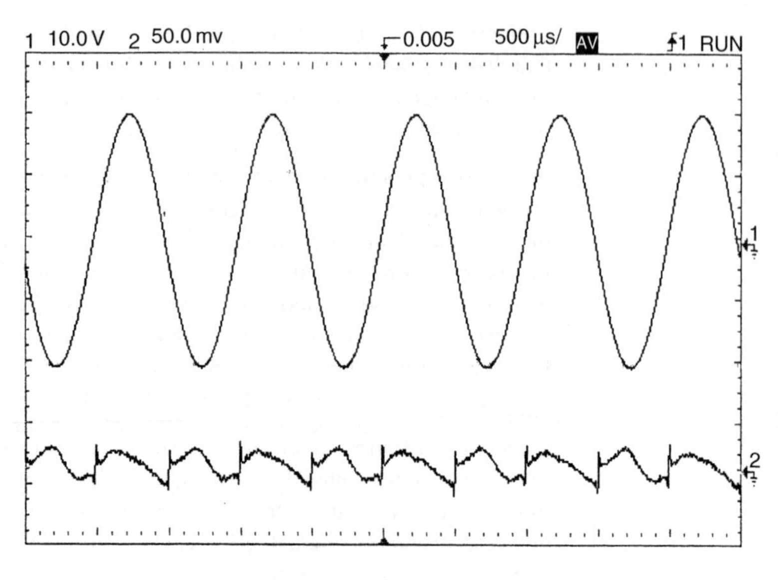

Figure 5.15 Large Signal Non-linearity, driving 50 W into 4 Ω and averaged 64 times

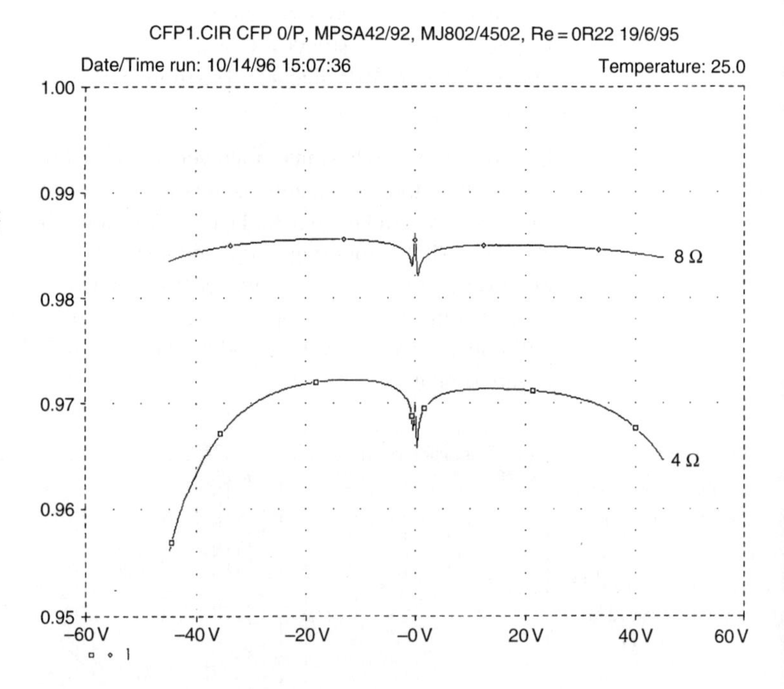

Figure 5.16 The incremental gain of a standard CFP output stage. The 4 Ω trace droops much more as the gain falls off at higher currents. PSpice simulation

As it happens, an 8 Ω nominal impedance is a reasonably good match for standard power BJTs, though 16 Ω might be better for minimizing LSN if loudspeaker technology permits. It is coincidental that an 8 Ω nominal impedance corresponds approximately with the heaviest load that can be driven without LSN appearing, as this value is a legacy from valve technology. LSN is an extra distortion component laid on top of others, and usually dominating them in amplitude, so it is obviously simplest to minimise the 8 Ω distortion first. 4 Ω effects can then be seen more or less in isolation when load impedance is reduced.

The typical result of 4 Ω loading was shown in Figure 5.14, for the modern MJ15024/25 complementary pair from Motorola. Figure 5.17 shows the same diagram for one of the oldest silicon complementary pairs, the 2N3055/2955. The 8 Ω distortion is similar for the different devices, but the 4 Ω THD is 3.0 times worse for the venerable 2N3055/2955. Such is progress.

Such experiments with different output devices throw useful light on the Blameless concept – from the various types tried so far it can be said that Blameless performance, whatever the output device type, should not exceed 0.001% at 1 kHz and 0.006% at 10 kHz, when driving 8 Ω. The components existed to build sub-0.001% THD amplifiers in mid-1969, but not the knowledge.

Low-impedance loads have other implications beyond worse THD. The requirements for sustained long-term 4 Ω operation are severe, demanding more heatsinking and greater power supply capacity. For economic reasons the peak/average ratio of music is usually fully exploited, though this can cause real problems on extended sinewave tests, such as the FTC 40%-power-for-an-hour preconditioning procedure.

The focus of this section is the extra distortion generated in the output stage itself by increased loading, but there are other ways in which linearity may be degraded by the higher currents flowing. Of the amplifier distortion mechanisms (see page 65), Distortions 1, 2, and 8 are unaffected by output stage current magnitudes. Distortion 4 might be expected to increase, as

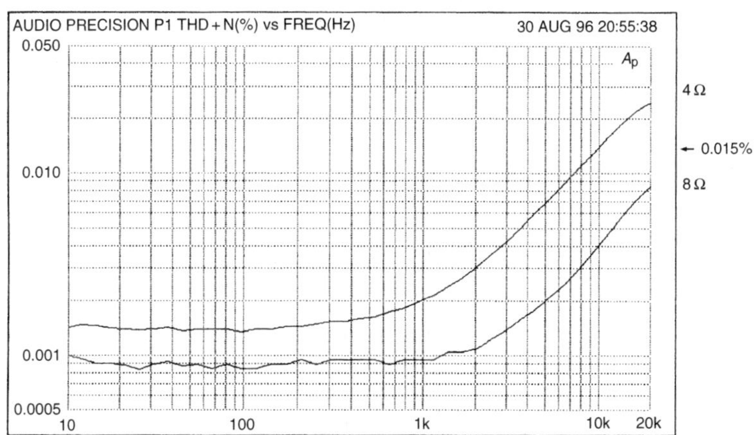

Figure 5.17
4 Ω distortion is 3× greater than 8 Ω for 2N3055/2955 output devices. Compare Figure 5.14

increased loading on the output stage is reflected in increased loading on the VAS. However, both the beta-enhanced EF and buffered-cascode methods of VAS linearisation deal effectively with sub-8 Ω loads, and this does not seem to be a problem.

When a 4 Ω load is driven, the current taken from the power supply is greater, potentially increasing the rail ripple, which could worsen Distortion 5. However, if the supply reservoir capacitances have been sized to permit greater power delivery, their increased capacitance reduces ripple again, so this effect tends to cancel out. Even if rail ripple doubles, the usual RC filtering of bias supplies should keep it out of the amplifier, preventing intrusion via the input pair tail, and so on.

Distortion 6 could worsen as the half-wave currents flowing in the output circuitry are twice as large, with no counteracting mechanism. Distortion 7, if present, will be worse due to the increased load currents flowing in the output stage wiring resistances.

Of those mechanisms above, Distortion 4 is inherent in the circuit configuration (though easily reducible below the threshold of measurement) while 5, 6, and 7 are topological, in that they depend on the spatial and geometrical relationships of components and wiring. The latter three distortions can therefore be completely eliminated in both theory and practice. This leaves only the LSN component, otherwise known as Distortion 3a, to deal with.

The load-invariant concept

In an ideal amplifier the extra LSN distortion component would not exist. Such an amplifier would give no more distortion into 4 Ω than 8, and could be called 'Load-Invariant to 4 Ω'. The minimum load qualification is required because it will be seen that the lower the impedance, the greater the difficulties in aspiring to Load-Invariance. I assume that we start out with an amplifier that is Blameless at 8 Ω; it would be logical but quite pointless to apply the term 'Load-Invariant' to an ill-conceived amplifier delivering 1% THD into both 8 and 4 Ω.

The LSN mechanism

When the load impedance is reduced, the voltage conditions are essentially unchanged. LSN is therefore clearly a current-domain effect, a function of the magnitude of the signal currents flowing in drivers and output devices.

A 4 Ω load doubles the output device currents, but this does not in itself generate significant extra distortion. The crucial factor appears to be that the current drawn from the drivers by the output device bases *more* than doubles, due to beta falloff in the output devices as collector current increases.

It is this *extra* increase of current that causes almost all the additional distortion. The exact details of this have not been completely clarified, but it seems that this 'extra current' due to beta falloff varies very non-linearly with output voltage, and combines with driver non-linearity to reinforce it rather than cancel. Beta-droop is ultimately due to high-level injection effects, which are in the province of semiconductor physics rather than amplifier design. Such effects vary greatly with device type, so when output transistors are selected, the likely performance with loads below 8 Ω must be considered.

There is good simulator evidence that LSN is entirely due to beta-droop causing extra current to be drawn from the drivers. To summarise:

- Simulated output stages with output devices modified to have no beta-droop (by increasing SPICE model parameter IKF) do not show LSN. It appears to be specifically that extra current taken due to beta-droop causes the extra non-linearity.
- Simulated output devices driven with zero-impedance voltage sources instead of the usual transistor drivers exhibit no LSN. This shows that LSN does not occur in the outputs themselves, and so it must be happening in the driver transistors.
- Output stage distortion can be treated as an error voltage between input and output. The double emitter-follower (EF) stage error is therefore: driver Vbe + output Vbe + Re drop. A simulated EF output stage with the usual drivers shows that it is primarily non-linearity increases in the driver Vbe rather than in the output Vbe, as load resistance is reduced. The voltage drop across the emitter resistors Re is essentially linear.

The knowledge that beta-droop caused by increased output device Ic is at the root of the problem leads to some solutions. First, the per-device Ic can be reduced by using parallel output devices. Alternatively Ic can be left unchanged and output device types selected for those with the least beta-droop.

Doubled output devices

LSN can be effectively reduced by doubling the output devices, when this is quite unnecessary for handling the rated power output. The falloff of beta depends on collector current, and if two output devices are connected in parallel, the collector current divides in two between them. Beta-droop is much reduced.

From the above evidence, I predicted that this doubling ought to reduce LSN – and when measured, indeed it does. Such reality checks must never be omitted when using circuit simulators. Figure 5.18 compares the 4 Ω THD at 60 W for single and double output devices, showing that doubling reduces distortion by about 1.9 times, which is a worthwhile improvement.

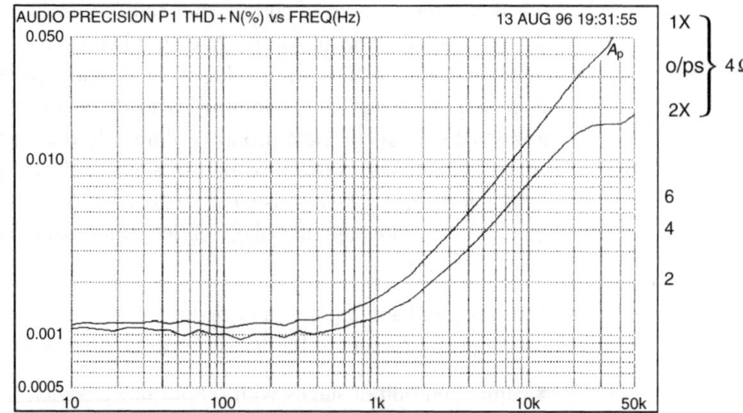

Figure 5.18
4 Ω distortion is reduced by 1.9× upon doubling standard (MJ15024/15025) output transistors. 30 W/8 Ω

The output transistors used for this test were modern devices, the Motorola MJ15024/15025. The much older 2N3055/2955 complementary pair give a similar halving of LSN when their number is doubled, though the initial distortion is three times higher into 4 Ω. 2N3055 specimens with an H suffix show markedly worse linearity than those without.

No explicit current-sharing components were added when doubling the devices, and this lack seemed to have no effect on LSN reduction. There was no evidence of current hogging, and it appears that the circuit cabling resistances alone were sufficent to prevent this.

Doubling the number of power devices naturally increases the power output capability, though if this is exploited LSN will tend to rise again, and you are back where you started. Opting for increased power output will also make it necessary to uprate the power supply, heatsinks, and so on. The essence of this technique is to use parallel devices to reduce distortion long before power handling alone compels you to do so.

Better output devices

The 2SC3281 2SA1302 complementary pair are plastic TO3P devices with a reputation in the hi-fi industry for being 'more linear' than the general run of transistors. Vague claims of this sort arouse the deepest of suspicions; compare the many assertions of superior linearity for power FETs, which is the exact opposite of reality. However, in this case the core of truth is that 2SC3281 and 2SA1302 show much less beta-droop than average power transistors. These devices were introduced by Toshiba; Motorola versions are MJL3281A, MJL1302A, also in TO3P package. Figure 5.19 shows beta-droop, for the various devices discussed here, and it is clear that more droop means more LSN.

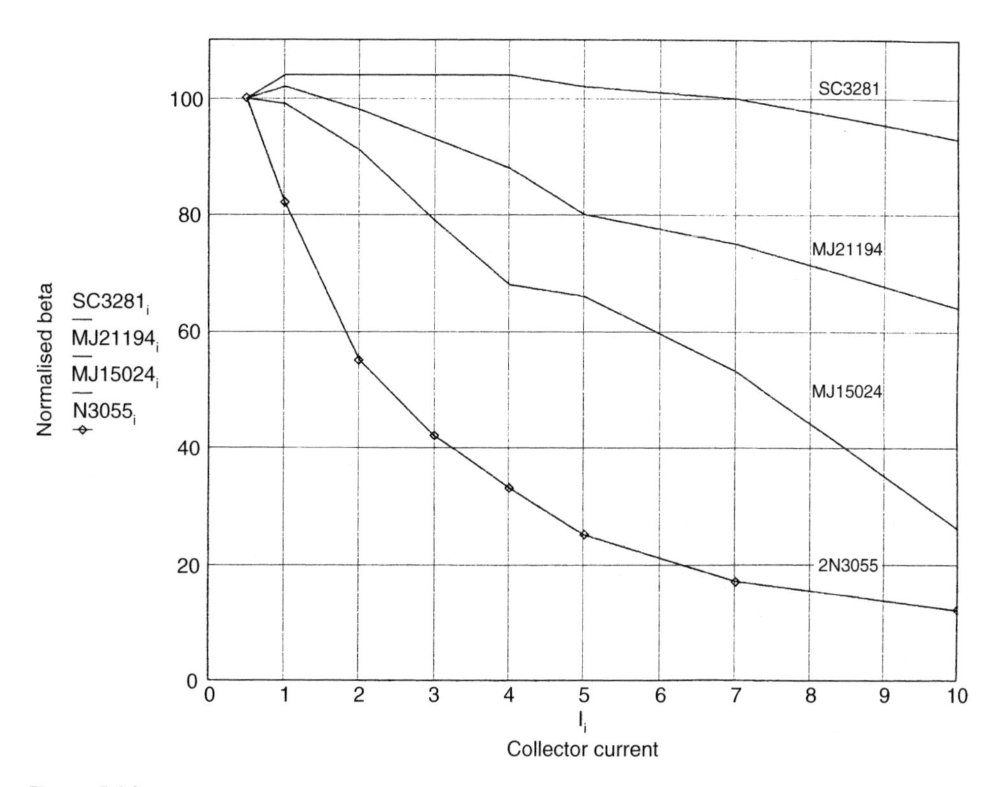

Figure 5.19

Power transistor beta falls as collector current increases. Beta is normalised to 100 at 0.5 A (from manufacturers' data sheets)

The 3281/1302 pair are clearly in a different class from conventional transistors, as they maintain beta much more effectively when collector current increases. There seems to be no special name for this class of BJTs, so I have called them 'sustained-beta' devices here.

The THD into 4 and 8 Ω for single 3281/1302 devices is shown in Figure 5.20. Distortion is reduced by about 1.4 times compared with the standard devices of Figure 5.14, over the range 2–8 kHz. Several pairs of 3281/1302 were tested and the 4 Ω improvement is consistent and repeatable.

The obvious next step is to combine these two techniques by using doubled sustained-beta devices. The doubled-device results are shown in Figure 5.21 where the distortion at 80 W/4 Ω (15 kHz) is reduced from 0.009% in Figure 5.20 to 0.0045%; in other words, halved. The 8 and 4 Ω traces are now very close together, the 4 Ω THD being only 1.2 times higher than in the 8 Ω case.

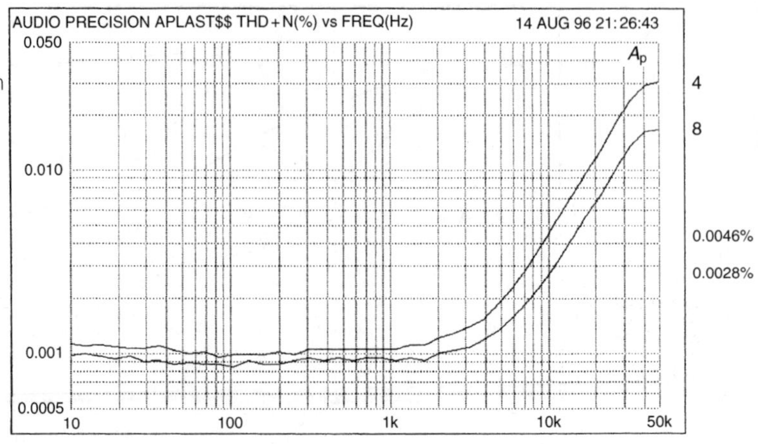

Figure 5.20
THD at 40 W/8 Ω and 80 W/4 Ω with single 3281/1302 devices

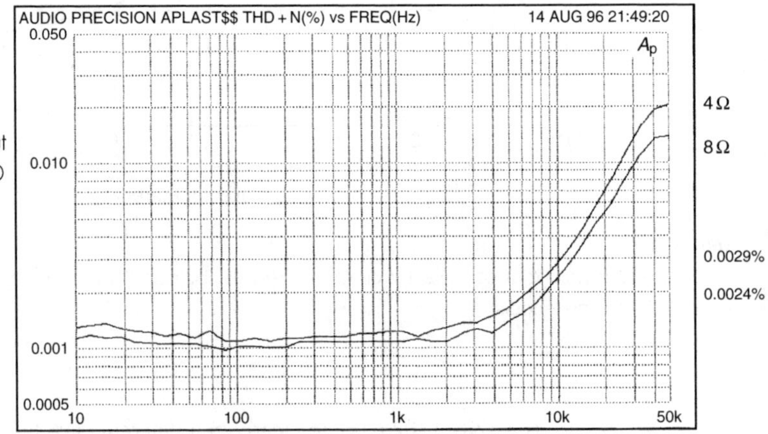

Figure 5.21
THD at 40 W/8 Ω and 80 W/4 Ω with doubled 3281/1302 output transistors. 4 Ω THD has been halved compared with Figure 5.12

There are other devices showing less beta-droop than standard. In a very quick survey I unearthed the MJ21193, MJ21194 pair (TO3 package) and the MJL21193, MJL21194 pair (TO3P package), both from Motorola. These devices show beta-maintenance intermediate between the 'super' 3281/1302 and 'ordinary' MJ15024/25, so it seemed likely that they would give less LSN than ordinary power devices, but more than the 3281/1302. This prediction was tested and duly fulfilled.

It could be argued that multiplying output transistors is an expensive way to solve a linearity problem. To give this perspective, in a typical stereo power amplifier the total cost including heatsink, metal work and mains transformer will only increase by about 5% when the output devices are doubled.

Feedforward diodes

The first technique I tried to reduce LSN was the addition of power diodes across OR22 output emitter resistors. The improvement was only significant for high power into sub-3 Ω loading, and was of rather doubtful utility for hi-fi. Feedforward diodes treat the symptoms (by attempting distortion cancellation) rather than the root cause, so it is not surprising this method is of limited effectiveness; see Figure 5.25.

It is my current practice to set the output emitter resistors Re at 0.1 Ω, rather than the more common OR22. This change both improves voltage-swing efficiency and reduces the extra distortion generated if the amplifier is erroneously biased into Class AB. As a result even low-impedance loads give a relatively small voltage drop across Re, which is insufficient to turn on a silicon power diode at realistic output levels.

Schottky diodes have a lower forward voltage drop and might be useful here. Tests with 50 A diodes have been made but have so far not been encouraging in the distortion reduction achieved. Suitable Schottky diodes cost at least as much as an output transistor, and two will be needed.

Trouble with triples

In electronics, as in many fields, there is often a choice between applying brawn (in this case multiple power devices) or brains to solve a given problem. The 'brains' option here would be a clever circuit configuration that reduced LSN without replication of expensive power silicon, and the obvious place to look is the output-triple approach. Note 'output triples' here refers to pre-driver, driver, and output device all in one local NFB loop, rather than three identical output devices in parallel, which I would call 'tripled outputs'. Getting the nomenclature right is a bit of a problem.

In simulation, output-triple configurations do reduce the gain-droop that causes LSN. There are many different ways to configure output triples, and they vary in their linearity and immunity to LSN. The true difficulty with this approach is that three transistors in a tight local loop are very prone to parasitic and local oscillations. This tendency is exacerbated by reducing the load impedances, presumably because the higher collector currents lead to increased device transconductance. This sort of instability can be very hard to deal with, and in some configurations appears almost insoluble. At present this approach has not been studied further.

Loads below 4 Ω

So far I have concentrated on 4 Ω loads; loudspeaker impedances often sink lower than this, so further tests were done at 3 Ω. One pair of

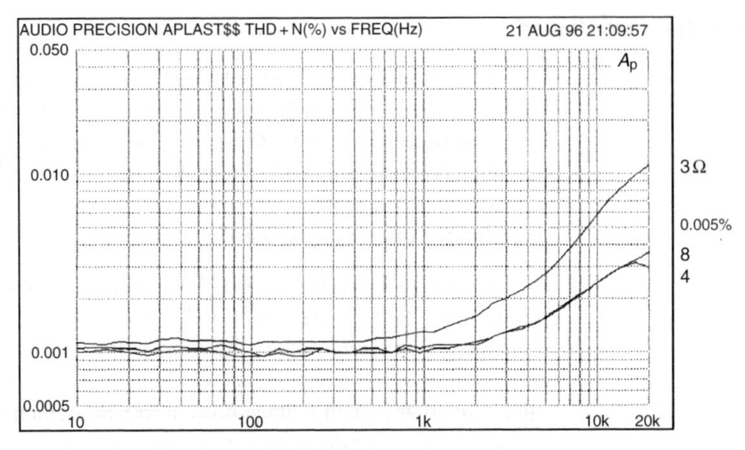

Figure 5.22
Distortion for 3, 4 and 8 Ω loads, single 3281/1302 devices. 20 W/8 Ω, 40 W/40 Ω and 60 W/3 Ω

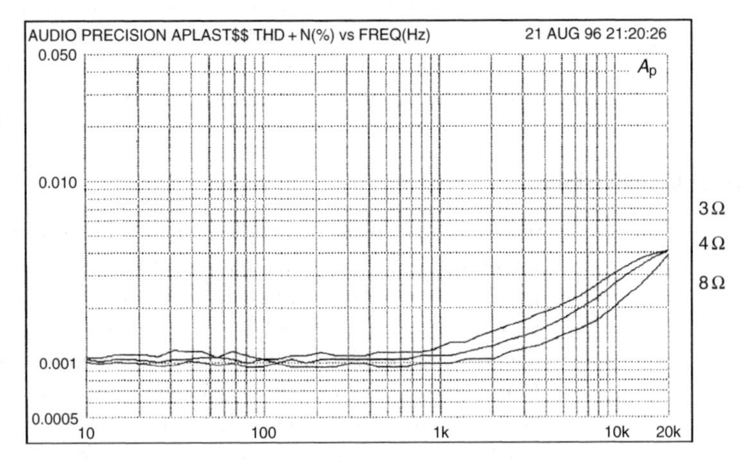

Figure 5.23
Distortion for 3, 4 and 8 Ω load, double 3281/1302 devices. Power as Figure 5.22

3281/1302 devices will give 50 W into 3 Ω for THD of 0.006% (10 kHz), see Figure 5.22. Two pairs of 3281/1302 reduce the distortion to 0.003% (10 kHz) as in Figure 5.23. This is an excellent result for such simple circuitry, and may well be a record for 3 Ω linearity.

It appears that whatever the device type, doubling the outputs halves the THD percentage for 4 Ω loading. This principle can be extended to 2 Ω operation, but tripled devices are required for sustained operation at significant powers. The resistive losses will be serious, so 2 Ω power output may be little greater than that into 4 Ω.

Better 8 Ω performance

It was not expected that the sustained-beta devices would also show lower crossover distortion at 8 Ω, but they do, and the effect is once more repeatable. It may be that whatever improves the beta characteristic also somewhat alters the turn-on law so that crossover distortion is reduced; alternatively traces of LSN, not visible in the THD residual, may have been eliminated. The latter is probably the more likely explanation.

The plot in Figure 5.23 shows the improvement over the MJ15024/25 pair; compare the 8 Ω line in Figure 5.14. The 8 Ω THD at 10 kHz is reduced from 0.003% to 0.002%, and with correct bias adjustment, the crossover artefacts are invisible on the 1 kHz THD residual. Crossover artefacts are only just visible in the 4 Ω case, and to get a feel for the distortion being produced, and to set the bias optimally, it is necessary to test at 5 kHz into 4 Ω.

A practical load-invariant design

Figure 5.24 is the circuit of a practical Load-Invariant amplifier designed for 8 Ω nominal loads with 4 Ω impedance dips; not for speakers that start out at 4 Ω nominal and plummet from there. The distortion performance is shown in Figures 5.21 and 5.22 for various fitments of output device. The supply voltage can be from ±20 to ±40 V; checking power capability for a given output device fit must be left to the constructor.

Apart from Load-Invariance, the design also incorporates two new techniques from the Thermal Dynamics section of this book.

The first technique greatly reduces time lag in the thermal compensation. With a CFP output stage, the bias generator aims to shadow driver junction temperature rather than the output junctions. A much faster response to power dissipation changes is obtained by mounting bias generator transistor TR8 on top of driver TR14, rather than on the other side of the heatsink. The driver heatsink mass is largely decoupled from the thermal compensation system, and the response is speeded up by at least two orders of magnitude.

The second innovation is a bias generator with an increased temperature coefficient, to reduce the static errors introduced by thermal losses between driver and sensor. The bias generator tempco is increased to −4.0 mV/°C. D5 also compensates for the effect of ambient temperature changes.

This design is not described in detail because it closely resembles the Blameless Class-B amp described on page 179. The low-noise feedback network is taken from the Trimodal amplifier on page 274; note the requirement for input bootstrapping if a 10k input impedance is required. Single-slope VI limiting is incorporated for overload protection, implemented by TR12, 13. The global NFB factor is once more a modest 30 dB at 20 kHz.

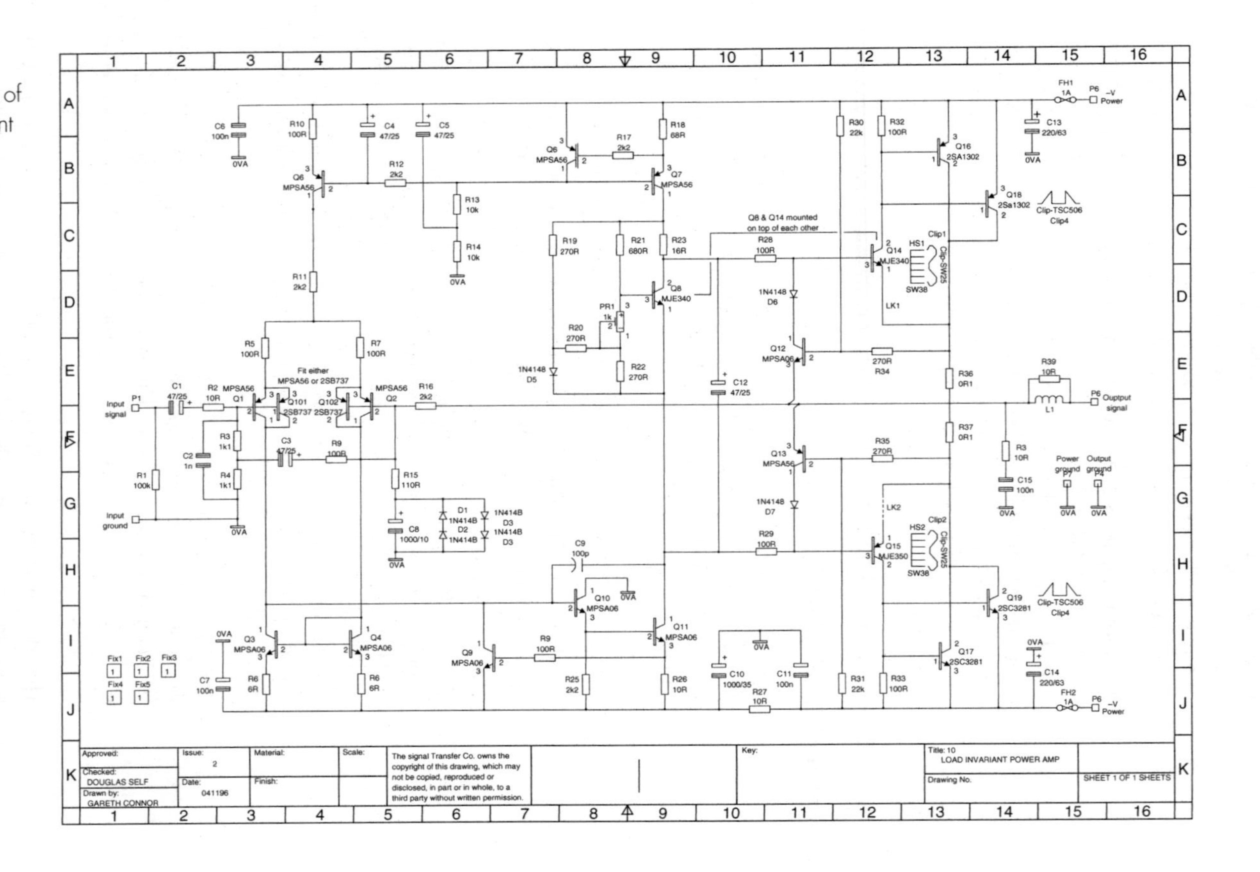

Figure 5.24 Circuit diagram of the Load-Invariant power amlifier

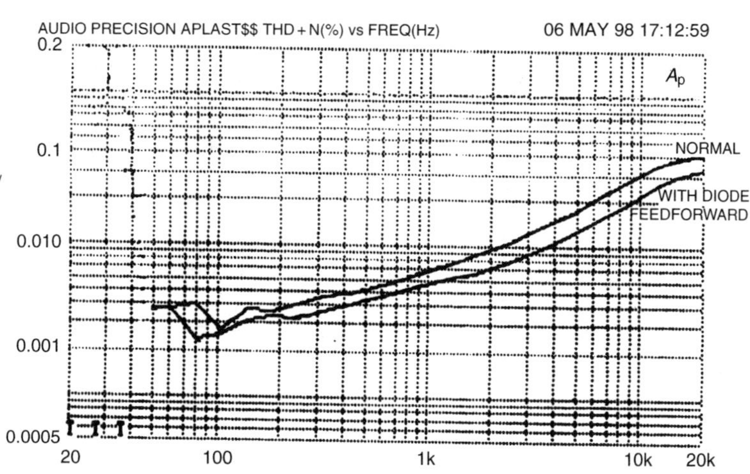

Figure 5.25 Simple diode feedforward reduces distortion with sub-8 Ω loads. Measured at 210 W into 2.7 Ω

The latest findings

I have recently done further experiments with multiple devices, using three, four, five and six in parallel. The 2SC2922/2SA1612 complementary pair were used. In this case the circuit used was somewhat different (see Figure 5.26). With a greater number of devices I was now more concerned about proper current sharing, and so each device has its own emitter resistor. This makes it look much more like a conventional paralleled output stage, which essentially it is. This time I tried both double and the triple-EF output configurations, as I wished to prove:

(a) that LSN theory worked for both of the common configurations EF and CFP – it does;
(b) that LSN theory worked for both double and triple versions of the EF output stage – it does.

For reasons of space only the triple-EF results are discussed here.

Figure 5.27 shows the measured THD results for one complementary pair of output devices in the triple-EF circuit of Fig 5.25. Distortion is slightly higher, and the noise floor relatively lower, than in the standard result (Fig 2 in Part 1) because of the higher output power of 50 W/8 Ω. Figure 5.28 shows the same except there are now two pairs of output devices. Note that THD has halved at both 8 and 4 Ω loads; this is probably due to the larger currents taken by 8 Ω loads at this higher power. Figure 5.29 shows the result for six devices; 8 Ω distortion has almost been abolished, and the 4 Ω result is almost as good. It is necessary to go down to a 2 Ω load to get the THD clear of the noise so it can be measured accurately. With six outputs, driving a substantial amount of power into this load is not a problem.

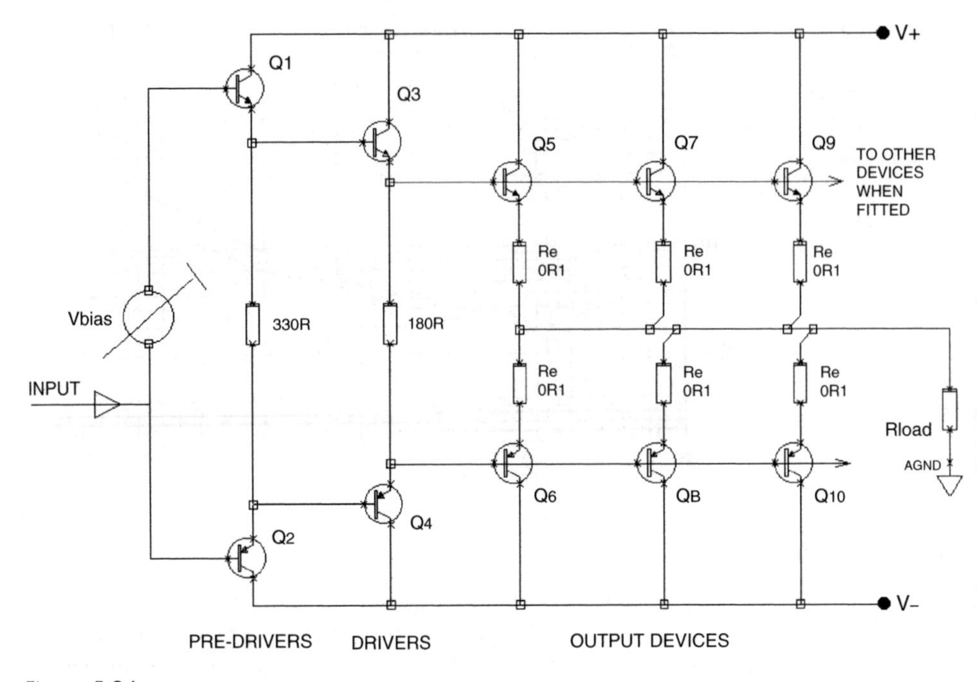

Figure 5.26
The triple-EF output stage used for the measurements described below. 'Triple' refers to the fact that there are three transistors from input to output, rather than the fact that there happen to be three output devices in parallel

On a practical note, the more output devices you have, the harder the amplifier may be to purge of parasitic oscillations in the output stage. This is presumably due to the extra raw transconductance available, and can be a problem even with the triple-EF circuit, which has no local NFB loops. I do not pretend to be able to give a detailed explanation of this effect at the moment.

Having demonstrated that sustained-beta output devices not only reduce LSN but also unexpectedly reduce crossover distortion, it seemed worth checking if using multiple output devices would give a similar reduction at light loading. I was rather surprised to find they did.

Adding more output devices in parallel, while driving an 8-Ω load, results in a steady reduction in distortion. Figures 5.27–5.29 show how this works in reality. The SPICE simulations in Figure 5.30 reveal that increasing the number N of output devices not only flattens the crossover gain wobble, but spreads it out over a greater width. This spreading effect is an extra bonus because it means that lower-order harmonics are generated, and at lower frequencies there will be more negative feedback to linearise them. (Bear in mind also that a triple-EF output has an inherently wider gain

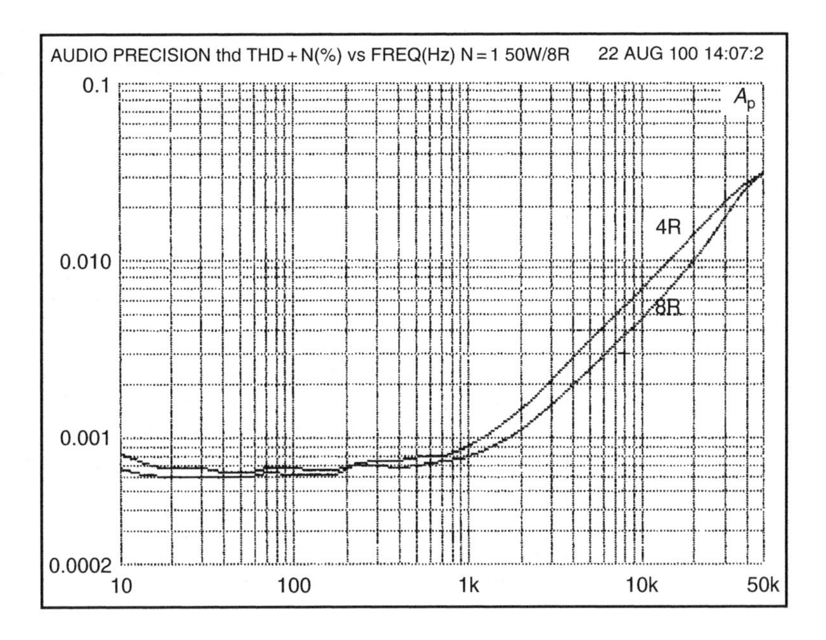

Figure 5.27 THD for one pair ($N = 1$) of output devices, at 50 W/8 R and 100 W/4 R

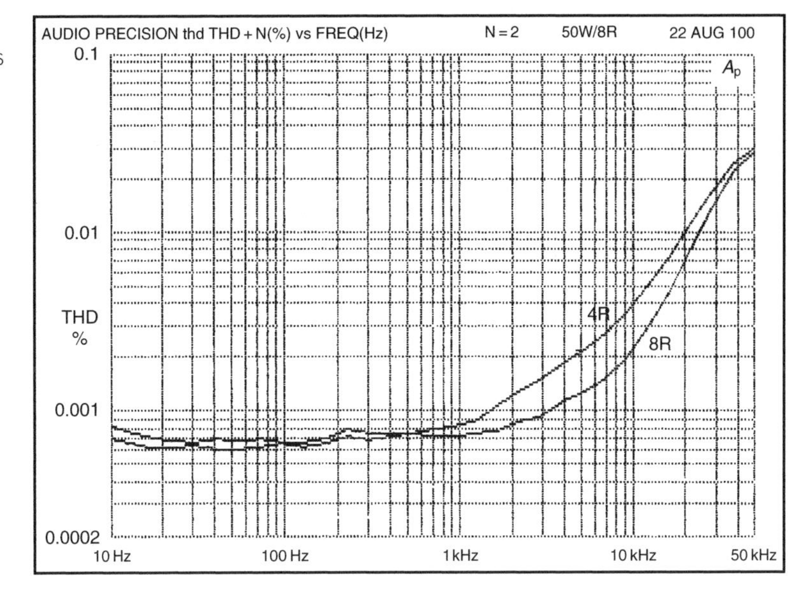

Figure 5.28 THD for two pairs ($N = 2$) of output devices, at 50 W/8 R and 100 W/4 R. A definite improvement

wobble than the double-EF.) Taking the gain wobble width as the voltage between the bottoms of the two dips, this appears to be proportional to N. The amount of gain wobble, as measured from top of the peak to bottom of the dips, appears to be proportional to 1/N.

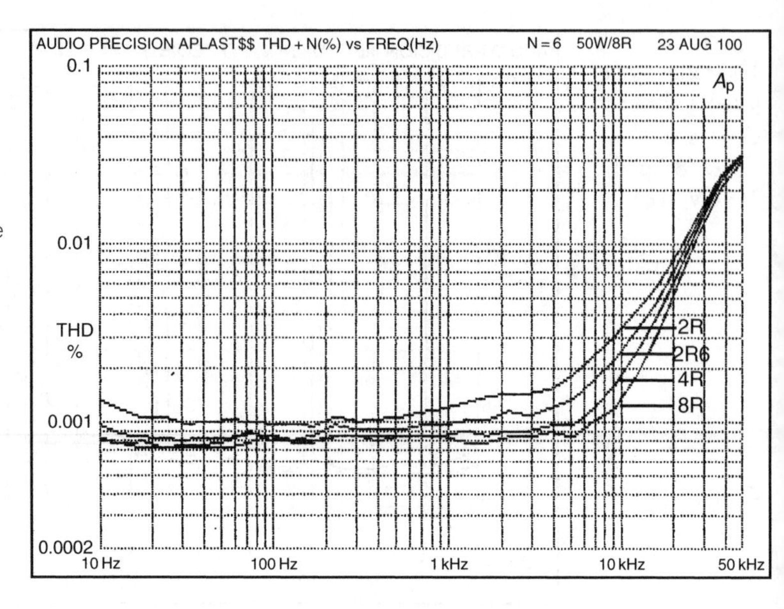

Figure 5.29
THD for six pairs ($N=6$) of output devices, at 50 W/8 R, 100 W/4 R, 200 W/2 R. Note very low distortion at 8 ohms

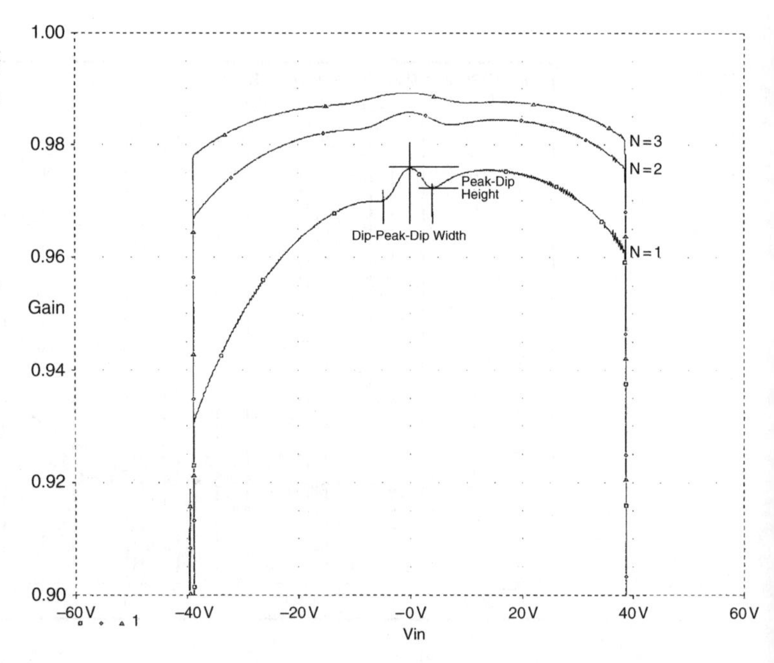

Figure 5.30
SPICE simulation of triple-EF output with $N=1$, 2 and 3. As N increases the crossover gain wobble becomes flatter and more spread out laterally

This makes sense. We know that crossover distortion increases with heavier loading, i.e., with greater currents flowing in the output devices, but under the same voltage conditions. It is therefore not surprising that reducing the device currents by using multiple devices has the same effect as reducing loading. If there are two output devices in parallel, each sees half the current variations, and crossover non-linearity is reduced. The voltage conditions are the same in each half and so are unchanged. This offers us the interesting possibility that crossover distortion – which has hitherto appeared inescapable – can be reduced to an arbitrary level simply by paralleling enough output transistors. To the best of my knowledge this is a new insight.

Summary

In conventional amplifiers, reducing the 8 Ω load to 4 Ω increases the THD by 2 to 3 times. The figure attained by this amplifier is 1.2 times, and the ratio could be made even closer to unity by tripling the output devices.

Crossover distortion (Distortion 3b)

In a field like Audio where consensus of any sort is rare, it is a truth universally acknowledged that crossover distortion is the worst problem that can afflict Class-B power amplifiers. The problem is the crossover region, where control of the output voltage must be handed over from one device to another. Crossover distortion is rightly feared as it generates unpleasant high-order harmonics, with at least the potential to increase in percentage as signal level falls.

The pernicious nature of crossover distortion is partly because it occurs over a small part of the signal swing, and so generates high-order harmonics. Worse still, this small range over which it does occur is at the zero-crossing point, so not only is it present at all levels and all but the lightest loads, but is generally believed to increase as output level falls, threatening very poor linearity at the modest listening powers that most people use.

There is a consensus that crossover caused the *transistor sound* of the 1960s, though to the best of my knowledge this has never actually been confirmed by the double-blind testing of vintage equipment.

The Vbe-Ic characteristic of a bipolar transistor is initially exponential, blending into linear as the internal emitter resistance re comes to dominate the transconductance. The usual Class-B stage puts two of these curves back-to-back, and Peter Blomley has shown[12] that these curves are non-conjugate, i.e., there is no way they can be shuffled about so they will sum to a completely linear transfer characteristic, whatever the offset between them imposed by the bias voltage. This can be demonstrated quickly and easily by SPICE simulation; see Figure 5.31. There is at first sight not much you can do except maintain the bias voltage, and hence quiescent current, at some optimal level for minimum gain deviation at crossover; quiescent-current control is a complex subject that could fill a big book in itself, and is considered in Chapter 12.

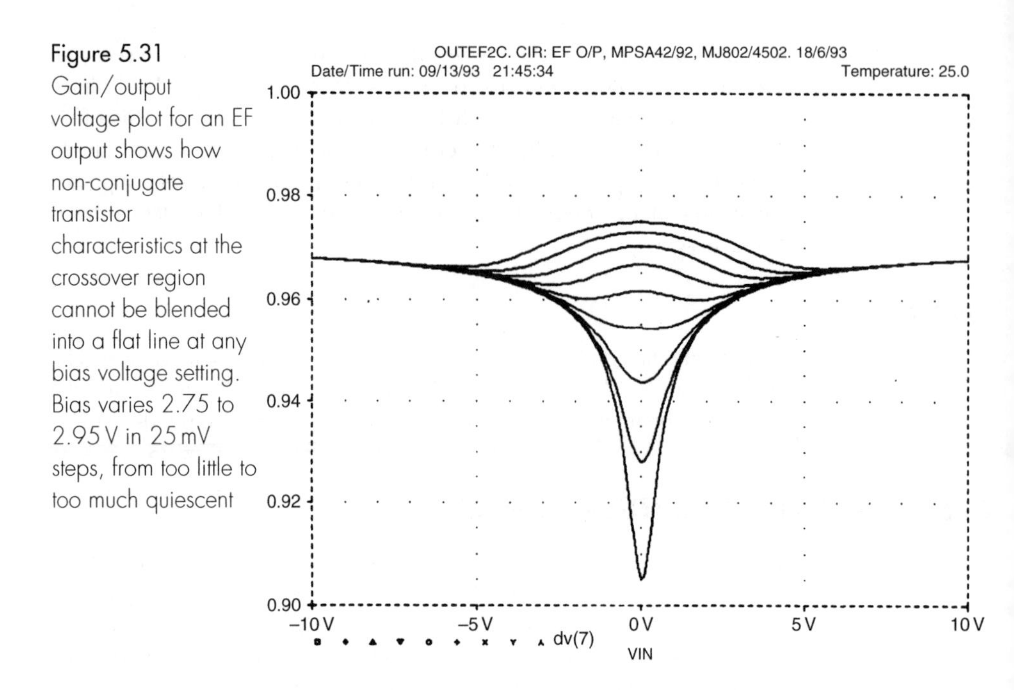

Figure 5.31 Gain/output voltage plot for an EF output shows how non-conjugate transistor characteristics at the crossover region cannot be blended into a flat line at any bias voltage setting. Bias varies 2.75 to 2.95 V in 25 mV steps, from too little to too much quiescent

It should be said that the crossover distortion levels generated in a Blameless amplifier can be very low up to around 1 kHz, being barely visible in residual noise and only measurable with a spectrum-analyser. As an instructive example, if a Blameless closed-loop Class-B amplifier is driven through a TL072 unity-gain buffer the added noise from this op-amp will usually submerge the 1 kHz crossover artefacts into the noise floor, at least as judged by the eye on the oscilloscope. (It is most important to note that Distortions 4, 5, 6 and 7 create disturbances of the THD residual at the zero-crossing point that can be easily mistaken for crossover distortion, but the actual mechanisms are quite different). However, the crossover distortion becomes obvious as the frequency increases, and the high-order harmonics benefit less from NFB.

It will be seen later that in a Blameless amplifier driving 8 Ω the overall linearity is dominated by crossover distortion, even with a well-designed and optimally biased output stage. There is an obvious incentive to minimise this distortion mechanism, but there seems no obvious way to reduce crossover gain deviations by tinkering with any of the relatively conventional stages considered so far.

Figure 5.32 shows the signal waveform and THD residual from a Blameless power amplifier with optimal Class-B bias. Output power was 25 W into 8 Ω, or 50 W into 4 Ω (i.e., the same output voltage) as appropriate, for all the residuals shown here. The figure is a record of a single sweep so the residual appears to be almost totally random noise; without the

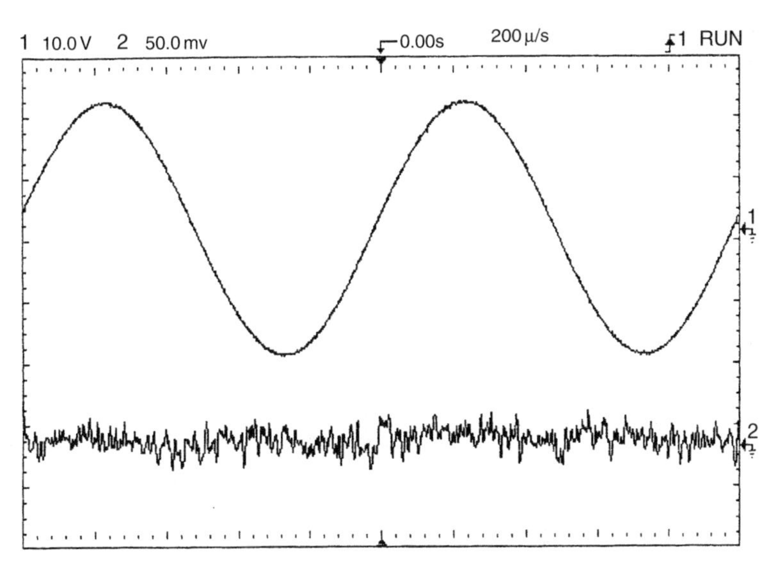

Figure 5.32
The THD residual from an optimally biased Blameless power amplifier at 1 kHz, 25 W/8 Ω is essentially white noise. There is some evidence of artefacts at the crossover point, but they are not measurable. THD 0.00097%, 80 kHz bandwidth

visual averaging that occurs when we look at an oscilloscope the crossover artefacts are much less visible than in real time.

In Figure 5.33, 64 times averaging is applied, and the disturbances around crossover become very clear. There is also revealed a low-order component at roughly 0.0004%, which is probably due to very small amounts of Distortion 6 that were not visible when the amplifier layout was optimised.

Figure 5.34 shows Class-B slightly underbiased to generate crossover distortion. The crossover spikes are very sharp, so their height in the residual

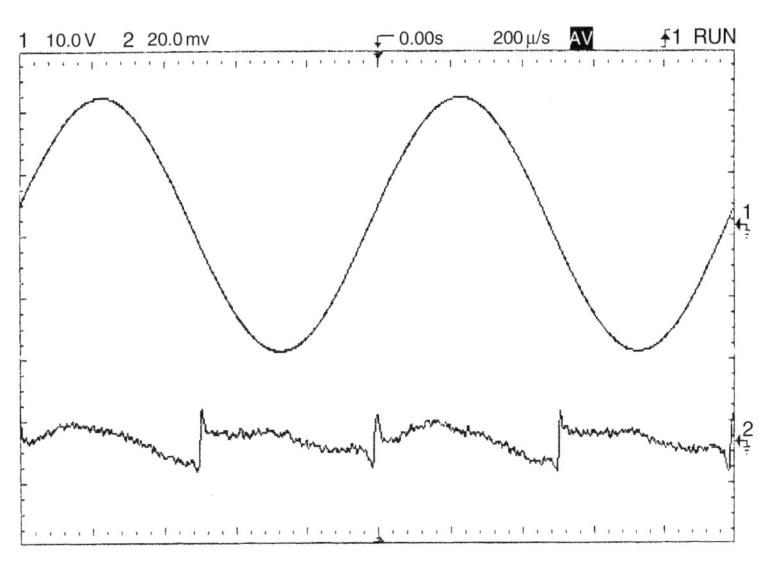

Figure 5.33
Averaging Figure 5.2 residual 64 times reduces the noise by 18 dB, and crossover discontinuities are now obvious. The residual has been scaled up by 2.5 times from Figure 5.2 for greater clarity

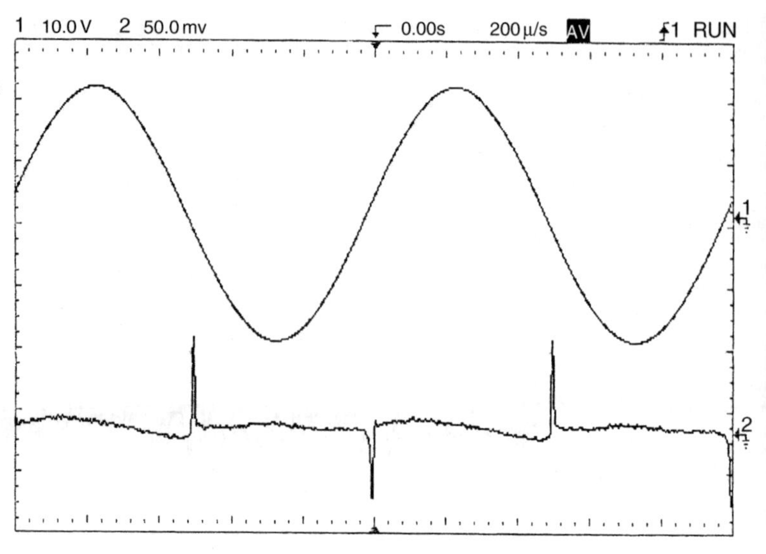

Figure 5.34
The results of mild underbias in Class-B

depends strongly on measurement bandwidth. Their presence warns immediately of underbiasing and avoidable crossover distortion.

In Figure 5.35 an optimally biased amplifier is tested at 10 kHz. The THD increases to approximately 0.004%, as the amount of global negative feedback is 20 dB less than at 1 kHz. The timebase is faster so crossover events appear wider than in Figure 5.34. The THD level is now higher and above the noise so the residual is averaged 8 times only. The measurement bandwidth is still 80 kHz, so harmonics above the eighth are now lost. This

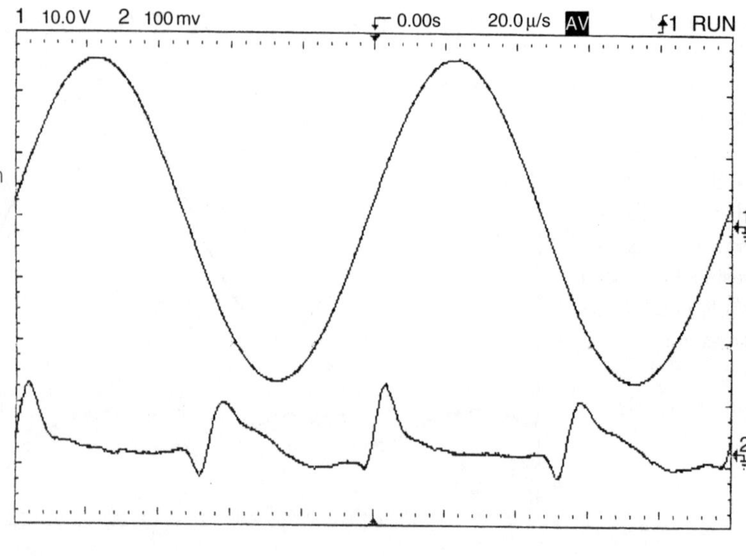

Figure 5.35
An optimally biased Blameless power amplifier at 10 kHz. THD approximately 0.004%, bandwidth 80 kHz. Averaged 8 times

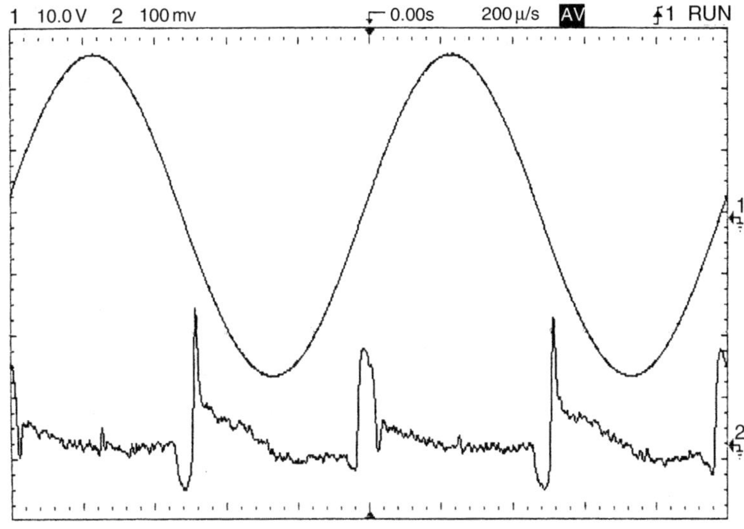

Figure 5.36
As Figure 5.6, but in 500 kHz bandwidth. The distortion products look quite different

is illustrated in Figure 5.36, which is Figure 5.35 rerun with a 500 kHz bandwidth. The distortion products now look much more jagged.

Figure 5.37 shows the gain-step distortion introduced by Class-AB. The undesirable edges in the residual are no longer in close pairs that partially cancel, but are spread apart on either side of the zero crossing. No averaging is used here as the THD is higher. See page 279 for more on Class-AB distortion.

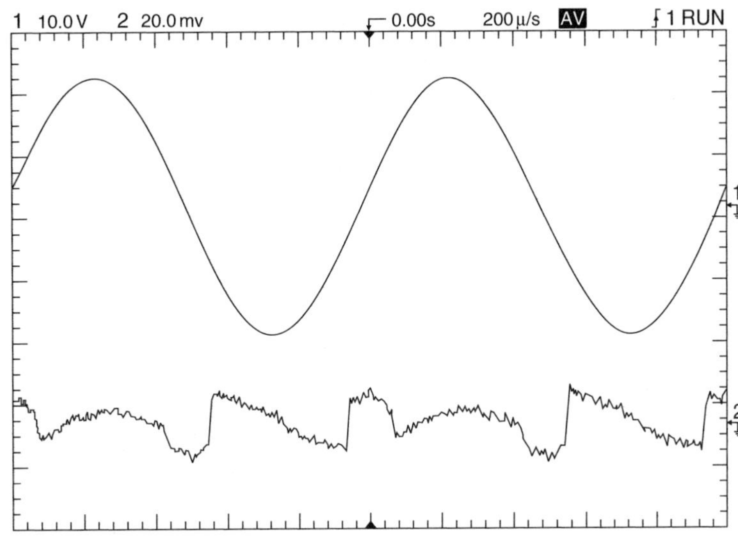

Figure 5.37
The gm-doubling distortion introduced by Class-AB. The edges in the residual are larger and no longer at the zero crossing, but displaced either side of it

It is commonplace in Audio to discover that a problem like crossover distortion has been written about and agonised over for decades, but the amount of technical investigation that has been done (or at any rate published) is disappointingly small. I had to do some basic investigations myself.

I first looked to see if crossover distortion really *did* increase with decreasing output level in a Blameless amplifier; to attempt its study with an amplifier contaminated with any of the avoidable distortion mechanisms is completely pointless. One problem is that a Blameless amplifier has such a low level of distortion at 1 kHz (0.001% or less) that the crossover artefacts are barely visible in circuit noise, even if low-noise techniques are used. The measured percentage level of the noise-plus-distortion residual is bound to rise with falling output, because the noise voltage remains constant; this is the lowest line in Figure 5.38. To circumvent this, the amplifier was deliberately underbiased by varying amounts to generate ample crossover spikes, on the assumption that any correctly adjusted amplifier should be less barbarous than this.

The answer from Figure 5.38 is that the THD percentage does increase as level falls, but relatively slowly. Both EF and CFP output stages give similar diagrams to Figure 5.38, and whatever the degree of underbias, THD increases by about 1.6 times as the output voltage is halved. In other words, reducing the output power from 25 W to 250 mW, which is pretty drastic, only increases THD % by six times, and so it is clear that the *absolute* (as opposed to percentage) THD level in fact falls slowly with amplitude, and therefore probably remains imperceptible. This is something of a relief; but crossover distortion remains a bad thing to have.

Distortion versus level was also investigated at high frequencies, i.e., above 1 kHz where there is more THD to measure, and optimal biasing can be used. Figure 5.39 shows the variation of THD with level for the EF stage at

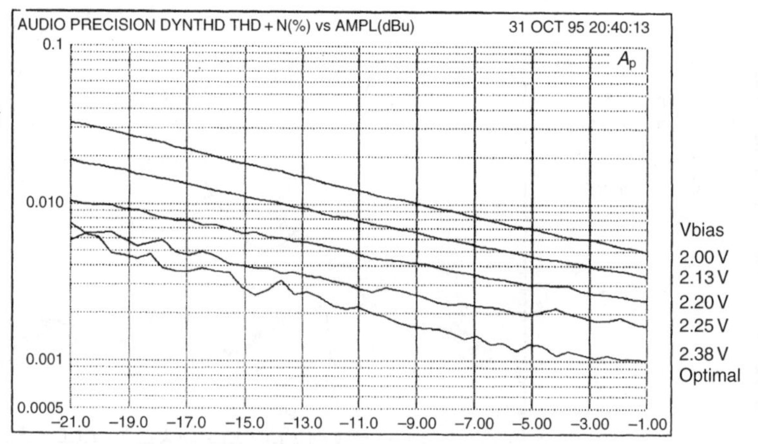

Figure 5.38 Showing how crossover distortion rises slowly as output power is reduced from 25 W to 250 mW (8 Ω) for optimal bias and increasingly severe underbias (upper lines). This is an EF type output stage. Measurement bandwidth 22 kHz

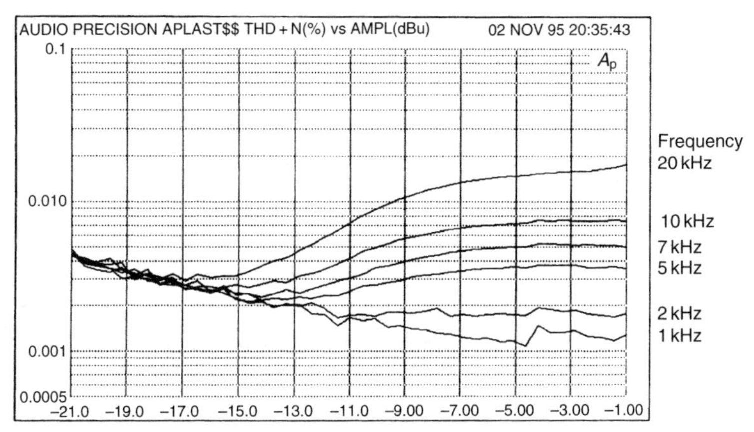

Figure 5.39 Variation of crossover distortion with output level for higher frequencies. Optimally biased EF output stage. Bandwidth 80 kHz

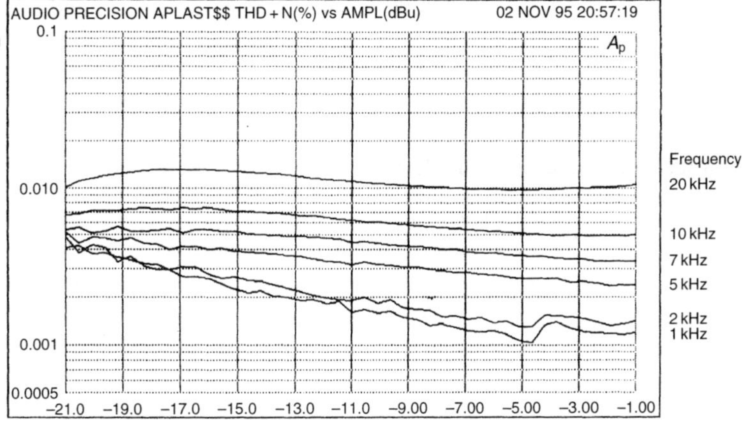

Figure 5.40 Variation of distortion with level for higher frequencies. Optimally biased CFP output stage. Bandwidth 80 kHz

a selection of frequencies; Figure 5.40 shows the same for the CFP. Neither shows a significant rise in percentage THD with falling level, though it is noticeable that the EF gives a good deal less distortion at lower power levels around 1 W. This is an unexpected observation, and possibly a new one.

To further get the measure of the problem, Figure 5.41 shows how HF distortion is greatly reduced by increasing the load resistance, providing further confirmation that almost all the 8 Ω distortion originates as crossover in the output stage.

Crossover distortion, unlike some more benign kinds of signal-warping, is unanimously agreed to be something any amplifier could well do without. The amount of crossover distortion produced depends strongly on optimal quiescent adjustment, so the thermal compensation used to stabilise this against changes in temperature and power dissipation must be accurate.

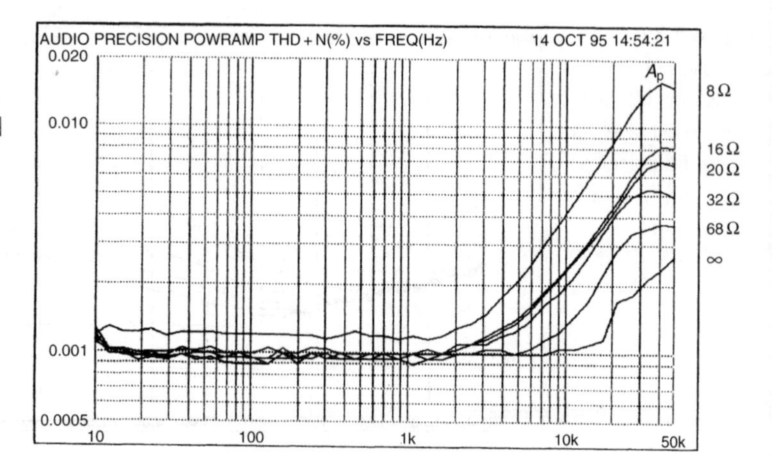

Figure 5.41
How crossover distortion is reduced with increasing load resistance. 20 W into 8 Ω, 80 kHz bandwidth

This section deals with the crossover region and its quiescent conditions, and the specific issues of the effectiveness of the thermal compensation for temperature effects is dealt with in detail in Chapter 12.

Output stage quiescent conditions

Figure 5.42 shows the two most common types of output stage: the Emitter-Follower (EF) and the Complementary-Feedback-Pair (CFP) configurations. The manifold types of output stage based on triples will have to be set aside for the moment. The two circuits shown have few components, and there are equally few variables to explore in attempting to reduce crossover distortion.

To get the terminology straight: here, as in my previous writings, Vbias refers to the voltage set up across the driver bases by the Vbe-multiplier bias generator, and is in the range 1–3 V for Class-B operation. Vq is the quiescent voltage across the two emitter resistors (hereafter Re) alone, and is between 5 and 50 mV, depending on the configuration chosen. Quiescent current Iq refers only to that flowing in the output devices, and does not include driver standing currents.

I have already shown that the two most common output configurations are quite different in behaviour, with the CFP being superior on most criteria. Table 5.2 shows that crossover gain variation for the EF stage is smoother (being some 20 times wider) but of four times higher amplitude than for the CFP version. It is not immediately obvious from this which stage will generate the least HF THD, bearing in mind that the NFB factor falls with frequency.

Table 5.2 also emphasises that a little-known drawback of the EF version is that its quiescent dissipation may be far from negligible.

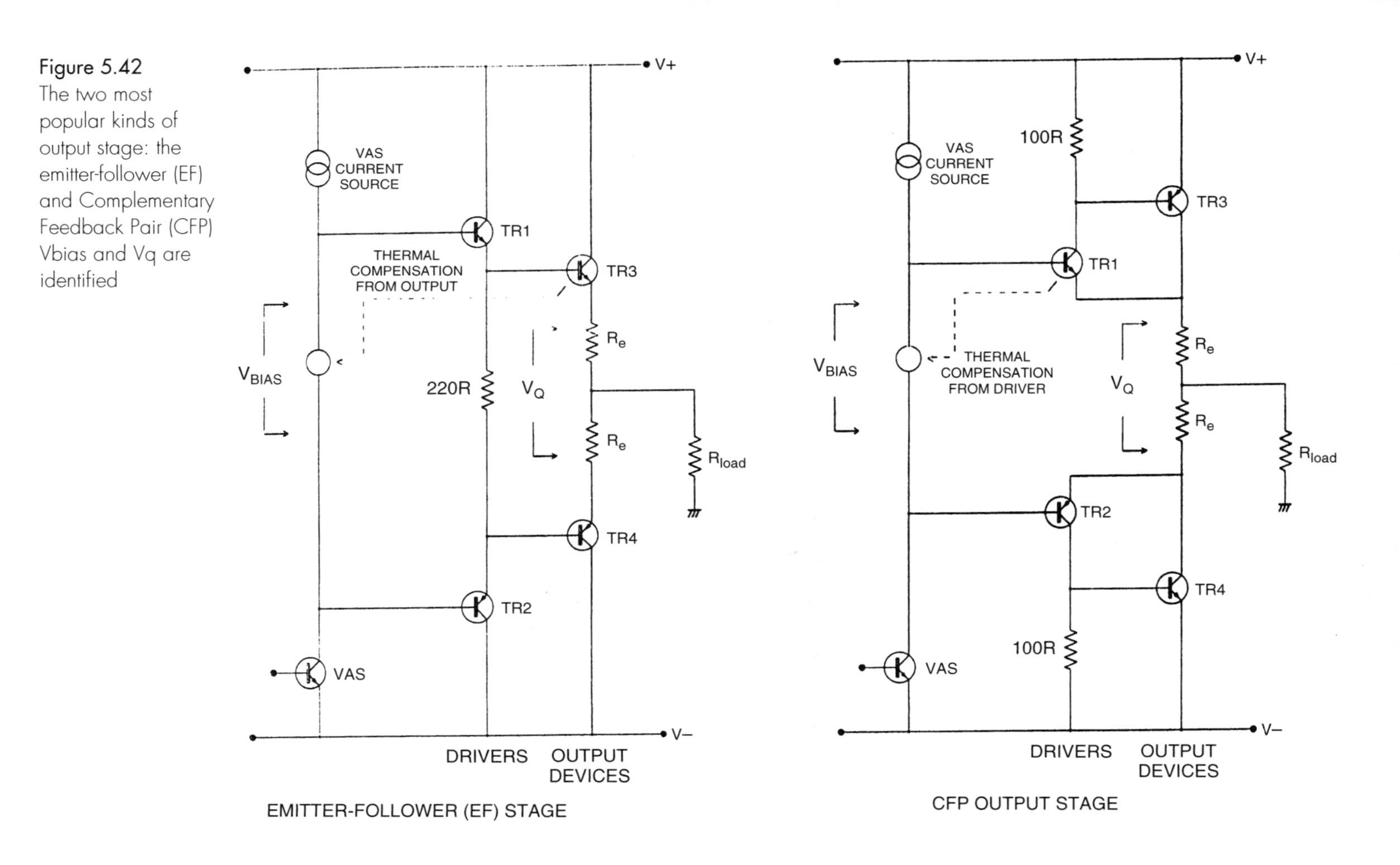

Figure 5.42 The two most popular kinds of output stage: the emitter-follower (EF) and Complementary Feedback Pair (CFP) Vbias and Vq are identified

Table 5.2 Quiescent conditions compared

	Emitter-follower	CFP
Vbias	2.930 V	1.297 V
Vq	50 mV	5 mV
Iq	114 mA	11 mA
Pq (per o/p device)	4.6 W	0.44 W
Average gain	0.968	0.971
Peak gain deviation from average	0.48%	0.13%
Crossover width*	±12 V	±0.6 V

(For Re = OR22, 8 Ω load, and ±40 V supply rails)
* Crossover-width is the central region of the output voltage range over which crossover effects are significant; I have rather arbitrarily defined it as the ± output range over which the incremental gain curves diverge by more than 0.0005 when Vbias is altered around the optimum value. This is evaluated here for an 8 Ω load only.

An experiment on crossover distortion

Looking hard at the two output stage circuit diagrams, intuition suggests that the value of emitter resistor Re is worth experimenting with. Since these two resistors are placed between the output devices, and alternately pass the full load current, it seems possible that their value could be critical in mediating the handover of output control from one device to the other. Re was therefore stepped from 0.1 to 0.47 Ω, which covers the practical range. Vbias was reoptimised at each step, though the changes were very small, especially for the CFP version.

Figure 5.43 shows the resulting gain variations in the crossover region for the EF stage, while Figure 5.44 shows the same for the CFP configuration. Table 5.3 summarises some numerical results for the EF stage, and Table 5.4 for the CFP.

There are some obvious features; first, Re is clearly not critical in value as the gain changes in the crossover region are relatively minor. Reducing the Re value allows the average gain to approach unity more closely, with a consequent advantage in output power capability. See page 276. Similarly, reducing Re widens the crossover region for a constant load resistance, because more current must pass through one Re to generate enough voltage-drop to turn off the other output device. This implies that as Re is reduced, the crossover products become lower-order and so of lower frequency. They should be better linearised by the frequency-dependent global NFB, and so overall closed-loop HF THD should be lower.

The simulated crossover distortion experiment described on page 113 showed that as the crossover region was made narrower, the distortion energy became more evenly spread over higher harmonics. A wider crossover region implies energy more concentrated in the lower harmonics,

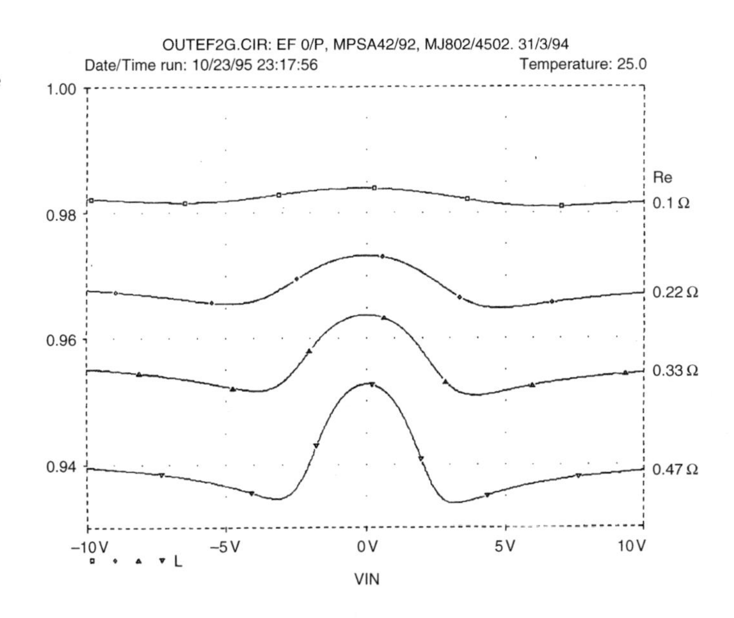

Figure 5.43
Output linearity of the EF output stage for emitter-resistance Re between 0.1 and 0.47 Ω

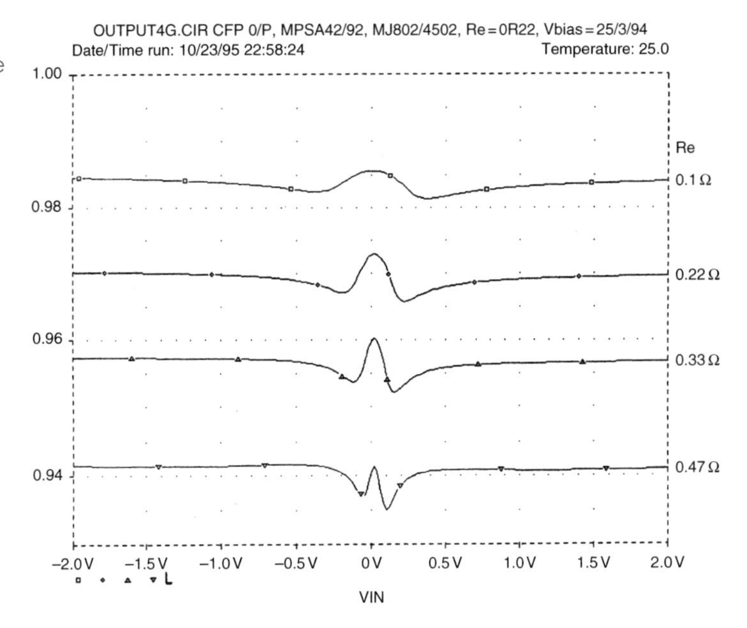

Figure 5.44
Output linearity of the CFP output stage for emitter-resistance Re between 0.1 and 0.47 Ω

which will receive the benefit of more negative feedback. However, if the region is made wider, but retains the same amount of gain deviation, it seems likely that the total harmonic energy is greater, and so there are two opposing effects to be considered.

This is partly confirmed by Figure 5.41, where measurements show that the THD reaches a very shallow minimum for Re = OR22, at any rate for

Table 5.3
Emitter-follower output (Type-1). Data for 8 Ω load and EF o/p stage

Re ohms	Optimal Vbias Volts	Optimal Vq mV	Iq mA	X-Width Volts	Average Gain ratio
0.1	2.86	42.6	215	18	0.982
0.22	2.87	46.2	107	12	0.968
0.33	2.89	47.6	74	9	0.955
0.47	2.93	54.8	59	7	0.939

As Re is varied, Vq varies by only 29%, while Iq varies by 365%

Table 5.4
CFP output. Data for 8 Ω load and CFP o/p stage

Re ohms	Optimal Vbias Volts	Optimal Vq mV	Iq mA	X-Width Volts	Average Gain ratio
0.1	1.297	3.06	15.3	1.0	0.983
0.22	1.297	4.62	11.5	0.62	0.971
0.33	1.297	5.64	8.54	0.40	0.956
0.47	1.298	7.18	7.64	0.29	0.941

that particular configuration, level, and load; this is consistent with two opposing effects. While the variation of THD with Re appears to be real, it is small, and I conclude that selecting *Re* = OR1 for maximum efficiency is probably the over-riding consideration. This has the additional benefit that if the stage is erroneously over-biased into Class AB, the resulting gm-doubling distortion will only be half as bad as if the more usual OR22 values had been used for *Re*.

It would be easy to assume that higher values of *Re* must be more linear, because of a vague feeling that *there is more local feedback* but this cannot be true as an emitter-follower already has 100% voltage feedback to its emitter, by definition. Changing the value of *Re* alters slightly the total resistive load seen by the emitter itself, and this does seem to have a small but measurable effect on linearity.

As *Re* is varied, *V*q varies by 230% while *I*q varies by 85%. However the absolute *V*q change is only 4 mV, while the sum of Vbe's varies by only 0.23%. This makes it pretty plain that the voltage domain is what counts, rather than the absolute value of *I*q.

The first surprise from this experiment is that in the typical Class-B output stage, quiescent current as such does not matter a great deal. This may be hard to believe, particularly after my repeated statements that quiescent conditions are critical in Class-B, but both assertions are true. The data for both the EF and CFP output stages show that changing Re alters the Iq considerably, but the optimal value of *V*bias and *V*q barely change.

The voltage across the transistor base-emitter junctions and *Re*'s seems to be what counts, and the actual value of current flowing as a result is not in itself of much interest. However, the Vbias setting remains critical for minimum distortion; once the *Re* value is settled at the design stage, the adjustment procedure for optimal crossover is just as before.

The irrelevance of quiescent current was confirmed by the Trimodal amplifier, which was designed after the work described here was done, and where I found that changing the output emitter resistor value *Re* over a 5:1 range required no alteration in *V*bias to maintain optimal crossover conditions.

The critical factor is therefore the voltages across the various components in the output stage. Output stages get hot, and when the junction temperatures change, both experiment and simulation show that if *V*bias is altered to maintain optimal crossover, *V*q remains virtually constant. This confirms the task of thermal compensation is solely to cancel out the Vbe changes in the transistors; this may appear to be a blinding glimpse of the obvious, but it was worth checking as there is no inherent reason why the optimal *V*q should not be a function of device temperature. Fortunately it is not, for thermal compensation that also dealt with a need for *V*q to change with temperature might be a good deal more complex.

*V*q as the critical quiescent parameter

The recognition that *V*q is the critical parameter has some interesting implications. Can we immediately start setting up amplifiers for optimal crossover with a cheap DVM rather than an expensive THD analyser? Setting up quiescent current with a milliammeter has often been advocated, but the direct measurement of this current is not easy. It requires breaking the output circuit so a meter can be inserted, and not all amplifiers react favourably to so rude an intrusion. (The amplifier must also have near-zero DC offset voltage to get any accuracy.) Measuring the total amplifier consumption is not acceptable because the standing-current taken by the small-signal and driver sections will, in the CFP case at least, swamp the quiescent current. It is possible to determine quiescent current indirectly from the *V*q drop across the *Re*'s (still assuming zero DC offset) but this can never give a very accurate current reading as the tolerance of low-value *Re*'s is unlikely to be better than ±10%.

However, if *V*q is the real quantity we need to get at, then *Re* tolerances can be blissfully ignored. This does not make THD analysers obsolete overnight. It would be first necessary to show that Vq was always a reliable indicator of crossover setting, no matter what variations occurred in driver or output transistor parameters. This would be a sizeable undertaking.

There is also the difficulty that real-life DC offsets are not zero, though this could possibly be side-stepped by measuring *V*q with the load disconnected. A final objection is that without THD analysis and visual examination of the residual, you can never be sure an amplifier is free from parasitic oscillations and working properly.

I have previously demonstrated that the distortion behaviour of a typical amplifier is quite different when driving 4 Ω rather than 8 Ω loads. This is because with the heavier load, the output stage gain-behaviour tends to be dominated by beta-loss in the output devices at higher currents, and consequent extra loading on the drivers, giving third-harmonic distortion. If this is to be reduced, which may be well worthwhile as many loudspeaker loads have serious impedance dips, then it will need to be tackled in a completely different way from crossover distortion.

It is disappointing to find that no manipulation of output-stage component values appears to significantly improve crossover distortion, but apart from this one small piece of (negative) information gained, we have in addition determined that:

1 quiescent current as such does not matter; *V*q is the vital quantity,
2 a perfect thermal compensation scheme, that was able to maintain *V*q at exactly the correct value, requires no more information than the junction temperatures of the driver and output devices. Regrettably none of these temperatures are actually accessible, but at least we know what to aim for.

Switching distortion (Distortion 3c)

This depends on several variables, notably the speed characteristics of the output devices and the output topology. Leaving aside the semiconductor physics and concentrating on the topology, the critical factor is whether or not the output stage can reverse-bias the output device base-emitter junctions to maximise the speed at which carriers are sucked out, so the device is turned off quickly. The only conventional configuration that can reverse-bias the output base-emitter junctions is the EF Type II, described on page 115.

A second influence is the value of the driver emitter or collector resistors; the lower they are the faster the stored charge can be removed. Applying these criteria can reduce HF distortion markedly, but of equal importance is that it minimises overlap of output conduction at high frequencies, which if unchecked results in an inefficient and potentially destructive increase in supply current[13]. To illustrate this, Figure 5.45 shows a graph of current consumption versus frequency for varying driver collector resistance, for a CFP type output.

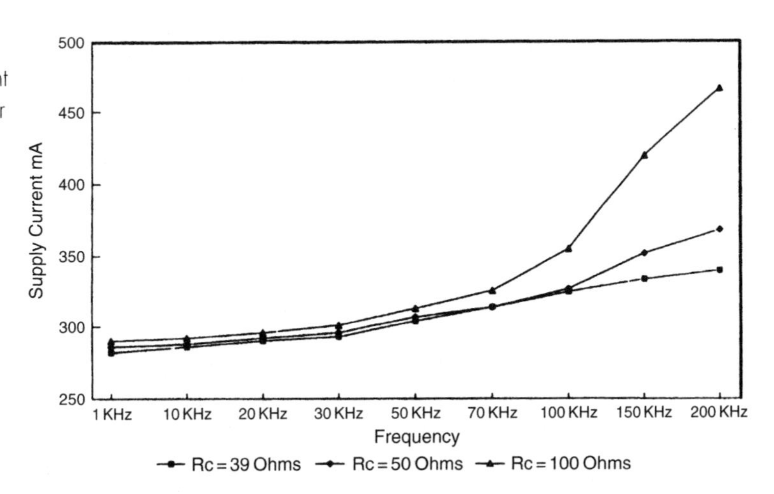

Figure 5.45 Power supply current versus frequency, for a CFP output with the driver collector resistors varied. There is little to be gained from reducing *Rc* below 50 Ω

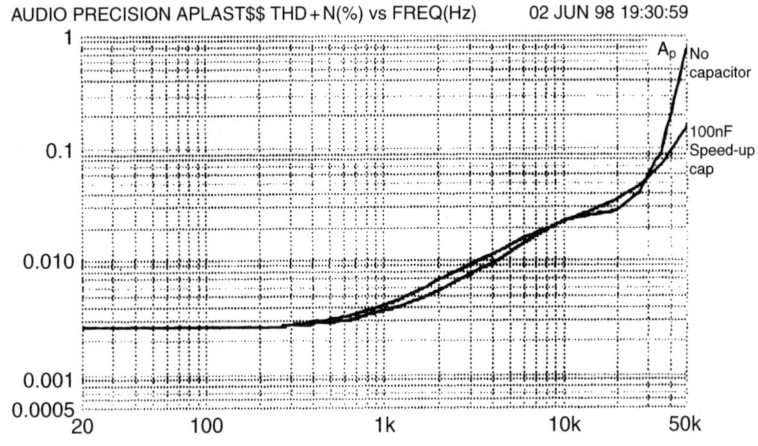

Figure 5.46 HF THD reduction by adding speedup capacitance across the common driver resistance of a Type II EF output stage

Figure 5.46 shows the reduction of HF THD by adding a speedup capacitor across the common driver resistor of an EF Type II. At LF the difference is small, but at 40 kHz THD is halved, indicating much cleaner switchoff. There is also a small benefit over the range 300 Hz–8 kHz.

Thermal distortion

Thermal distortion is that caused by cyclic temperature changes at signal frequency, causing corresponding modulation of device parameters. While it is certainly a real problem in IC op-amps, which have input and output devices in very close thermal proximity, the situation in a normal discrete-component power amplifier is quite different, and thermal distortion cannot be detected. Having studied in detail distortion mechanisms that are all too real, it comes as some relief to find that one prospective distortion is illusory. Some writers appear to take it as given that such a distortion mechanism exists in power amplifiers, but having studied the subject in some depth I have yet to see the effect, and quite frankly I do not think it exists.

While now and again there have been odd mentions of thermal distortion in power amps in some of the hi-fi press, you will never find:

1 any explanation of how it might work,
2 any estimate of the magnitude of the effect,
3 a circuit that will demonstrate its production.

In the usual absence of specific theories, one can only assume that the alleged mechanism induces parameter changes in semiconductors whose power dissipation varies over a cycle. If this were to happen, it would presumably manifest itself as a rise in second or third harmonic distortion at very low frequencies, but this simply does not happen. The largest effects would be expected in Class-B output stages where dissipation varies wildly over a cycle; the effect is still wholly absent.

One reason for this may be that drivers and output devices have relatively large junctions with high thermal inertia – a few seconds with a hammer and chisel revealed that an MJE340 driver has a chip with four times the total area of a TL072. Given this thermal mass, parameters presumably cannot change much even at 10 Hz. Low frequencies are also where the global NFB factor is at its maximum; it is perfectly possible to design an amplifier with 100 dB of feedback at 10 Hz, though much more modest figures are sufficient to make distortion unmeasurably low up to 1 kHz or so. Using my design methodology a Blameless amplifier can be straightforwardly designed to produce less than 0.0006% THD at 10 Hz (150 W/8 Ω) without even considering thermal distortion; this suggests that we have here a non-problem.

I accept that it is not uncommon to see amplifier THD plots that rise at low frequencies; but whenever I have been able to investigate this, the LF rise could be eliminated by attending to either defective decoupling or feedback-capacitor distortion. Any thermal distortion must be at a very low level as it is invisible at 0.0006%; remember that this is the level of a THD reading that is visually pure noise, though there are real amplifier distortion products buried in it.

I have therefore done some deeper investigation by spectrum analysis of the residual, which enables the harmonics to be extracted from the noise. The test amplifier was an optimally biased Class-B machine very similar to that on Figure 6.16, except with a CFP output. The Audio Precision oscillator is very, very clean but this amplifier tests it to its limits, and so Table 5.5 below shows harmonics in a before-and-after-amplifier comparison. The spectrum analyser bandwidth was 1 Hz for 10 Hz tests, and 4.5 Hz for 1 kHz, to discriminate against wideband noise.

Table 5.5 Relative amplitute of distortion harmonics

	10 Hz AP out (%)	Amp out (%)	1 kHz AP out (%)	Amp out (%)
Fundamental	0.00013	0.00031	0.00012	0.00035
Second	0.00033	0.00092	0.00008	0.00060
Third	0.00035	0.00050	0.000013	0.00024
Fourth	<0.000002	0.00035	<0.000008	0.00048
Fifth	<0.00025	<0.00045	0.000014	0.00024
Sixth	<0.000006	0.00030	0.000008	0.00021
Seventh	<0.000006	<0.00008	0.000009	0.00009
Eighth	<0.000003	0.00003	0.000008	0.00016
Ninth	<0.000004	0.00011	0.000007	<0.00008
AP THD reading (80 kHz bandwidth)	0.00046	0.00095	0.00060	0.00117

NB: The rejection of the fundamental is not perfect, and this is shown as it contributes to the THD figure.

This further peeling of the distortion onion shows several things; that the AP is a brilliant piece of machinery, and that the amplifier is really quite linear too. However there is nothing resembling evidence for thermal distortion effects.

As a final argument, consider the distortion residual of a slightly under-biased power-amp, using a CFP output configuration so that output device junction temperatures do not affect the quiescent current; it therefore depends only on the driver temperatures. When the amplifier is switched on and begins to apply sinewave power to a load, the crossover spikes (generated by the deliberate underbiasing) will be seen to slowly shrink in height over a couple of minutes as the drivers warm up. This occurs even with the usual temperature compensation system, because of the delays and losses in heating up the Vbe-multiplier transistor.

The size of these crossover spikes gives in effect a continuous readout of driver temperature, and the slow variations that are seen imply time-constants measured in tens of seconds or more; this must mean a negligible response at 10 Hz.

There is no doubt that long-term thermal effects can alter Class-B amplifier distortion, because as I have written elsewhere, the quiescent current setting is critical for the lowest possible high-frequency THD. However this is strictly a slow (several minutes) phenomenon, whereas enthusiasts for thermal distortion are thinking of the usual sort of per-cycle distortion.

The above arguments lead me to conclude that thermal distortion as usually described does not exist at a detectable level.

Thermal distortion in a power amp IC

Audio writers sometimes speculate about 'thermal distortion'. This is assumed to be caused by cyclic temperature changes at signal frequency, causing modulation of transistor parameters. It is undoubtedly a real problem in power ICs, which have input and output devices in close thermal proximity on the same piece of sillcon, but in a discrete-component power amplifier there is no such thermal coupling, and no such distortion.

Thermal non-linearities would presumably appear as second or third harmonic distortion rising at low frequencies, and the largest effects should be in Class-B output stages where dissipation varies greatly over a cycle. There is absolutely no such effect to be seen in discrete-component power amplifiers.

But thermal distortion certainly does exist in IC power amplifiers. Figure 5.47 is a distortion plot for the Philips TDA 1522Q power amp IC, which I believe shows the effect. The power level was 4.4 W into 8 Ω, 8 W into 4 Ω. As is usual for such amplifiers, the distortion is generally high, but drops into a notch at 40 Hz; the only feasible explanation for this is cancellation of distortion products from two separate distortion sources. At frequencies below this notch there is second-harmonic distortion rising at 12 dB/octave as frequency falls. The LF residual looks quite different from the midband distortion, which was a mixture of second and third harmonic plus crossover spikes.

The THD figure falls above 10 kHz because of the 80 kHz bandwidth limitation on the residual, and the high-order nature of the harmonics that make up crossover distortion.

All other possible sources of an LF distortion rise, such as inadequate decoupling, were excluded. There was no output capacitor to introduce non-linearity.

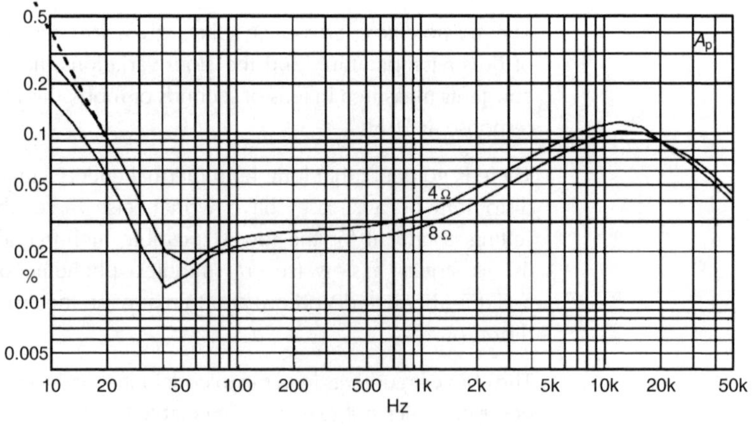

Figure 5.47
Distortion plot for the Phllips TDA1522Q IC. Power out was 4.4 W rms into 8 Ω, 8 W rms into 4 Ω. The dotted line shows a 12 dB/octave slope

It seems pretty clear that the steep rise here is due to thermal distortion, in the form of feedback from the power output stage to earlier parts of the amplifier – probably the input stage. As would be expected, the effect is greater with a heavier load which causes more heating; in fact halving the load doubles the THD reading below the 40 Hz notch.

Selecting an output stage

Even if we stick to the most conventional of output stages, there are still an embarrassingly large number to choose form. The cost of a complementary pair of power FETs is currently at least twice that of roughly equivalent BJTs, and taken with the poor linearity and low efficiency of these devices, the use of them may require a marketing rather than a technical motivation.

Turning to BJTs, I conclude that there are the following candidates for *Best Output Stage*:

1 the Emitter-Follower Type II output stage is the best at coping with switchoff distortion but the quiescent-current stability may be doubtful,
2 the CFP topology has good quiescent stability and low LSN; its worst drawback is that reverse-biasing the output bases for fast switchoff is impossible without additional HT rails,
3 the quasi-complementary-with-Baxandall-diode stage comes close to mimicking the EF-type stages in linearity, with a potential for cost-saving on output devices. Quiescent stability is not as good as the CFP.

Closing the loop: distortion in complete amplifiers

In Chapter 4 it was shown how relatively simple design rules could ensure that the THD of the small-signal stages alone could be reduced to less than 0.001% across the audio band, in a thoroughly repeatable fashion, and without using frightening amounts of negative feedback. Combining this sub-system with one of the more linear output stages described in Chapter 4, such as the CFP version which gives 0.014% THD open-loop, and bearing in mind that ample NFB is available, it seems we have all the ingredients for a virtually distortionless power amplifier. However, life is rarely so simple . . . .

(Note – the AP plots in Figures 5.5–5.7 were taken at 100 W rms into 8 Ω, from an amplifier with an input error of −70 dB at 10 kHz and a c/l gain of 27 dB, giving a feedback factor of 43 dB at this frequency. This is well above the dominant-pole frequency and so the NFB factor is dropping at 6 dB/octave and will be down to 37 dB at 20 kHz. My experience suggests that this is about as much feedback as is safe for general hi-fi usage, assuming an output inductor to improve stability with capacitative loads. Sadly, published data on this touchy topic seems to be non-existent).

Figure 5.48 shows the distortion performance of such a closed-loop amplifier with an EF output stage, Figure 5.49 showing the same with a CFP output stage. Figure 5.50 shows the THD of a quasi-complementary stage with Baxandall diode[14]. In each case Distortion 1, Distortion 2 and Distortion 4–Distortion 7 have been eliminated, by methods described in past and future chapters, to make the amplifier Blameless.

It will be seen at once that these amplifiers are not distortionless, though the performance is markedly superior to the usual run of hardware. THD in the LF region is very low, well below a noise floor of 0.0007%, and the usual rise below 100 Hz is very small indeed. However, above 2 kHz, THD rises with frequency at between 6 and 12 dB/octave, and the distortion residual

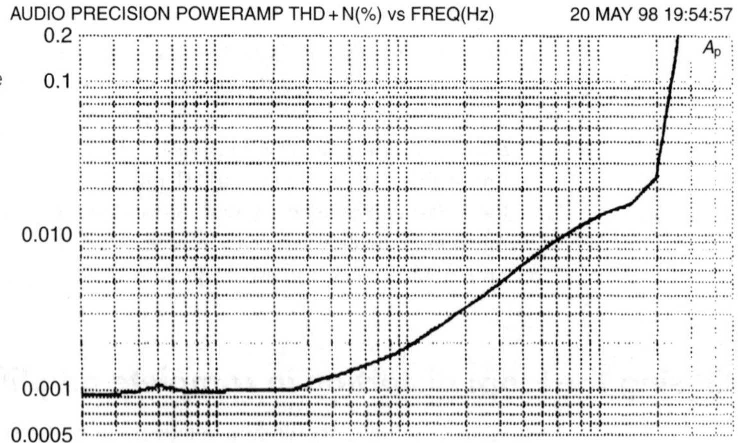

Figure 5.48 Closed-loop amplifier performance with Emitter-Follower output stage. 100 W into 8 Ω

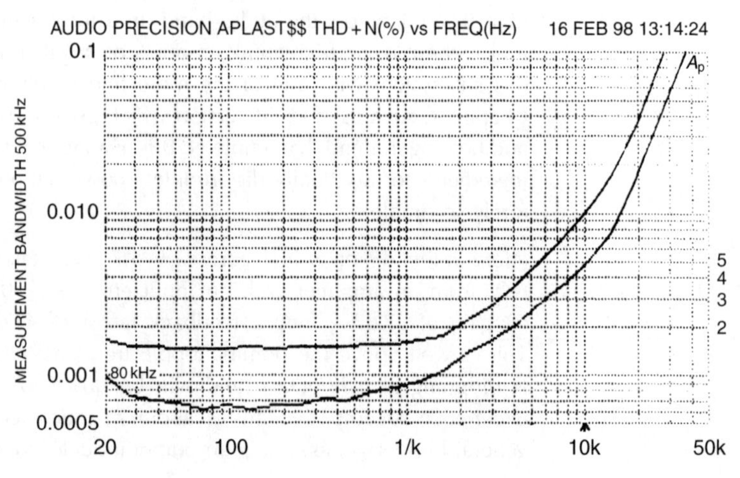

Figure 5.49 Closed-loop amplifier performance with CFP output. 100 W into 8 Ω

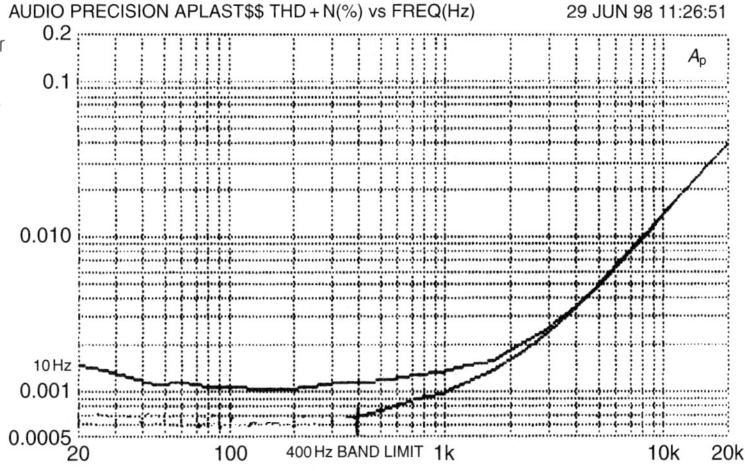

Figure 5.50 Closed-loop amplifier performance; quasi-complementary output stage with Baxandall diode. 100 W into 8 Ω

Table 5.6 Summary of closed-loop amp performance

	1 kHz (%)	10 kHz (%)
EF	0.0019	0.013
CFP	0.0008	0.005
Quasi Bax	0.0015	0.015

in this region is clearly time-aligned with the crossover region, and consists of high-order harmonics rather than second or third. It is intriguing to note that the quasi-Baxandall output gives about the same HF THD as the EF topology, which confirms the statement on page 122 that the addition of a Baxandall diode essentially turns a conventional quasi-complementary stage with serious crossover asymmetry into a reasonable emulation of a complementary EF stage. There is less HF THD with a CFP output; this cannot be due to large-signal non-linearity as this is negligible with an 8 Ω load for all three stages, and so it must be due to high-order crossover products. (See Table 5.6.)

The distortion figures given in this book are rather lower than usual. I would like to emphasise that these are not freakish or unrepeatable figures; they are the result of attending to all of the major sources of distortion, rather than just one or two. I have at the time of writing personally built 12 models of the CFP version, and performance showed little variation.

Here the closed-loop distortion is much greater than that produced by the small-signal stages alone; however if the input pair is badly designed its HF distortion can easily exceed that caused by the output stage.

Our feedback-factor here is a minimum of 70× across the band (being much higher at LF) and the output stages examined above are mostly capable of less than 0.1% THD open-loop. It seems a combination of these should yield a closed loop distortion at least 70 times better, i.e., below 0.001% from 10 Hz to 20 kHz. This happy outcome fails to materialise, and we had better find out why....

First, when an amplifier with a frequency-dependent NFB factor generates distortion, the reduction is not that due to the NFB factor at the fundamental frequency, but the amount available at the frequency of the harmonic in question. A typical amplifier with o/l gain rolling-off at 6 dB/octave will be half as effective at reducing fourth-harmonic distortion as it is at reducing the second harmonic. LSN is largely third (and possibly second) harmonic, and so NFB will deal with this effectively. However, both crossover and switchoff distortions generate high-order harmonics significant up to at least the nineteenth and these receive much less linearissation. As the fundamental moves up in frequency the harmonics do too, and benefit from even less feedback. This is the reason for the *differentiated* look to many distortion residuals; higher harmonics are emphasised at the rate of 6 db/octave.

Here is a real example of the inability of NFB to cure all possible amplifier ills. To reduce this HF distortion we must reduce the crossover gain-deviations of the output stage before closing the loop. There seems no obvious way to do this by minor modifications to any of the conventional output stages; we can only optimise the quiescent current.

As I stated on page 34, Class AB is generally not a Good Thing, as it gives more distortion than Class B, rather than less, and so will not help us. Figure 5.51 makes this very clear for the closed-loop case; Class-AB clearly gives the worst performance. (As before, the AB quiescent was set for 50:50 m/s ratio of the gm-doubling artefacts on the residual.)

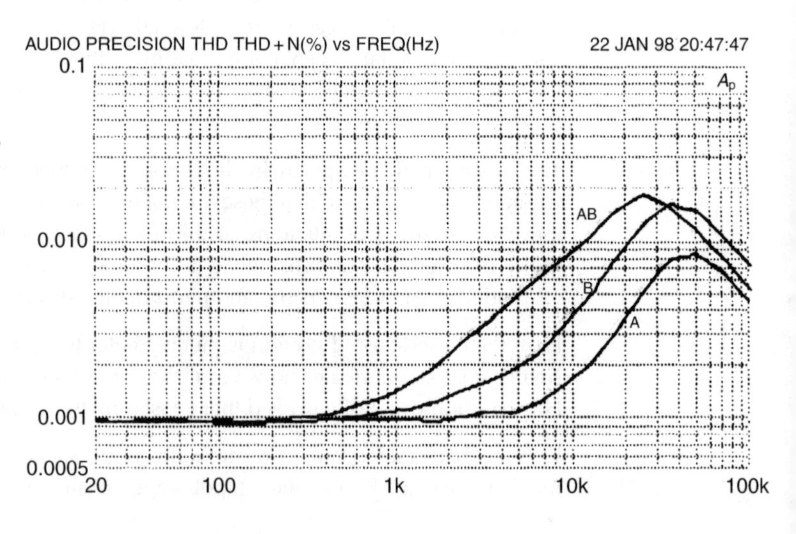

Figure 5.51 Closed-loop CFP amp. Setting quiescent for Class AB gives more HF THD than either Class A or B

Conclusions

1 Class-AB is best avoided. Use pure Class-A or B, as AB will always have more distortion than either,

2 FET outputs offer freedom from some BJT problems, but in general have poorer linearity and cost more,

3 The distortion generated by a Blameless amplifier driving an 8 Ω load is almost wholly due to the effects of crossover and switching distortion. This does not hold for 4 Ω or lower loads, where third harmonic on the residual shows the presence of large-signal non-linearity, caused by beta-loss at high output currents.

References

1. Mann *The Texan 20 + 20 Watt Stereo Amplifier* Practical Wireless, May 1972, p. 48 (Output stage with gain).
2. Takahashi *Design and Construction of High Slew Rate Amplifiers* Preprint No. 1348 (A-4) for 60th AES Convention 1978 (Class-B small-signal stages).
3. Hawksford *Distortion Correction in Audio Power Amplifiers* JAES, Jan/Feb 1981, p. 27 (Error-correction).
4. Walker, P *Current-Dumping Audio Amplifier* Wireless World 1975, pp. 560–562.
5. Blomley *New Approach To Class-B* Wireless World, Feb 1971, p. 57 and March 1971, pp. 127–131.
6. Otala *An Audio Power Amplifier for Ultimate Quality Requirements* IEEE Trans on Audio and Electroacoustics, Dec 1973, p. 548.
7. Lin, H Electronics, Sept 1956, pp. 173–175 (Quasi-comp).
8. Baxandall, P *Symmetry in Class B* Letters, Wireless World, Sept 1969, p. 416 (Baxandall diode).
9. Gray and Meyer *Analysis and Design of Analog Integrated Circuits*, Wiley 1984, p. 172.
10. Crecraft et al *Electronics* pub. Chapman and Hall 1993, p. 538.
11. Oliver *Distortion In Complementary-Pair Class-B Amps* Hewlett-Packard Journal, Feb 1971, p. 11.
12. Blomley, P *New Approach To Class-B* Wireless World, Feb 1971, p. 57.
13. Alves, J *Power Bandwidth Limitations in Audio Amplifiers* IEEE Trans on Broadcast and TV, March 1973, p. 79.
14. Baxandall, P *Symmetry in Class B* Letters, Wireless World, Sept 1969, p. 416 (Baxandall diode).

6

The output stage II

Distortion number 4: VAS loading distortion

Distortion 4 is that which results from the loading of the Voltage Amplifier Stage (VAS) by the non-linear input impedance of a Class-B output stage. This was looked at in Chapter 4 from the point of view of the VAS, where it was shown that since the VAS provides all the voltage gain, its collector impedance tends to be high. This renders it vulnerable to non-linear loading unless it is buffered or otherwise protected.

The VAS is routinely (though usually unknowingly) linearised by applying local negative-feedback via the dominant-pole Miller capacitor *C*dom, and this is a powerful argument against any other form of compensation. If VAS distortion still adds significantly to the amplifier total, then the local open-loop gain of the VAS stage can be raised to increase the local feedback factor. The obvious method is to raise the impedance at the VAS collector, and so the gain, by cascoding. However, if this is done without buffering the output stage loading will render the cascoding almost completely ineffective. Using a VAS-buffer eliminates this problem.

As explained in Chapter 4, the VAS collector impedance, while high at LF compared with other circuit nodes, falls with frequency as soon as *C*dom takes effect, and so Distortion 4 is usually only visible at LF. It is also often masked by the increase in output stage distortion above dominant-pole frequency P1 as the amount of global NFB reduces.

The fall in VAS impedance with frequency is demonstrated in Figure 6.1, obtained from the Spice conceptual model in Chapter 4, with values appropriate to real life. The LF impedance is basically that of the VAS collector resistance, but halves with each octave once P1 is reached. By 3 kHz the impedance is down to 1 kΩ, and still falling. Nevertheless, it usually remains high enough for the input impedance of a Class-B output stage to significantly degrade linearity, the actual effect being shown in Figure 6.2.

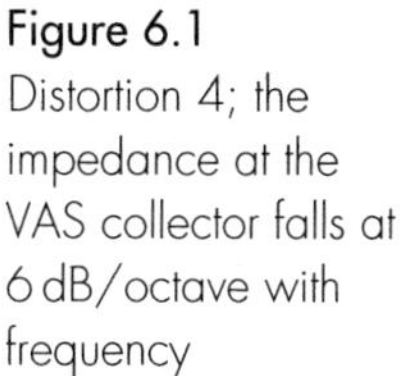

Figure 6.1
Distortion 4; the impedance at the VAS collector falls at 6 dB/octave with frequency

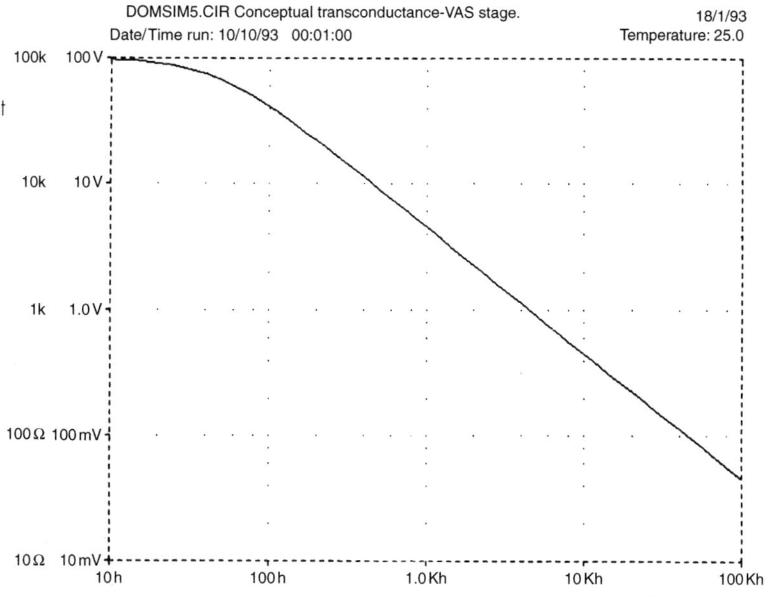

Figure 6.2
Distortion 4 in action; the lower trace shows the result of its elimination by the use of a VAS-buffer

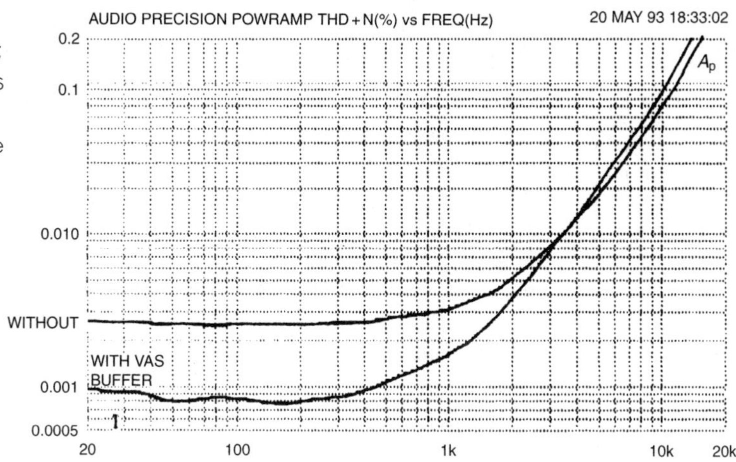

In Chapter 4, it was shown that as an alternative to cascoding, an effective means of linearising the VAS is to add an emitter-follower within the VAS local feedback loop, increasing the local NFB factor by raising effective beta rather than the collector impedance. As well as good VAS linearity, this establishes a much lower VAS collector impedance across the audio band, and is much more resistant to Distortion 4 than the cascode version. VAS buffering is not required, so this method has a lower component

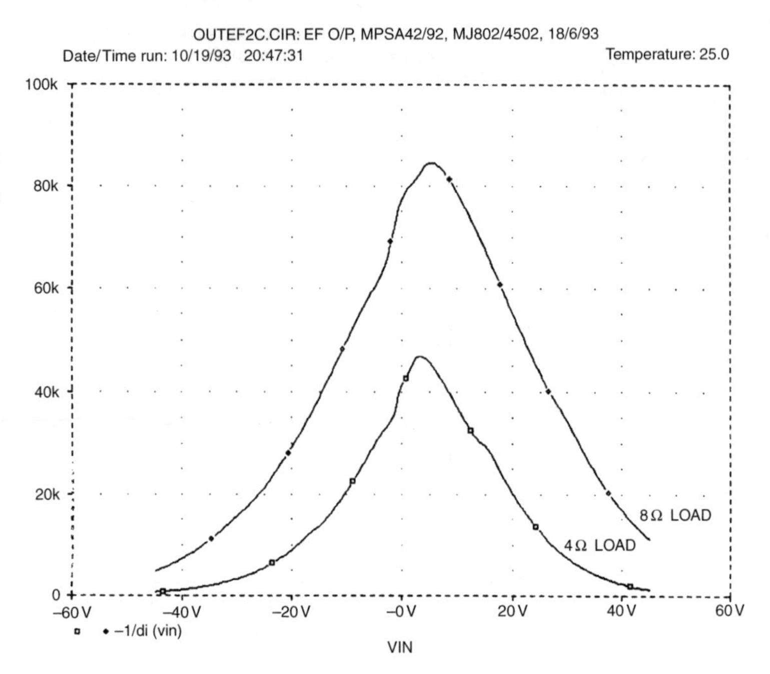

Figure 6.3
Distortion 4 and its root cause; the nonlinear input impedance of an EF Class-B output stage

count. The only drawback is a greater tendency to parasitic oscillation near negative clipping, when used with a CFP output stage.

Figure 6.3 confirms that the input impedance of a conventional EF Type I output stage is highly non-linear; the data is derived from a SPICE output stage simulation with optimal Iq. Even with an undemanding 8 Ω load, the impedance varies by 10:1 over the output voltage swing. The Type II EF output (using a shared drive emitter resistance) has a 50% higher impedance around crossover, but the variation ratio is rather greater. CFP output stages have a more complex variation that includes a precipitous drop to less than 20 kΩ around the crossover point. With all types under-biasing produces additional sharp impedance changes at crossover.

Distortion number 5: rail decoupling distortion

Almost all amplifiers have some form of rail decoupling apart from the main reservoir capacitors; this is usually required to guarantee HF stability. Standard decoupling arrangements include small to medium-size electrolytics (say 10–470 μF) connected between each rail and ground, and an inevitable consequence is that rail-voltage variations cause current to flow into the ground connection chosen. This is just one mechanism that defines the Power Supply Rejection Ratio (PSRR) of an amplifier, but it is one that can seriously damage linearity.

If we use an unregulated power supply (and there are almost overwhelming reasons for using such a supply, detailed in Chapter 8) comprising transformer, bridge rectifier, and reservoir capacitors, then these rails have a non-zero AC impedance and their voltage variations will be due to amplifier load currents as well as 100 Hz ripple. In Class-B, the supply-rail currents are halfwave-rectified sine pulses with strong harmonic content, and if they contaminate the signal then distortion is badly degraded; a common route for interaction is via decoupling grounds shared with input or feedback networks, and a separate decoupler ground is usually a complete cure. This point is easy to overlook, and attempts to improve amplifier linearity by labouring on the input pair, VAS, etc., are doomed to failure unless this distortion mechanism is eliminated first. As a rule it is simply necessary to take the decoupling ground separately back to the ground star point, as shown in Figure 6.4. (Note that the star-point A is defined on a short spur from the heavy connection joining the reservoirs; trying to use B as the star point will introduce ripple due to the large reservoir-charging current pulses passing through it.)

Figure 6.5 shows the effect on an otherwise Blameless amplifier handling 60 W/8 Ω, with 220 µF rail decoupling capacitors; at 1 kHz distortion has increased by more than ten times, which is quite bad enough. However, at 20 Hz the THD has increased at least 100-fold, turning a very good amplifier into a profoundly mediocre one with one misconceived connection.

When the waveform on the supply rails is examined, the 100 Hz ripple amplitude will usually be found to exceed the pulses due to Class-B signal current, and so some of the *distortion* on the upper curve of the plot is actually due to ripple injection. This is hinted at by the phase-crevasse at 100 Hz, where the ripple happened to partly cancel the signal at the instant of measurement. Below 100 Hz the curve rises as greater demands are made on the reservoirs, the signal voltage on the rails increases, and more distorted current is forced into the ground system.

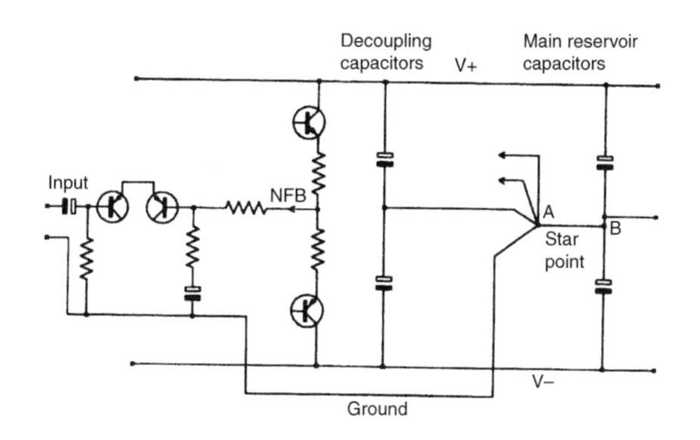

Figure 6.4
Distortion 5; The correct way to route decouple grounding to the star-point

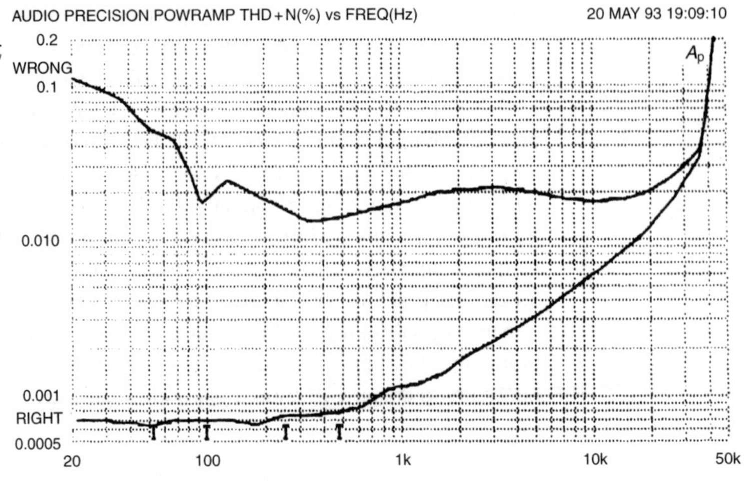

Figure 6.5
Distortion 5 in action; The upper trace was produced simply by taking the decoupler ground from the star-point and connecting it via the input ground line instead

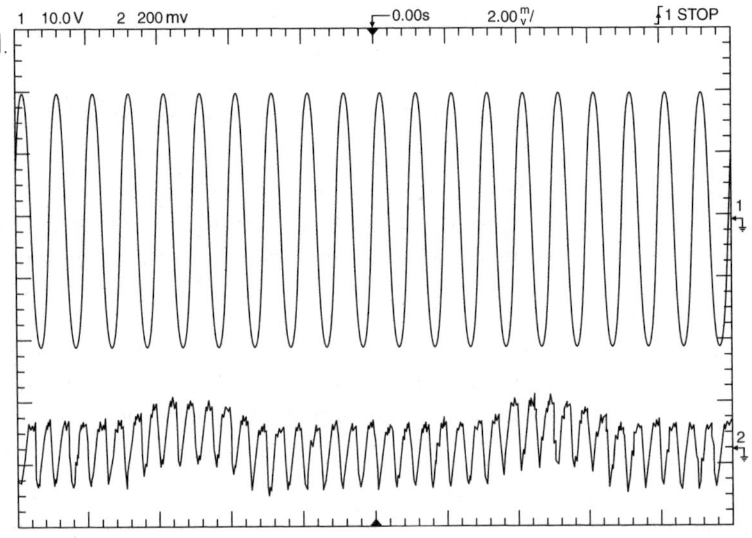

Figure 6.6
Distortion 5 revealed. Connecting the rail decoupler to input ground increases THD eight-fold from 0.00097% to 0.008%, mostly as second harmonic. 100 Hz ripple is also visible. No averaging

Figure 6.6 shows a typical Distortion 5 residual, produced by deliberately connecting the negative supply-rail decoupling capacitor to the input ground instead of properly giving it its own return to the far side of the star-point. THD increased from 0.00097% to 0.008%, appearing mostly as second harmonic. Distortion 5 is usually easy to identify as it is accompanied by 100 Hz power-supply ripple; Distortions 6 and 7 introduce no extra ripple. The ripple contamination here – the the two humps at the bottom – is significant and contributes to the THD reading.

As a general rule, if an amplifier is made free from ripple injection under drive conditions, demonstrated by a THD residual without ripple components, there will be no distortion from the power-supply rails, and the complications and inefficiencies of high-current rail regulators are quite unnecessary.

There has been much discussion of PSRR-induced distortion in the literature recently, e.g., Greg Ball[1]. I part company with some writers at the point where they assume a power amplifier is likely to have 25 dB PSRR, making an expensive set of HT regulators the only answer. Greg Ball also initially assumes that a power amp has the same PSRR characteristics as an op-amp, i.e., falling steadily at 6 dB/octave. There is absolutely no need for this to be so, given a little RC decoupling, and Ball states at the end of his article that *a more elegant solution... is to depend on a high PSRR in the amplifier proper*. This issue is dealt with in detail in Chapter 8.

Distortion number 6: induction distortion

The existence of this distortion mechanism, like Distortion 5, stems directly from the Class-B nature of the output stage. With a sine input, the output hopefully carries a good sinewave, but the supply-rail currents are halfwave-rectified sine pulses, which will readily crosstalk into sensitive parts of the circuit by induction. This is very damaging to the distortion performance, as Figure 6.7 shows.

The distortion signal may intrude into the input circuitry, the feedback path, or even the cables to the output terminals. The result is a kind of sawtooth on the distortion residual that is very distinctive, and a large extra distortion component that rises at 6 dB/octave with frequency.

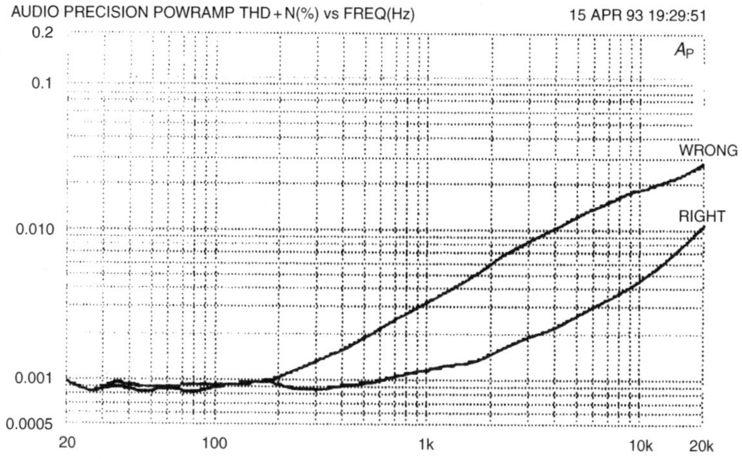

Figure 6.7
Distortion 6 exposed. The upper trace shows the effects of Class-B rail induction into signal circuitry

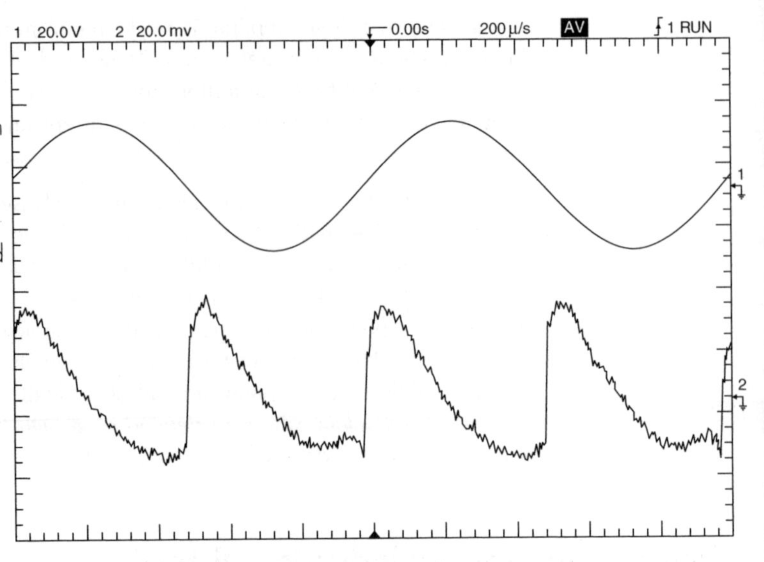

Figure 6.8 Distortion 6. Induction of half-wave signal from the negative supply rail into the NFB line increases THD to 0.0021%. Averaged 64 times

A Distortion 6 residual is displayed in Figure 6.8. The V-supply rail was routed parallel to the negative-feedback line to produce this diagram. THD is more than doubled, but is still relatively low at 0.0021%. 64-times averaging is used. Distortion 6 is easily identified if the DC supply cables are movable, for altering their run will strongly affect the quantity generated.

This inductive effect appears to have been first publicised by Cherry[2], in a paper that deserves more attention. The effect has however been recognised and avoided by some practitioners for many years[3]. However, having examined many power amplifiers with varying degrees of virtue, I feel that this effect, being apparently unknown to most designers, is probably the most widespread cause of unnecessary distortion.

The contribution of Distortion 6 can be reduced below the measurement threshold by taking sufficient care over the layout of supply-rail cabling relative to signal leads, and avoiding loops that will induce or pick up magnetic fields. I wish I could give precise rules for layout that would guarantee freedom from the problem, but each amplifier has its own physical layout, and the cabling topology has to take this into account. However, here are some guidelines:

First, implement rigorous minimisation of loop area in the input and feedback circuitry; keeping each signal line as close to its ground return as possible. Second, minimise the ability of the supply wiring to establish magnetic fields in the first place; third, put as much distance between these two areas as you can. Fresh air beats shielding on price every time.

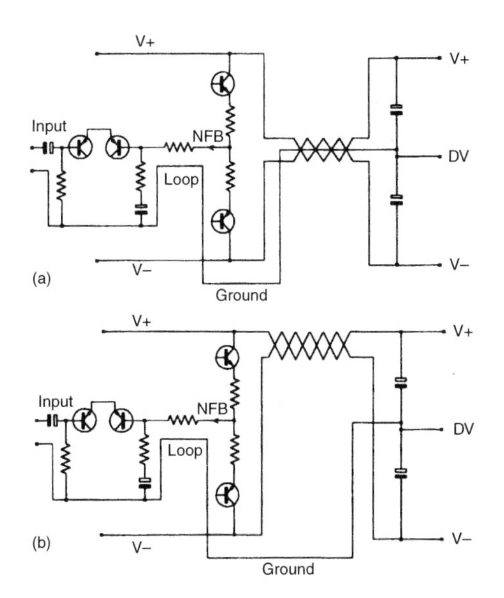

Figure 6.9
Distortion 6; Countermeasures against the induction of distortion from the supply rails. 6.7b is usually more effective

Figure 6.9 shows one straightforward approach to solving the problem; the supply and ground wires are tightly twisted together to reduce radiation. In practice this does not seem to effective, for reasons that are not wholly clear, but seem to involve the difficulty of ensuring exactly equal coupling between three twisted conductors. In Figure 6.9, the supply rails are twisted together but kept well away from the ground return; this will allow field generation, but if the currents in the two rails butt together to make a nice sinewave at the output, then they should do the same when the magnetic fields from each rail sum. There is an obvious risk of interchannel crosstalk if this approach is used in a stereo amplifier, but it does deal effectively with the induced-distortion problem in some layouts.

Distortion number 7: NFB takeoff point distortion

It has become a tired old truism that negative feedback is a powerful technique, and like all such, must be used with care if you are to avoid tweeter-frying HF instability.

However, there is another and much more subtle trap in applying global NFB. Class-B output stages are a maelstrom of high-amplitude halfwave-rectified currents, and if the feedback takeoff point is in slightly the wrong place, these currents contaminate the feedback signal, making it an inaccurate representation of the output voltage, and hence introducing distortion; Figure 6.10 shows the problem. At the current levels in question, all wires and PCB tracks must be treated as resistances, and it follows that point C is not at the same potential as point D whenever TR1 conducts. If feedback is taken from D, then a clean signal will be established here, but the signal

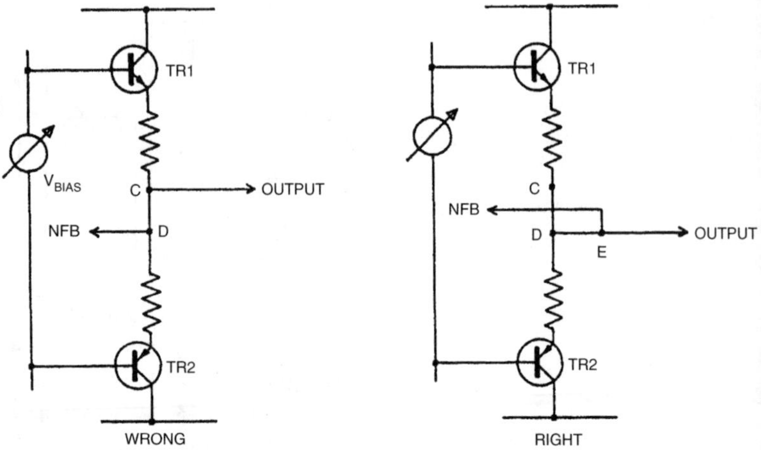

Figure 6.10
Distortion 7; Wrong and right ways of arranging the critical negative-feedback takeoff point

at output point C will have a half-wave rectified sinewave added to it, due to the resistance C–D. The actual output will be distorted but the feedback loop will do nothing about it as it does not know about the error.

Figure 6.11 shows the practical result for an amplifier driving 100 W into 8 Ω, with the extra distortion interestingly shadowing the original curve as it rises with frequency. The resistive path C–D that did the damage was a mere 6 mm length of heavy-gauge wirewound resistor lead.

Figure 6.12 shows a THD residual for Distortion 7, introduced by deliberately taking the NFB from the wrong point. The THD rose from 0.00097% to 0.0027%, simply because the NFB feed was taken from the wrong end of the leg of one of the output emitter resistors *R*e. Note this is not the wrong

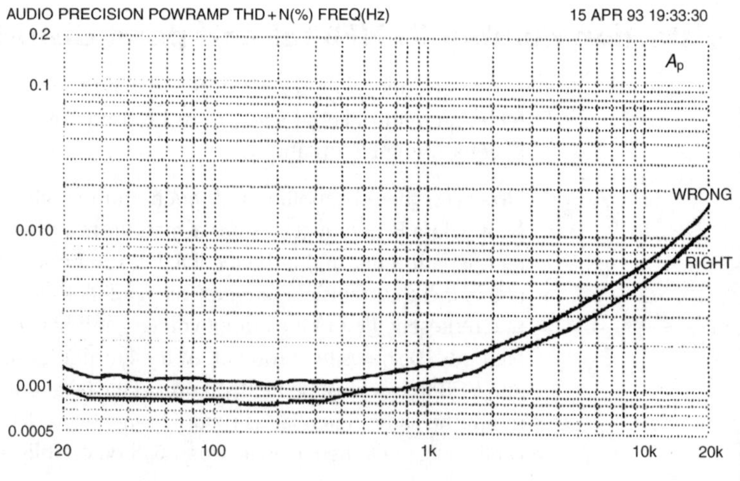

Figure 6.11
Distortion 7 at work; the upper (WRONG) trace shows the result of a mere 6 mm of heavy-gauge wire between the output and the feedback point

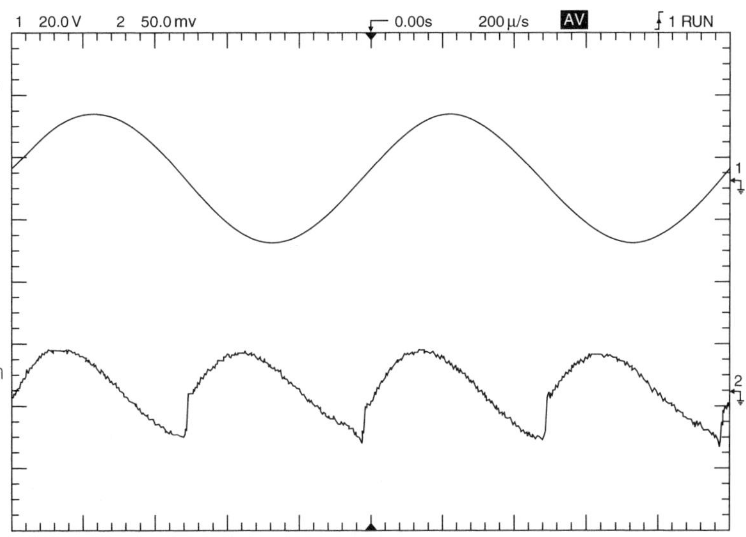

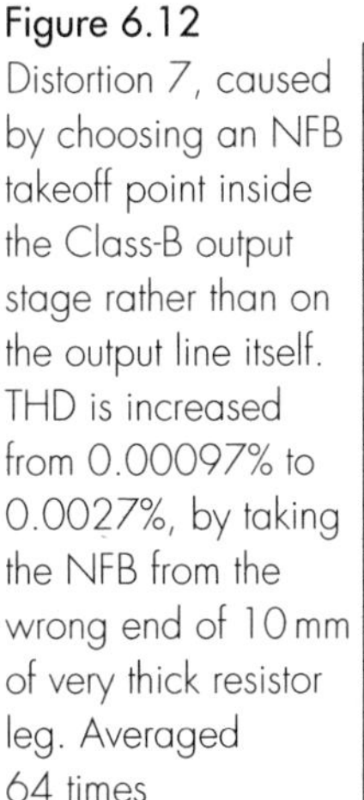
Figure 6.12 Distortion 7, caused by choosing an NFB takeoff point inside the Class-B output stage rather than on the output line itself. THD is increased from 0.00097% to 0.0027%, by taking the NFB from the wrong end of 10 mm of very thick resistor leg. Averaged 64 times

side of the resistor, or the distortion would have been gross, but a mere 10 mm along a very thick resistor leg from the actual output junction point.

Of the distortions that afflict generic Class-B power amplifiers, 5, 6 and 7 all look rather similar when they appear in the THD residual, which is perhaps not surprising since all result from adding half-wave disturbances to the signal.

To eliminate this distortion is easy, once you are alert to the danger. Taking the NFB feed from D is not advisable as D is not a mathematical point, but has a physical extent, inside which the current distribution is unknown. Point E on the output line is much better, as the half-wave currents do not flow through this arm of the circuit.

Distortion number 8: capacitor distortion

When I wrote the original series on amplifier distortion[4], I listed seven types of distortion that defined an amplifier's linearity. The number has grown to eight, with the addition of electrolytic capacitor distortion. This has nothing to do with Subjectivist hypotheses about mysterious non-measurable effects; this phenomenon is all too real, though for some reason it seems to be almost unknown amongst audio designers.

Standard aluminium electrolytics generate distortion whenever they are used so a significant AC voltage develops across them; this is usually when they are used for coupling and DC blocking, whilst driving a significant resistive load. Figure 6.13 is the test circuit; Figure 6.14 shows the resulting distortion for a 47 µF 25 V capacitor driving +20 dBm (7.75 V rms) into

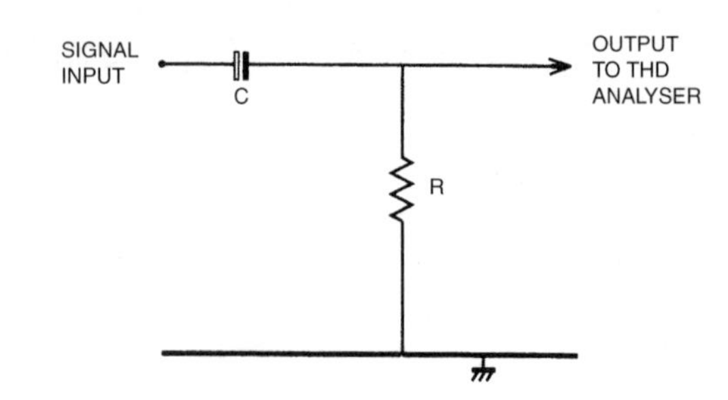

Figure 6.13
A very simple circuit to demonstrate electrolytic capacitor distortion. Measurable distortion begins at 100 Hz

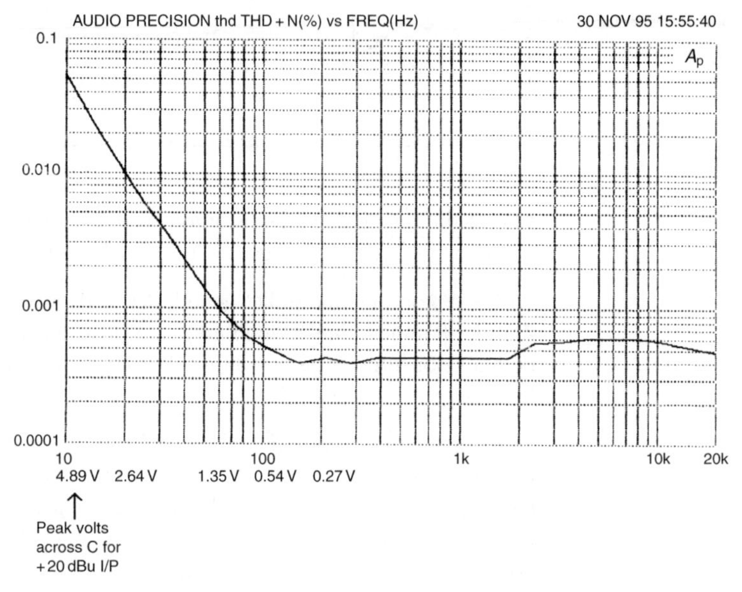

Figure 6.14
Capacitor distortion versus frequency, showing the rapid rise in THD once the distortion threshold is reached

a 680 Ω load, while Figure 6.15 shows how the associated LF roll-off has barely begun. The distortion is a mixture of second and third harmonic, and rises rapidly as frequency falls, at something between 12 and 18 dB/octave.

The great danger of this mechanism is that serious distortion begins while the response roll-off is barely detectable; here the THD reaches 0.01% when the response has only fallen by 0.2 dB. The voltage across the capacitor is 2.6 V peak, and this voltage is a better warning of danger than the degree of roll-off.

Further tests showed that the distortion roughly triples as the applied voltage doubles; this factor seems to vary somewhat between different capacitor rated voltages.

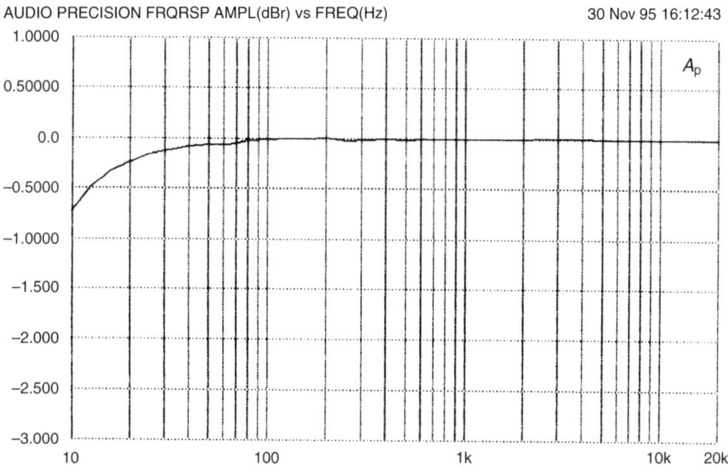

Figure 6.15
The small amount of LF roll-off associated with the distortion rise in Figure 6.11

The mechanism by which capacitors generate this distortion is unclear. Dielectric absorption appears to be ruled out as this is invariably (and therefore presumably successfully) modelled by adding linear components, in the shape of resistors and capacitors, to the basic capacitor model. Reverse biasing is not the problem, for capacitors DC biased by up to +15 V show slightly increased, not reduced distortion. Non-polarized electrolytics show the same effect but at a much greater AC voltage, typically giving the same distortion at one-tenth the frequency of a conventional capacitor with the same time-constant; the cost and size of these components generally rules out their use to combat this effect. Usually the best solution is simply to keep increasing the capacitor value until the LF distortion rise disappears off the left of the THD graph. Negligible roll-off in the audio band is not a sufficient criterion.

Electrolytics are therefore best reserved for DC filtering, and for signal coupling where the AC voltage across them will be negligible. If a coupling capacitor does have AC voltage across it, and drives the usual resistive load, then it must be acting as a high-pass filter. This is never good design practice, because electrolytics have large tolerances and make inaccurate filters; it is now clear they generate distortion as well.

It is therefore most undesirable to define the lower bandwidth limit simply by relying on the high-pass action of electrolytics and circuit resistances; it should be done with a non-electrolytic capacitor, made as large as possible economically in order to reduce the value of the associated resistance and so keep down circuit impedances, thus minimising the danger of noise and crosstalk.

Capacitor distortion in power amplifiers is most likely to occur in the feedback network blocking capacitor (assuming a DC-coupled amplifier).

The input blocking capacitor usually feeds a high impedance, but the feedback arm must have the lowest possible resistances to minimise both noise and DC offset. The feedback capacitor therefore tends to be relatively large, and if it is not quite large enough the THD plot of the amplifier will show the characteristic kick up at the LF end. An example of this is dealt with in detail on page 90.

It is common for amplifiers to show a rise in distortion at the LF end, but there is no reason why this should ever occur. Capacitor distortion is usually the reason, but Distortion 5 (Rail Decoupling Distortion) can also contribute. They can be distinguished because Distortion 5 typically rises by only 6 dB/octave as frequency decreases, rather than 12–18 dB/octave.

Amplifiers with AC-coupled outputs are now fairly rare, possibly because distortion in the output capacitor is a major problem, occurring in the mid-band as well as at LF. See page 44 for details.

Design example: a 50 W Class-B amplifier

Figure 6.16 shows a design example of a Class-B amplifier, intended for domestic hi-fi applications. Despite its relatively conventional appearance, the circuit parameters selected give much better than a conventional distortion performance; this is potentially a Blameless design, but only if due care is given to wiring topology and physical layout will this be achieved.

With the supply voltages and values shown it gives 50 W into 8 Ω, for 1 V rms input. In earlier chapters, I have used the word *Blameless* to describe amplifiers in which all distortion mechanisms, except the apparently unavoidable ones due to Class-B, have been rendered negligible. This circuit has the potential to be Blameless (as do we all) but achieving this depends on care in cabling and layout. It does not aim to be a cookbook project; for example, overcurrent and DC-offset protection are omitted.

In Chapter 11, output topologies are examined, and the conclusion drawn that power-FETs were disappointingly expensive, inefficient, and non-linear. Therefore, it is bipolars. The best BJT configurations were the Emitter-Follower Type II, with least output switchoff distortion, and the Complementary-Feedback Pair (CFP), giving the best basic linearity.

The output configuration chosen is the Emitter-Follower Type II, which has the advantage of reducing switchoff non-linearities (Distortion 3c) due to the action of R15 in reverse-biasing the output base-emitter junctions as they turn off. A possible disadvantage is that quiescent stability might be worse than for the CFP output topology, as there is no local feedback loop to servo out Vbe variations in the hot output devices. Domestic ambient temperature changes will be small, so that adequate quiescent stability can be attained by suitable heatsinking and thermal compensation.

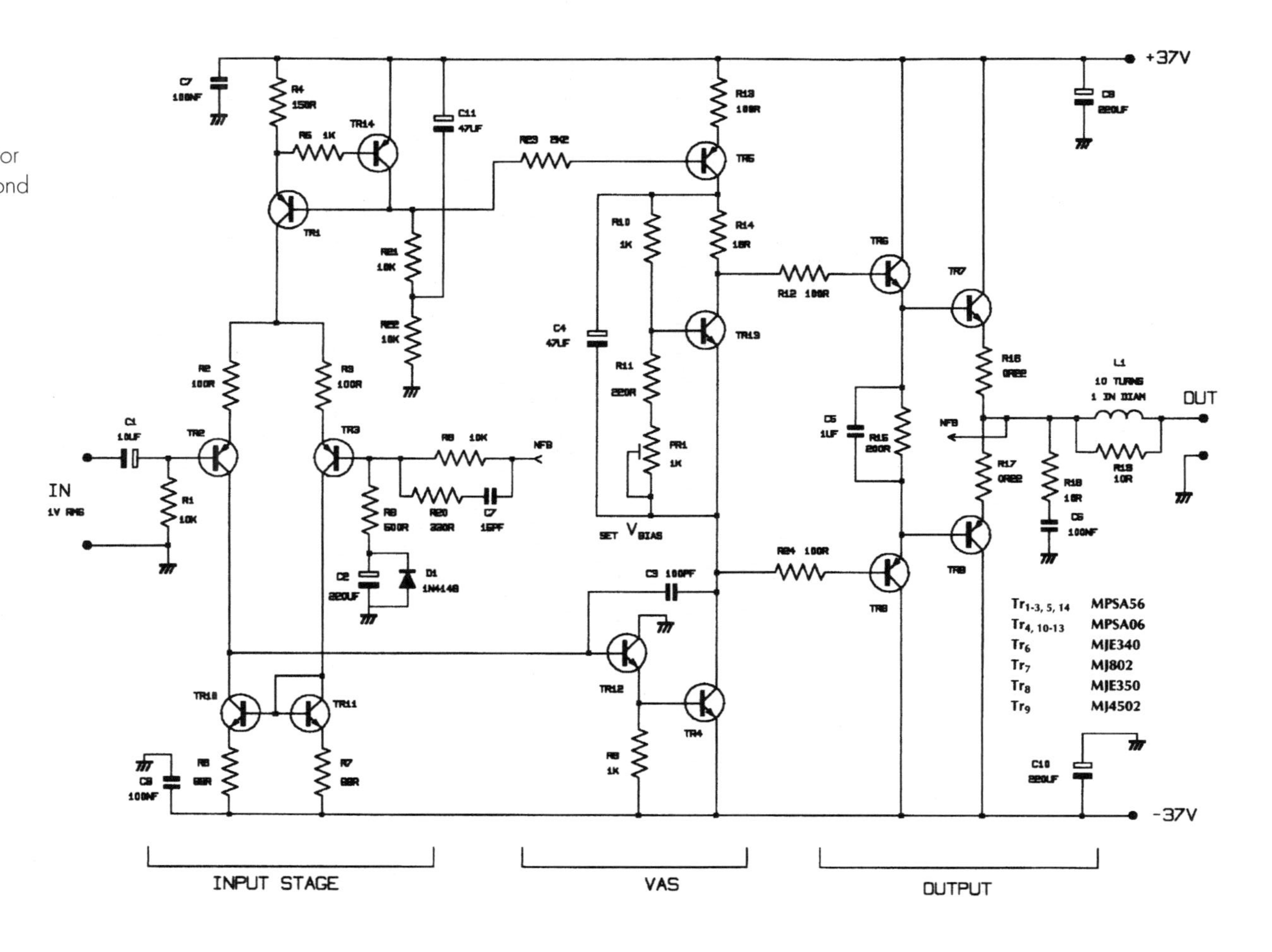

Figure 6.16
50 W Class-B amplifier circuit diagram. Transistor numbers correspond with the generic amplifier in Chapter 3

A global NFB factor of 30 dB at 20 kHz was chosen, which should give generous HF stability margins. The input stage (current-source TR1 and differential pair TR2, 3) is heavily degenerated by R2, 3 to delay the onset of third-harmonic Distortion 1, and to assist this the contribution of transistor internal *re* variation is minimised by using the unusually high tail current of 4 mA. TR11, 12 form a degenerated current-mirror that enforces accurate balance of the TR2, 3 collector currents, preventing the generation of second-harmonic distortion. Tail source TR1, 14 has a basic PSRR 10 dB better than the usual two-diode version, though this is academic when C11 is fitted.

Input resistor R1 and feedback arm R8 are made equal and kept as low as possible consistent with a reasonably high input impedance, so that base current mismatch caused by beta variations will give a minimal DC offset; this does not affect TR2–TR3 Vbe mismatches, which appear directly at the output, but these are much smaller than the effects of Ib. Even if TR2, 3 are high voltage types with low beta, the output offset should be within ±50 mV, which should be quite adequate, and eliminates balance presets and DC servos. A low value for R8 also gives a low value for R9, which improves the noise performance.

The value of C2 shown (220 µF) gives an LF roll-off with R9 that is −3 dB at 1.4 Hz. The aim is not an unreasonably extended sub-bass response, but to prevent an LF rise in distortion due to capacitor non-linearity; 100 µF degraded the THD at 10 Hz from less than 0.0006% to 0.0011%, and I judge this unacceptable aesthetically if not audibly. Band-limiting should be done earlier, with non-electrolytic capacitors. Protection diode D1 prevents damage to C2 if the amplifier suffers a fault that makes it saturate negatively; it looks unlikely but causes no measurable distortion[5]. C7 provides some stabilising phase-advance and limits the closed-loop bandwidth; R20 prevents it upsetting TR3.

The VAS stage is enhanced by an emitter-follower inside the Miller-compensation loop, so that the local NFB that linearises the VAS is increased by augmenting total VAS beta, rather than by increasing the collector impedance by cascoding. This extra local NFB effectively eliminates Distortion 2 (VAS non-linearity). Further study has shown that thus increasing VAS beta gives a much lower collector impedance than a cascode stage, due to the greater local feedback, and so a VAS-buffer to eliminate Distortion 4 (loading of VAS collector by the non-linear input impedance of the output stage) appears unnecessary. *C*dom is relatively high at 100 pF, to swamp transistor internal capacitances and circuit strays, and make the design predictable. The slew-rate calculates as 40 V/µsec. The VAS collector-load is a standard current source, to avoid the uncertainties of bootstrapping.

Since almost all the THD from a blameless amplifier is crossover, keeping the quiescent conditions optimal is essential. Quiescent stability requires the bias generator to cancel out the Vbe variations of four junctions in series; those of two drivers and of two output devices. Bias generator TR8 is the standard Vbe-multiplier, modified to make its voltage more stable against variations in the current through it. These occur because the biasing of TR5 does not completely reject rail variations; its output current also drifts initially due to heating and changes in TR5 Vbe. Keeping Class-B quiescent stable is hard enough at the best of times, and so it makes sense to keep these extra factors out of the equation. The basic Vbe-multiplier has an incremental resistance of about 20 Ω; in other words its voltage changes by 1 mV for a 50 μA drift in standing current. Adding R14 converts this to a gently peaking characteristic that can be made perfectly flat at one chosen current; see Figure 6.17. Setting R14 to 22 Ω makes the voltage peak at 6 mA, and standing current now must deviate from this value by more than 500 μA for a 1 mV bias change. The R14 value needs to be altered if TR15 is run at a different current; for example, 16 Ω makes the voltage peak at 8 mA instead. If TO3 outputs are used the bias generator should be in contact with the top or can of one of the output devices, rather than the heatsink, as this is the fastest and least attenuated source for thermal feedback.

The output stage is a standard double emitter-follower apart from the connection of R15 between the driver emitters without connection to the output rail. This gives quicker and cleaner switchoff of the outputs at high frequencies; switchoff distortion may significantly degrade THD

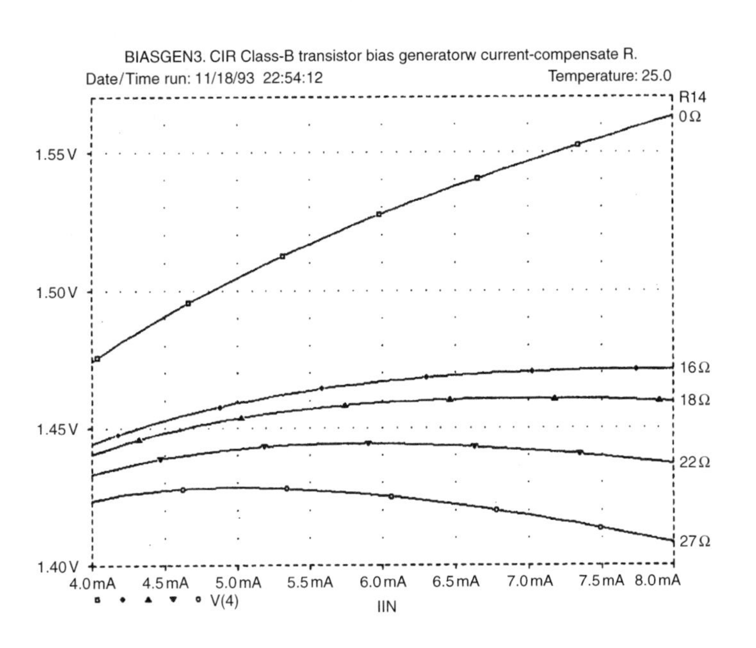

Figure 6.17 SPICE plot of the voltage-peaking behaviour of a current-compensated bias generator

from 10 kHz upwards, dependent on transistor type. Speedup capacitor C4 noticeable improves the switchoff action. C6, R18 form the Zobel network (sometimes confusingly called a Boucherot cell) while L1, damped by R19, isolates the amplifier from load capacitance.

Figure 6.18 shows the 50 W/8 Ω distortion performance; about 0.001% at 1 kHz, and 0.006% at 10 kHz (See Table 6.1). The measurement bandwidth makes a big difference to the appearance, because what little distortion is present is crossover-derived, and so high-order. It rises at 6 dB/octave, at the rate the feedback factor falls, and it is instructive to watch the crossover glitches emerging from the noise, like Grendel from the marsh, as the test frequency increases above 1 kHz. There is no precipitous THD rise in the ultrasonic region.

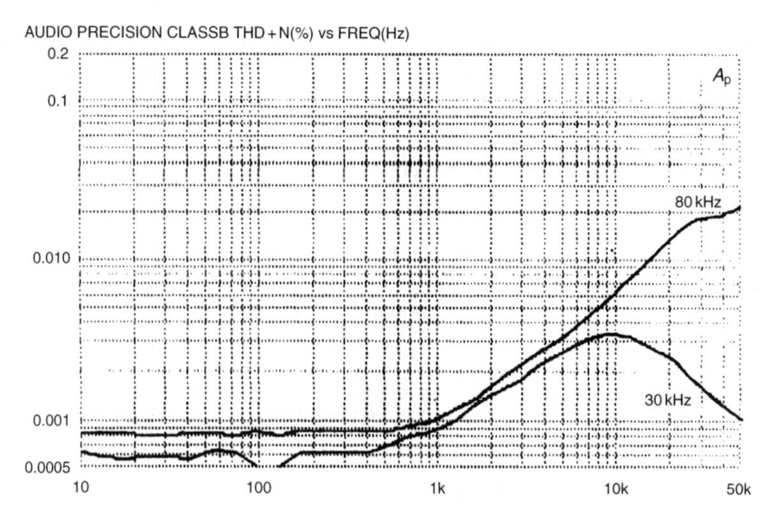

Figure 6.18
Class-B amplifier: THD performance at 50 W/8 Ω; measurement bandwidths 30 kHz and 80 kHz

Table 6.1
Class-B amplifier performance

Power output	50 W rms into 8 Ω
Distortion	Below 0.0006% at 1 kHz and 50 W/8 Ω
	Below 0.006% at 10 kHz
Slew-rate	Approximately 35 V/μsec
Noise	91 dBu at the output
EIN	117 dBu (referred to input)
Freq Response	+0, −0.5 dB over 20 Hz–20 kHz

(Most of the AP plots in this book were obtained from an amplifier similar to Figure 6.16, though with higher supply rails and so greater power capability. The main differences were the use of a cascode-VAS with a buffer, and a CFP output to minimise distracting quiescent variations. Measurements at powers above 100 W/8 Ω used a version with two paralleled output devices.)

The zigzags on the LF end of the plot are measurement artefacts, apparently caused by the Audio Precision system trying to winkle out distortion from visually pure white noise. Below 700 Hz the residual was pure noise with a level equivalent to approximately 0.0006% (yes, three zeros) at 30 kHz bandwidth; the actual THD here must be microscopic. This performance can only be obtained if all seven of the distortion mechanisms are properly addressed; Distortions 1–4 are determined by the circuit design, but the remaining three depend critically on physical layout and grounding topology.

It is hard to beat a well-gilded lily, and so Figure 6.19 shows the startling results of applying 2-pole compensation to the basic amplifier; C3 remains 100 pF, while CP2 was 220 pF and Rp 1 k (see Figure 7.1d, page 188). The extra global NFB does its work extremely well, the 10 kHz THD dropping to 0.0015%, while the 1 kHz figure can only be guessed at. There were no unusual signs of instability, but as always unusual compensation schemes require careful testing. It does appear that a Blameless amplifier with 2-pole compensation takes us close to the long-sought goal of the Distortionless Amplifier.

The basic Blameless EF amplifier was experimentally rebuilt with three alternative output stages; the simple quasi-complementary, the quasi-Baxandall, and the CFP. The results for both single and two-pole compensation are shown in Figures 6.20, 6.21, and 6.22. The simple quasi-complementary generates more crossover distortion, as expected, and the quasi-Baxandall version is not a lot better, probably due to remaining asymmetries around the crossover region. The CFP gives even lower distortion than the original EF-II output, with Figure 6.19 showing only the result for single-pole compensation; in this case the improvement with two-pole was marginal and the trace is omitted for clarity.

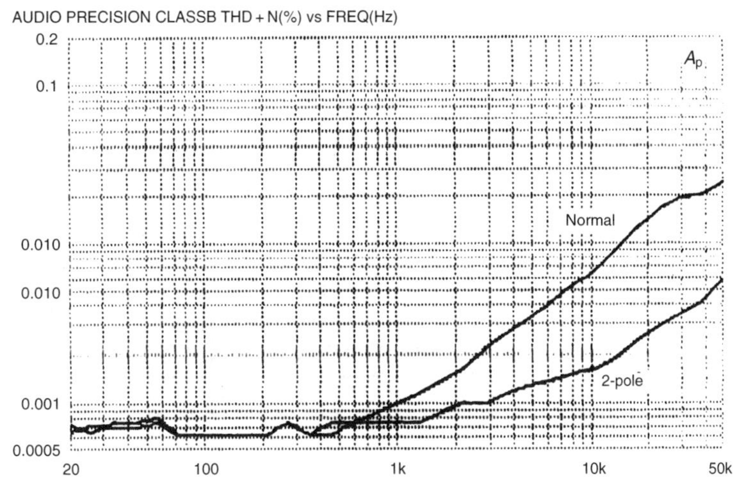

Figure 6.19 The dramatic THD improvement obtained by converting the Class-B amplifier to 2-pole compensation

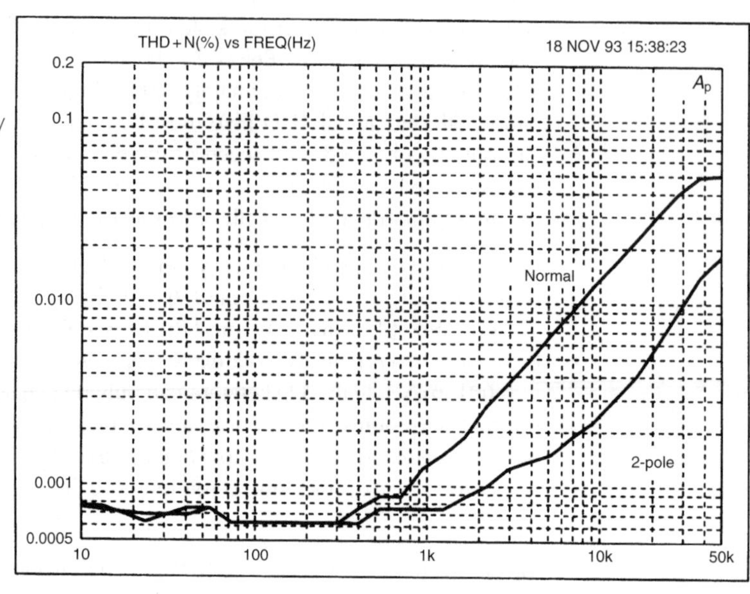

Figure 6.20 Class-B amplifier with simple quasi-complementary output. Lower trace is for two-pole compensation

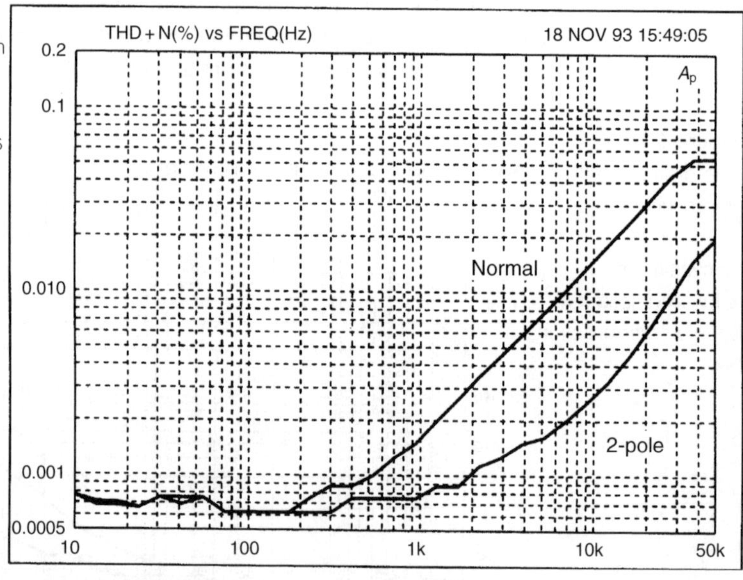

Figure 6.21 Class-B amplifier with quasi-comp plus Baxandall diode output. Lower trace is the two-pole case

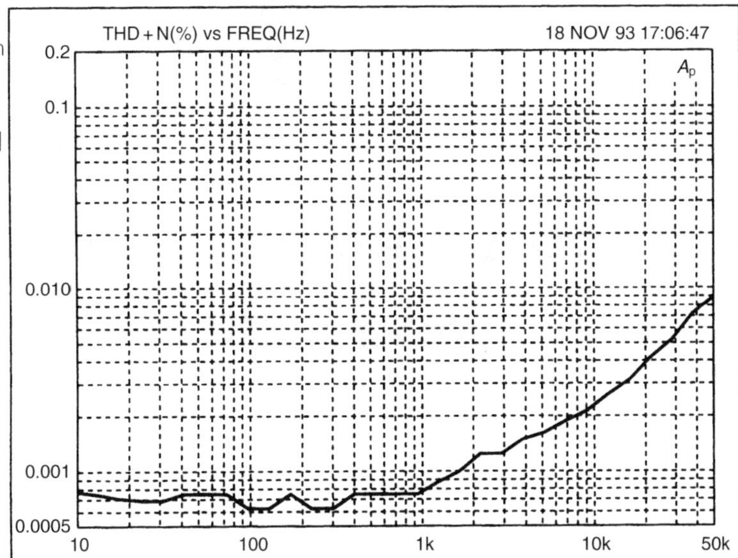

Figure 6.22
Class-B amplifier with Complementary-Feedback Pair (CFP) output stage. Normal compensation only

References

1. Ball, G *Distorting Power Supplies* Electronics & Wireless World, Dec 1990, p. 1084.
2. Cherry, *A New Distortion Mechanism in Class-B Amplifiers* JAES, May 1981, p. 327.
3. Baxandall, P *Private communication*, 1995.
4. Self, D *Distortion In Power Amplifiers* Series in Electronics & Wireless World, Aug 93 to March 94.
5. Self, D *An Advanced Preamplifier* Wireless World, Nov 1976, p. 43.

7

Compensation, slew-rate, and stability

Frequency compensation in general

The compensation of an amplifier is the tailoring of its open-loop gain and phase characteristics so that is dependably stable when the global feedback loop is closed.

It must be said straight away that *compensation* is a thoroughly misleading word to describe the subject of this chapter. It implies that one problematic influence is being balanced out by another opposing force, when in fact it means the process of tailoring the open-loop gain and phase of an amplifier so that it is satisfactorily stable when the global feedback loop is closed. The derivation of the word is historical, going back to the days when all servomechanisms were mechanical, and usually included an impressive Watt governor pirouetting on top of the machinery.

An amplifier requires compensation because its basic open-loop gain is still high at frequencies where the internal phase-shifts are reaching 180°. This turns negative feedback into positive at high frequencies, and causes oscillation, which in audio amplifiers can be very destructive. The way to prevent this is to ensure that the loop gain falls to below unity before the phase-shift reaches 180°; oscillation therefore cannot develop. Compensation is therefore vital simply because it makes the amplifier stable; there are other considerations, however, because the way in which the compensation is applied has a major effect on the closed-loop distortion behaviour.

The distortion performance of an amplifier is determined not only by open-loop linearity, but also the negative feedback factor applied when the loop is closed; in most practical circumstances doubling the NFB factor halves the distortion. So far I have assumed that open-loop gain falls at

6 dB/octave due to a single dominant pole, with the amount of NFB permissible at HF being set by the demands of HF stability. We have seen that this results in the distortion from a Blameless amplifier consisting almost entirely of crossover artefacts, because of their high-order and hence high frequency. Audio amplifiers using more advanced compensation are rather rare. However, certain techniques do exist, and are described later.

This book sticks closely to conventional topologies, because even apparently commonplace circuitry has proven to have little-known aspects, and to be capable of remarkable linearity. This means the classical three-stage architecture circuit with transconductance input, transimpedance VAS, and unity-gain output stage. Negative feedback is applied globally, but is smoothly transferred by *C*dom to be local solely to the VAS as frequency increases. Other configurations are possible; a two-stage amplifier with transconductance input and unity-gain output is an intriguing possibility – this is common in CMOS op-amps – but is probably ill-suited to power-amp impedances. Another architecture with a voltage-gain input stage is described in Chapter 11, and see Otala[1] for an eccentric four-stage amplifier with a low open-loop gain of 52 dB (due to the dogged use of local feedback) and only 20 dB of global feedback. Most of this chapter relates only to the conventional three-stage structure.

Dominant-pole compensation

Dominant-pole compensation is the simplest kind, though its action is subtle. Simply take the lowest pole to hand (P1), and make it dominant, i.e., so much lower in frequency than the next pole P2 that the total loop-gain (i.e., the open-loop gain as reduced by the attenuation in the feedback network) falls below unity before enough phase-shift accumulates to cause HF oscillation. With a single pole, the gain must fall at 6 dB/octave, corresponding to a constant 90° phase shift. Thus the phase margin will be 90°, giving good stability.

Figure 7.1a shows the traditional Miller method of creating a dominant pole. The collector pole of TR4 is lowered by adding the external Miller-capacitance *C*dom to that which unavoidably exists as the internal *C*bc of the VAS transistor. However, there are some other beneficial effects; *C*dom causes *pole-splitting*, in which the pole at TR2 collector is pushed up in frequency as P1 is moved down – most desirable for stability. Simultaneously the local NFB through *C*dom linearises the VAS.

Assuming that input-stage transconductance is set to a plausible 5 mA/V, and stability considerations set the maximal 20 kHz open-loop gain to 50 dB, then from Equations 3.1–3.3 on pages 63 and 64, *C*dom must be 125 pF. This is more than enough to swamp the internal capacitances of the VAS transistor, and is a practical real-life value.

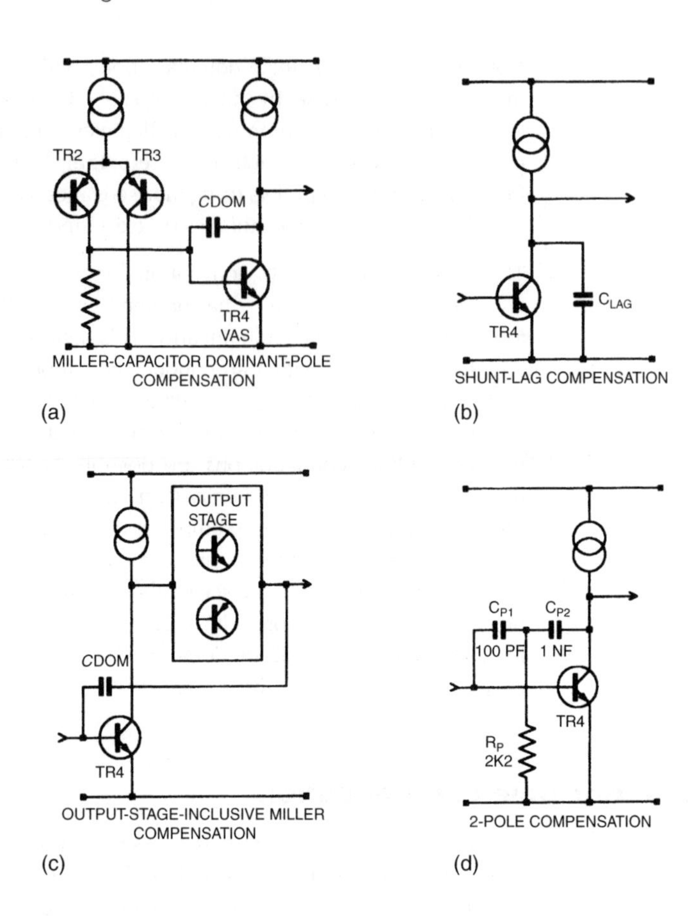

Figure 7.1 (a) The traditional Miller method of making a dominant pole. (b) Shunt compensation shows a much less satisfactory method – the addition of capacitance to ground from the VAS collector. (c) Inclusive Miller compensation. (d) Two-pole compensation

The peak current that flows in and out of this capacitor for an output of 20 V rms at 20 kHz, is 447 μA. Since the input stage must sink *C*dom current while the VAS collector-load sources it, and likewise the input stage must source it while the VAS sinks it, there are four possible ways in which slewrate may be limited by inadequate current capacity; if the input stage is properly designed then the usual limiting factor is VAS current-sourcing. In this example a peak current of less than 0.5 mA should be easy to deal with, and the maximum frequency for unslewed output will be comfortably above 20 kHz.

Lag compensation

Figure 7.1b shows a much less satisfactory method of compensation – the addition of capacitance to ground from the VAS collector. This is usually called shunt or lag compensation, and as Peter Baxandall[2] aptly put it, '*The technique is in all respects sub-optimal*'. We have already seen on page 104 that loading the VAS collector resistively to ground is a very poor option for reducing LF open-loop gain, and a similar argument shows that capacitative loading to ground for compensation purposes is an even worse idea. To reduce open-loop gain at 20 kHz to 50 dB as before, the shunt capacitor *C*lag must be 43.6 nF, which is a whole different order of things from 125 pF. The current in and out of *C*lag at 20 V rms, 20 kHz, is 155 mA peak, which is going to require some serious electronics to provide it. This important result is yielded by simple calculation, confirmed by Spice simulation. The input stage no longer constrains the slew-rate limits, which now depend entirely on the VAS.

A VAS working under these conditions will have poor linearity. The *I*c variations in the VAS, caused by the heavy extra loading, produce more distortion and there is no local NFB through a Miller capacitor to correct it. To make matters worse, the dominant pole P1 will probably need to be set to a lower frequency than for the Miller case, to maintain the same stability margins, as there is now no pole-splitting action to increase the frequency of the pole at the input-stage collector. Hence *C*lag may have to be even larger than 43 nF, requiring even higher peak currents.

Takahashi[3] has produced a fascinating paper on this approach, showing one way of generating the enormous compensation currents required for good slew-rates. The only thing missing is an explanation of why shunt compensation was chosen in the first place.

Including the output stage: inclusive Miller compensation

Miller-capacitor compensation elegantly solves several problems at once, and the decision to adopt it is simple. However the question of whether to include the output stage in the Miller feedback loop is less easy. Such inclusion (see Figure 7.1c) presents the alluring possibility that local feed-back could linearise both the VAS and the output stage, with just the input stage left out in the cold as frequency rises and global NFB falls. This idea is most attractive as it would greatly increase the total feedback available to linearise a distortive Class-B output stage.

There is certainly some truth in this, as I have shown[4], where applying *C*dom around the output as well as the VAS reduced the peak (not rms) 1 kHz THD from 0.05% to 0.02%. However I must say that the output stage was deliberately under-biased to induce crossover spikes, because with optimal bias the improvement, although real, was too small to be either convincing or worthwhile. A vital point is that this demonstration used a model amplifier with TO-92 *output* transistors, because in my experience the technique just does not work well with real power bipolars, tending to intractable HF oscillation. There is evidence that inclusive compensa-tion, when it can be made stable, is much less effective at dealing with ordinary crossover distortion than with the spikes produced by deliberate under-biasing.

The use of local NFB to linearise the VAS demands a tight loop with minimal extra phase-shift beyond that inherent in the Cdom dominant pole. It is permissible to insert a cascode or a small-signal emitter-follower into this local loop, but a sluggish output stage seems to be pushing luck too far; the output stage poles are now included in the loop, which loses its dependable HF stability. Bob Widlar[5] stated that output stage behaviour must be well-controlled up to 100 MHz for the technique to be reliable; this would appear to be virtually impossible for discrete power stages with varying loads.

While I have so far not found *Inclusive Miller-compensation* to be useful myself, others may know different; if anyone can shed further light I would be most interested.

Nested feedback loops

Nested feedback is a way to apply more NFB around the output stage without increasing the global feedback factor. The output has an extra voltage gain stage bolted on, and a local feedback loop is closed around these two stages. This NFB around the composite output bloc reduces output stage distortion and increase frequency response, to make it safe to include in the global NFB loop.

Suppose that bloc A1 (Figure 7.2a) is a Distortionless small-signal amplifier providing all the open-loop gain and so including the dominant pole. A3 is a unity-gain output stage with its own main pole at 1 MHz and distortion of 1% under given conditions; this 1 MHz pole puts a firm limit on the amount of global NFB that can be safely applied. Figure 7.2b shows a nested-feedback version; an extra gain-bloc A2 has been added, with local feedback around the output stage. A2 has the modest gain of 20 dB so there is a good chance of stability when this loop is closed to bring the gain of A3 + A2 back to unity. A2 now experiences 20 dB of NFB, bringing the distortion down to 0.1%, and raising the main pole to 10 MHz, which should allow the application of 20 dB more global NFB around the overall loop that includes A1. We have thus decreased the distortion that exists before global NFB is applied, and simultaneously increased the amount of NFB that can be safely used, promising that the final linearity could be very good indeed. For another theoretical example see Pernici et al.[6]

Real-life examples of this technique in power amps are not easy to find, but it is widely used in op-amps. Many of us were long puzzled by the way that the much-loved 5534 maintained such low THD up to high frequencies. Contemplation of its enigmatic entrails appears to reveal a three-gain-stage design with an inner Miller loop around the third stage, and an outer

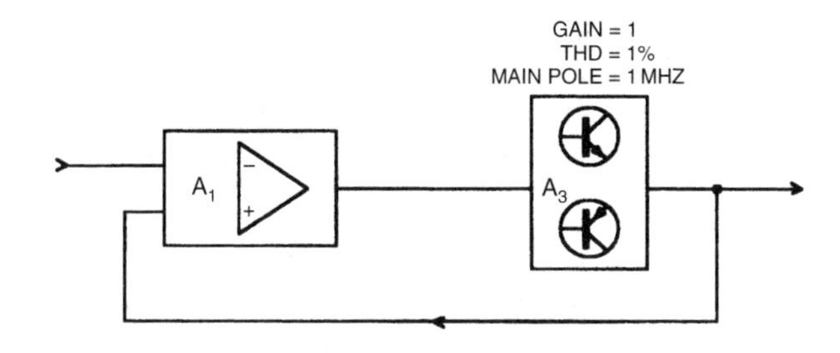

Figure 7.2a
Normal single-loop global negative feedback

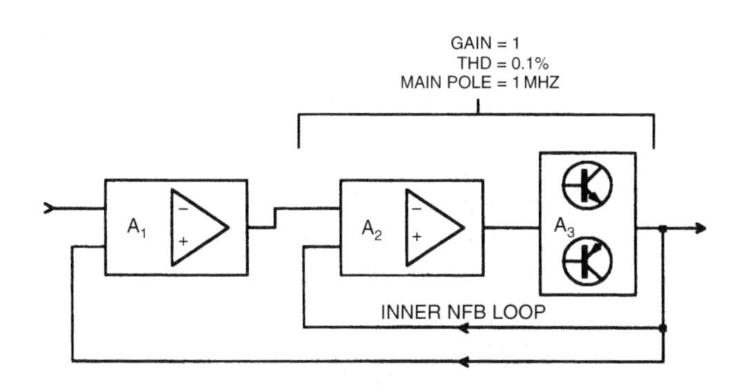

Figure 7.2b
Nested feedback

Miller loop around the second and third stages; global NFB is then applied externally around the whole lot. Nested Miller compensation has reached its apotheosis in CMOS op-amps – the present record appears[7] to be three nested Miller loops plus the global NFB; do not try this one at home. More details on the theory of nested feedback can be found in Scott and Spears[8].

Two-pole compensation

Two-pole compensation is well-known as a technique for squeezing the best performance from an op-amp[9],[10], but it has rarely been applied to power amplifiers; the only example I know is found in Widlar[5]. An extra HF time constant is inserted in the Cdom path, giving an open-loop gain curve that initially falls at almost 12 dB/octave, but which gradually reverts to 6 dB/octave as frequency continues to increase. This reversion is arranged to happen well before the unity loop-gain line is reached, and so stability should be the same as for the conventional dominant-pole scheme, but with increased negative feedback over part of the operational frequency range. The faster gain roll-off means that the maximum amount of feedback can be maintained up to a higher frequency. There is no measurable mid-band peak in the closed-loop response.

It is right to feel nervous about any manoeuvre that increases the NFB factor; power amplifiers face varying conditions and it is difficult to be sure that a design will always be stable under all circumstances. This makes designers rather conservative about compensation, and I approached this technique with some trepidation. However, results were excellent with no obvious reduction in stability. See Figure 7.4 for the happy result of applying this technique to the Class-B amplifier seen in Figure 7.5.

The simplest way to implement two-pole compensation is shown in Figure 7.1d, with typical values. Cp1 should have the same value as it would for stable single-pole compensation, and Cp2 should be at least twice as big; *R*p is usually in the region 1 k–10 k. At intermediate frequencies Cp2 has an impedance comparable with *R*p, and the resulting extra time-constant causes the local feedback around the VAS to increase more rapidly with frequency, reducing the open-loop gain at almost 12 dB/octave. At HF the impedance of *R*p is high compared with Cp2, the gain slope asymptotes back to 6 dB/octave, and then operation is the same as conventional dominant-pole, with *C*dom equal to the series capacitance combination. So long as the slope returns to 6 dB/octave before the unity loop-gain crossing occurs, there seems no obvious reason why the Nyquist stability should be impaired. Figure 7.3 shows a simulated two-pole open-loop gain plot for realistic component values; Cp2 should be at least twice Cp1 so the gain falls back to the 6 dB/octave line before the unity loop-gain line is crossed. The potential feedback factor has been increased by more than 20 dB from 3 kHz to 30 kHz, a region where THD tends to increase due to falling NFB. The open-loop gain peak at 8 kHz

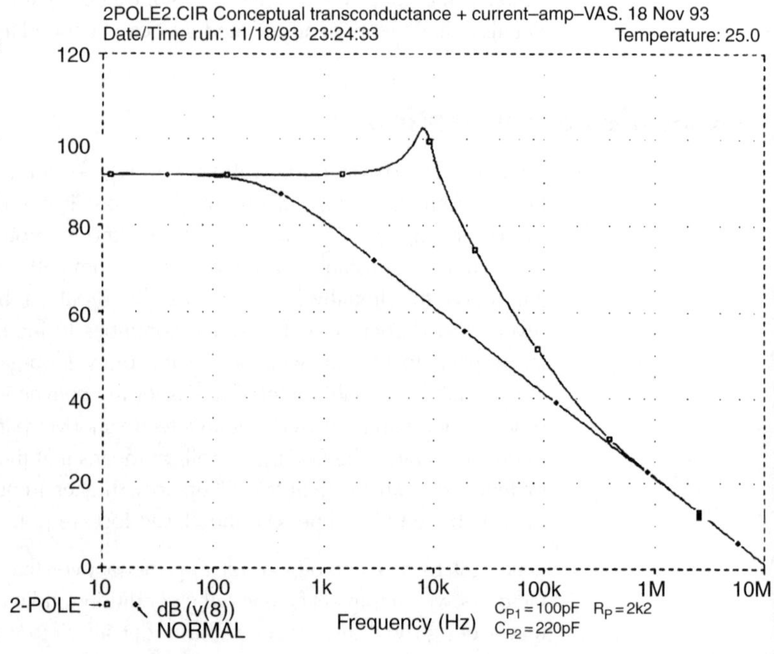

Figure 7.3
The open-loop gain plot for two-pole compensation with realistic component values

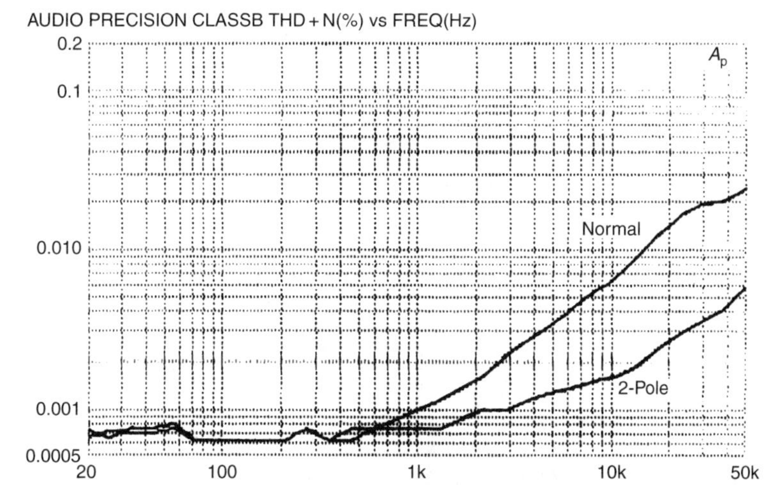

Figure 7.4
Distortion reduction with two-pole compensation

looks extremely dubious, but I have so far failed to detect any resulting ill-effects in the closed-loop behaviour.

There is however a snag to the approach shown here, which reduces the linearity improvement. Two-pole compensation may decrease open-loop linearity at the same time as it raises the feedback factor that strives to correct it. At HF, Cp2 has low impedance and allows Rp to directly load the VAS collector to ground; as we have seen, this worsens VAS linearity.

However, if Cp2 and Rp are correctly proportioned the overall reduction in distortion is dramatic and extremely valuable. When two-pole compensation was added to the amplifier circuit shown in Figure 7.5, the crossover glitches on the THD residual almost disappeared, being partially replaced by low-level second harmonic which almost certainly results from VAS loading. The positive slew-rate will also be slightly reduced.

This looks like an attractive technique, as it can be simply applied to an existing design by adding two inexpensive components. If Cp2 is much larger than Cp1, then adding/removing Rp allows instant comparison between the two kinds of compensation. Be warned that if an amplifier is prone to HF parasitics then this kind of compensation may worsen them.

Output networks

The usual output networks for a power amplifier are shown in Figure 7.6, with typical values. They comprise a shunt Zobel network, for stability into inductive loads, and a series output inductor/damping resistor for stability into capacitive loads.

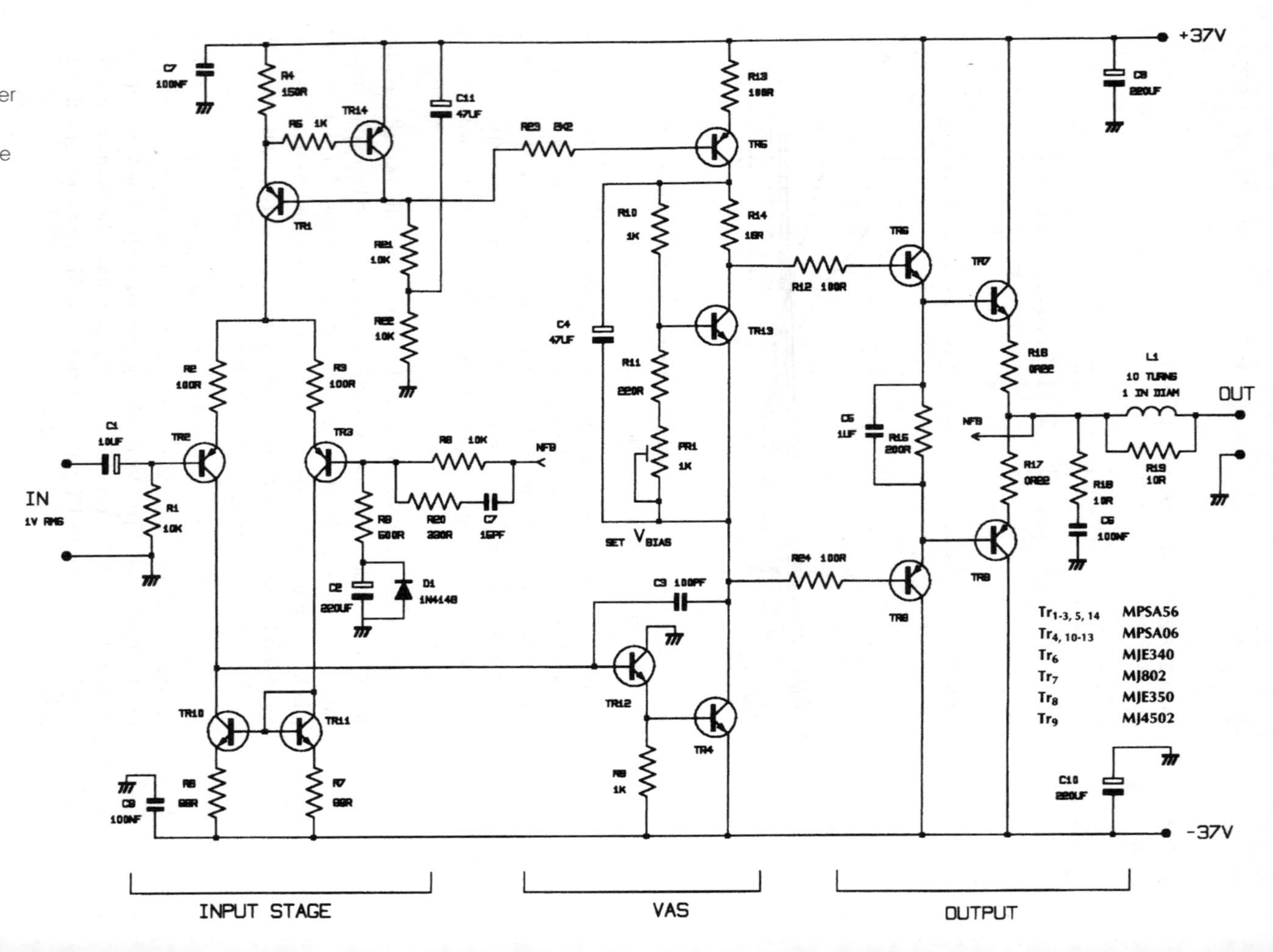

Figure 7.5
The Class-B amplifier from Chapter 6. At the simplest level the maximum slew-rate is defined by the current source TR1 and the value of C_{dom}

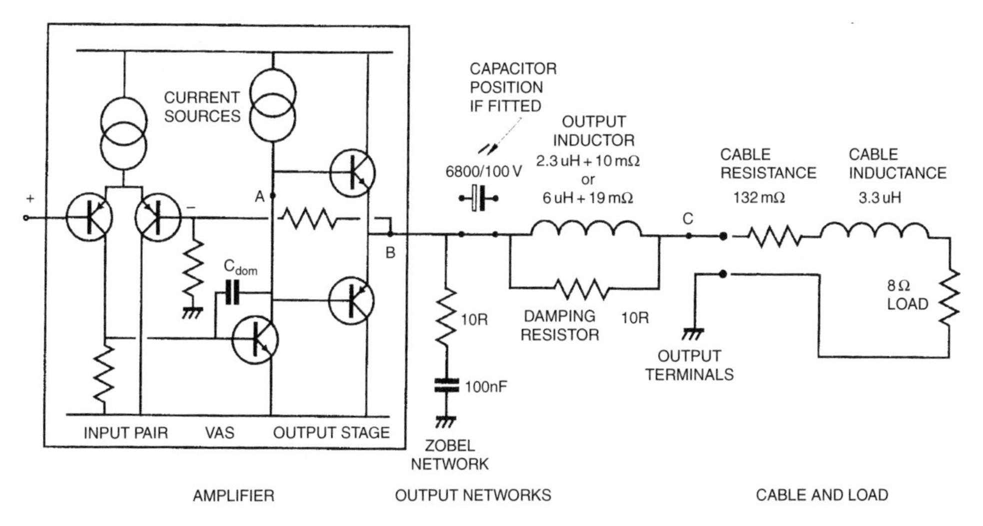

Figure 7.6
The amplifier-cable-speaker system. Simplified amplifier with Zobel network and damped output inductor, and a resistive load. Cable resistance and inductance values are typical for a 5 m length

Amplifier output impedance

The main effect of output impedance is usually thought to be its effect on Damping Factor. This is wrong, as explained in Chapter 1. Despite this demonstration of its irrelevance, I will refer to Damping Factor here, to show how an apparently impressive figure dwindles as more parts of the speaker-cable system are included.

Figure 7.6 shows a simplified amplifier with Zobel network and series output inductor, plus simple models of the connecting cable and speaker load. The output impedance of a solid-state amplifier is very low if even a modest amount of global NFB is used. I measured a Blameless Class-B amplifier similar to Figure 7.5 with the usual NFB factor of 29 dB at 20 kHz, increasing at 6 dB/octave as frequency falls. Figure 7.7 shows the output impedance at point B before the output inductor, measured by injecting a 10 mA signal current into the output via a 600 Ω resistance.

The low-frequency output impedance is approximately 9 mΩ (an 8 Ω Damping Factor of 890). To put this into perspective, one metre of thick 32/02 equipment cable (32 strands of 0.2 mm diameter) has a resistance of 16.9 mΩ. The internal cabling resistance in an amplifier can equal or exceed the output impedance of the amplifier itself at LF.

Output impedance rises at 6 dB/octave above 3 kHz, as global NFB falls off, reaching 36 mΩ at 20 kHz. The 3 kHz break frequency does not correspond

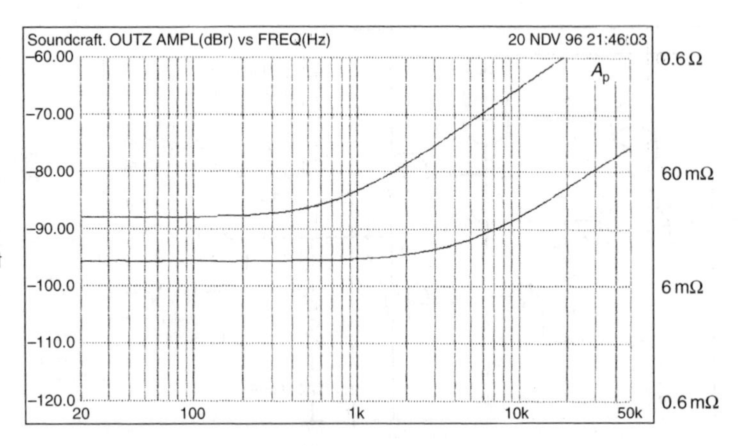

Figure 7.7 Output impedance of a Blameless amplifier, with and without 6 μH output inductor. Adding the inductor (upper trace) increases both the flat LF output impedance, due to its series resistance, and the rising HF impedance

with the amplifier dominant pole frequency, which is much lower at around 10 Hz.

The closed-loop output impedance of any amplifier is set by the open-loop output impedance and the negative feedback factor. The output impedance is not simply the output impedance of the output stage alone, because the latter is driven from the VAS, so there is a significant and frequency-varying source impedance at point A in Figure 7.6.

When the standard EF and CFP stages are *driven from a zero-impedance source*, in both cases the raw output impedance is in the region of 150–180 mΩ. This assumes the emitter resistors *R*e are 0.1 Ω. Increasing *R*e to 0.22 Ω increases output impedance to the range 230–280 mΩ, showing that these resistors in fact make up most of the output impedance. The output devices and drivers have little influence.

If the average open-loop output impedance is 200 mΩ, and the NFB factor at 20 kHz is 29 dB, or 28 times, we would expect a closed-loop output impedance of approximately 200/28, which is 7 mΩ. Since it is actually about 33 mΩ at this frequency, there is clearly more going on than simple theory implies. In a real amplifier the output stage is not driven from a zero impedance, but a fairly high one that falls proportionally with frequency; for my Blameless Class-B design it falls from 3 kΩ at 1 kHz to about 220 Ω at 20 kHz. A 220 Ω source impedance produces an open-loop output impedance of about 1 Ω, which when reduced by a factor of 28 when global feedback is applied, gives 35 mΩ. This is close to the value measured at 20 kHz at point B in Figure 7.6.

All of these measured closed-loop output impedances are very low compared with the other impedances in the amp-cable-speaker system. It would appear they can in most cases be ignored.

The Blameless amplifier design shown on page 179 has an output inductor of approximately 6 μH; the aim is absolutely guaranteed stability into all capacitative loads, and the inductance is therefore at the high end of the permissible range. This is limited by the HF roll-off into the lowest load resistance to be driven. This substantial component comprises 20 turns of 1.5 mm dia. copper wire, wound to 1 in. dia., and has a DC resistance of 19 mΩ. This small extra resistance raises the flat section of the impedance plot to 24 mΩ, and in fact dominates the LF output impedance as measured at the amplifier terminals (point C). It also sharply reduces the notional Damping Factor from 890 to 330.

Naturally the inductance of the coil pushes the rising portion of the impedance curve higher. The output impedance now starts to rise from 700 Hz, still at 6 dB per octave, reaching 0.6 Ω at 20 kHz. See Figure 7.7.

Minimising amplifier output impedance

This issue is worth considering, not because it optimises speaker dynamics, which it does not, but because it minimises frequency response variations due to varying speaker impedance. There is also, of course, specmanship to be considered.

It is clear from Figure 7.7 that the output impedance of a generic amplifier will very probably be less than the inductor resistance, so the latter should be attended to first. Determine the minimum output inductance for stability with capacitive loads, because lower inductance means fewer turns of wire and less resistance. Some guidance on this is given in the next section. Note, however, that the inductance of the usual single-layer coil varies with the square of the number of turns, so halving the inductance only reduces the turns, and hence the series resistance, by root-two. The coil wire must be as thick as the cost/quality tradeoffs allow.

It is also desirable to minimise the resistance of the amplifier internal wiring, and to carefully consider any extra resistance introduced by output relays, speaker switching, etc. When these factors have been reduced as far as cost and practicality allow, it is likely that the output impedance of the actual amplifier will still be the smallest component of the total.

Zobel networks

All power amplifiers except for the most rudimentary kinds include a Zobel network in their arrangements for stability. This simple but somewhat enigmatic network comprises a resistor and capacitor in series from the amplifier output rail to ground. It is always fitted on the inside (i.e., upstream) of the output inductor, though a few designs have a second Zobel network after the output inductor; the thinking behind this latter approach is obscure. The resistor approximates to the expected load impedance, and is usually between 4.7 and 10 Ω. The capacitor is almost invariably 100 nF, and these convenient values and their constancy in the face of changing amplifier design might lead one to suppose that they are not critical; in fact experiment suggests that the real reason is that the traditional values are just about right.

The function of the Zobel network (sometimes also called a Boucherot cell) is rarely discussed, but is usually said to prevent too inductive a reactance being presented to the amplifier output by a loudspeaker voice-coil, the implication being that this could cause HF instability. It is intuitively easy to see why a capacitative load on an amplifier with a finite output resistance could cause HF instability by introducing extra lagging phase-shift into the global NFB loop, but it is less clear why an inductive load should be a problem; if a capacitive load reduces stability margins, then it seems reasonable that an inductive one would increase them.

At this point I felt some experiments were called for, and so I removed the standard 10 Ω/0.1 μF Zobel from a Blameless Class-B amplifier with CFP output and the usual NFB factor of 32 dB at 20 kHz. With an 8 Ω resistive load the THD performance and stability were unchanged. However, when a 0.47 mH inductor was added in series, to roughly simulate a single-unit loudspeaker, there was evidence of local VHF instability in the output stage; there was certainly no Nyquist instability of the global NFB loop.

I also attempted to reduce the loading placed on the output by the Zobel network. However, increasing the series resistance to 22 Ω still gave some evidence of stability problems, and I was forced to the depressing conclusion that the standard values are just about right. In fact, with the standard 10 Ω/0.1 μF network the extra loading placed on the amplifier at HF is not great; for a 1 V output at 10 kHz the Zobel network draws 6.3 mA, rising to 12.4 mA at 20 kHz, compared with 125 mA drawn at all frequencies by an 8 Ω resistor. These currents can be simply scaled up for realistic output levels, and this allows the Zobel resistor power rating to be determined. Thus an amplifier capable of 20 V rms output must have a Zobel resistor capable of sustaining 248 mA rms at 20 kHz, dissipating 0.62 W; a 1 W component could be chosen.

In fact, the greatest stress is placed on the Zobel resistor by HF instability, as amplifier oscillation is often in the range 50–500 kHz. It should therefore be chosen to withstand this for at least a short time, as otherwise faultfinding becomes rather fraught; ratings in the range 3 to 5 W are usual.

To conclude this section, there seems no doubt that a Zobel network is required with any load that is even mildly inductive. The resistor can be of an ordinary wire-wound type, rated to 5 W or more; this should prevent its burn-out under HF instability. A wire-wound resistor may reduce the effectiveness of the Zobel at VHF, but seems to work well in practice; the Zobel still gives effective stabilisation with inductive loads.

Output inductors

Only in the simplest kinds of power amplifier is it usual for the output stage to be connected directly to the external load. Direct connection is generally only feasible for amplifiers with low feedback factors, which have large safety margins against Nyquist instability caused by reactive loads.

For many years designers have been wary of what may happen when a capacitive load is connected to their amplifiers; a fear that dates back to the introduction of the first practical electrostatic loudspeaker from Quad Acoustics, which was crudely emulated by adding a 2 μF capacitor in parallel to the usual 8 Ω resistive test load. The real load impedance presented by an electrostatic speaker is far more complex than this, largely as a result of the step-up transformer required to develop the appropriate drive voltages, but a 2 μF capacitor alone can cause instability in an amplifier unless precautions are taken.

When a shunt capacitor is placed across a resistive load in this way, and no output inductor is fitted, it is usually found that the value with the most destabilising effect is nearer 100 nF than 2 μF.

The most effective precaution against this form of instability is a small air-cored inductor in series with the amplifier output. This isolates the amplifier from the shunt capacitance, without causing significant losses at audio frequencies. The value is normally in the region 1–7 μH, the upper limit being set by the need to avoid significant HF roll-off into a 4 Ω load. If 2 Ω loads are contemplated then this limit must be halved.

It is usual to test amplifier transient response with a square-wave while the output is loaded with 8 Ω and 2 μF in parallel to simulate an electrostatic loudspeaker, as this is often regarded as the most demanding condition. However, there is an inductor in the amplifier output, and when there is significant capacitance in the load they resonate together, giving a peak in the frequency response at the HF end, and overshoot and ringing on fast edges.

This test therefore does not actually examine amplifier response at all, for the damped ringing that is almost universally seen during these capacitive loading tests is due to the output inductor resonating with the test load capacitance, and has nothing whatever to do with amplifier stability. The ringing is usually around 40 kHz or so, and this is much too slow to be blamed on any normally compensated amplifier. The output network adds ringing to the transient response even if the amplifier itself is perfect.

It is good practice to put a low-value damping resistor across the inductor; this reduces the Q of the output LC combination on capacitive loading, and thus reduces overshoot and ringing.

If a power amplifier is deliberately provoked by shorting out the output inductor and applying a capacitive load, then the oscillation is usually around 100–500 kHz, and can be destructive of the output transistors if allowed to persist. It is nothing like the neat ringing seen in typical capacitive load tests. In this case there is no such thing as *nicely damped ringing* because damped oscillation at 500 kHz probably means you are one bare step away from oscillatory disaster.

Attempts to test this on the circuit of Figure 7.5 were frustrated because it is actually rather resistant to capacitance-induced oscillation, probably because the level of global feedback is fairly modest. 100 nF directly across the output induced damped ringing at 420 kHz, while 470 nF gave ringing at 300 kHz, and 2 μF at 125 kHz.

While the 8 Ω/2 μF test described above actually reveals nothing about amplifier transient response, it is embedded in tradition, and it is too optimistic to expect its doubtful nature to be universally recognised. Minimising output ringing is of some commercial importance; several factors affect it, and can be manipulated to tidy up the overshoot and avoid deterring potential customers:

- The output inductance value. Increasing the inductance with all other components held constant reduces the overshoot and the amount of response peaking, but the peak moves downward in frequency so the rising response begins to invade the audio band. See Figures 7.8 and 7.9.
- The value of the damping resistor across the output coil. Reducing its value reduces the Q of the output LC tuned circuit, and so reduces overshoot and ringing. The resistor is usually 10 Ω, and can be a conventional wirewound type without problems due to self-inductance; 10 Ω reduces the overshoot from 58% without damping to 48%, and much reduces ringing. Response peaking is reduced with only a slight effect on frequency. See Figures 7.10 and 7.11. The damping resistor can in fact be reduced as to low as 1 Ω, providing the amplifier stability into capacitance remains dependable, and this reduces the transient overshoot further from 48% to 19%, and eliminates ringing altogether; there is just a single overshoot. Whether this is more visually appealing to the potential customer is an interesting point.
- The load capacitance value. Increasing this with the shunt resistor held at 8 Ω gives more overshoot and lower frequency ringing that decays more slowly. The response peaking is both sharper and lower in frequency, which is not a good combination. However, this component is part of the standard test load and is outside the designer's control. See Figures 7.12 and 7.13.

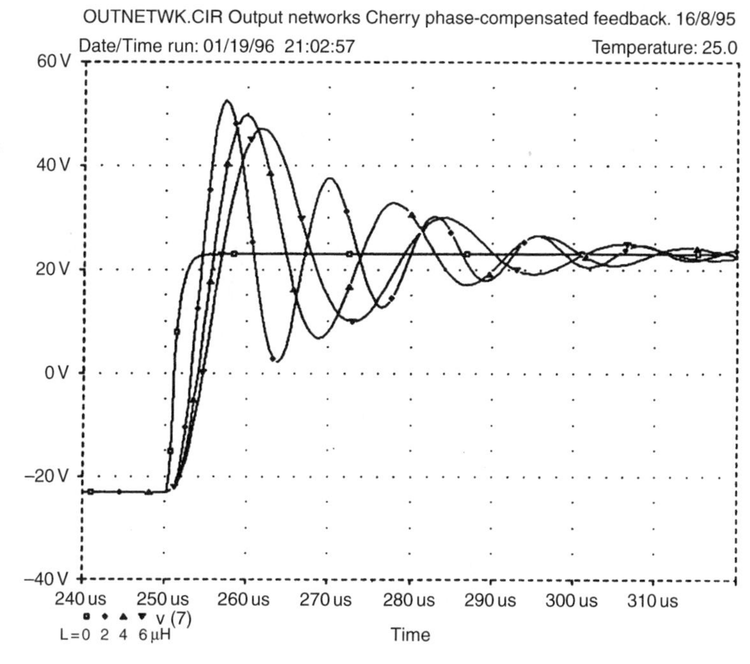

Figure 7.8
Transient response with varying output inductance; increasing L reduces ringing frequency, without much effect on overshoot. Input risetime 1 µsec

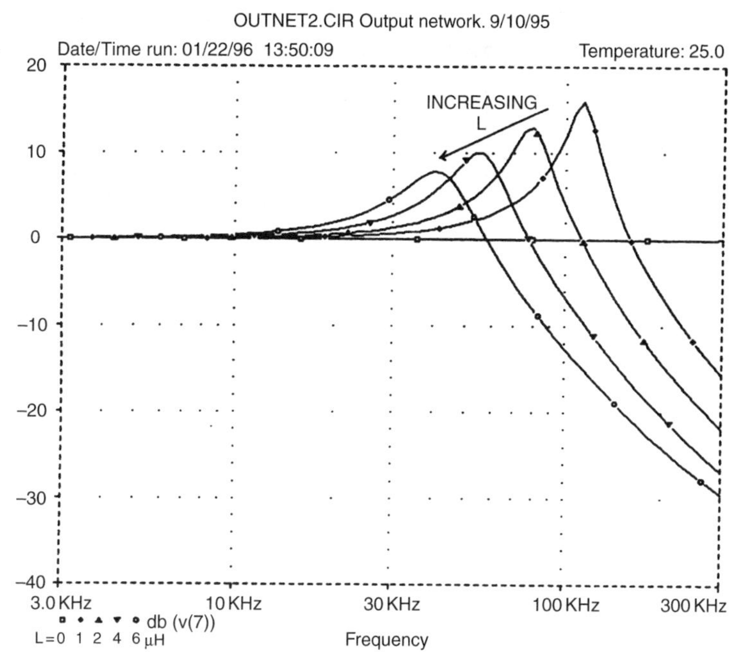

Figure 7.9
Increasing the output inductance reduces frequency response peaking and lowers its frequency

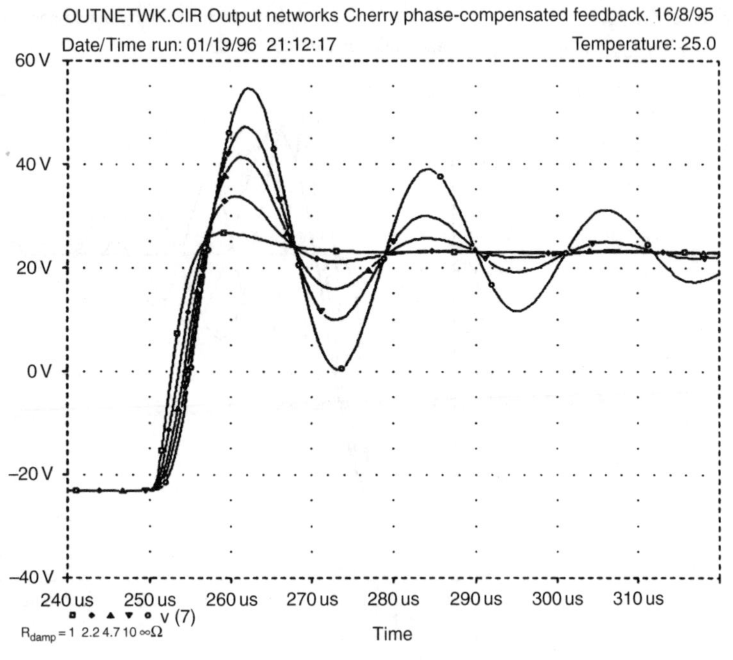

Figure 7.10
The effect of varying the damping resistance on transient response. 1 Ω almost eliminates overshoot

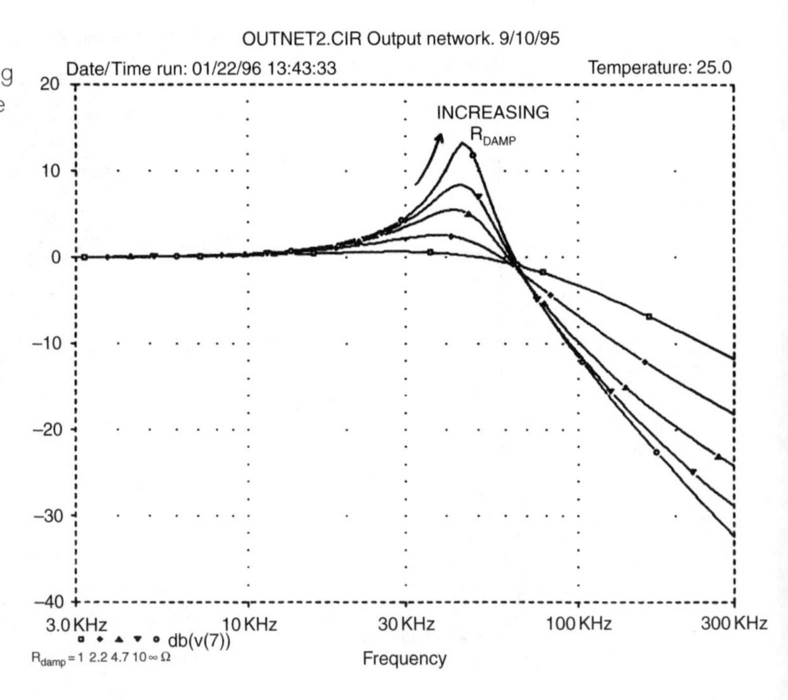

Figure 7.11
The effect of varying damping resistance on frequency response. Lower values reduce the peaking around 40 kHz

Figure 7.12
Increasing the load capacitance increases the transient overshoot, while lowering its frequency

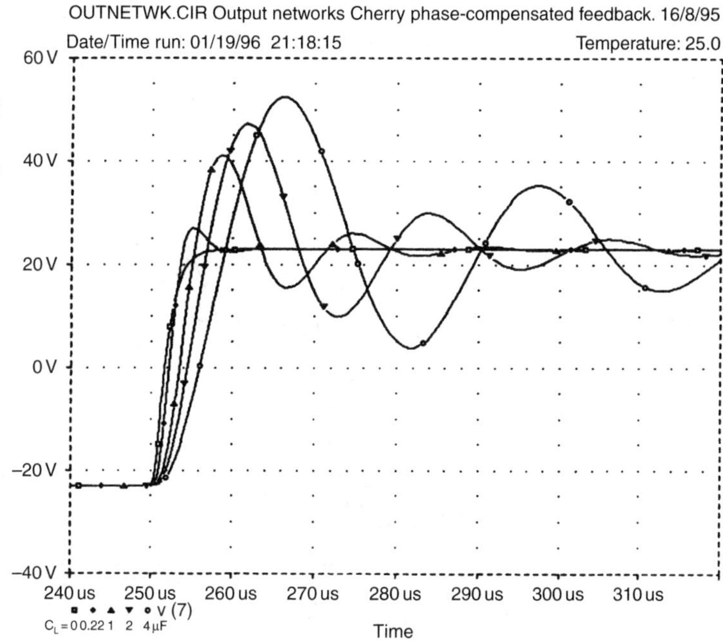

Figure 7.13
Increasing the load capacitance increases frequency response peaking and lowers its frequency

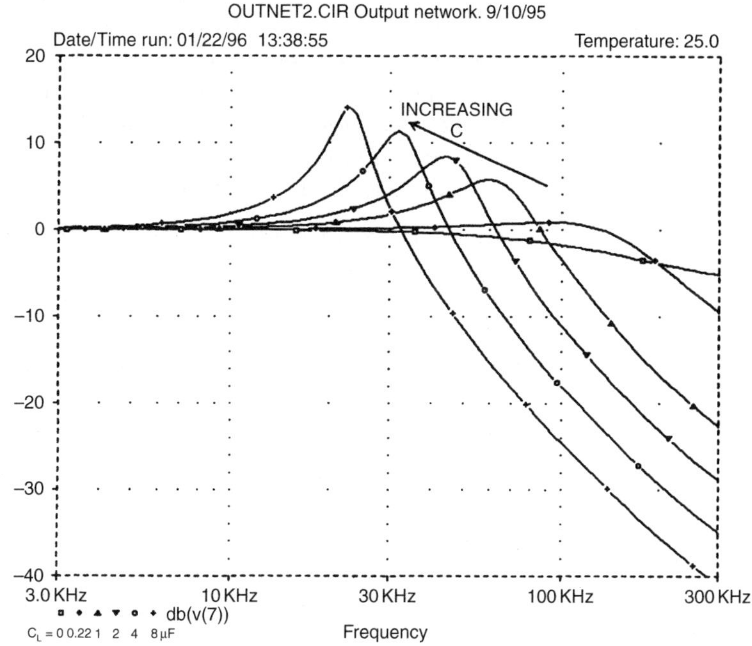

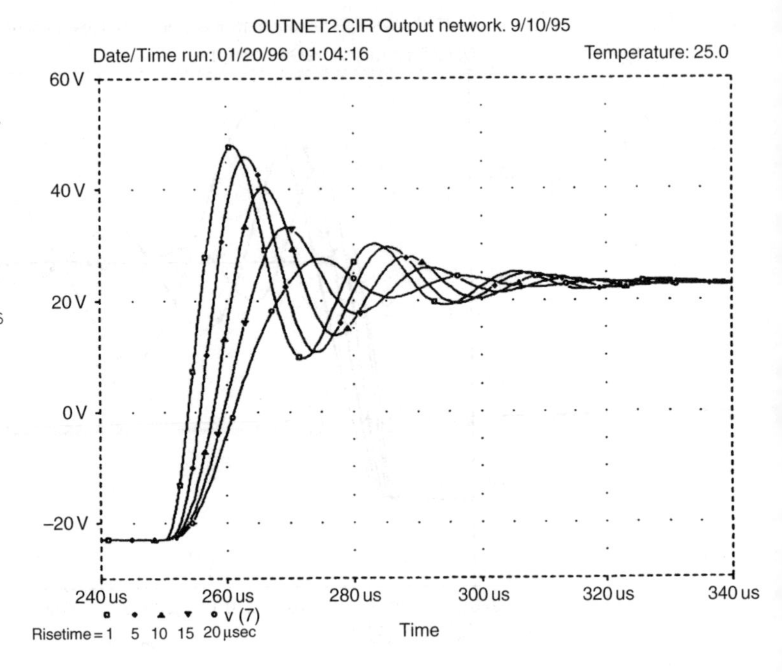

Figure 7.14
The most important factor in transient response is actually the rise-time of the square-wave input, especially for overshoot percentage. The ringing frequency is unaffected

- In actual fact, by far the most important factor affecting overshoot and ringing is the rise-time of the applied square wave. This is yet another rather important audio fact that seems to be almost unknown. Figure 7.14 shows how the overshoot given by the circuit in Figure 7.6 is 51% for a 1 µsec rise-time, but only 12% for a 20 µsec rise-time. It is clear that the *transient response* measured in this test may depend critically on the details of the testgear and the amplifier slew-rate, and can be manipulated to give the result you want.

An output inductor should be air-cored to eliminate the possibility of extra distortion due to the saturation of magnetic materials. Ferrite-based VHF chokes give stable operation, but their linearity must be considered dubious. In the 1970s there was a fashion for using one of the big power-supply electrolytics as a coil-former, but this is not a good idea. The magnetic characteristics of the capacitor are unknown, and its lifetime may be reduced by the heat dissipated in the coil winding resistance.

The resistance of an air-cored 7 µH coil made from 20 turns of 1.5 mm diameter wire (this is quite a substantial component 3 cm in diameter and 6 cm long) is enough to cause a measurable power loss into a 4 Ω load, and to dominate the output impedance as measured at the amplifier terminals. The coil wire should therefore be as thick as your cost/quality tradeoffs allow.

The power rating for the damping resistor is assessed as follows. For a resistive 8 Ω load the voltage across the output inductor increases slowly with frequency, and the damping resistor dissipation only reaches 1.2 mW at 20 kHz for 1 V rms output. This assumes a normal 10 Ω damping resistor; if the value is reduced to 1 Ω to eliminate ringing into capacitive loads, as described above, then the dissipation is ten times as great at 12 mW.

A much greater potential dissipation occurs when the load is the traditional 8 Ω/2 μF combination. The voltage across the output inductor peaks as it resonates with the load capacitance, and the power dissipated in a 10 Ω damping resistor at resonance is 0.6 W for 1 V rms. This is however at an ultrasonic frequency (around 50 kHz with a 7 μH inductor) and is a fairly sharp peak, so there is little chance of musical signals causing high dissipation in the resistor in normal use. However, as for the Zobel network, some allowance must be made for sinewave testing and oscillatory faults, so the damping resistor is commonly rated at between 1 and 5 W. An ordinary wirewound component works well with no apparent problems due to self-inductance.

The output inductor value

As mentioned above, the output inductor for all my designs started out at 20 turns and approximately 6 μH. In later tests the inductor was cut in half, now measuring 2.3 μH inductance and 10.1 mΩ DC resistance; this component was stable for all capacitor values, but has not had rigorous testing with real loudspeakers. It does now look more like an 'average' amplifier inductor, rather than an oversised one.

An alternative method of stabilisation is a series resistor instead of the inductor. Even with 100 nF loading, a OR1 wirewound output resistor completely removed ringing on the amplifier output. This is cheaper, but obviously less efficient than an inductor, as 100 mΩ of extra resistance has been introduced instead of 10 mΩ with the new 2.3 μH inductor. The Damping Factor with OR1 cannot exceed 80. A more important objection is that the 4 Ω output power appears to be significantly reduced – a 200 W/4 Ω amplifier is reduced to a 190 W unit, which does not look so good in the specs, even though the reduction in perceived loudness is negligible.

Cable effects

Looking at the amplifier-cable-load system as a whole, the amplifier and cable impedances have the following effects with an 8 Ω resistive load:

- A constant amplitude loss due to the cable resistance forming a potential divider with the 8 Ω load. The resistive component from the amplifier output is usually negligible.

- A high-frequency roll-off due to the cable inductance forming an LR lowpass filter with the 8 Ω load. The amplifier's output inductor (to give stability with capacitative loads) adds directly to this to make up the total series inductance. The shunt capacitance of any normal speaker cable is trivially small, and can have no significant effect on frequency response or anything else.

The main factors in speaker cable selection are therefore series resistance and inductance. If these parameters are below 100 mΩ and 3 μH, any effects will be imperceptible. This can be met by 13 A mains cable, especially if all three conductors are used.

If the amplifier is connected to a typical loudspeaker rather than a pure resistance the further effects are:

- The frequency response of the voltage at the loudspeaker terminals shows small humps and dips as the uneven speaker impedance loads the series combination of amplifier output impedance and cable resistance.
- The variable loading affects the amplifier distortion performance. HF crossover distortion reduces as load resistance increases above 8 Ω; even 68 Ω loading increases HF distortion above the unloaded condition. For heavier loading than 8 Ω, crossover may continue to increase, but this is usually masked by the onset of Large Signal Non-linearity[16].
- Severe dips in impedance may activate the overload protection circuitry unexpectedly. Signal amplitudes are higher at LF so impedance dips here are potentially more likely to draw enough current to trigger protection.

Crosstalk in amplifier output inductors

When designing a stereo power amplifier, the issue of interchannel crosstalk is always a concern. Now that amplifiers with up to seven channels for home theatre are becoming more common, the crosstalk issue is that much more important, if only because the channels are likely to be more closely packed. Here I deal with one aspect of it. Almost all power amplifiers have output coils to stabilise them against capacitative reactances, and a question often raised is whether inductive coupling between the two is likely to degrade crosstalk. It is sometimes suggested that the coils – which are usually in solenoid form, with length and diameter of the same order – should be mounted with their axes at right angles rather than parallel, to minimise coupling. But does this really work?

I think I am pretty safe in saying there is no published work on this, so it was time to make some. The coil coupling could no doubt be calculated (though not by me) but as often in the glorious pursuit of electronics, it was quicker to measure it.

The coils I used were both of 14 turn of 1 mm diameter copper wire, overall length 22 mm and diameter 20 mm. This has an inductance of about 2 μH, and is pretty much an 'average' output coil, suitable for stabilising amplifiers up to about 150 W/8 Ω. Different coils will give somewhat different results, but extrapolation to whatever component you are using should be straightforward; for example, twice the turns on both coils means four times the coupling.

Figure 7.15 shows the situation in a stereo power amplifier. The field radiated due to the current in Coil A is picked up by Coil B and a crosstalk voltage added to the output signal at B.

Figure 7.16 shows the experimental setup. Coil A is driven from a signal generator with a source impedance of 50 Ω, set to 5 V rms. Virtually all of this is dropped across the source resistance, so Coil A is effectively driven with a constant current of 100 mA rms.

Figure 7.17 shows the first result, taken with the coils coaxial and the ends touching. (This proved, as expected, to be the worst case for coupling.) The crosstalk rises at 6 dB/octave, because the voltage induced in Coil B is proportional to the rate of change of flux, and the magnitude of peak flux is fixed. This is clearly not the same as conventional transformer action, where

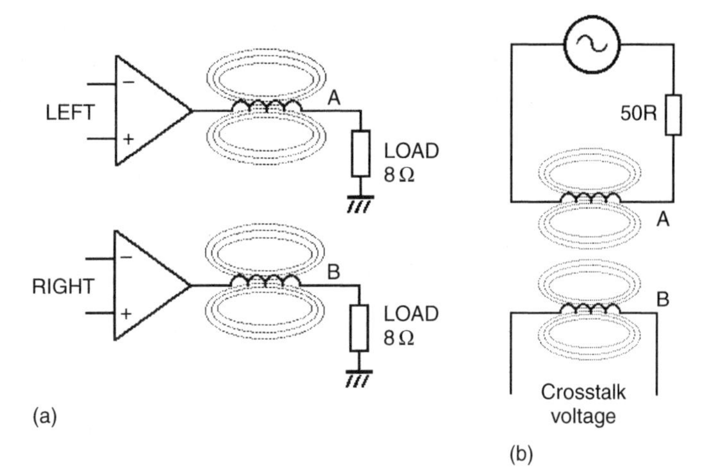

Figure 7.15
(a) The coupling of output coils in a stereo power amplifier. (b) The experimental circuit. The 'transmitting' Coil A is driven with an effectively constant current, and the voltage across the 'receiving' Coil B measured

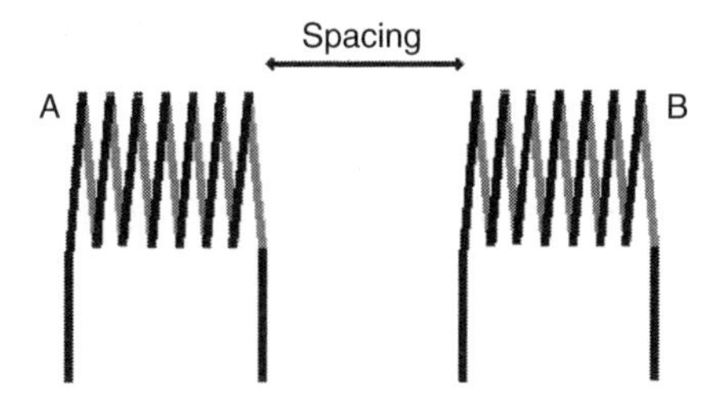

Figure 7.16
The physical coil configuration for the measurement of coaxial coils

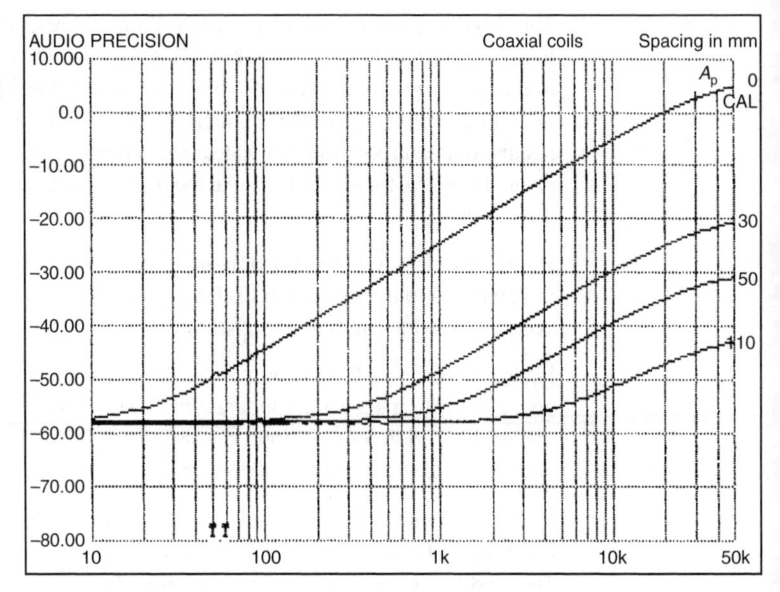

Figure 7.17 Crosstalk versus spacing for coaxial coils

the frequency response is flat. In a transformer the primary inductance is much greater than the circuit series impedance, so the magnetic flux that couples with the secondary halves when the input frequency doubles, and the voltage induced in the secondary is constant. The crosstalk at 20 kHz was taken as the 0 dB reference. This represented 2.4 mV rms across Coil B. 100 mA rms in Coil A corresponds to 800 mV rms across an 8 Ω load, so this gives a final crosstalk figure from channel to channel of −54 dB at 20 kHz. It carries on deteriorating above 20 kHz but no one can hear it. All crosstalk figures given below are at 20 kHz.

The coils were then separated 10 mm at a time, and with each increment the crosstalk dropped by 10 dB, as seen in Figure 7.17. At 110 mm spacing, which is quite practical for most designs, the crosstalk had fallen by 47 dB from the reference case, giving an overall crosstalk of 54 + 47 = 101 dB total. This is a very low level, and at the very top of the audio band. At 1 kHz, where the ear is much more sensitive, the crosstalk will be some 25 dB less, which brings it down to −126 dB total which I can say with some confidence is not going to be a problem. This is obtained with what looks like the least favourable orientation of coils. Coil–coil coupling is −32 dB at 50 mm, and the figure at this spacing will be used to compare the configurations.

The next configuration tested was that of Figure 7.18, where the coils have parallel axes but are displaced to the side. The results are in Figure 7.19; the crosstalk is now −38 dB at 50 mm. With each 10 mm spacing increment the crosstalk dropped by 7 dB. This setup is worse than the crossed-axis version but better than the coaxial one.

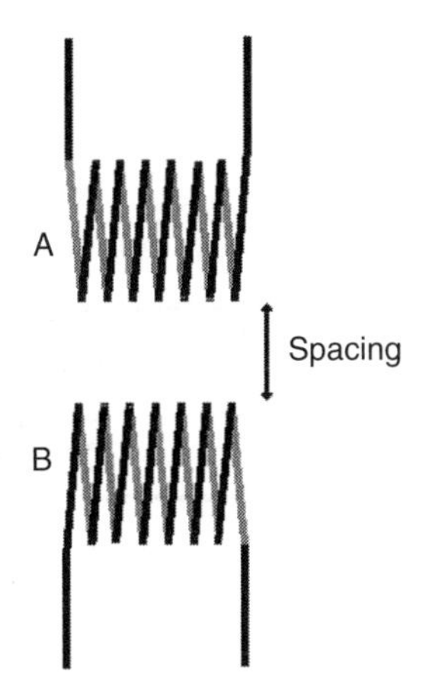

Figure 7.18
The coil configuration for non-coaxial parallel-axis coils

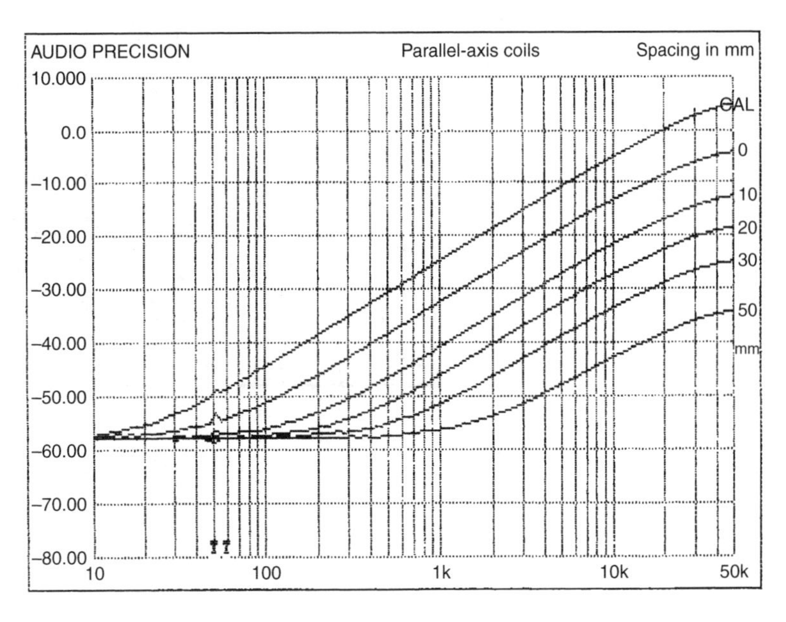

Figure 7.19
Crosstalk versus spacing for parallel-axis coils

The final configurations had the axes of the coils at 90°; the crossed-axis condition. The base position is with the corners of the coils touching; see Figure 7.20. When the coil is in the position X, still touching, crosstalk almost vanishes as there is a cancellation null. With the coils so close, this is a very sharp null and exploiting it in quantity production is quite impractical. The slightest deformation of either coil ruins the effect. Moving the Coil A away from B again gives the results in Figure 7.21. The crosstalk is now −43 dB at 50 mm, only an improvement of 11 dB over the coaxial case; turning coils around is clearly not as effective as might be supposed. This time, with each 10 mm spacing increment the crosstalk dropped by 8 dB rather than 10 dB.

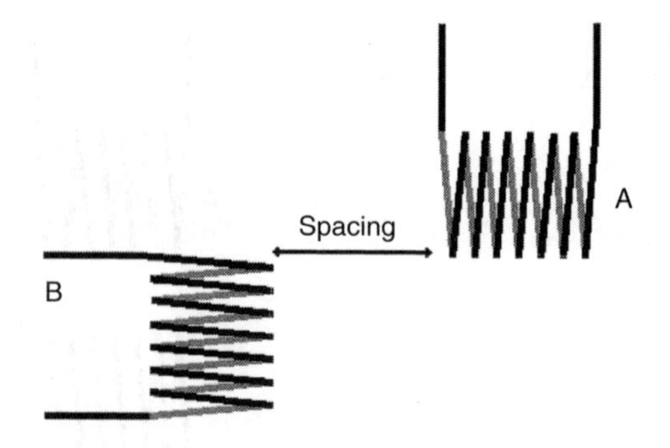

Figure 7.20
The coil configuration for crossed-axis measurements

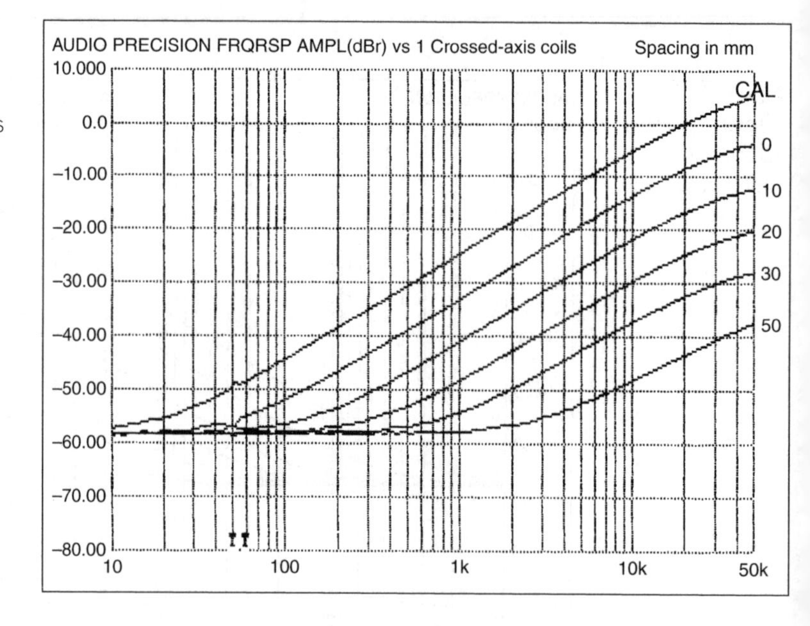

Figure 7.21
Crosstalk versus spacing for crossed-axis coils

The obvious next step is to try combining distance with cancellation as in Figure 7.22. This can give a good performance even if a large spacing is not possible. Figure 7.23 shows that careful coil positioning can give crosstalk better than −60 dB (−114 dB total) across the audio band, although the spacing is only 20 mm. The other curves show the degradation of performance when the coil is misaligned by moving it bodily sideways by 1, 2, 3 and 4 mm; just a 2 mm error has worsened crosstalk by 20 dB at 20 kHz. Obviously in practice the coil PCB hole will not move – but it is very possible that coils will be bent slightly sideways in production.

Figure 7.24 gives the same results for a 50 mm spacing, which can usually be managed in a stereo design. The null position once more just gives the

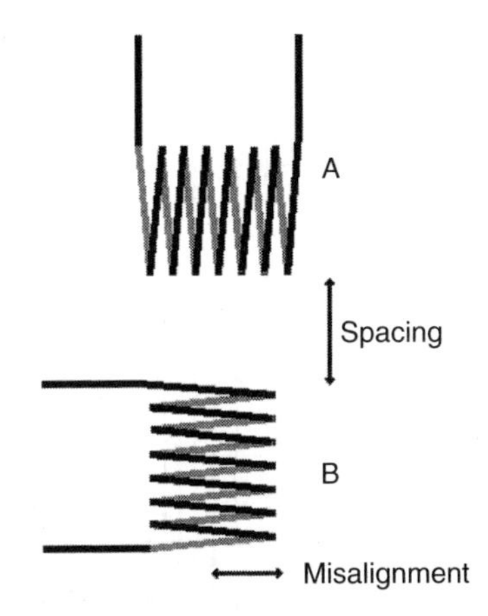

Figure 7.22
The coil configuration for crossed-axis with cancellation

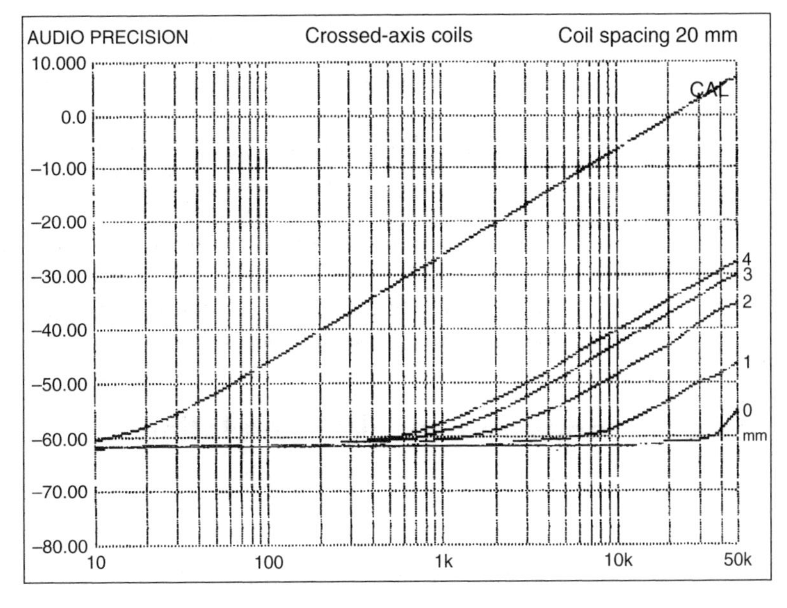

Figure 7.23
Crosstalk versus alignment for crossed-axis coils spaced at 20 mm, using cancellation

noise floor across the band, and a 2 mm misalignment now only worsens things by about 5 dB. This is definitely the best arrangement if the spacing is limited.

Conclusions

Coil orientation can help. Simply turning one coil through 90° gives an improvement of only 11 dB, but if it is aligned to cancel out the coupling, there is a big improvement. See how −38 dB in Figure 7.19 becomes −61 dB in Figure 7.24 at 20 kHz. On a typical stereo amplifier PCB, the coils are likely to be parallel – probably just for the sake of appearance – but their

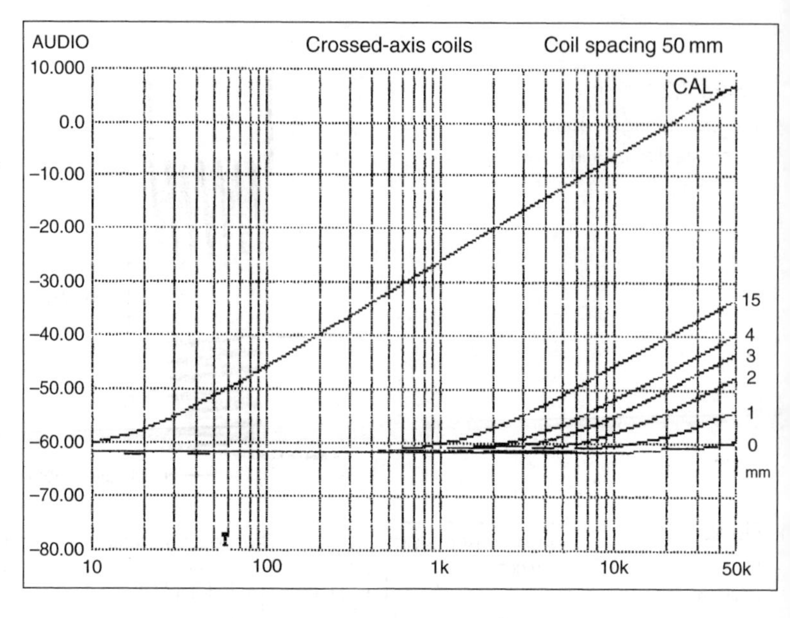

Figure 7.24 Crosstalk versus alignment for crossed-axis coils spaced at 50 mm, using cancellation

spacing is unlikely to be less than 50 mm unless the output components have been deliberately grouped together. As with capacitative crosstalk, physical distance is cheaper than anything else, and if the results are not good enough, use more of it. In this case the overall crosstalk at 20 kHz will be $54+38=-92$ dB total, which is probably already well below other forms of interchannel crosstalk. A quick quarter-turn of the coil improves this to at least -114 dB. It should do.

Reactive loads and speaker simulation

Amplifiers are almost universally designed and tested running into a purely resistive load, although they actually spend their working lives driving loudspeakers, which contain both important reactive components and also electromechanical resonances. At first sight this is a nonsensical situation; however, testing into resistive loads is neither naive nor an attempt to avoid the issue of real loads; there is in fact little alternative.

Loudspeakers vary greatly in their design and construction, and this is reflected in variations in the impedance they present to the amplifier on test. It would be necessary to specify a *standard speaker* for the results from different amplifiers to be comparable. Second, loudspeakers have a notable tendency to turn electricity into sound, and the sinewave testing of a 200 W amplifier would be a demanding experience for all those in earshot; soundproof chambers are not easy or cheap to construct. Third, such a standard test speaker would have to be capable of enormous power-handling if it were to be able to sustain long-term testing at high power;

loudspeakers are always rated with the peak/average ratio of speech and music firmly in mind, and the lower signal levels at high frequencies are also exploited when choosing tweeter power ratings. A final objection is that loudspeakers are not noted for perfect linearity, especially at the LF end, and if the amplifier does not have a very low output impedance this speaker non-linearity may confuse the measurement of distortion. Amplifier testing would demand a completely different sort of loudspeaker from that used for actually listening to music; the market for it would be very, very small, so it would be expensive.

Resistive loads

Amplifiers are normally developed through 8 and 4 Ω testing, though intermediate values such as 5.66 Ω (the geometric mean of 8 and 4) are rarely explored considering how often they occur in real use. This is probably legitimate in that if an amplifier works well at 8 and 4 Ω it is most unlikely to give trouble at intermediate loadings. In practice few nominal 8 Ω speakers have impedance dips that go below 5 Ω, and design to 4 Ω gives a safety margin, if not a large one.

The most common elaboration on a simple resistive load is the addition of 2 μF in parallel with 8 Ω to roughly simulate an electrostatic loudspeaker; this is in fact not a particularly reactive load, for the impedance of a 2 μF capacitor only becomes equal to the resistance at 9.95 kHz, so most of the audio band is left undisturbed by phase shift. This load is in fact a worse approximation to a moving-coil speaker than is a pure resistance.

Modelling real loudspeaker loading

The impedance curve of a real loudspeaker may be complex, with multiple humps and dips representing various features of the speaker. The resonance in the bass driver unit will give a significant hump in LF impedance, with associated phase changes. Reflex (ported enclosure) designs have a characteristic double-hump in the LF, with the middle dip corresponding to the port tuning. The HF region is highly variable, and depends in a complicated fashion on the number of drive units, and their interactions with the crossover components.

Connection of an amplifier to a typical speaker impedance rather than a resistance has several consequences:

- The frequency response, measured in terms of the voltage across the loudspeaker terminals, shows small humps and bumps due to the uneven impedance loading the series combination of amplifier output impedance and connecting cable resistance.

- Severe dips in impedance may activate the overload protection circuitry prematurely. This has to be looked at in terms of probability, because a high amplitude in a narrow frequency band may not occur very often, and if it does it may be so brief that the distortion generated is not perceptible. Amplitudes are higher at LF and so impedance dips here are potentially more serious.
- The variable loading affects the distortion performance.

Figure 7.25 shows how the HF crossover distortion varies with load resistance for loads lighter than those usually considered. Even 68 Ω loading increases HF distortion.

Figure 7.26 shows an electrical model of a single full-range loudspeaker unit. While a single-driver design is unlikely to be encountered in hi-fi applications, many PA, disco and sound reinforcement applications use full-range drive units, for which this is a good model. Rc and Lc represent the resistance and inductance of the voicecoil. Lr and Cr model the electromechanical resonance of the cone mass with the suspension compliance

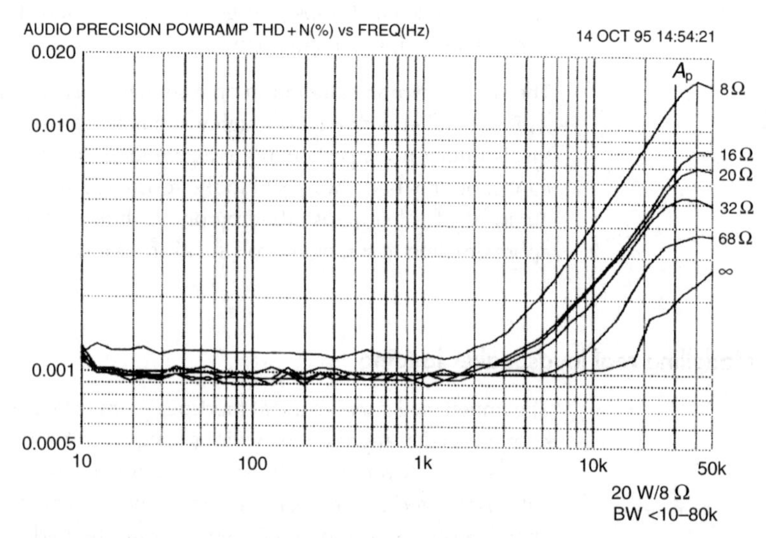

Figure 7.25
The reduction of HF THD as resistive amplifier loading is made lighter than 8 Ω

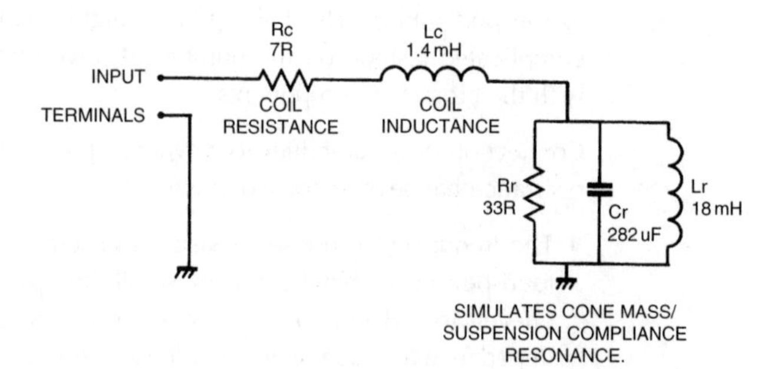

Figure 7.26
Electrical model of a single speaker unit in a sealed enclosure

and air-spring of the enclosure, with Rr setting the damping; these last three components have no physical existence, but give the same impedance characteristics as the real resonance.

The input impedance magnitude this network presents to an amplifier is shown in Figure 7.27. The peak at 70 Hz is due to the cone resonance; without the sealed enclosure, the restoring force on the cone would be less and the free-air resonance would be at a lower frequency. The rising impedance above 1 kHz is due to the voicecoil inductance Lc.

When the electrical model of a single-unit load replaces the standard 8 Ω resistive load, something remarkable happens; HF distortion virtually disappears, as shown in Figure 7.28. This is because a Blameless amplifier driving 8 Ω only exhibits crossover distortion, increasing with frequency as the NFB factor falls, and the magnitude of this depends on the current drawn from the output stage; with an inductive load this current falls at high frequencies.

Most hi-fi amplifiers will be driving two-way or three-way loudspeaker systems, and four-way designs are not unknown. This complicates the impedance characteristic, which in a typical two-way speaker looks something like Figure 7.29, though the rise above 10 kHz is often absent. The bass resonance remains at 70 Hz as before, but there are two drive units, and hence two resonances. There is also the considerable complication of a crossover network to direct the HF to the tweeter and the LF to the

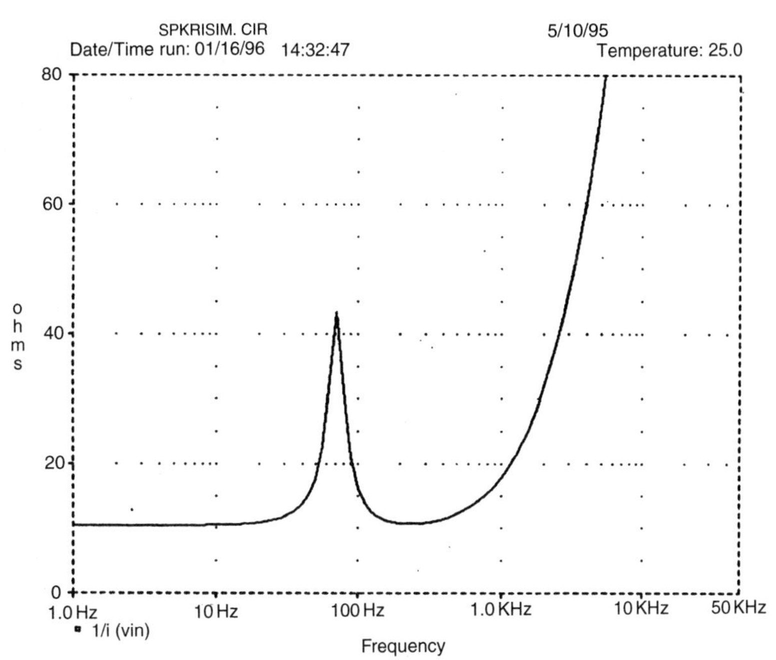

Figure 7.27
Input impedance of single speaker unit

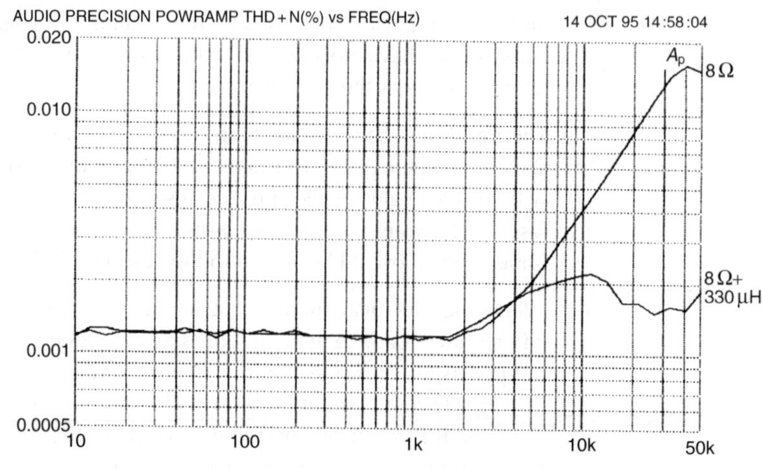

Figure 7.28 The reduction of HF THD with an inductive load; adding 330 μH in series with the 8 Ω reduces the 20 kHz THD by more than four times

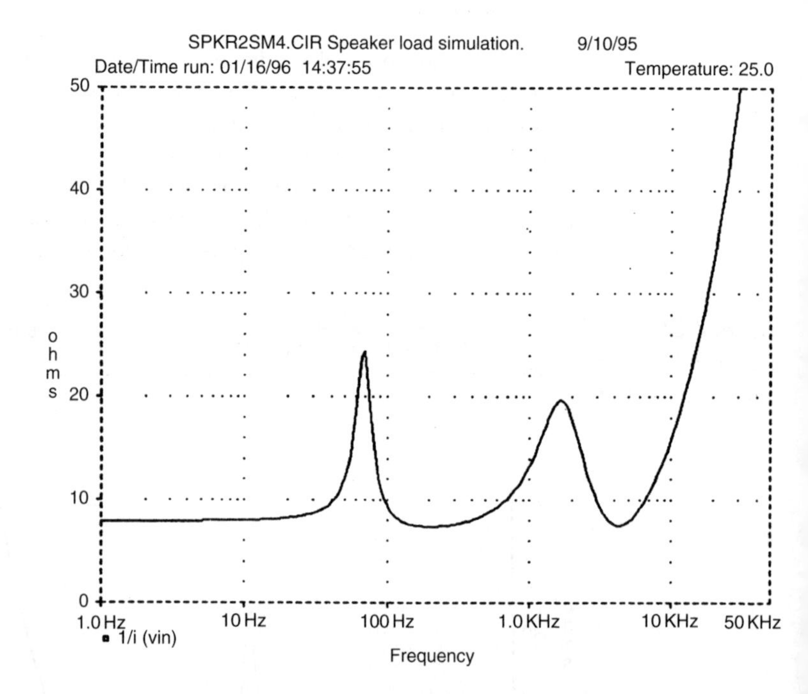

Figure 7.29 The circuit of the 2-way speaker model

low-frequency unit, and this adds several extra variables to the situation. In a bass reflex design the bass resonance hump may be supplemented by another LF resonant peak due to the port tuning. An attempt at a representative load simulator for a two-way infinite-baffle loudspeaker system is shown in Figure 7.30. This assumes a simple crossover network without compensation for rising tweeter coil impedance, and is partially based on a network proposed by Ken Kantnor in Atkinson[11].

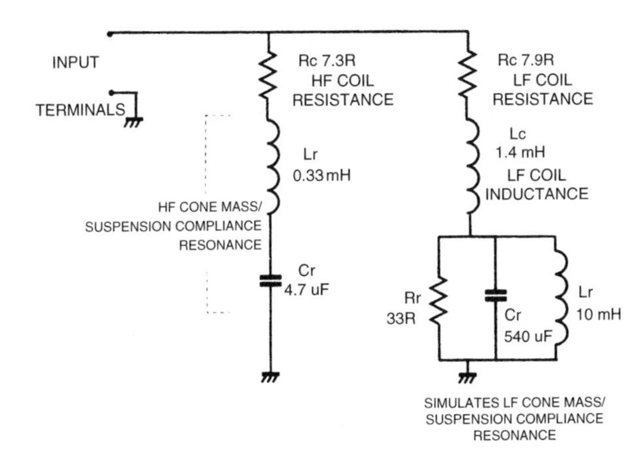

Figure 7.30
The circuit of the 2-way speaker model

Some loudspeaker crossover designs include their own Zobel networks, typically placed across the tweeter unit, to compensate for the HF rise in impedance due to the voicecoil inductance. If these Zobels are placed there to terminate the crossover circuitry in a roughly resistive load, then the loudspeaker designer has every right to do it; electroacoustic design is quite difficult enough without adding extra restrictions. However, if they are incorporated simply to make the impedance curve look tidier, and allow a claim that the load has been made easier for the amplifier to drive, then this seems misguided. The actual effect is the opposite; a typical amplifier has no difficulty driving an inductive reactance, and the HF crossover distortion can be greatly reduced when driving a load with an impedance that rises above the nominal value at HF.

This is only an introduction to the huge subject of real amplifier loads. More detailed information is given in Benjamin[12].

Loudspeaker loads and output stages

There is a common assumption that any reactive load is more difficult for an amplifier to drive than a purely resistive one; however, it is devoutly to be wished that people would say what they mean by 'difficult'. It could mean that stability margins are reduced, or that the stresses on the output devices are increased. Both problems can exist, but I suspect that this belief is rooted in anthropomorphic thinking. It is easy to assume that if a signal is more complex to contemplate, it is harder for an amplifier to handle. This is not, however, true; it is not necessary to understand the laws of physics to obey them. Everything does anyway.

When solid-state amplifiers show instability it is always at ultrasonic frequencies, assuming we are not grappling with some historical curiosity that has AC coupling in the forward signal path. It never occurs in the middle of the audio band although many loudspeakers have major convulsions in their impedance curves in this region. Reactive loading can and does imperil stability at high frequencies unless precautions are taken, usually in the form of an output inductor. It does not cause oscillation or ringing mid-band.

Reactive loads do increase output device stresses. In particular peak power dissipation is increased by the altered voltage/current phase relationships in a reactive load.

Single-speaker load

Considering a single speaker unit with the equivalent circuit of Figure 7.26, the impedance magnitude never falls below the 8 Ω nominal value, and is much greater in some regions; this suggests the overall amplifier power dissipation would be less than for an 8 Ω resistive load.

Unfortunately this is not so; the voltage/current phase relationship brought about by the reactive load is a critical factor. When a pure resistance is driven, the voltage across the output device falls as the current through it rises, and they never reach a maximum at the same time. See Figure 7.31, for Class-B with an 8 Ω resistive load. The instantaneous power is the product of instantaneous current and voltage drop, and in Class-B has a

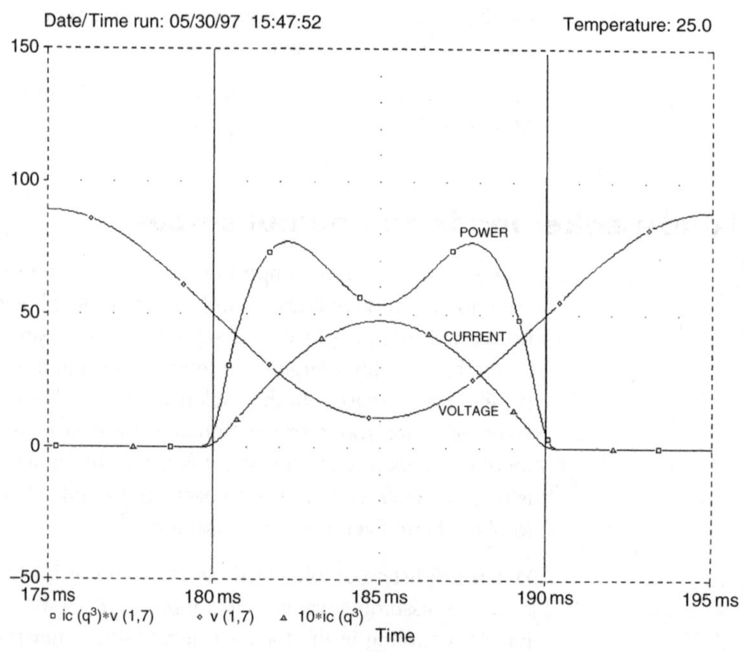

Figure 7.31 Instantaneous Vce, Ic, and Pdiss in an output transistor driving 8 Ω to 40 V peak at 50 Hz, from ±50 V rails. Device dissipation peaks twice at 77 watts in each half-cycle

characteristic two-horned shape, peaking twice at 77 W during its conducting half-cycle.

When the single-speaker load is driven at 50 Hz, the impedance is a mix of resistive and inductive, at $8.12 + 3.9\,j\Omega$. Therefore the current phase-lags the voltage, altering the instantaneous product of voltage and power to that shown in Figure 7.32. The average dissipation over the Class-B half-cycle is slightly reduced, but the peak instantaneous power increases by 30% due to the voltage/current phase shift. This could have serious results on amplifier reliability if not considered at the design stage. Note that this impedance is equivalent *at 50 Hz only* to 8.5 Ω in series with 10.8 mH. Trying to drive this replacement load at any other frequency, or with a non-sine waveform, would give completely wrong results. Not every writer on this topic appears to appreciate this.

Similarly, if the single-speaker load is driven at 200 Hz, on the other side of the resonance peak, the impedance is a combination of resistive and capacitative at $8.4 - 3.4\,j\Omega$ and the current leads the voltage. This gives much the same result as Figure 7.32, except that the peak power now occurs in the first part of the half-cycle. The equivalent load *at 200 Hz only* is 10.8 Ω in parallel with 35 μF.

When designing output stages, there are four electrical quantities to accommodate within the output device ratings; peak current, average current, peak power and average power. (Junction temperatures must of course also

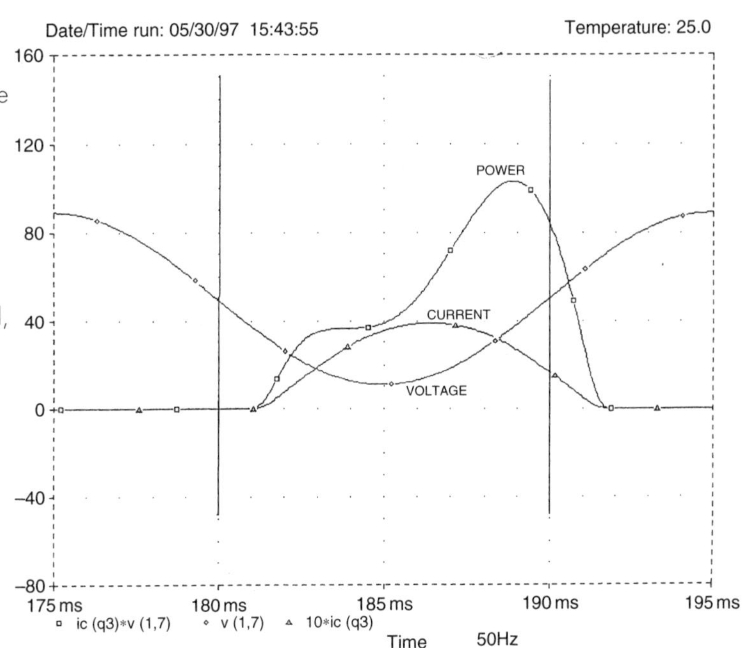

Figure 7.32
As Figure 7.31, but driving 50 Hz into the single-speaker load. At this frequency the load is partly inductive so current lags voltage and the instantaneous power curve is asymmetrical, peaking higher at 110 watts towards the end of the half-cycle

be considered at some point.) The critical quantities for semiconductor safety in amplifiers are usually the peak instantaneous values; for heatsink design average power is what counts, while for the power supply average current is the significant quantity.

To determine the effect of real speaker loads on device stress I simulated an EF output stage driving a single-speaker load with a 40 V peak sinewave, powered from ±50 V rails. The load was as Figure 7.26 except for a reduction in the voicecoil inductance to 0.1 mH; the resulting impedance curve is shown in Figure 7.33. Transient simulations over many cycles were done for 42 spot frequencies from 20 Hz to 20 kHz, and the peak and average quantities recorded and plotted. Many cycles must be simulated as the bass resonance in the impedance model takes time to reach steady state when a sinewave is abruptly applied; not everyone writing on this topic appears to have appreciated this point.

Steady sinewave excitation was used as a practical approach to simulation and testing, and does not claim to be a good approximation to music or speech. Arbitrary non-cyclic transients could be investigated by the same method, but the number of waveform possibilities is infinite. It would also be necessary to be careful about the initial conditions.

Figures 7.33, 7.34 and 7.35 are the distilled results of a very large number of simulations. Figure 7.34 shows that the gentle foothills of the impedance peak at bass resonance actually increase the peak instantaneous power stress on the output devices by 30%, despite the reduced current drawn.

The most dangerous regions for the amplifier are the sides of a resonance hump where the phase shift is the greatest. Peak dissipation only falls below

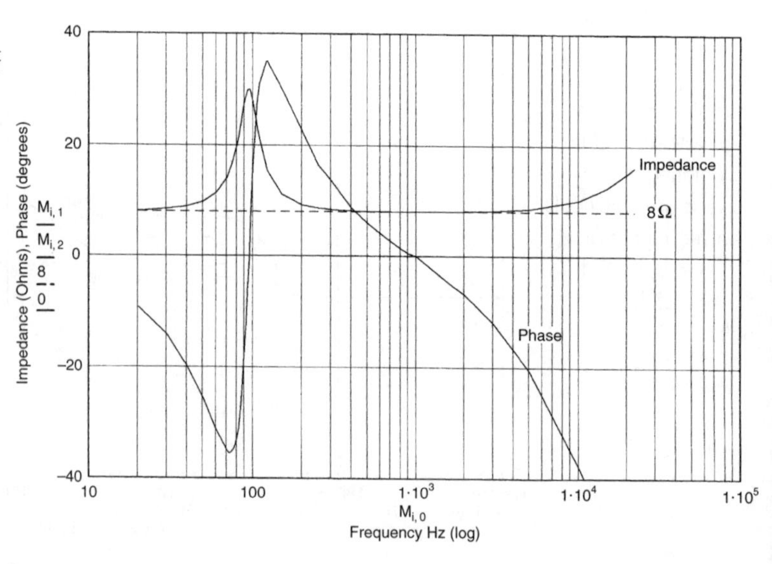

Figure 7.33
Impedance curve of the single-speaker model. The dotted line is 8 Ω resistive

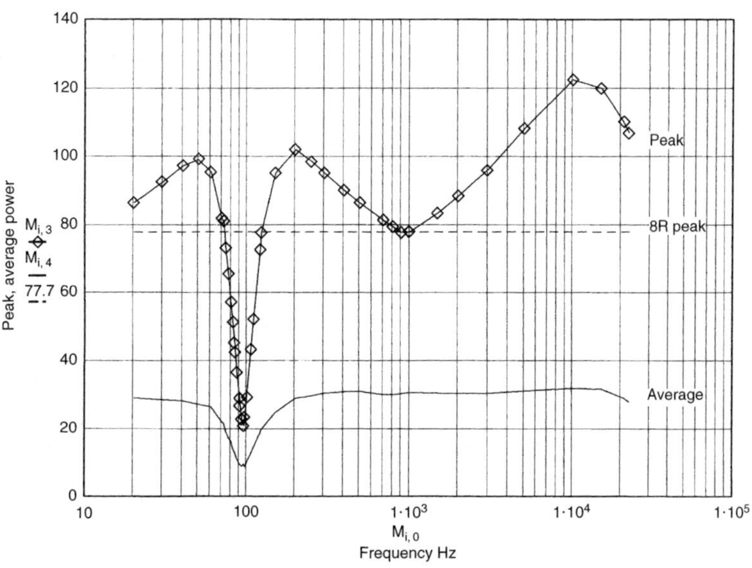

Figure 7.34
Peak and average output device power dissipation driving the single-unit speaker impedance as Figure 7.33. The dotted line is peak power for 8 Ω resistive

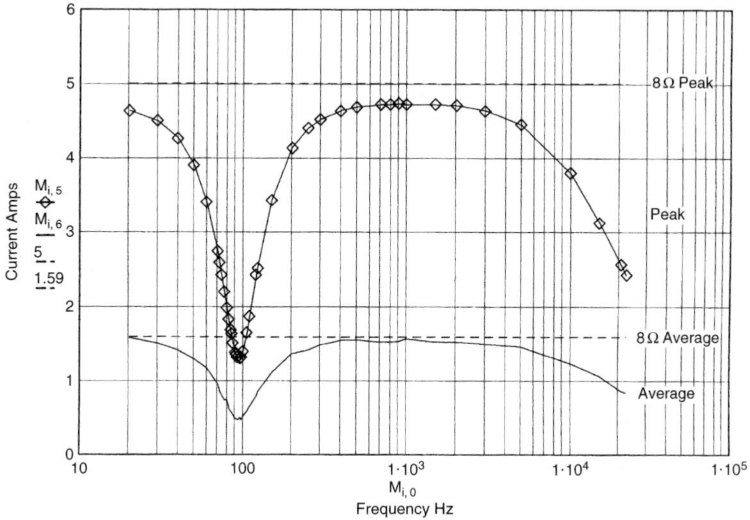

Figure 7.35
Peak and average output device current driving the single-unit speaker impedance. Dotted lines are peak and average current into 8 Ω

that for an 8 Ω resistor (shown dotted) around the actual resonance peak, where it drops quickly to a quarter of the resistive case.

Likewise, the increase in impedance at the HF end of the spectrum, where voicecoil inductance is significant, causes a more serious rise in peak dissipation to 50% more than the resistive case. The conclusion is that for peak power, the phase angle is far more important than the impedance magnitude.

The effects on the average power dissipation, and on the peak and average device current in Figure 7.35, are more benign. With this type of load network, all three quantities are reduced when the speaker impedance increases, the voltage/current phase shifts having no effect on the current.

Two-way speaker loads

The impedance plot for the simulated two-way speaker load of Figure 7.29 is shown in Figure 7.36 at 59 spot frequencies. The curve is more complex and shows a dip below the nominal impedance as well as peaks above; this is typical of multi-speaker designs. An impedance dip causes the maximum output device stress as it combines increased current demand with phase shifts that increase peak instantaneous dissipation.

In Figure 7.37 the impedance rise at bass resonance again causes increased peak power dissipation due to phase shifts; the other three quantities are reduced. In the HF region there is an impedance dip at 6 kHz which nearly doubles peak power dissipation on its lower slopes, the effect being greater because both phase-shift and increased current demand are acting. The actual bottom of the dip sharply reduces peak power where the phase angle passes through zero, giving the notch effect at the top of the peak.

Average power (Figure 7.37) and peak and average current (Figure 7.38) are all increased by the impedance dip, but to a more modest extent.

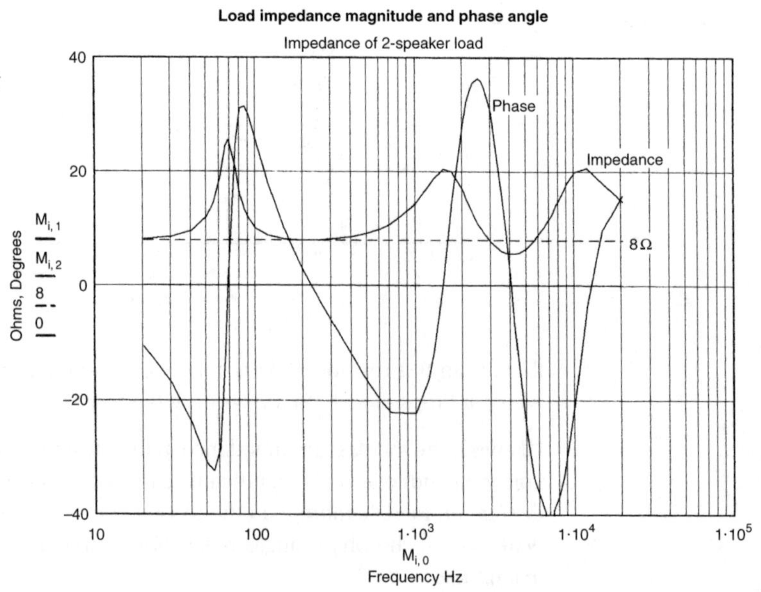

Figure 7.36
Impedance curve of model of the two-unit speaker model in Figure 7.30. Dotted line is 8 Ω resistive

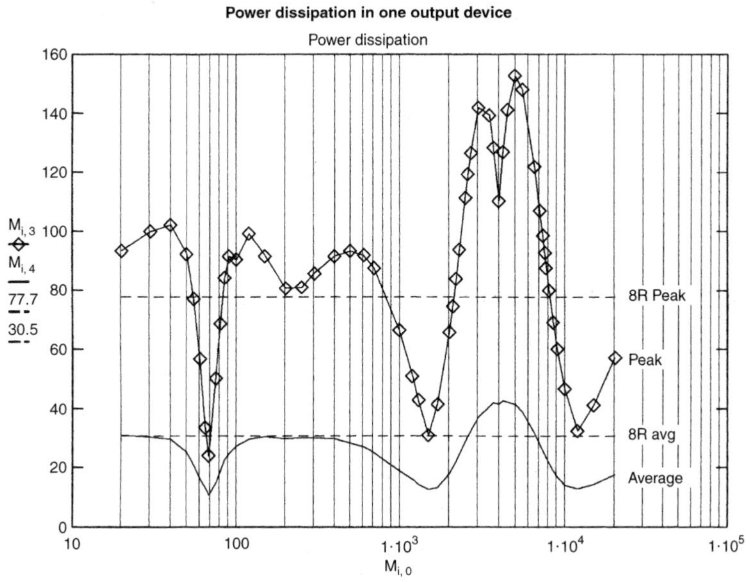

Figure 7.37
Peak and average output device power dissipation driving the two-way speaker model. Dotted lines are peak and average for 8 Ω

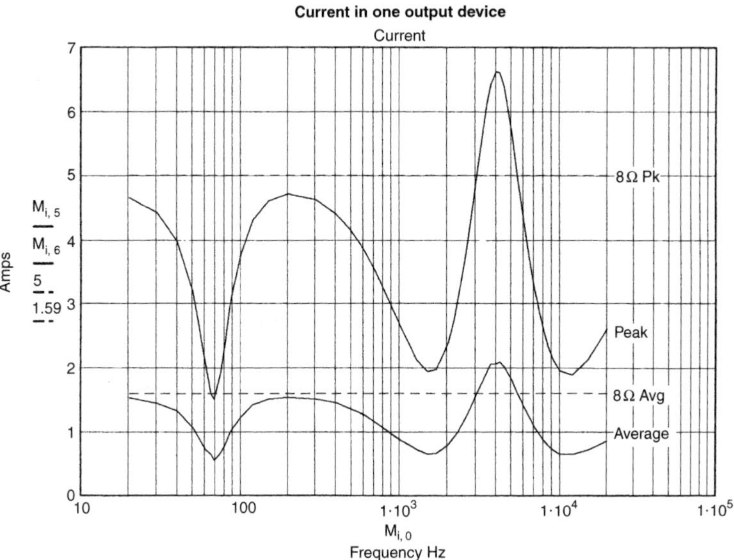

Figure 7.38
Peak and average output device current driving two-way speaker impedance as Figure 7.13. Dotted lines are peak and average for 8 Ω

Peak power would appear to be the critical quantity. Power device ratings often allow the power and second-breakdown limits (and sometimes the bondwire current limit also) to be exceeded for brief periods. If you attempt to exploit these areas in an audio application, you are living very dangerously, as the longest excursion specified is usually 5 msec, and a half-cycle at 20 Hz lasts for 25 msec.

From this it can be concluded that a truly 'difficult' load impedance is one with lots of small humps and dips giving significant phase shifts and increased peak dissipation across most of the audio band. Impedance dips cause more stress than peaks, as might be expected. Low impedances at the high-frequency end (above 5 kHz) are particularly undesirable as they will increase amplifier crossover distortion.

Enhanced loudspeaker currents

When amplifier current capability and loudspeaker loading are discussed it is often said that it is possible to devise special waveforms that cause a loudspeaker to draw more transient current than would at first appear to be possible. This is perfectly true. The issue was raised by Otala et al[13], and expanded on in Otala and Huttunen[14]. The effect was also demonstrated by Cordell[15].

The effect may be demonstrated with the electrical analogue of a single speaker unit as shown in Figure 7.26. Rc is the resistance of the voice-coil and Lc its inductance. Lr and Cr model the cone resonance, with Rr controlling its damping. These three components simulate the impedance characteristics of the real electromechanical resonance. The voicecoil inductance is 0.29 mH, and its resistance 6.8 Ω, typical for a 10 inch bass unit of 8 Ω nominal impedance. Measurements on this circuit cannot show an impedance below 6.8 Ω at any frequency, and it is easy to assume that the current demands can therefore never exceed those of a 6.8 Ω resistance. This is not so.

The secret of getting unexpectedly high currents flowing is to make use of the energy stored in the circuit reactances. This is done by applying an asymmetrical waveform with transitions carefully timed to match the speaker resonance. Figure 7.39 shows PSpice simulation of the currents drawn by the circuit of Figure 7.26. The rectangular waveform is the current in a reference 8 Ω resistance driven with the same waveform. A ±10 V output limit is used here for simplicity but this could obviously be much higher, depending on the amplifier rail voltages.

At the start of the waveform at A, current flows freely into Cr, reducing to B as the capacitance charges. Current is also slowly building up in Lr, causing the total current drawn to increase again to C. A positive transition to the opposite output voltage then takes the system to point D; this is not the same state as at A because energy has been stored in Lr during the long negative period.

A carefully timed transition is then made at E, at the lowest point in this part of the curve. The current change is the same amplitude as at D, but it starts off from a point where the current is already negative, so the final

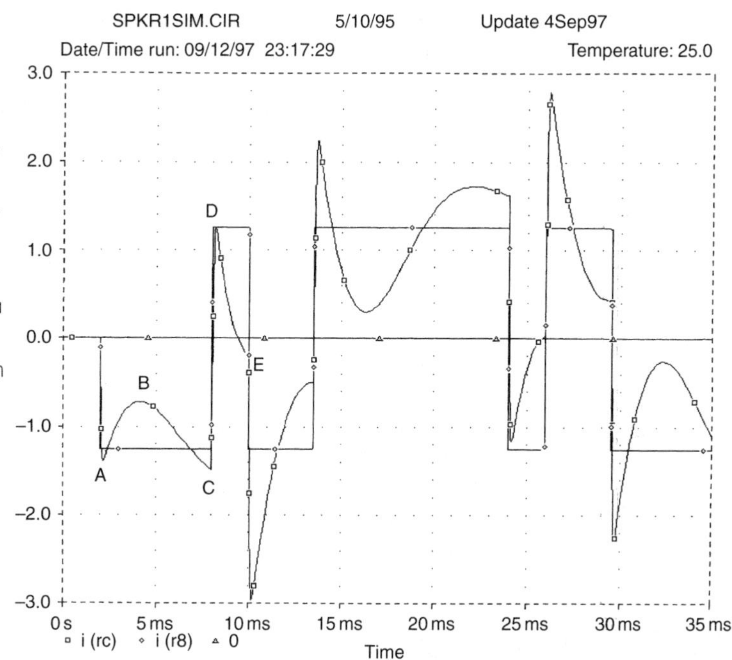

Figure 7.39 An asymmetrical waveform to generate enhanced speaker currents. The sequence ABCDE generates a negative current spike; to the right, the inverse sequence produces a positive spike. The rectangular waveform is the current through an 8 Ω resistive load

peak goes much lower to 2.96 amps, 2.4 times greater than that drawn by the 8 Ω resistor. I call this the Current Timing Factor, or CTF.

Otala and Huttunen[14] show that the use of multi-way loudspeakers, and more complex electrical models, allows many more degrees of freedom in maximising the peak current. They quote a worst case CTF of 6.6 times. An amplifier driving 50 W into 8 Ω must supply a peak current into an 8 Ω resistance of 3.53 amps; amplifiers are usually designed to drive 4 Ω or lower to allow for impedance dips and this means the peak current capability must be at least 7.1 amps. However, a CTF of implies that the peak capability should be at least 23 amps. This peak current need only be delivered for less than a millisecond, but it could complicate the design of protection circuitry.

The vital features of the provocative waveform are the fast transitions and their asymmetrical timing. The optimal transition timing for high currents varies with the speaker parameters. The waveform in Figure 7.39 uses ramped transitions lasting 10 μsec; if these transitions are made longer the peak currents are reduced. There is little change up to 100 μsec, but with transitions lengthened to 500 μsec the CTF is reduced from 2.4 to 2.1.

Without doing an exhaustive survey, it is impossible to know how many power amplifiers can supply six times the nominal peak current required.

I suspect there are not many. Is this therefore a neglected cause of real audible impairment? I think not, because:

1 Music signals do not contain high-level rectangular waveforms, nor trapezoidal approximations to them. A useful investigation would be a statistical evaluation of how often (if ever) waveforms giving significant peak current enhancement occur. As an informal test, I spent some time staring at a digital scope connected to general-purpose rock music, and saw nothing resembling the test waveform. Whether the asymmetrical timings were present is not easy to say; however, the large-amplitude vertical edges were definitely not.
2 If an amplifier does not have a huge current-peak capability, then the overload protection circuitry will hopefully operate. If this is of a non-latching type that works cleanly, the only result will be rare and very brief periods of clipping distortion when the loudspeaker encounters a particularly unlucky waveform. Such infrequent transient distortion is known to be inaudible and this may explain why the current enhancement effect has attracted relatively little attention so far.

Amplifier instability

Amplifier instability refers to unwanted oscillations at either HF or LF. The latter is very rare in solid-state amplifiers, though still very much an issue for valve designers. Instability has to be taken very seriously, because it may not only destroy the amplifier that hosts it, but also damage the loudspeakers.

Instability at middle frequencies such as 1 kHz is virtually impossible unless you have a very eccentric design with roll-offs and phase-shifts in the middle of the audio band.

HF instability

HF instability is probably the most difficult problem that may confront the amplifier designer, and there are several reasons for this:

1 The most daunting feature of HF oscillation is that under some circumstances it can cause the destruction of the amplifier in relatively short order. It is often most inadvisable to let the amplifier sit there oscillating while you ponder its shortcomings.

BJT amplifiers will suffer overheating because of conduction overlap in the output devices; it takes time to clear the charge carriers out of the device junctions. Some designs deal with this better than others, but it is still true that subjecting a BJT design to prolonged sinewave testing above 20 kHz should be done with great caution. Internal oscillations may of course have much higher frequencies than this, and in some cases the output devices may be heated to destruction in a few seconds. The resistor in the Zobel network will probably also catch fire.

FET amplifiers are less vulnerable to this overlap effect, due to their different conduction mechanism, but show a much greater tendency to parasitic oscillation at high frequencies, which can be equally destructive. Under high-amplitude oscillation plastic-package FETs may fail explosively; this is usually a prompt failure within a second or so and leaves very little time to hit the off switch.

2 Various sub-sections of the amplifier may go into oscillation on their own account, even if the global feedback loop is stable against Nyquist oscillation. Even a single device may go into parasitic oscillation (e.g., emitter-followers fed from inappropriate source impedances) and this is usually at a sufficiently high frequency that it either does not fight its way through to the amplifier output, or does not register on a 20 MHz scope. The presence of this last kind of parasitic is usually revealed by excessive and unexpected non-linearity.
3 Another problem with HF oscillation is that it cannot in general be modelled theoretically. The exception to this is global Nyquist oscillation (i.e., oscillation around the main feedback loop because the phase-shift has become too great before the loop gain has dropped below unity) which can be avoided by calculation, simulation, and design. The forward-path gain and the dominant pole frequency are both easy to calculate, though the higher pole frequencies that cause phase-shift to accumulate are usually completely mysterious; to the best of my knowledge virtually no work has been done on the frequency response of audio amplifier output stages. Design for Nyquist stability therefore reduces to deciding what feedback factor at 20 kHz will give reliable stability with various resistive and reactive loads, and then apportioning the open-loop gain between the transconductance of the input stage and the transresistance of the VAS.

The other HF oscillations, however, such as parasitics and other more obscure oscillatory misbehaviour, seem to depend on various unknown or partly known second-order effects that are difficult or impossible to deal with quantitatively and are quite reasonably left out of simulator device models. This means we are reduced to something not much better than trial-and-error when faced with a tricky problem.

The CFP output stage has two transistors connected together in a very tight 100% local feedback loop, and there is a clear possibility of oscillation inside this loop. When it happens, this tends to be benign, at a relatively high frequency (say 2–10 MHz) with a clear association with one polarity of half-cycle.

LF instability

Amplifier instability at LF (motorboating) is largely a thing of the past now that amplifiers are almost invariably designed with DC-coupling throughout the forward and feedback paths. The theoretical basis for it is exactly as for HF Nyquist oscillation; when enough phase-shift accumulates at a given frequency, there will be oscillation, and it does not matter if that frequency is 1 Hz or 1 MHz.

At LF things are actually easier, because all the relevant time-constants are known, or can at least be pinned down to a range of values based on electrolytic capacitor tolerances, and so the system is designable. The techniques for dealing with almost any number of LF poles and zeros were well-known in the valve era, when AC coupling between stages was usually unavoidable, because of the large DC voltage difference between the anode of one stage and the grid of the next.

Oscillation at LF is unlikely to be provoked by awkward load impedances. This is not true at HF, where a capacitative load can cause serious instability. However, this problem at least is easily handled by adding an output inductor.

Speed and slew-rate in audio amplifiers

It seems self-evident that a fast amplifier is a better thing to have than a slow one, but – what is a fast amplifier? Closed-loop bandwidth is not a promising yardstick; it is virtually certain that any power amplifier employing negative feedback will have a basic closed-loop frequency response handsomely in excess of any possible aural requirements, even if the overall system bandwidth is defined at a lower value by earlier filtering.

There is always a lot of loose talk about the importance of an amplifier's open-loop bandwidth, much of it depressingly ill-informed. I demonstrated[16] that the frequency of the dominant pole P1 that sets the open-loop bandwidth is a variable and rather shifty quantity that depends on transistor beta and other ill-defined parameters. (I also showed how it can be cynically manipulated to make it higher by reducing open-loop gain below P1.) While P1 may vary, the actual gain at HF (say 20 kHz) is thankfully a much more dependable figure that is set only by frequency, input stage transconductance, and the value of Cdom[17]. It is this which is the meaningful figure in describing the amount of NFB that an amplifier enjoys.

The most meaningful definition of an amplifier's *speed* is its maximal slew-rate. The minimum slew-rate for a 100 W/8 Ω amplifier to cleanly reproduce a 20 kHz sinewave is easily calculated as 5.0 V/μsec; so 10 V/μsec is adequate for 400 W/8 Ω, a power level that takes us somewhat out of the realms of domestic hi-fi. A safety-margin is desirable, and if we make this a bare factor of two then it could be logically argued that 20 V/μsec is enough for any hi-fi application; there is in fact a less obvious but substantial safety-margin already built in, as 20 kHz signals at maximum level are mercifully rare in music; the amplitude distribution falls off rapidly at higher frequencies.

Firm recommendations on slew-rate are not common; Peter Baxandall made measurements of the slew-rate produced by vinyl disc signals, and concluded that they could be reproduced by an amplifier with a slew limit corresponding to maximum output at 2.2 kHz. For the 100 W amplifier this corresponds to 0.55 V/μsec[18].

Nelson Pass made similar tests, with a moving-magnet (MM) cartridge, and quoted a not dissimilar maximum of 1 V/μsec at 100 W. A moving-coil (MC) cartridge doubled this to 2 V/μsec, and Pass reported[19] that the absolute maximum possible with a combination of direct-cut discs and MC cartridges was 5 V/μsec at 100 W. This is comfortably below the 20 V/μsec figure arrived at theoretically above; Pass concluded that even if a generous 10:1 factor of safety was adopted, 50 V/μsec would be the highest speed ever required from a 100 W amplifier.

However, in the real world we must also consider The Numbers Game; if all else is equal then the faster amplifier is the more saleable. As an example of this, it has been recently reported in the hi-fi press that a particular 50 W/8 Ω amplifier has been upgraded from 20 V/μsec to 40 V/μsec[20] and this is clearly expected to elicit a positive response from intending purchasers. This report is exceptional, for equipment reviews in the hi-fi press do not usually include slew-rate measurements. It is therefore difficult to get a handle on the state of the art, but a trawl through the accumulated data of years shows that the most highly specified equipment usually plumps for 50 V/μsec – slew-rates always being quoted in suspiciously round numbers. There was one isolated claim of 200 V/μsec, but I must admit to doubts about the reality of this.

The Class-B amplifier shown in Figure 7.5 is that already described in Chapter 6; the same component numbers have been preserved. This generic circuit has many advantages, though an inherently good slew performance is not necessarily one of them; however, it remains the basis for the overwhelming majority of amplifiers so it seems the obvious place to start. I have glibly stated that its slew-rate calculated at 40 V/μsec, which by the above arguments is more than adequate. However, let us assume that a major improvement in slew-rate is required to counter the propaganda of the Other Amplifier Company down the road, and examine how it might be done. As in so many areas of life, things will prove much more complicated than expected.

The basics of amplifier slew-limiting

At the simplest level, slew-rate in a conventional amplifier configuration like Figure 7.5 depends on getting current in and out of Cdom (C3) with the convenient relation:

$$\text{Slew-rate} = \text{I}/\text{Cdom V}/\mu\text{sec}, \quad \text{for I in } \mu\text{A, Cdom in pF} \qquad \text{Equation 7.1}$$

The maximum output frequency for a given slew-rate and voltage is:

$$\text{Freq max} = S_R/(2 \times \pi \times \text{Vpk}) = S_R/(2 \times \pi \times \sqrt{2} \times \text{V rms}) \qquad \text{Equation 7.2}$$

So, for example, with a slew-rate of 20 V/µsec the maximum freq at which 35 V rms can be sustained is 64 kHz, and if Cdom is 100 pF then the input stage must be able to source and sink 2 mA peak. Likewise, a sinewave of given amplitude and frequency has a maximum slew-rate (at zero-crossing) of:

$$\begin{aligned} \text{SR of sinewave} &= \text{dV}/\text{dt} = \varphi_{\text{max}}\text{Vpk} \\ &= 2 \times \pi \times \text{freq} \times \text{Vpk} \qquad \text{Equation 7.3} \end{aligned}$$

For Figure 7.5, our slew-rate equation yields 4000/100, or about 40 V/µsec, as quoted above, if we assume (as all textbooks do) that the only current-limitation is the tail-source of the input pair. If this differential pair has a current-mirror collector load – and there are pressing reasons why it should – then almost the full tail-current is available to service Cdom. This seems very simple – to increase slew-rate increase the tail-current. But . . .

The tail-current is not the only limit on the slew current in Cdom. (This point was touched on by Self[21].) Figure 7.40 shows the current paths for positive and negative slew-limit, and it can be seen at once that the positive current can only be supplied by the VAS current-source load. This will reduce the maximum positive rate, causing slew asymmetry, if the VAS current-source cannot supply as much current as the tail source. In contrast, for negative slewing TR4 can turn on as much as required to sink the Cdom current, and the VAS collector load is not involved.

In most designs the VAS current-source value does not appear to be an issue, as the VAS is run at a higher current than the input stage to ensure enough pull-up current for the top half of the output stage; however it will transpire that the VAS source can still cause problems.

Slew-rate measurement techniques

Directly measuring the edge-slopes of fast square waves from a scope screen is not easy, and without a delayed timebase it is virtually impossible. A much easier (and far more accurate) method is to pass the amplifier output through a suitably-scaled differentiator circuit; slew-rate then becomes

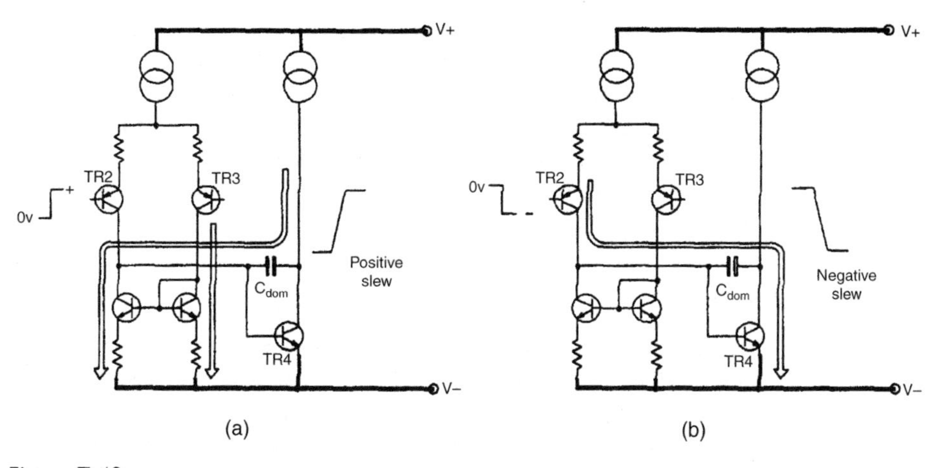

Figure 7.40
(a) The current path for positive slewing. At the limit all of the slewing current has to pass through the current-mirror, TR2 being cut off. (b) The current path at negative slew limit. TR2 is saturated and the current-mirror is cut off

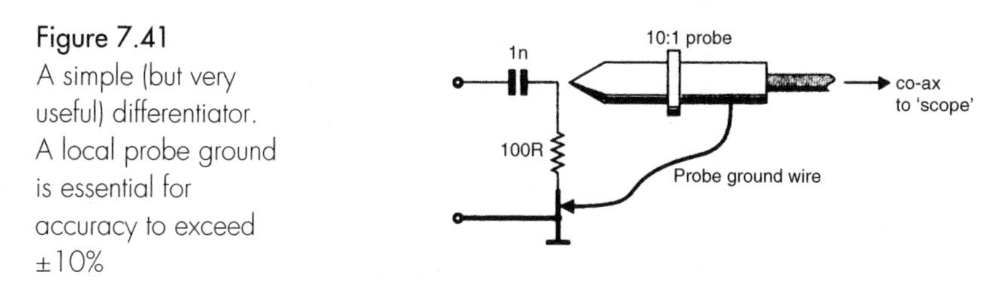

Figure 7.41
A simple (but very useful) differentiator. A local probe ground is essential for accuracy to exceed ±10%

simple amplitude, which is much easier to read from a graticule. The circuit in Figure 7.41 gives a handy 100 mV output for each V/μsec of slew; the RC time-constant must be very short for reasonable accuracy. The differentiator was driven directly by the amplifier, and *not* via an output inductor. Be aware that this circuit needs to be coupled to the scope by a proper ×10 probe; the capacitance of plain screened cable gives serious underreadings. We are dealing here with sub-microsecond pulse techniques, so bear in mind that waveform artefacts such as ringing are as likely to be due to test cabling as to the amplifier.

Applying a fast-edged square wave to an amplifier does not guarantee that it will show its slew-rate limits. If the error voltage so generated is not enough to saturate the input stage then the output will be an exponential response, without non-linear effects. For most of the tests described here, the amplifier had to be driven hard to ensure that the true slew-limits were revealed; this is due to the heavy degeneration that reduces the transconductance of the input pair. Degeneration increases the error voltage required for saturation, but does not directly alter slew limits.

Running a slew test on the circuit of Figure 7.5, with an 8 Ω load, sharply highlights the inadequacies of simple theory. The differentiator revealed asymmetrical slew-rates of +21 V/μsec up and −48 V/μsec down, which is both a letdown and a puzzle considering that the simple theory promises 40 V/μsec. To get results worse than theory predicts is merely the common lot of the engineer; to simultaneously get results that are *better* is grounds for the gravest suspicions.

Improving the slew-rate

Looking again at Figure 7.5, the VAS current-source value is apparently already bigger than required to source the current *C*dom requires when the input stage is sinking hard, so we confidently decrease R4 to 100 R (to match R13) in a plausible attempt to accelerate slewing. With considerable disappointment we discover that the slew-rate only changes to +21 V/μsec, −62 V/μsec; the negative rate still exceeds the new theoretical value of 60 V/μsec. Just what is wrong here? Honesty compels us to use the lower of the two figures in our ads (doesn't it?) and so the priority is to find out why the positive slewing is so feeble.

At first it seems unlikely that the VAS current source is the culprit, as with equal-value R4 and R13, the source should be able to supply all the input stage can sink. Nonetheless, we can test this cherished belief by increasing the VAS source current while leaving the tail-current at its original value. We find that R4 = 150 R, R13 = 68 R gives +23 V/μsec, −48 V/μsec, and this small but definite increase in positive rate shows clearly there is something non-obvious going on in the VAS source.

(This straightforward method of slew acceleration by increasing standing currents means a significant increase in dissipation for the VAS and its current source. We are in danger of exceeding the capabilities of the TO92 package, leading to a cost increase. The problem is less in the input stage, as dissipation is split between at least three devices.)

Simulating slew-limiting

When circuits turn truculent, it's time to simplify and simulate. The circuit was reduced to a *model* amplifier by replacing the Class-B output stage with a small-signal Class-A emitter follower; this was then subjected to some brutally thorough PSPICE simulation, which revealed the various mechanisms described below.

Figure 7.42 shows the positive-going slew of this model amplifier, with both the actual output voltage and its differential, the latter suitably scaled by dividing by 10^6 so it can be read directly in V/μsec from the same plot. Figure 7.43 shows the same for the negative-going slew. The plots are done for a series of changes to the resistors R4, 23 that set the standing currents.

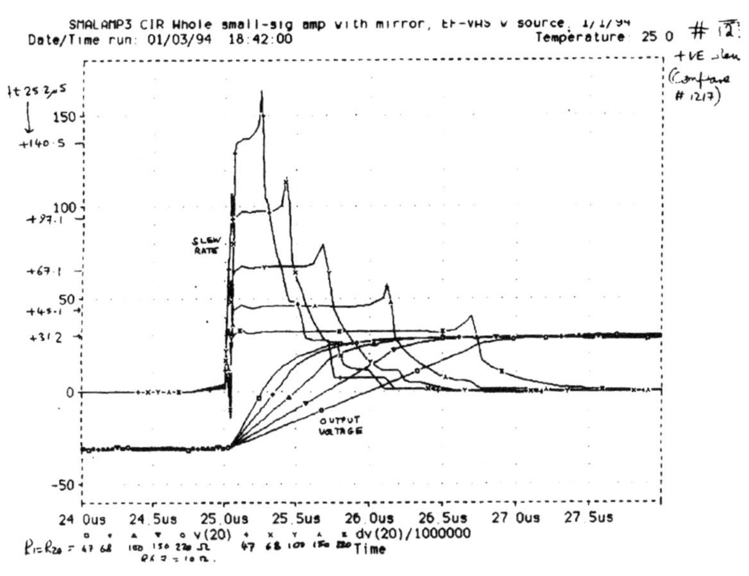

Figure 7.42
Positive slewing of simulated model amplifier. The lower traces show the amplifier output slewing from −30 to +30 V while the upper traces are the scaled differentiation

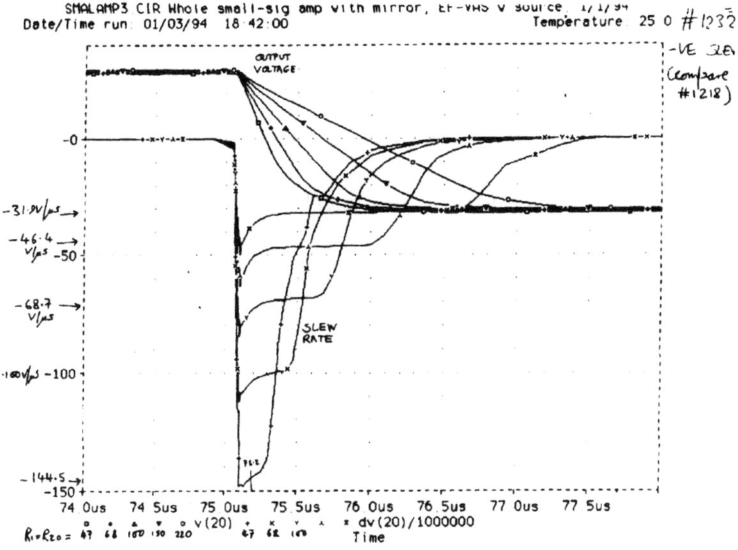

Figure 7.43
Negative slewing of simulated model amplifier. Increasing the slew-rate limit causes a larger part of the output transient to become exponential, as the input pair spends less time saturated. Thus the differential trace has a shorter flat period

Several points need to be made about these plots; first, the slew-rates shown for the lower R4, 23 values are not obtainable in the real amplifier with output stage, for reasons that will emerge. Note that almost imperceptible wobbles in the output voltage put large spikes on the plot of the slew-rate, and it is unlikely that these are being simulated accurately, if only because circuit strays are neglected. To get valid slew-rates, read the flat portions of the differential plots.

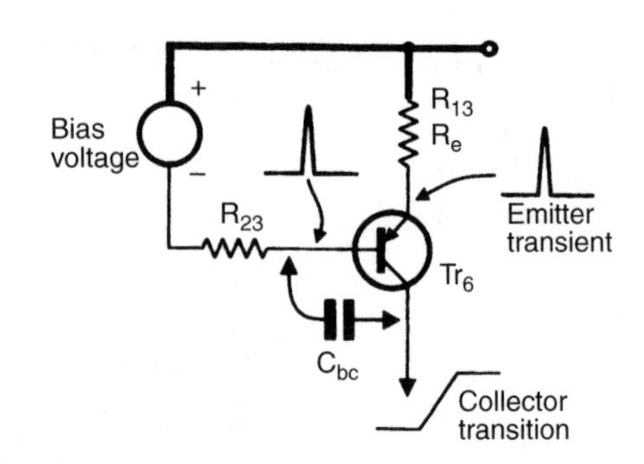

Figure 7.44
One reason why simple theory fails. Fast positive edges on the collector of the VAS source TR6 couple through the internal Cbc to momentarily reduce standing current

Using this method, the first insight into slew-rate asymmetry was obtained. At audio frequencies, a constant current-source provides a fairly constant current and that is the end of the matter, making it the usual choice for the VAS collector load; as a result its collector is exposed to the full output swing and the full slew-rate. When an amplifier slews rapidly, there is a transient feedthrough from the collector to the base (see Figure 7.44) via the collector-base capacitance. If the base voltage is not tightly fixed then fast positive slewing drives the base voltage upwards, reducing the voltage on the emitter and hence the output current. Conversely, for negative slew the current-source output briefly increases; see Erdi[22]. In other words, fast positive slewing itself reduces the current available to implement it.

Having discovered this hidden constraint, the role of isolation resistor R23 immediately looks suspect. Simulation confirms that its presence worsens the feedthrough effect by increasing the impedance of the reference voltage fed to TR5 base. As is usual, the input-stage tail-source TR1 is biased from the same voltage as TR5; this minor economy complicates things significantly, as the tail current also varies during fast transients, reducing for positive slew, and increasing for negative.

Slewing limitations in real life

Bias isolation resistors are not unique to the amplifier of Figure 7.5; they are very commonly used. For an example taken at random, see Meyer[23]. My own purpose in adding R23 was not to isolate the two current sources from each other at AC (something it utterly fails to do) but to aid fault-finding. Without this resistor, if the current in either source drops to zero (e.g., if TR1 fails open-circuit) then the reference voltage collapses, turning off both sources, and it can be time-consuming to determine which has died and which has merely come out in sympathy. Accepting this, we return to the original Figure 7.5 values and replace R23 with a link; the measured slew-rates at once improve from +21, −48 to +24, −48 (from here on the V/μsec is omitted). This is already slightly faster than our first attempt at acceleration, without the thermal penalties of increasing the VAS standing current.

The original amplifier used an active tail-source, with feedback control by TR14; this was a mere whim, and a pair of diodes gave identical THD figures. It seems likely that reconfiguring the two current-sources so that the VAS source is the active one would make it more resistant to feedthrough, as the current-control loop is now around TR5 rather than TR1, with feedback applied directly to the quantity showing unwanted variations (see Figure 7.45). There is indeed some improvement, from +24, −48 to +28, −48.

This change seems to work best when the VAS current is increased, and R4 = 100 R, R13 = 68 R now gives us +37, −52, which is definite progress on the positive slewing. The negative rate has also slightly increased, indicating that the tail-current is still being increased by feedthrough effect.

It seems desirable to minimise this transient feedthrough, as it works against us just at the wrong time. One possibility would be a cascode transistor to shield TR5 collector from rapid voltage changes; this would require more biasing components and would reduce the positive output swing, albeit only slightly.

Since it is the VAS current-source feedthrough capacitance that causes so much grief, can we turn it against itself, so that an abrupt voltage transition increases the current available to sustain it, rather than reducing it? Oh yes we can, for if a small capacitance Cs is added between TR5 collector (carrying the full voltage swing) and the sensing point A of the active tail source, then as the VAS collector swings upward, the base of TR14 is also driven positive, tending to turn it off and hence increasing the bias applied to VAS source TR5 via R21. This technique is highly effective, but it smacks of positive feedback and should be used with caution; Cs must be kept small. I found 7.5 pF to be the highest value usable without degrading the amplifier's HF stability.

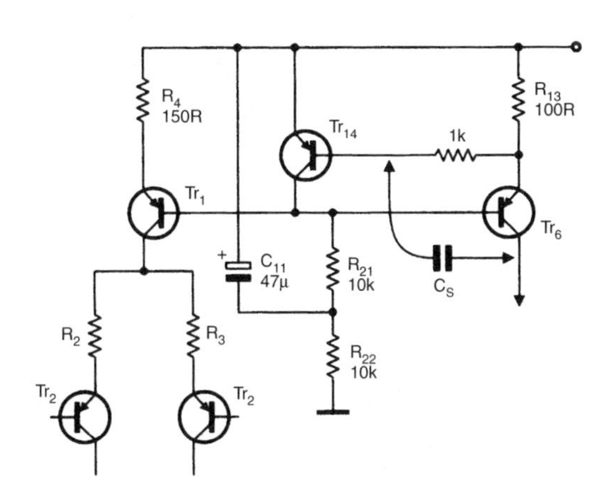

Figure 7.45
A modified biasing system that makes TR6 current the controlled variable, and reduces the feedthrough effect

With R4 = 100, R13 = 68 adding Cs = 6 pF takes us from +37, −52 to +42, −43; and the slew asymmetry that has dogged this circuit from the start has been corrected. Fine adjustment of this capacitance is needful if good slew-symmetry is demanded.

Some additional complications

Some other unsuspected effects were uncovered in the pursuit of speed; it is not widely known that slew-rate is affected both by output loading and the output stage operating class. For example, above we have noted that R4 = 100, R13 = 68 yields +37, −52 for Class-B and an 8 Ω load. With 4 Ω loading this changes to +34, −58, and again the loss in positive speed is the most significant. If the output stage is biased into Class-A (for an 8 Ω load) then we get +35, −50. The explanation is that the output stage, despite the cascading of drivers and output devices, draws significant current from the VAS stage. The drivers draw enough base current in the 4 Ω case to divert extra current from Cdom and current is in shortest supply during positive slew. The effect in Class-A is more severe because the output device currents are always high, the drivers requiring more base current even when quiescent, and again this will be syphoned off from the VAS collector.

Speeding-up this amplifier would be easier if the Miller capacitor Cdom was smaller. Does it really need to be that big? Well yes, because if we want the NFB factor to be reasonably low for dependable HF stability, the HF loop gain must be limited. Open-loop gain above the dominant pole frequency P1 is the product of input stage gm with the value of Cdom, and the gm is already as low as it can reasonably be made by emitter degeneration. Emitter resistors R2, 3 at 100 Ω are large enough to mildly compromise the input offset voltage, because the tail current splits in two through a pair of resistors that are unlikely to be matched to better than 1%, and noise performance is also impaired by this extra resistance in the input pair emitters. Thus for a given NFB factor at 20 kHz, Cdom is fixed.

Despite these objections, the approach was tested by changing the distribution of open-loop gain between the input stage and the VAS. R2, 3 were increased from 100 R to 220 R, and Cdom reduced to 66 pF; this does not give exactly the same NFB factor, but in essence we have halved the transconductance of the input stage, while doubling the gain of the VAS. This gain-doubling allows Cdom to be reduced to 66 pF without reduction of stability margins.

With R4 = 100, R13 = 68 as before, the slew-rate is increased to +50, −50 with Cs = 6 pF to maintain slewing symmetry. This is a 25% increase in speed rather than the 50% that might be expected from simple theory, and indicates that other restrictions on speed still exist; in fact PSPICE showed there are several.

One of these restrictions is as follows; when slewing positively, TR4 and TR12 must be turned off as fast as possible, by pulling current out of Cdom. The input pair therefore causes TR10 to be turned on by an increasing voltage across TR11 and R7. As TR10 turns on, its emitter voltage rises due to R6, while at the same time the collector voltage must be pulled down to near the –ve rail to turn off Q4. In the limit TR10 runs out of Vce, and is unable to pull current out of Cdom fast enough. The simplest way to reduce this problem is to reduce the resistors R6, 7 that degenerate the current-mirror. This risks HF distortion variations due to input-pair Ic imbalance, but values down to 12 Ω have given acceptable results. Once more it is the positive rate that suffers.

Another way to reduce the value needed for Cdom is to lower the loop-gain by increasing the feedback network attenuation, or in other words, to run the amplifier at a higher closed-loop gain. This might be no bad thing; the current *standard* of 1 V for full output is (I suspect) due to a desire for low closed-loop gain in order to maximise the NFB factor, so reducing distortion. I recall JLH advocating this strategy back in 1974. However, we must take the world as we find it, and so I have left closed-loop gain alone. We could of course attenuate the input signal so it can be amplified more, though I have an uneasy feeling about this sort of thing; amplifying in a pre-amp then attenuating in the power amp implies a headroom bottleneck, if such a curdled metaphor is permissible. It might be worth exploring this approach; this amplifier has good open-loop linearity and I do not think excessive THD would be a problem.

Having previously spent some effort on minimising distortion, we do not wish to compromise the THD of a Blameless amplifier. Mercifully, none of the modifications set out here have any significant effect on overall THD, though there may be minor variations around 10–20 kHz.

Further improvements and other configurations

The results I have obtained in my attempts to improve slewing are not exactly stunning at first sight; however they do have the merit of being as grittily realistic as I can make them. I set out in the belief that enhancing slew-rate would be fairly simple; the very reverse has proved to be the case. It may well be that other VAS configurations, such as the push-pull VAS examined in Self[16], will prove more amenable to design for rapid slew-rates; however such topologies have other disadvantages to overcome.

Stochino in a fascinating paper[24] has presented a topology, which, although a good deal more complex than the conventional arrangement, claims to make slew-rates up to 400 V/μsec achievable.

References

1. Otala, M *An Audio Power Amplifier for Ultimate Quality Requirements* IEEE Trans on Audio and Electroacoustics, Vol AU-21, No. 6, Dec 1973.
2. Baxandall *Audio Power Amplifier Design: Part 4* Wireless World, July 1978, p. 76.
3. Takahashi et al *Design and Construction of High Slew-Rate Amplifiers* AES 60th Convention, Preprint No. 1348 (A-4) 1978.
4. Self *Crossover Distortion and Compensation* Letters, Electronics and Wireless World, Aug 1992, p. 657.
5. Widlar, *A Monolithic Power Op-Amp* IEEE J Solid-State Circuits, Vol 23, No 2, April 1988.
6. Bonello *Advanced Negative Feedback Design for High Performance Amplifiers* AES 67th Convention, Preprint No. 1706 (D-5) 1980.
7. Pernici et al *A CMOS Low-Distortion Amplifier with Double-Nested Miller Compensation* IEEE J. Solid-State Circuits, July 1993, p. 758.
8. Scott and Spears *On The Advantages of Nested Feedback Loops* JAES Vol 39, March 1991, p. 115.
9. National Semi *Fast Compensation Extends Power Bandwidth* Linear Brief 4, NatSem Linear Apps Handbook, 1991.
10. Feucht *Handbook of Analog Circuit Design* Academic Press 1990, p. 264.
11. Atkinson, J *Review of Krell KSA-50S Power Amplifier* Stereophile, Aug 1995, p. 168.
12. Benjamin, E *Audio Power Amplifiers for Loudspeaker Loads* JAES Vol 42, Sept 1994, p. 670.
13. Otala et al *Input Current Requirements of High-Quality Loudspeaker Systems* AES preprint #1987 (D7) for 73rd Convention, March 1983.
14. Otala and Huttunen *Peak Current Requirement of Commercial Loudspeaker Systems* JAES, June 1987, p. 455. See Ch. 12, p. 294.
15. Cordell, R *Interface Intermodulation in Amplifiers* Wireless World, Feb 1983, p. 32.
16. Self, D *Distortion In Power Amplifiers, Part 3* Electronics World+WW, Oct 1993, p. 824.
17. Self, D *Ibid Part 1* Electronics World+WW, Aug 1993, p. 631.
18. Baxandall, P *Audio Power Amplifier Design* Wireless World, Jan 1978, p. 56.
19. Pass, N *Linearity, Slew rates, Damping, Stasis and*... Hi-Fi News and RR, Sept 1983, p. 36.
20. Hughes, J *Arcam Alpha5/Alpha6 Amplifier Review* Audiophile, Jan 1994, p. 37.
21. Self, D *Distortion In Power Amplifiers, Part 7* Electronics World+WW, Feb 1994, p. 138.
22. Erdi, G *A 300 v/uS Monolithic Voltage Follower* IEEE J. of Solid-State Circuits, Dec 1979, p. 1062.
23. Meyer, D *Assembling a Universal Tiger* Popular Electronics, Oct 1970.
24. Stochino, G *Ultra-Fast Amplifier* Electronics World+WW, Oct 1995, p. 835.

8

Power supplies and PSRR

Power supply technologies

There are three principal ways to power an amplifier:

1 a simple unregulated power supply consisting of transformer, rectifiers, and reservoir capacitors;
2 a linear regulated power supply;
3 a switch-mode power supply.

It is immediately obvious that the first and simplest option will be the most cost-effective, but at a first glance it seems likely to compromise noise and ripple performance, and possibly interchannel crosstalk. It is therefore worthwhile to examine the pros and cons of each technology in a little more detail.

Simple unregulated power supplies

Advantages

- Simple, reliable, and cheap. (Relatively speaking – the traditional copper and iron mains transformer will probably be the most expensive component in the amplifier.)
- No possibility of instability or HF interference from switching frequencies.
- The amplifier can deliver higher power on transient peaks, which is just what is required.

Disadvantages

- Significant ripple is present on the DC output and the PSRR of the amplifier will need careful attention.

- The mains transformer will be relatively heavy and bulky.
- Transformer primary tappings must be changed for different countries and mains voltages.
- The absence of switch-mode technology does not mean total silence as regards RF emissions. The bridge rectifier will generate bursts of RF at a 100 Hz repetition rate as the diodes turn off. This worsens with increasing current drawn.

Linear regulated power supplies

Advantages

- Can be designed so that virtually no ripple is present on the DC output (in other words the ripple is below the white noise the regulator generates) allowing relaxation of amplifier supply-rail rejection requirements. However, you can only afford to be careless with the PSRR of the power amp if the regulators can maintain completely clean supply-rails in the face of sudden current demands. If not, there will be interchannel crosstalk unless there is a separate regulator for each channel. This means four for a stereo amplifier, making the overall system very expensive.
- A regulated output voltage gives absolutely consistent audio power output in the face of mains voltage variation.
- The possibility exists of electronic shutdown in the event of an amplifier DC fault, so that an output relay can be dispensed with. However, this adds significant circuitry, and there is no guarantee that a failed output device will not cause a collateral failure in the regulators which leaves the speakers still in jeopardy.

Disadvantages

- Complex and therefore potentially less reliable. The overall amplifier system is at least twice as complicated. The much higher component-count must reduce overall reliability, and getting it working in the first place will take longer and be more difficult. For an example circuit see Sinclair[1]. If the power amplifier fails, due to an output device failure, then the regulator devices will probably also be destroyed, as protecting semiconductors with fuses is a very doubtful business; in fact it is virtually impossible. The old joke about the transistors protecting the fuse is not at all funny to power-amplifier designers, because this is precisely what happens. Electronic overload protection for the regulator sections is therefore essential to avert the possibility of a domino-effect failure, and this adds further complications, as it will probably need to be some sort of foldback protection characteristic if the regulator transistors are to have a realistic prospect of survival.

- Comparatively expensive, requiring at least two more power semiconductors, with associated control circuitry and over-current protection. These power devices in turn need heatsinks and mounting hardware, checking for shorts in production, etc.
- Transformer tappings must still be changed for different mains voltages.
- IC voltage regulators are usually ruled out by the voltage and current requirements, so it must be a discrete design, and these are not simple to make bulletproof. Cannot usually be bought in as an OEM item, except at uneconomically high cost.
- May show serious HF instability problems, either alone or in combination with the amplifiers powered. The regulator output impedance is likely to rise with frequency, and this can give rise to some really unpleasant sorts of HF instability. Some of my worst amplifier experiences have involved (very) conditional stability in such amplifiers.
- The amplifier can no longer deliver higher power on transient peaks.
- The overall power dissipation for a given output is considerably increased, due to the minimum voltage-drop though the regulator system.
- The response to transient current demands is likely to be slow, affecting slewing behaviour.

Switch-mode power supplies

Advantages

- Ripple can be considerably lower than for unregulated power supplies, though never as low as a good linear regulator design. 20 mV pk–pk is typical.
- There is no heavy mains transformer, giving a considerable saving in overall equipment weight. This can be important in PA equipment.
- Can be bought in as an OEM item; in fact this is virtually compulsory as switch-mode design is a specialised job for experts.
- Can be arranged to shutdown if amplifier develops a dangerous DC offset.
- Can be specified to operate properly, and give the same audio output without adjustment, over the entire possible worldwide mains-voltage range, which is normally taken as 90–260 V.

Disadvantages

- A prolific source of high-frequency interference. This can be extremely difficult to eradicate entirely from the audio output.
- The 100 Hz ripple output is significant, as noted above, and will require the usual PSRR precautions in the amplifiers.

- Much more complex and therefore less reliable than unregulated supplies. Dangerous if not properly cased, as high DC voltage is present.
- The response to transient current demands is likely to be relatively slow.

On perusing the above list, it seems clear that regulated supplies for power amplifiers are a Bad Thing. Not everyone agrees with me; see for example Linsley-Hood[2]. Unfortunately he does not adduce any evidence to support his case.

The usual claim is that linear regulated supplies give *tighter bass*; advocates of this position are always careful not to define *tighter bass* too closely, so no-one can disprove the notion. If the phrase means anything, it presumably refers to changes in the low-frequency transient response; however since no such changes can be detected, this appears to be simply untrue. If properly designed, all three approaches can give excellent sound, so it makes sense to go for the easiest solution; with the unregulated supply the main challenge is to keep the ripple out of the audio, which will be seen to be straightforward if tackled logically. The linear regulated approach presents instead the challenge of designing not one but two complex negative-feedback systems, close-coupled in what can easily become a deadly embrace if one of the partners shows any HF instability. As for switch-mode supplies, their design is very much a matter for specialists.

The generic amplifier designs examined in this book have excellent supply-rail rejection, and so a simple unregulated supply is perfectly adequate. The use of regulated supplies is definitely unnecessary, and I would recommend strongly against their use. At best, you have doubled the amount of high-power circuitry to be bought, built, and tested. At worst, you could have intractable HF stability problems, peculiar slew-limiting, and some expensive device failures.

Design considerations for power supplies

A typical unregulated power supply is shown in Figure 8.1. This is wholly conventional in concept, though for optimal hum performance the wiring topology and physical layout need close attention, and this point is rarely made.

For amplifiers of moderate power the total reservoir capacitance per rail usually ranges from 4700 to 20,000 μF, though some designs have much more. Ripple current ratings must be taken seriously. It is often claimed that large amounts of reservoir capacitance give *firmer bass*; this is untrue for all normal amplifier designs below clipping.

I do not propose to go through the details of designing a simple PSU, as such data can be found in standard textbooks, but I instead offer some hints

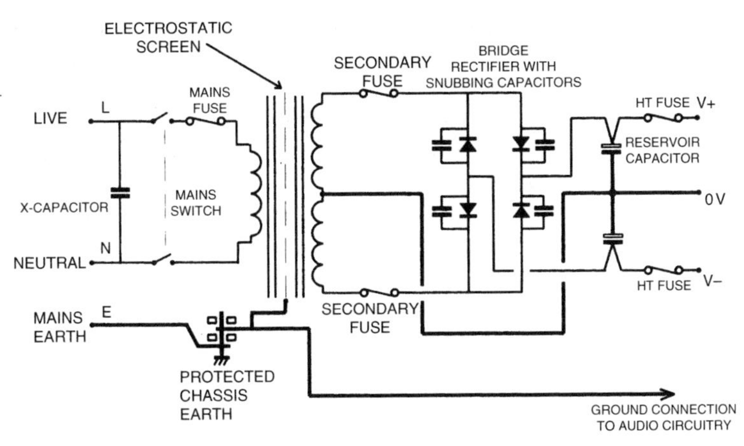

Figure 8.1
A simple unregulated power supply, including rectifier-snubbing and X-capacitor.

and warnings that are either rarely published or are especially relevant to audio amplifier design.

Mains transformers

The mains transformer will normally be either the traditional E-&-I frame type, or a toroid. The frame type is used where price is more important than compactness or external field, and vice-versa. There are various other types of transformer, such as C-core, or R-core, but they do not seem to be able to match the low external field of the toroid, while being significantly more expensive than the frame type.

The external field of a frame transformer can be significantly reduced by specifying a hum strap, or *belly-band* as it is sometimes rather indelicately called. This is a wide strip of copper that forms a closed circuit around the outside of the core and windings, so it does not form a shorted turn in the main transformer flux. Instead it intersects with the leakage flux, partially cancelling it.

The design of the mains transformer for a given voltage at a given current is simple in principle, but in practice always seems to involve a degree of trial and error. The main reason for this is that the voltage developed on the reservoir capacitors depends on losses that are not easily predicted, and this is inherent in any rectifier circuit where the current flows only in short sharp peaks at the crest of the AC waveform.

First the voltage developed depends on the transformer regulation, i.e., the amount the voltage drops as more current is drawn. (The word *regulation* in this context has nothing to do with negative-feedback voltage control – unfortunate and confusing, but there it is.) Transformer manufactures are usually reluctant to predict anything more than a very approximate figure for this.

Voltage losses also depend strongly on the peak amplitude of the charging pulses from the rectifier to the reservoir; these peaks cause voltage drops in the AC wiring, transformer winding resistances, and rectifiers that are rather larger than might be expected. Unfortunately the peak current value is poorly defined, by wiring resistance and transformer leakage reactance (a parameter that transformer manufacturers are even more reluctant to predict) and any calculations are so rough that they are really valueless. There may also be uncertainties in the voltage efficiency of the amplifier itself, and there are so many variables that it is only realistic to expect to try two or three transformer designs before the exact output power required is obtained.

Since most amplifiers are intended to reproduce music and speech, with high peak-to-average power ratios, they will operate satisfactorily with transformers rated to supply only 70% of the current required for extended sinewave operation, and in a competitive market the cost savings are significant. Trouble comes when the amplifiers are subjected to sinewave testing, and a transformer so rated will probably fail from internal overheating, though it may take an hour or more for the temperatures to climb high enough. The usual symptom is breakdown of the interlayer winding insulation, the resultant shorted turns causing the primary mains fuse to blow. This process is usually undramatic, without visible transformer damage or the evolution of smoke, but it does of course ruin an expensive component.

To prevent such failures when a mains transformer is deliberately underrated, some form of thermal cutout is essential. Self-resetting cutouts based on snap-action bimetal discs are physically small enough to be buried in the outer winding layers and work very well. They are usually chosen to act at 100°C or 110°C, as transformer materials are usually rated to 120°C unless special construction is required.

If the primary side of the mains transformer has multiple taps for multicountry operation, remember that some of the primary wiring will carry much greater currents at low voltage tappings; the mains current drawn on 90 V input will be nearly 3 times that at 240 V, for the same power out.

Fusing and rectification

The rectifier (almost always a packaged bridge) must be generously rated to withstand the initial current surge as the reservoirs charge from empty on switch-on. Rectifier heatsinking is definitely required for most sizes of amplifier; the voltage drop in a silicon rectifier may be low (1 V per diode is a good approximation for rough calculation) but the current pulses are large and the total dissipation is significant.

Reservoir capacitors must have the incoming wiring from the rectifier going directly to the capacitor terminals; likewise the outgoing wiring to the HT rails must leave from these terminals. In other words, do not run a tee off to the cap, because if you do its resistance combined with the high-current charging pulses adds narrow extra peaks to the ripple crests on the DC output and may worsen the hum/ripple level on the audio.

The cabling to and from the rectifiers carry charging pulses that have a considerably higher peak value than the DC output current. Conductor heating is therefore much greater due to the higher value of I-squared-R. Heating is likely to be especially severe if connectors are involved. Fuseholders may also heat up and consideration should be given to using heavy-duty types. Keep an eye on the fuses; if the fusewire sags at turn-on, or during transients, the fuse will fail after a few dozen hours, and the rated value needs to be increased.

When selecting the value of the mains fuse in the transformer primary circuit, remember that toroidal transformers take a large current surge at switch-on. The fuse will definitely need to be of the slow-blow type.

The bridge rectifier must be adequately rated for long-term reliability, and it needs proper heat-sinking.

RF emissions from bridge rectifiers

Bridge rectifiers, even the massive ones intended solely for 100 Hz power rectification, generate surprising quantities of RF. This happens when the bridge diodes turn off; the charge carriers are swept rapidly from the junction and the current flow stops with a sudden jolt that generates harmonics well into the RF bands. The greater the current, the more RF produced, though it is not generally possible to predict how steep this increase will be. The effect can often be heard by placing a transistor radio (long or medium wave) near the amplifier mains cable. It is the only area in a conventional power amplifier likely to give trouble in EMC emissions testing[3].

Even if the amplifier is built into a solidly grounded metal case, and the mains transformer has a grounded electrostatic screen, RF will be emitted via the live and neutral mains connections. The first line of defence against this is usually four *snubbing* capacitors of approximately 100 nF across each diode of the bridge, to reduce the abruptness of the turn-off. If these are to do any good, it is vital that they are all as close as possible to the bridge rectifier connections. (Never forget that such capacitors must be of the type intended to withstand continuous AC stress.)

The second line of defence against RF egress is an X-capacitor wired between Live and Neutral, as near to the mains inlet as possible (see Figure 8.1). This is usually only required on larger power amplifiers of 300 W total and above. The capacitor must be of the special type that can withstand direct mains connection. 100 nF is usually effective; some safety standards set a maximum of 470 nF.

Power supply-rail rejection in amplifiers

The literature on power amplifiers frequently discusses the importance of power-supply rejection in audio amplifiers, particularly in reference to its possible effects on distortion[4].

I hope I have shown in Chapters 5 and 6 that regulated power supplies are just not unnecessary for an exemplary THD performance. I want to confirm this by examining just how supply-rail disturbances insinuate themselves into an amplifier output, and the ways in which this rail-injection can be effectively eliminated. My aim is not just the production of hum-free amplifiers, but also to show that there is nothing inherently mysterious in power-supply effects, no matter what Subjectivists may say on the subject.

The effects of inadequate power-supply rejection ratio (PSRR) in a typical Class-B power amplifier with a simple unregulated supply, may be two-fold:

1. a proportion of the 100 Hz ripple on the rails will appear at the output, degrading the noise/hum performance. Most people find this much more disturbing than the equivalent amount of distortion,
2. the rails also carry a signal-related component, due to their finite impedance. In a Class-B amplifier this will be in the form of half-wave pulses, as the output current is drawn from the two supply-rails alternately; if this enters the signal path it will degrade the THD seriously.

The second possibility, the intrusion of distortion by supply-rail injection, can be eliminated in practice, at least in the conventional amplifier architecture so far examined. The most common defect seems to be misconnected rail bypass capacitors, which add copious ripple and distortion into the signal if their return lines share the signal ground; this was denoted No. 5 (Rail Decoupling Distortion) on my list of distortion mechanisms in Chapter 3.

This must not be confused with distortion caused by *inductive* coupling of halfwave supply currents into the signal path – this effect is wholly unrelated and is completely determined by the care put into physical layout; I labelled this Distortion No. 6 (Induction Distortion).

Assuming the rail bypass capacitors are connected correctly, with a separate ground return, ripple and distortion can only enter the amplifier directly through the circuitry. It is my experience that if the amplifier is made ripple-proof under load, then it is proof against distortion-components from the rails as well; this bold statement does however require a couple of qualifications:

First, the output must be ripple-free *under load*, i.e., with a substantial ripple amplitude on the rails. If a Class-B amplifier is measured for ripple output when quiescent, there will be a very low amplitude on the supply-rails and the measurement may be very good; but this gives no assurance that hum will not be added to the signal when the amplifier is operating and drawing significant current from the reservoir capacitors. Spectrum analysis could be used to sort the ripple from the signal under drive, but it is simpler to leave the amplifier undriven and artificially provoke ripple on the HT rails by loading them with a sizeable power resistor; in my work I have standardised on drawing 1 A. Thus one rail at a time can be loaded; since the rail rejection mechanisms are quite different for V+ and V−, this is a great advantage.

Drawing 1 A from the V− rail of the typical power amplifier in Figure 8.2 degraded the measured ripple output from −88 dBu (mostly noise) to −80 dBu.

Second, I assume that any rail filtering arrangements will work with constant or increasing effectiveness as frequency increases; this is clearly true for resistor-capacitor (RC) filtering, but is by no means certain for *electronic*

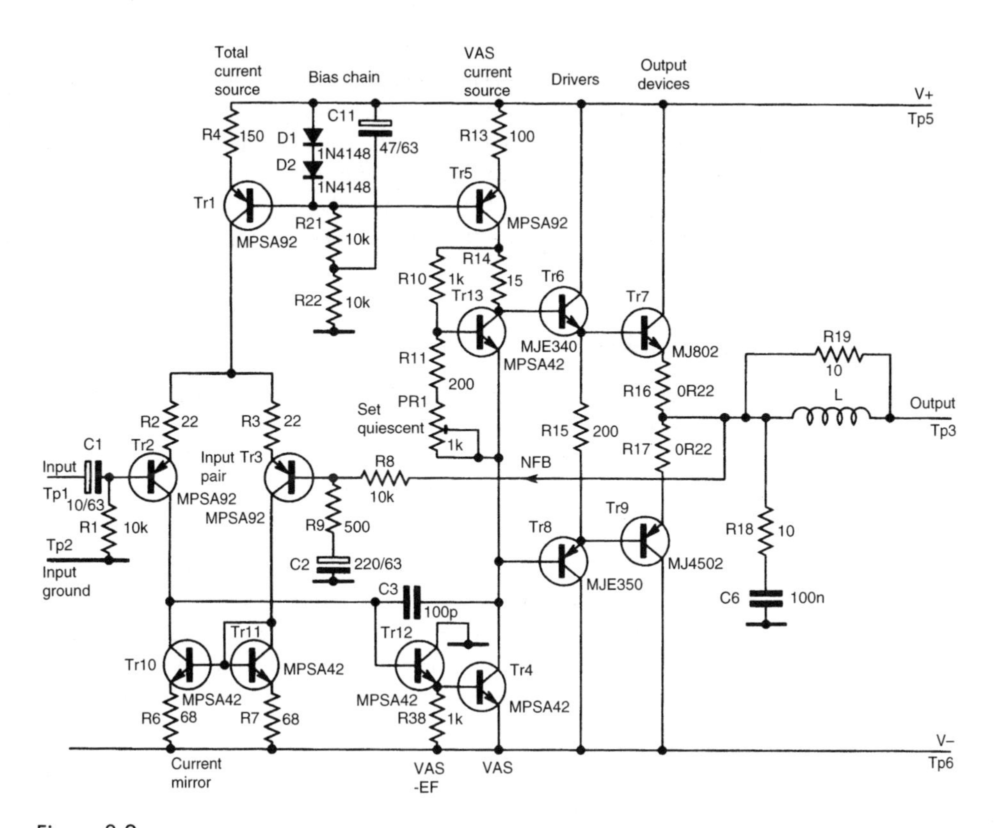

Figure 8.2
Diagram of a generic power amplifier, with diode biasing for input tail and VAS source

decoupling such as the NFB current-source biasing used in the design in Chapter 6. (These will show declining effectiveness with frequency as internal loop-gains fall.) Thus, if 100 Hz components are below the noise in the THD residual, it can usually be assumed that disturbances at higher frequencies will also be invisible, and not contributing to the total distortion.

To start with some hard experimental facts, I took a power amplifier – similar to Figure 8.2 – powered by an unregulated supply on the same PCB (the significance of this proximity will become clear in a moment) driving 140 W rms into 8 Ω at 1 kHz. The PSU was a conventional bridge rectifier feeding 10,000 μF reservoir capacity per rail.

The 100 Hz rail ripple under these conditions was 1 V pk–pk. Superimposed on this were the expected halfwave pulses at signal frequency; measured at the PCB track just before the HT fuse, their amplitude was about 100 mV peak-peak. This doubled to 200 mV on the downstream side of the fuse – the small resistance of a 6.3 A slow-blow fuse is sufficient to double this aspect of the PSRR problem, and so the fine details of PCB layout and PSU wiring could well have a major effect. (The 100 Hz ripple amplitude is of course unchanged by the fuse resistance.)

It is thus clear that improving the *transmitting* end of the problem is likely to be difficult and expensive, requiring extra-heavy wire, etc., to minimise the resistance between the reservoirs and the amplifier. It is much cheaper and easier to attack the *receiving* end, by improving the power-amp's PSRR. The same applies to 100 Hz ripple; the only way to reduce its amplitude is to increase reservoir capacity, and this is expensive.

A design philosophy for rail rejection

First, ensure there is a negligible ripple component in the noise output of the quiescent amplifier. This should be pretty simple, as the supply ripple will be minimal; any 50 Hz components are probably due to magnetic induction from the transformer, and must be removed first by attention to physical layout.

Second, the THD residual is examined under full drive; the ripple components here are obvious as they slide evilly along the distortion waveform (assuming that the scope is synchronised to the test signal). As another general rule, if an amplifier is made visually free of ripple-synchronous artefacts on the THD residual, then it will not suffer detectable distortion from the supply-rails.

PSRR is usually best dealt with by RC filtering in a discrete-component power amplifier. This will however be ineffective against the sub-50 Hz VLF signals that result from short-term mains voltage variations being reflected in the HT rails. A design relying wholly on RC filtering might have low AC ripple figures, but would show irregular jumps and twitches of the THD residual; hence the use of constant-current sources in the input tail and VAS to establish operating conditions more firmly.

The standard op-amp definition of PSRR is the dB loss between each supply-rail and the effective differential signal at the inputs, giving a figure independent of closed-loop gain. However, here I use the dB loss between rail and output, in the usual non-inverting configuration with a C/L gain of 26.4 dB. This is the gain of the amplifier circuit under consideration, and allows dB figures to be directly related to testgear readings.

Looking at Figure 8.2, we must assume that any connection to either HT rail is a possible entry point for ripple injection. The PSRR behaviour for each rail is quite different, so the two rails are examined separately.

Positive supply-rail rejection

The V+ rail injection points that must be eyed warily are the input-pair tail and the VAS collector load. There is little temptation to use a simple resistor tail for the input; the cost saving is negligible and the ripple performance inadequate, even with a decoupled mid-point. A practical value for such a tail-resistor would be 22 k, which in SPICE simulation gives a low-frequency PSRR of −120 dB for an undegenerated differential pair with current-mirror.

Replacing this tail resistor with the usual current source improves this to −164 dB, assuming the source has a clean bias voltage. The improvement of 44 dB is directly attributable to the greater output impedance of a current source compared with a tail resistor; with the values shown this is 4.6 M, and 4.6 M/22 k is 46 dB, which is a very reasonable agreement. Since the rail signal is unlikely to exceed +10 dBu, this would result in a maximum output ripple of −154 dBu.

The measured noise floor of a real amplifier (ripple excluded) was −94.2 dBu (EIN = −121.4 dBu) which is mostly Johnson noise from the emitter degeneration resistors and the global NFB network. The tail ripple contribution would be therefore 60 dB below the noise, where I think it is safe to neglect it.

However, the tail-source bias voltage in reality will not be perfect; it will be developed from V+, with ripple hopefully excluded. The classic method is a pair of silicon diodes; LED biasing provides excellent temperature compensation, but such accuracy in setting DC conditions is probably unnecessary. It may be desirable to bias the VAS collector current-source from the same voltage, which rules out anything above a volt or two. A 10 V zener might be appropriate for biasing the tail-source (given suitable precautions against noise generation) but this would seriously curtail the positive VAS voltage swing.

The negative-feedback biasing system used in the design in Chapter 6 provides a better basic PSRR than diodes, at the cost of some beta-dependence. It is not quite as good as an LED, but the lower voltage generated is more suitable for biasing a VAS source. These differences become academic if the bias chain mid-point is filtered with 47 μF to V+, as Table 8.1 shows; this is C11 in Figure 8.2.

As another example, the Figure 8.2 amplifier with diode-biasing and no bias chain filtering gives an output ripple of −74 dBu; with 47 μF filtering this improves to −92 dBu, and 220 μF drops the reading into limbo below the noise floor.

Figure 8.3 shows PSpice simulation of Figure 8.2, with a 0 dB sinewave superimposed on V+ only. A large Cdecouple (such as 100 μF) improves LF PSRR by about 20 dB, which should drop the residual ripple below the noise. However, there remains another frequency-insensitive mechanism at about −70 dB. The study of PSRR greatly resembles the peeling of onions, because there is layer after layer, and often tears. There also remains an

Table 8.1

	No decouple (dB)	Decoupled with 47 μF (dB)
2 diodes	−65	−87
LED	−77	−86
NFB low-beta	−74	−86
NFB high-beta	−77	−86

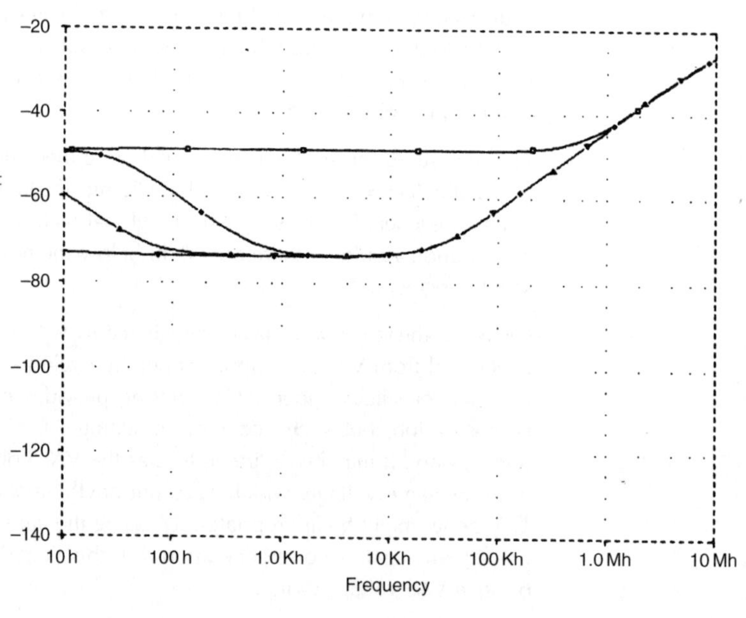

Figure 8.3
Positive-rail rejection; decoupling the tail current-source bias chain R21, R22 with 0, 1, 10 and 100 μF

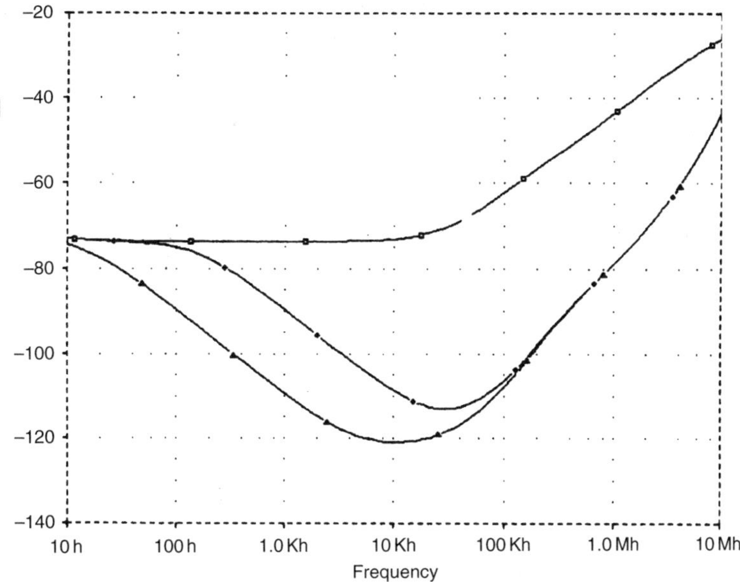

Figure 8.4 Positive-rail rejection; with input-stage supply-rail RC filtered with 100 Ω and 0, 10 and 100 μF. Same scale as Figure 8.3

HF injection route, starting at about 100 kHz in Figure 8.3, which is quite unaffected by the bias-chain decoupling.

Rather than digging deeper into the precise mechanisms of the next layer, it is simplest to RC filter the V+ supply to the input pair only (it makes very little difference if the VAS source is decoupled or not) as a few volts lost here are of no consequence. Figure 8.4 shows the very beneficial effect of this at middle frequencies, where the ear is most sensitive to ripple components.

Negative supply-rail rejection

The V– rail is the major route for injection, and a tough nut to analyse. The well-tried Wolf-Fence approach is to divide the problem in half, and in this case, the Fence is erected by applying RC filtering to the small-signal section (i.e., input current-mirror and VAS emitter) leaving the unity-gain output stage fully exposed to rail ripple. The output ripple promptly disappears, indicating that our wolf is getting in via the VAS or the bottom of the input pair, or both, and the output stage is effectively immune. We can do no more fencing of this kind, for the mirror has to be at the same DC potential as the VAS. SPICE simulation of the amplifier 1 with 1 V (0 dBV) AC signal on V– gives the PSRR curves in Figure 8.5, with *C*dom stepped in value. As before there are two regimes, one flat at –50 dB, and one rising at 6 dB per octave, implying at least two separate injection mechanisms. This suspicion is powerfully reinforced because as *C*dom is increased, the HF PSRR around 100 kHz improves to a maximum and then degrades again;

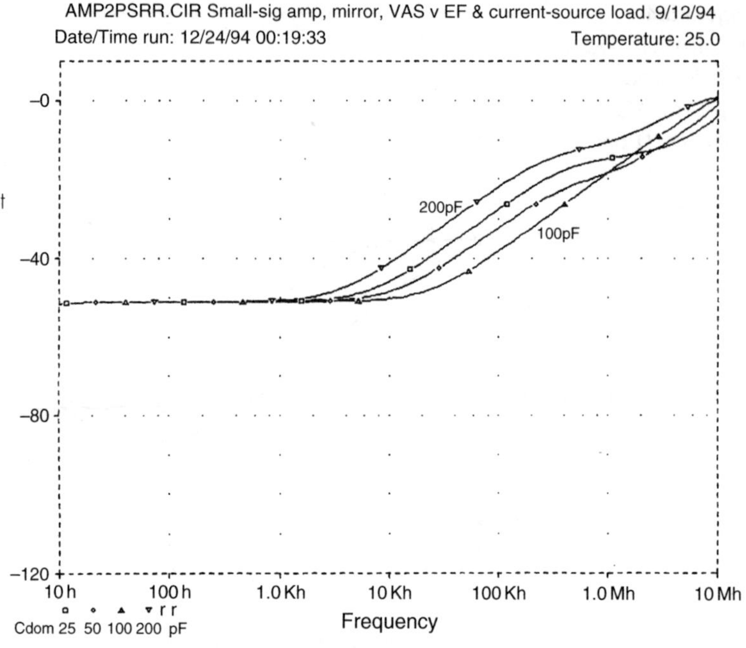

Figure 8.5 Negative-rail rejection varies with Cdom in a complex fashion; 100 pF is the optimal value. This implies some sort of concellation effect

i.e., there is an optimum value for Cdom at about 100 pF, indicating some sort of cancellation effect. (In the V+ case, the value of Cdom made very little difference.)

A primary LF ripple injection mechanism is Early Effect in the input-pair transistors, which determines the −50 dB LF floor of Curve 1 in Figure 8.7, for the standard input circuit (as per Figure 8.5 with Cdom = 100 pF).

To remove this effect, a cascode structure can be added to the input stage, as in Figure 8.6. This holds the Vce of the input pair at a constant 5 V, and gives Curve 2 in Figure 8.7. The LF floor is now 30 dB lower, although HF PSRR is slightly worse. The response to Cdom's value is now monotonic; simply a matter of more Cdom, less PSRR. This is a good indication that one of two partly cancelling injection mechanisms has been deactivated.

There is a deep subtlety hidden here. It is natural to assume that Early effect in the input pair is changing the signal current fed from the input stage to the VAS, but it is not so; this current is in fact completely unaltered. What *is* changed is the integrity of the feedback subtraction performed by the input pair; modulating the Vce of TR1, TR2 causes the output to alter at LF by global feedback action. Varying the amount of Early effect in TR1, TR2 by modifying VAF (Early intercept voltage) in the PSpice transistor model alters the floor height for Curve 1; the worst injection is with the lowest VAF (i.e., Vce has maximum effect on Ic) which makes sense.

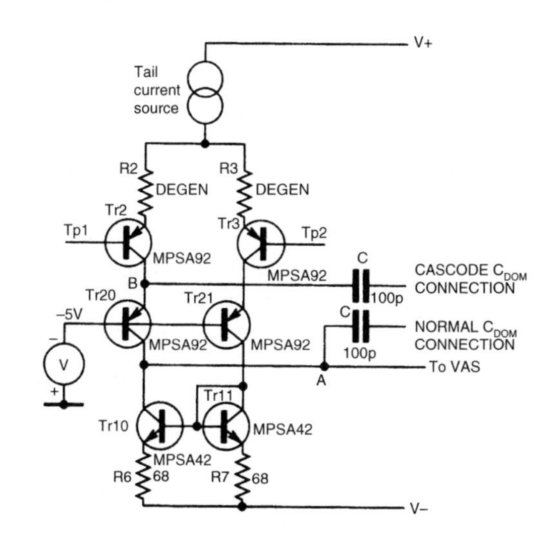

Figure 8.6
A cascoded input stage; Q21, Q22 prevent AC on V– from reaching TR2, TR3 collectors, and improve LF PSRR. B is the alternative C_{dom} connection point for cascode compensation

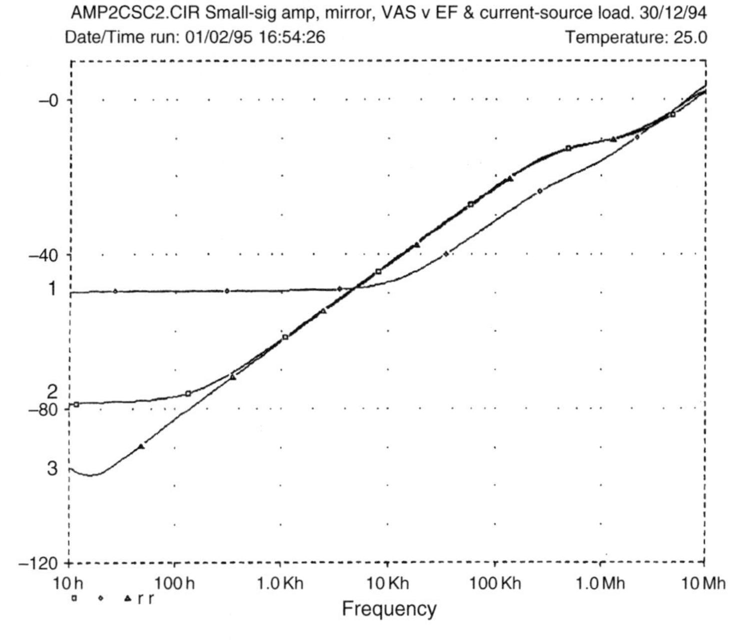

Figure 8.7
Curve 1 is negative-rail PSRR for the standard input. Curve 2 shows how cascoding the input stage improves rail rejection. Curve 3 shows further improvement by also decoupling TR12 collector to V–

We still have a LF floor, though it is now at –80 rather than –50 dB. Extensive experimentation showed that this is getting in via the collector supply of TR12, the VAS beta-enhancer, modulating Vce and adding a signal to the inner VAS loop by early effect once more. This is easily squished by decoupling TR12 collector to V–, and the LF floor drops to about –95 dB, where I think we can leave it for the time being (Curve 3 in Figure 8.7).

Having peeled two layers from the LF PSRR onion, something needs to be done about the rising injection with frequency above 100 Hz. Looking again at Figure 8.2, the VAS immediately attracts attention as an entry route. It is often glibly stated that such stages suffer from ripple fed in directly through Cdom, which certainly looks a prime suspect, connected as it is from V– to the VAS collector. However, this bald statement is untrue. In simulation it is possible to insert an ideal unity-gain buffer between the VAS collector and Cdom, without stability problems (A1 in Figure 8.8) and this absolutely prevents direct signal flow from V– to VAS collector through Cdom; the PSRR is completely unchanged.

Cdom has been eliminated as a direct conduit for ripple injection, but the PSRR remains very sensitive to its value. In fact the NFB factor available is the determining factor in suppressing V– ripple-injection, and the two quantities are often numerically equal across the audio band.

The conventional amplifier architecture we are examining inevitably has the VAS sitting on one supply-rail; full voltage swing would otherwise be impossible. Therefore the VAS input must be referenced to V–, and it is very likely that this change-of-reference from ground to V– is the basic source of injection. At first sight, it is hard to work out just what the VAS collector signal *is* referenced to, since this circuit node consists of two transistor collectors facing each other, with nothing to determine where it sits; the answer is that the global NFB references it to ground.

Consider an amplifier reduced to the conceptual model in Figure 8.9, with a real VAS combined with a perfect transconductance stage G, and unity-gain buffer A1. The VAS beta-enhancer TR12 must be included, as it proves to have a powerful effect on LF PSRR.

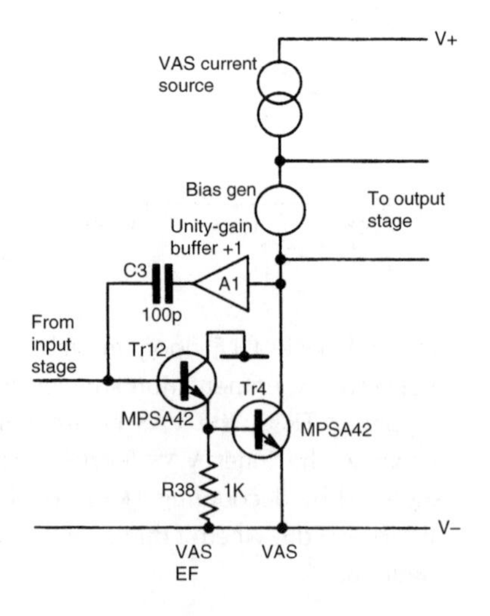

Figure 8.8
Adding a Cdom buffer A1 to prevent any possibility of signal entering directly from the V– rail

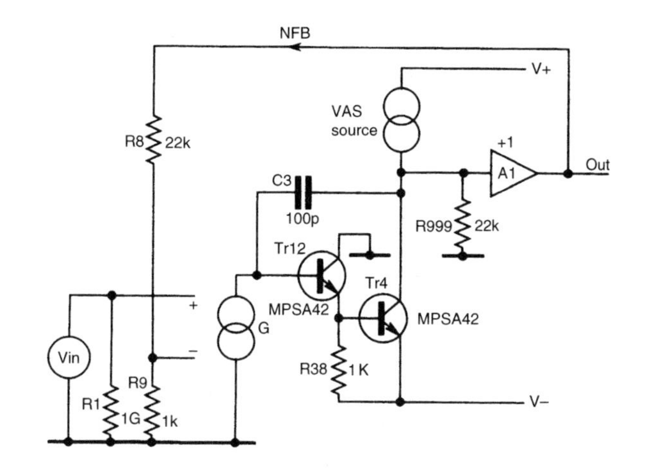

Figure 8.9 A conceptual SPICE model for V– PSRR, with only the VAS made from real components. R999 represents VAS loading

To start with, the global NFB is temporarily removed, and a DC input voltage is critically set to keep the amplifier in the active region (an easy trick in simulation). As frequency increases, the local NFB through *C*dom becomes steadily more effective, and the impedance at the VAS collector falls. Therefore the VAS collector becomes more and more closely bound to the AC on V–, until at a sufficiently high frequency (typically 10 kHz) the PSRR converges on 0 dB, and everything on the V– rail couples straight through at unity gain, as shown in Figure 8.10.

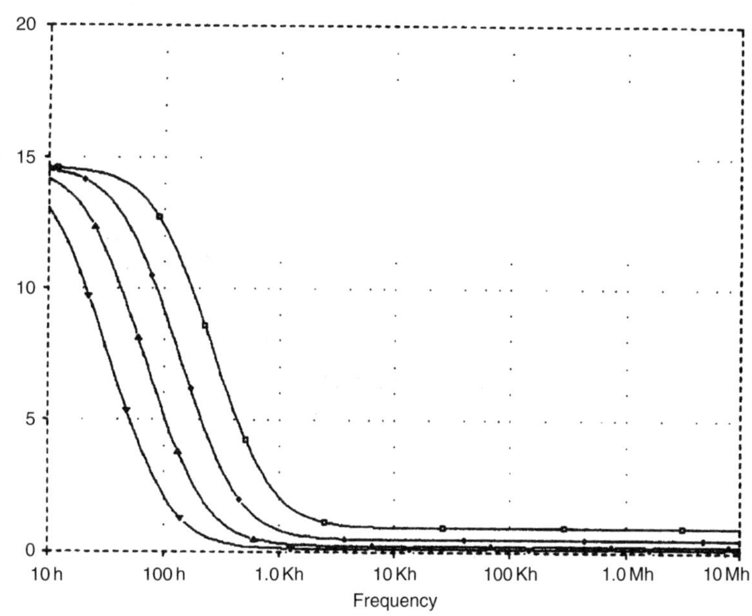

Figure 8.10 Open-loop PSRR from the model in Figure 8.8, with *C*dom value stepped. There is actual gain below 1 kHz

There is an extra complication here; the TR12/TR4 combination actually shows *gain* from V–to the output at low frequencies; this is due to Early effect, mostly in TR12. If TR12 was omitted the LF open-loop gain drops to about −6 dB.

Reconnecting the global NFB, Figure 8.11 shows a good emulation of the PSRR for the complete amplifier in Figure 8.7. The 10–15 dB open-loop-gain is flattened out by the global NFB, and no trace of it can be seen in Figure 8.11.

Now the NFB attempts to determine the amplifier output via the VAS collector, and if this control was perfect the PSRR would be infinite. It is not, because the NFB factor is finite, and falls with rising frequency, so PSRR deteriorates at exactly the same rate as the open-loop gain falls. This can be seen on many op-amp spec sheets, where V– PSRR falls off from the dominant-pole frequency, assuming conventional op-amp design with a VAS on V–.

Clearly a high global NFB factor at LF is vital to keep out V– disturbances. In Chapter 4, I rather tendentiously suggested that apparent open-loop bandwidth could be extended quite remarkably (without changing the amount of NFB at HF where it matters) by reducing LF loop gain; a high-value resistor RNFB in parallel with Cdom works the trick. What I did not say was that a high global NFB factor at LF is also invaluable for keeping the hum out; a point overlooked by those advocating low NFB factors as a matter of faith rather than reason.

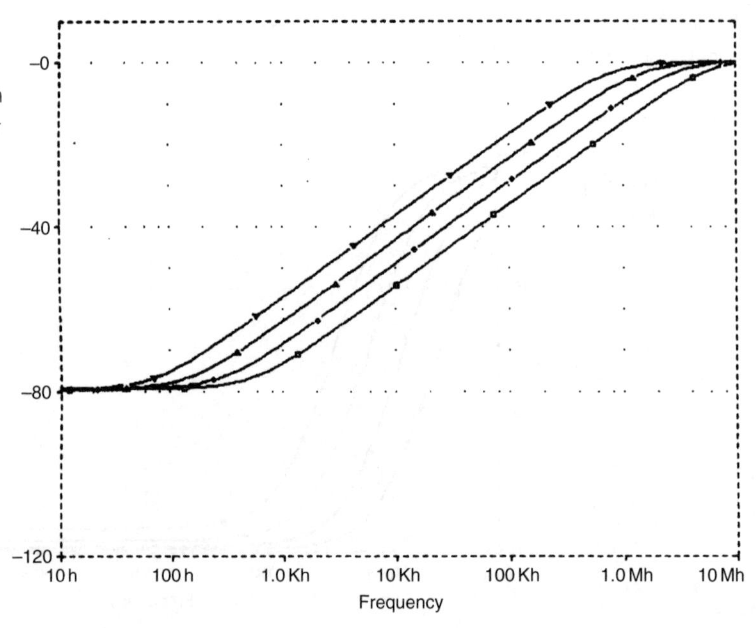

Figure 8.11 Closed-loop PSRR from Figure 8.9, with C_{dom} stepped to alter the closed-loop NFB factor

Table 8.2 shows how reducing global NFB by decreasing the value of RNFB degraded ripple rejection in a real amplifier.

Having got to the bottom of the V– PSRR mechanism, in a just world our reward would be a new and elegant way of preventing such ripple injection. Such a method indeed exists, though I believe it has never before been applied to power amplifiers[5],[6]. The trick is to change the reference, as far as Cdom is concerned, to ground. Figure 8.6 shows that cascode-compensation can be implemented simply by connecting Cdom to point B rather than the usual VAS base connection at A. Figure 8.12 demonstrates that this is effective, PSRR at 1 kHz improving by about 20 dB.

Elegant or not, the simplest way to reduce ripple below the noise floor still seems to be brute-force RC filtering of the V– supply to the input mirror and VAS, removing the disturbances before they enter. It may be crude, but it is effective, as shown in Figure 8.13. Good LF PSRR requires a large RC time-constant, and the response at DC is naturally unimproved, but

Table 8.2

RNFB	Ripple Out (dBu)
None	83.3
470 k	85.0
200 k	80.1
100 k	73.9

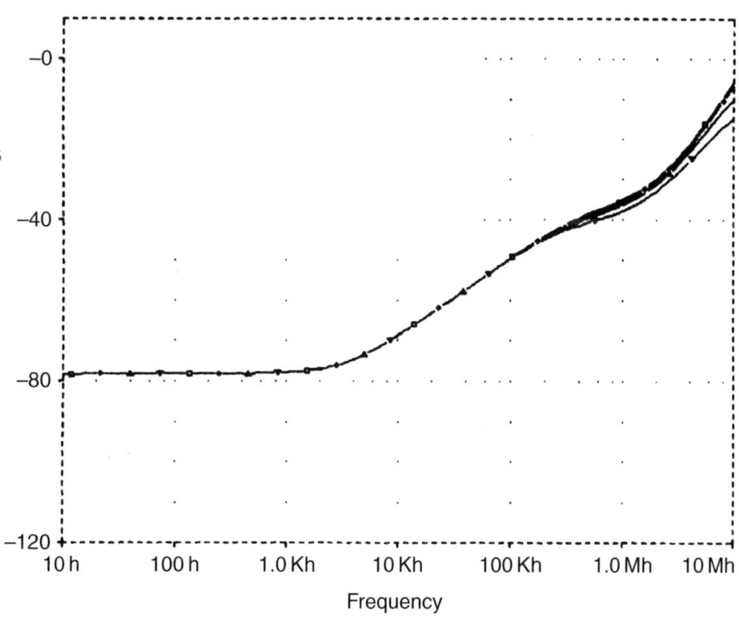

Figure 8.12
Using an input cascode to change the reference for Cdom. The LF PSRR is unchanged, but extends much higher in frequency. (Compare Curve 2 in Figure 8.7.) Note that Cdom value now has little effect

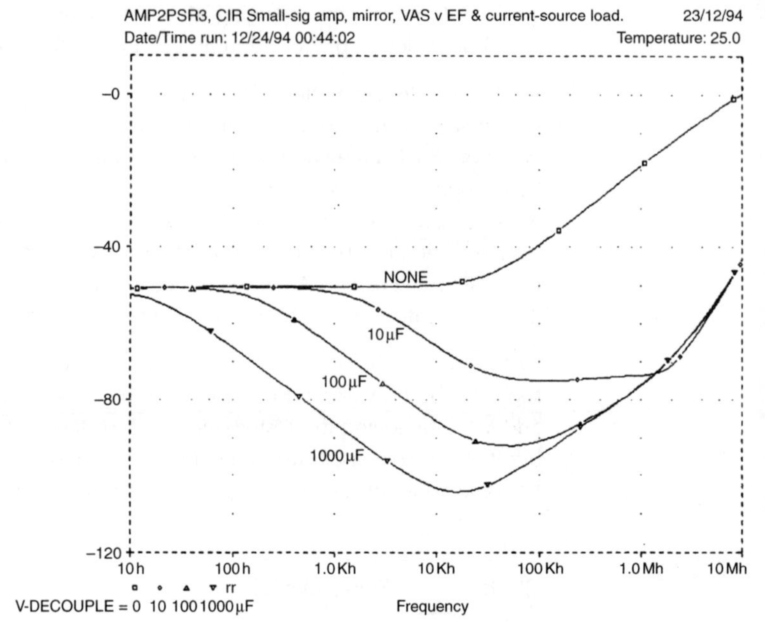

Figure 8.13
RC filtering of the V– rail is effective at medium frequencies, but less good at LF, even with 100 µF of filtering. R = 10 Ω

the real snag is that the necessary voltage drop across R directly reduces amplifier output swing, and since the magic number of watts available depends on voltage squared, it can make a surprising difference to the raw commercial numbers (though not, of course, to perceived loudness). With the circuit values shown 10 Ω is about the maximum tolerable value; even this gives a measurable reduction in output. The accompanying C should be at least 220 µF, and a higher value is desirable if every trace of ripple is to be removed.

References

1. Sinclair (ed) *Audio and Hi-Fi Handbook* pub. Newnes 1993, p. 541.
2. Linsley-Hood, J *Evolutionary Audio. Part 3* Electronics World, Jan 1990, p. 18.
3. Williams, T *EMC For Product Designers* pub. Newnes (Butterworth-Heinemann) 1992, ISBN 0 7506 1264 9, p. 106.
4. Ball, G *Distorting Power Supplies* EW+WW, Dec 90, p. 1084.
5. Ribner and Copeland *Design Techniques for Cascoded CMOS Opamps* IEEE J. Solid-State Circuits, Dec 1984, p. 919.
6. Ahuja, B K *Improved Frequency Compensation Technique for CMOS Opamps.* IEEE J. Solid-State Circuits, Dec 1983, pp. 629–633.

9

Class-A power amplifiers

An introduction to class-A

The two salient facts about Class-A amplifiers are that they are inefficient, and that they give the best possible distortion performance. They will never supplant Class-B amplifiers; but they will always be around.

The quiescent dissipation of the classic Class-A amplifier is equal to twice the maximum output power, making massive power outputs impractical, if only because of the discomfort engendered in the summer months. However, the nature of human hearing means that the power of an amplifier must be considerably increased to sound significantly louder. Doubling the sound pressure level (SPL) is not the same as doubling subjective loudness, the latter being measured in Sones rather than dB above threshold, and it appears that doubling subjective loudness requires nearer a 10 dB rather than 6 dB rise in SPL[1]. This implies amplifier power must be increased something like ten-fold, rather than merely quadrupled, to double subjective loudness. Thus a 40 W Class-B amplifier does not sound much larger than its 20 W Class-A cousin.

There is an attractive simplicity and purity about Class A. Most of the distortion mechanisms studied so far stem from Class B, and we can thankfully forget crossover and switchoff phenomena (Distortions 3b, 3c), non-linear VAS loading (Distortion 4), injection of supply-rail signals (Distortion 5), induction from supply currents (Distortion 6), and erroneous feedback connections (Distortion 7). Beta-mismatch in the output devices can also be ignored.

The only real disadvantage of Class-A is inefficiency, so inevitably efforts have been made to compromise between A and B. As compromises go, traditional Class-AB is not a happy one (see Chapters 5 and 6) because when the AB region is entered the step-change in gain generates significantly greater high-order distortion than that from optimally biased Class-B. However, a well-designed AB amplifier does give pure Class-A performance below the AB threshold, something a Class-B amp cannot do.

Another possible compromise is the so-called non-switching amplifier, with its output devices clamped to always pass a minimum current. However, it is not obvious that a sudden halt in current-change as opposed to complete turn-off makes a better crossover region. Those residual oscillograms that have been published seem to show that some kind of discontinuity still exists at crossover[2].

One potential problem is the presence of maximum ripple on the supply-rails at zero signal output; the PSRR must be taken seriously if good noise and ripple figures are to be obtained. This problem is simply solved by the measures proposed for Class-B designs in Chapter 8.

Class-A configurations and efficiency

There is a canonical sequence of efficiency in Class-A amplifiers. The simplest version is single-ended and resistively loaded, as at Figure 9.1a. When it sinks output current, there is an inevitable voltage drop across the emitter resistance, limiting the negative output capability, and resulting in an efficiency of 12.5% (erroneously quoted in at least one textbook as 25%, apparently on the grounds that power not dissipated in silicon does not count). This would be of purely theoretical interest – and not much of that – except that a single-ended design by Fuller Audio has recently appeared. This reportedly produces a 10 W output for a dissipation of 120 W, with output swing predictably curtailed in one direction[3].

A better method – Constant-current Class-A – is shown in Figure 9.1b. The current sunk by the lower constant-current source is no longer related to the voltage across it, and so the output voltage can approach the negative rail with a practicable quiescent current. (Hereafter shortened to Iq). Maximum efficiency is doubled to 25% at maximum output; for an example with 20 W output (and a big fan) see Nelson[4]. Some versions (Krell) make the current-source value switchable, controlling it with a kind of noise-gate.

Push-pull operation once more doubles full-power efficiency, getting us to a more practical 50%; most commercial Class-A amplifiers have been of this type. Both output halves now swing from zero to twice the Iq, and least voltage corresponds with maximum current, reducing dissipation. There is also the intriguing prospect of cancelling the even-order harmonics generated by the output devices.

Push-pull action can be induced in several ways. Figures 9.1c, d show the lower constant current-source replaced by a voltage-controlled current-source (VCIS). This can be driven directly by the amplifier forward path,

Figure 9.1
The canonical sequence of Class-A configurations. c, d and e are push-pull variants, and achieve 50% efficiency. e is simply a Class-B stage with higher Vbias

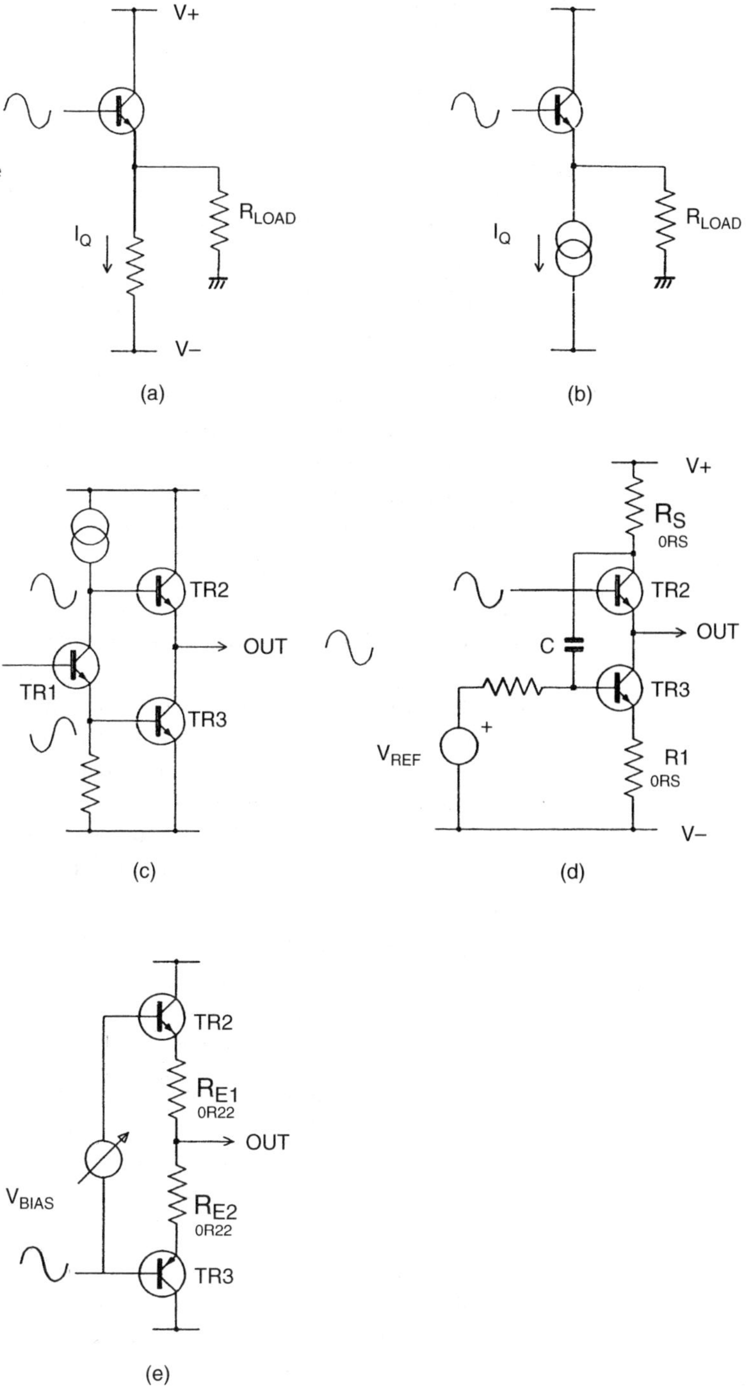

as in Figure 9.1c[5], or by a current-control negative-feedback loop, as at Figure 9.1d[6]. The first of these methods has the drawback that the stage generates gain, phase-splitter TR1 doubling as the VAS; hence there is no circuit node that can be treated as the input to a unity-gain output stage, making the circuit hard to analyse, as VAS distortion cannot be separated from output stage non-linearity. There is also no guarantee that upper and lower output devices will be driven appropriately for Class-A; in Linsley-Hood[5] the effective quiescent varies by more than 40% over the cycle.

The second push-pull method in Figure 9.1d is more dependable, and I have designed several versions that worked well. The disadvantage with the simple version shown is that a regulated supply is required to prevent rail ripple from disrupting the current-loop control. Designs of this type have a limited current-control range – in Figure 9.1d TR3 cannot be turned on any further once the upper device is fully off – so the lower VCIS will not be able to respond to an unforeseen increase in the output loading. In this event there is no way of resorting to Class-AB to keep the show going and the amplifier will show some form of asymmetrical hard clipping.

The best push-pull stage seems to be that in Figure 9.1e, which probably looks rather familiar. Like all the conventional Class-B stages examined in Chapters 5 and 6, this one will operate effectively in pure push-pull Class-A if the quiescent bias voltage is sufficiently increased; the increment over Class-B is typically 700 mV, depending on the value of the emitter resistors. For an example of high-biased Class B see Nelson-Jones[7]. This topology has the great advantage that, when confronted with an unexpectedly low load impedance, it will operate in Class-AB. The distortion performance will be inferior not only to Class-A but also to optimally biased Class-B, once above the AB transition level, but can still be made very low by proper design.

The push-pull concept has a maximum efficiency of 50%, but this is only achieved at maximum sinewave output; due to the high peak/average ratio of music, the true average efficiency probably does not exceed 10%, even at maximum volume before obvious clipping.

Other possibilities are signal-controlled variation of the Class-A amplifier rail voltages, either by a separate Class-B amplifier or by a modulated switch-mode supply. Both approaches are capable of high power output, but involve extensive extra circuitry, and present some daunting design problems.

A Class-B amplifier has a limited voltage output capability, but is flexible about load impedances; more current is simply turned on when required. However, Class-A has also a current limitation, after which it enters Class AB, and so loses its *raison d'être*. The choice of quiescent value has a major effect on thermal design and parts cost; so Class-A design demands a very clear idea of what load impedance is to be driven in pure A before we begin. The calculations to determine the required Iq are straightforward, though lengthy if supply ripple, Vce(sat)s, and Re losses, etc. are all considered, so I just give the results here. (An unregulated supply with 10,000 μF reservoirs is assumed.)

A 20 W/8 Ω amplifier will require rails of approximately ±24 V and a quiescent of 1.15 A. If this is extended to give roughly the same voltage swing into 4 Ω, then the output power becomes 37 W, and to deliver this in Class-A the quiescent must increase to 2.16 A, almost doubling dissipation. If however full voltage swing into 6 Ω will do (which it will for many reputable speakers) then the quiescent only needs to increase to 1.5 A; from here on I assume a quiescent of 1.6 A to give a margin of safety.

Output stages in Class-A

I consider here only the increased-bias Class-B topology, because it is probably the best approach, effectively solving the problems presented by the other methods. Figure 9.2 shows a Spice simulation of the collector currents in the output devices versus output voltage, and also the sum of these currents. This sum of device currents is in principle constant in Class-A, though it need not be so for low THD; the output signal is the difference of device currents, and is not inherently related to the sum. However, a large

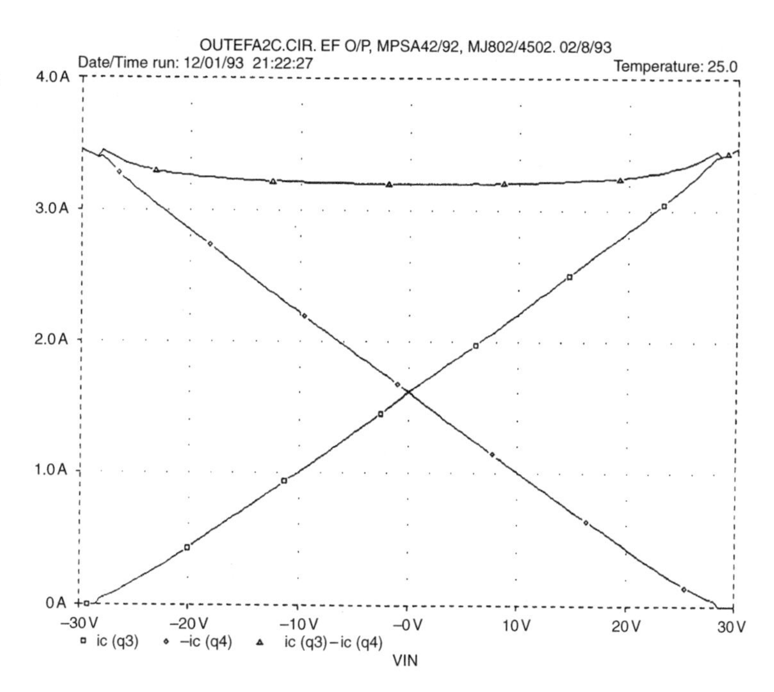

Figure 9.2
How output device current varies in push-pull Class-A. The sum of the currents is near-constant, simplifying biasing

deviation from this constant-sum condition means increased inefficiency, as the stage must be conducting more current than it needs to for some part of the cycle.

The constancy of this sum-of-currents is important because it shows that the voltage measured across Re1 and Re2 together is also effectively constant so long as the amplifier stays in Class-A. This in turn means that quiescent current can be simply set with a constant-voltage bias generator, in very much the same way as Class-B.

Figures 9.3, 9.4 and 9.5 show Spice gain plots for open-loop output stages, with 8 Ω loading and 1.6 A quiescent; the circuitry is exactly as for Class-B in Chapter 6. The upper traces show Class-A gain, and the lower traces optimal-bias Class-B gain for comparison. Figure 9.3 shows an emitter-follower output, Figure 9.4 a simple quasi-complementary stage, and Figure 9.5 a CFP output.

We would expect Class-A stages to be more linear than B, and they are. (Harmonic and THD figures for the three configurations, at 20 V Pk, are shown in Table 9.1.) There is absolutely no gain wobble around 0 V, as in Class-B, and push-pull Class-A really can and does cancel even-order distortion.

It is at once clear that the emitter-follower has more gain variation, and therefore worse linearity, than the CFP, while the quasi-comp circuit shows

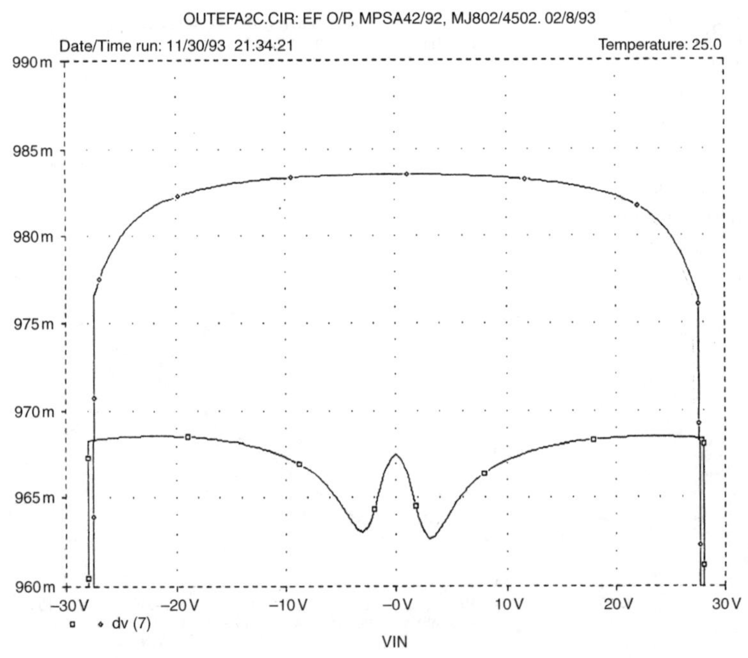

Figure 9.3 Gain linearity of the Class-A emitter-follower output stage. Load is 8 Ω, and quiescent current (Iq) is 1.6 A

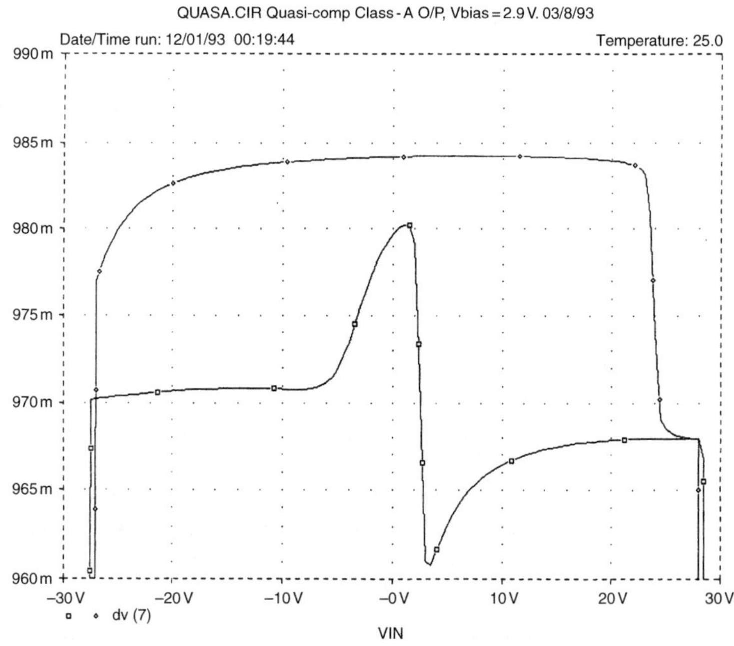

Figure 9.4 Gain linearity of the Class-A quasi-complementary output stage. Conditions as Figure 9.3

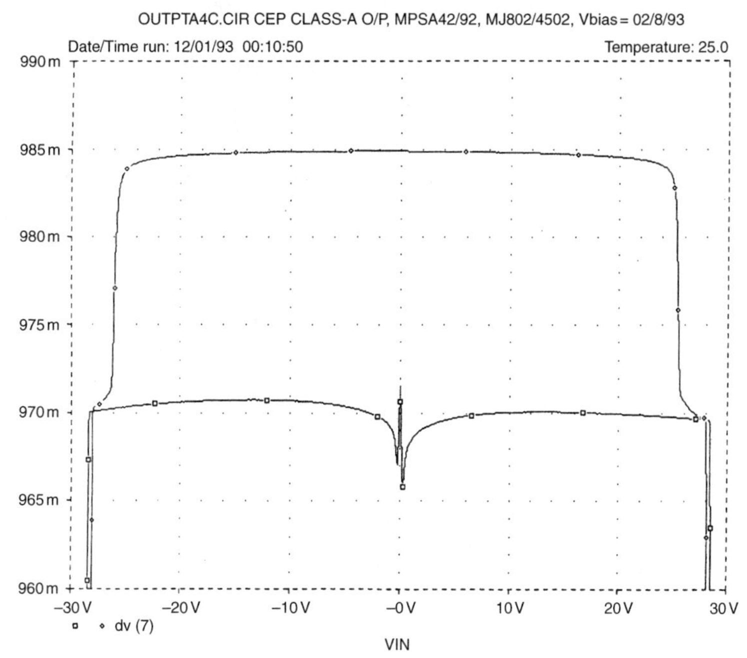

Figure 9.5 Gain linearity of the Class-A CFP output stage

Table 9.1

Harmonic	Emitter Follower (%)	Quasi-Comp (%)	CFP Output (%)
Second	0.00012	0.0118	0.00095
Third	0.0095	0.0064	0.0025
Fourth	0.00006	0.0011	0.00012
Fifth	0.00080	0.00058	0.00029
THD	0.0095	0.0135	0.0027

(THD is calculated from the first nine harmonics, though levels above the fifth are very small)

an interesting mix of the two. The more curved side of the quasi gain plot is on the −ve side, where the CFP half of the quasi circuit is passing most of the current; however we know by comparing Figure 9.3 and Figure 9.5 that the CFP is the more linear structure. Therefore it appears that the shape of the gain curve is determined by the output half that is turning off, presumably because this shows the biggest *gm* changes. The CFP structure maintains *gm* better as current decreases, and so gives a flatter gain curve with less rounding of the extremes.

The gain behaviour of these stages is reflected in their harmonic generation; Table 9.1 reveals that the two symmetrical topologies give mostly odd-order harmonics, as expected. The asymmetry of the quasi-comp version causes a large increase in even-order harmonics, and this is reflected in the higher THD figure. Nonetheless all the THD figures are still 2 to 3 times lower than for their Class-B equivalents.

This modest factor of improvement may seem a poor return for the extra dissipation of Class-A, but not so. The crucial point about the distortion from a Class-A output stage is not just that it is low in magnitude, but that it is low-order, and so benefits much more from the typical NFB factor that falls with frequency than does high-order crossover distortion.

The choice of Class-A output topology is now simple. For best performance, use the CFP; apart from greater basic linearity, the effects of output device temperature on Iq are servoed-out by local feedback, as in Class B. For utmost economy, use the quasi-complementary with two NPN devices; these need only a low Vce(max) for a typical Class-A amp, so here is an opportunity to recoup some of the money spent on heatsinking. The rules here are somewhat different from Class-B; the simple quasi-complementary configuration gives first-class results with moderate NFB, and adding a Baxandall diode to simulate a complementary emitter-follower stage gives little improvement in linearity. See however Nelson-Jones[7] for an example of its use.

It is sometimes assumed that the different mode of operation of Class-A makes it inherently short-circuit proof. This may be true with some configurations, but the high-biased type studied here will continue delivering current in time-honoured Class-B fashion until it bursts, and overload protection seems to be no less essential.

Quiescent current control systems

Unlike Class-B, precise control of quiescent current is not required to optimise distortion; for good linearity there just has to be enough of it. However, the Iq must be under some control to prevent thermal runaway, particularly if the emitter-follower output is used. A badly designed quiescent controller can ruin the linearity, and careful design is required. There is also the point that a precisely held standing-current is considered the mark of a well-bred Class-A amplifier; a quiescent that lurches around like a drunken sailor does not inspire confidence.

Straightforward thermal compensation with a Vbe-multiplier bias generator works[8], and will prevent thermal runaway. However, unlike Class-B, Class-A gives the opportunity of tightly controlling Iq by negative feedback. This is profoundly ironic because now that we can precisely control Iq, it is no longer critical. Nevertheless it seems churlish to ignore the opportunity, and so feedback quiescent control will be examined.

There are two basic methods of feedback current-control. In the first, the current in one output device is monitored, either by measuring the voltage across one emitter-resistor (Rs in Figure 9.6a), or by a collector sensing resistor; the second method monitors the sum of the device currents, which as described above, is constant in Class-A.

The first method as implemented in Figure 9.6a[7] compares the Vbe of TR4 with the voltage across Rs, with filtering by RF, CF. If quiescent is excessive, then TR4 conducts more, turning on TR5 and reducing the bias voltage between points A and B. In Figure 9.6b, which uses the VCIS approach, the voltage across collector sensing resistor Rs is compared with Vref by TR4, the value of Vref being chosen to allow for TR4 Vbe[9]. Filtering is once more by RF, CF.

For either Figure 9.6a or b, the current being monitored contains large amounts of signal, and must be low-pass filtered before being used for control purposes. This is awkward as it adds one more time-constant to worry about if the amplifier is driven into asymmetrical clipping. In the case of collector-sensing there are unavoidable losses in the extra sense resistor. It is also my experience that imperfect filtering causes a serious rise in LF distortion.

The Better Way is to monitor current in *both* emitter resistors; as explained above, the voltage across both is very nearly constant, and in practice filtering is unnecessary. An example of this approach is shown in Figure 9.6c,

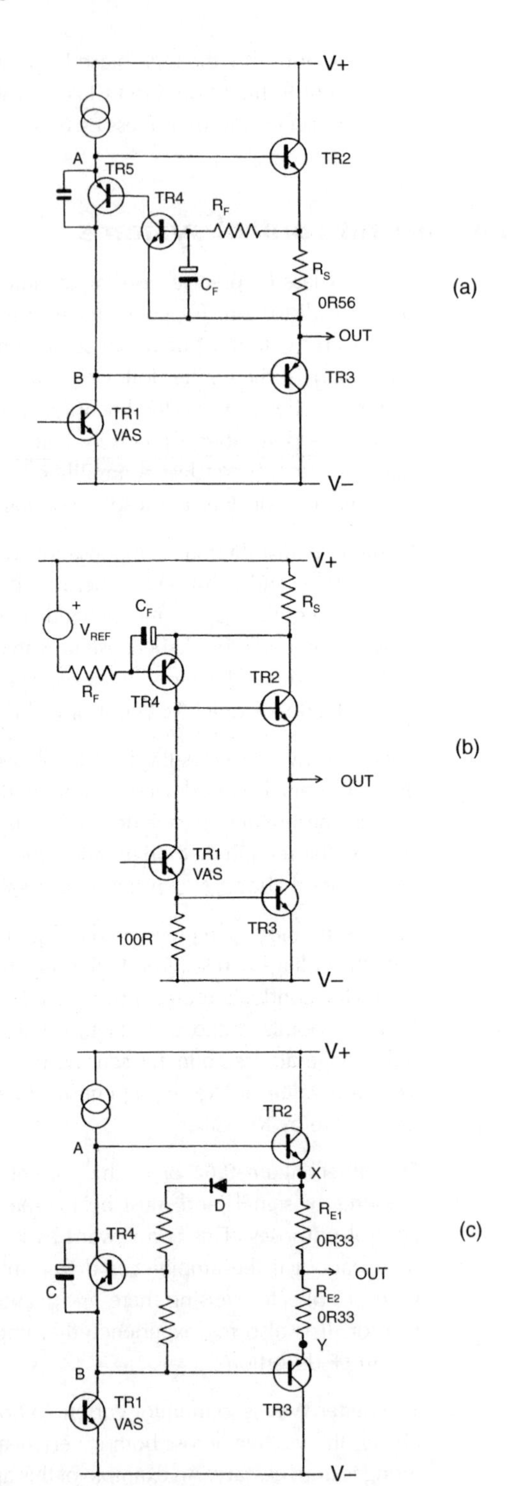

Figure 9.6 Current-control systems. Only that at c avoids the need to low-pass filter the control signal; C simply provides feedforward to speed up signal transfer to TR2

Table 9.2
Iq change per degree C

	Changing TR7 temp only (%)	Changing Global temp (%)
Quasi + Vbe-mult	+0.112	−0.43
Pass: as Figure 9.6c	+0.0257	−14.1
Pass: no diode D	+0.0675	−10.7
New system:	+0.006	−0.038

(assuming OR22 emitter resistors and 1.6 A Iq)

based on a concept originated by Nelson Pass[10]. Here TR4 compares its own Vbe with the voltage between X and B; excessive quiescent turns on TR4 and reduces the bias directly. Diode D is not essential to the concept, but usefully increases the current-feedback loop-gain; omitting it more than doubles Iq variation with TR7 temperature in the Pass circuit.

The trouble with this method is that TR3 Vbe directly affects the bias setting, but is outside the current-control loop. A multiple of Vbe is established between X and B, when what we really want to control is the voltage between X and Y. The temperature variations of TR4 and TR3 Vbe partly cancel, but only partly. This method is best used with a CFP or quasi output so that the difference between Y and B depends only on the driver temperature, which can be kept low. The *reference* is TR4 Vbe, which is itself temperature-dependent; even if it is kept away from the hot bits it will react to ambient temperature changes, and this explains the poor performance of the Pass method for global temp changes (Table 9.2).

A novel quiescent current controller

To solve this problem, I would like to introduce the novel control method in Figure 9.7. We need to compare the floating voltage between X and Y with a fixed reference, which sounds like a requirement for two differential amplifiers. This can be reduced to one by sitting the reference Vref on point Y; this is a very low-impedance point and can easily swallow a reference current of 1 mA or so. A simple differential pair TR15, 16 then compares the reference voltage with that at point Y; excess quiescent turns on TR16, causing TR13 to conduct more and reducing the bias voltage.

The circuitry looks enigmatic because of the high-impedance of TR13 collector would seem to prevent signal from reaching the upper half of the output stage; this is in essence true, but the vital point is that TR13 is part of an NFB loop that establishes a voltage at A that will keep the bias voltage between A and B constant. This comes to the same thing as maintaining a constant Vbias across TR5. As might be imagined, this loop does not shine at transferring signals quickly, and this duty is done by feed-forward

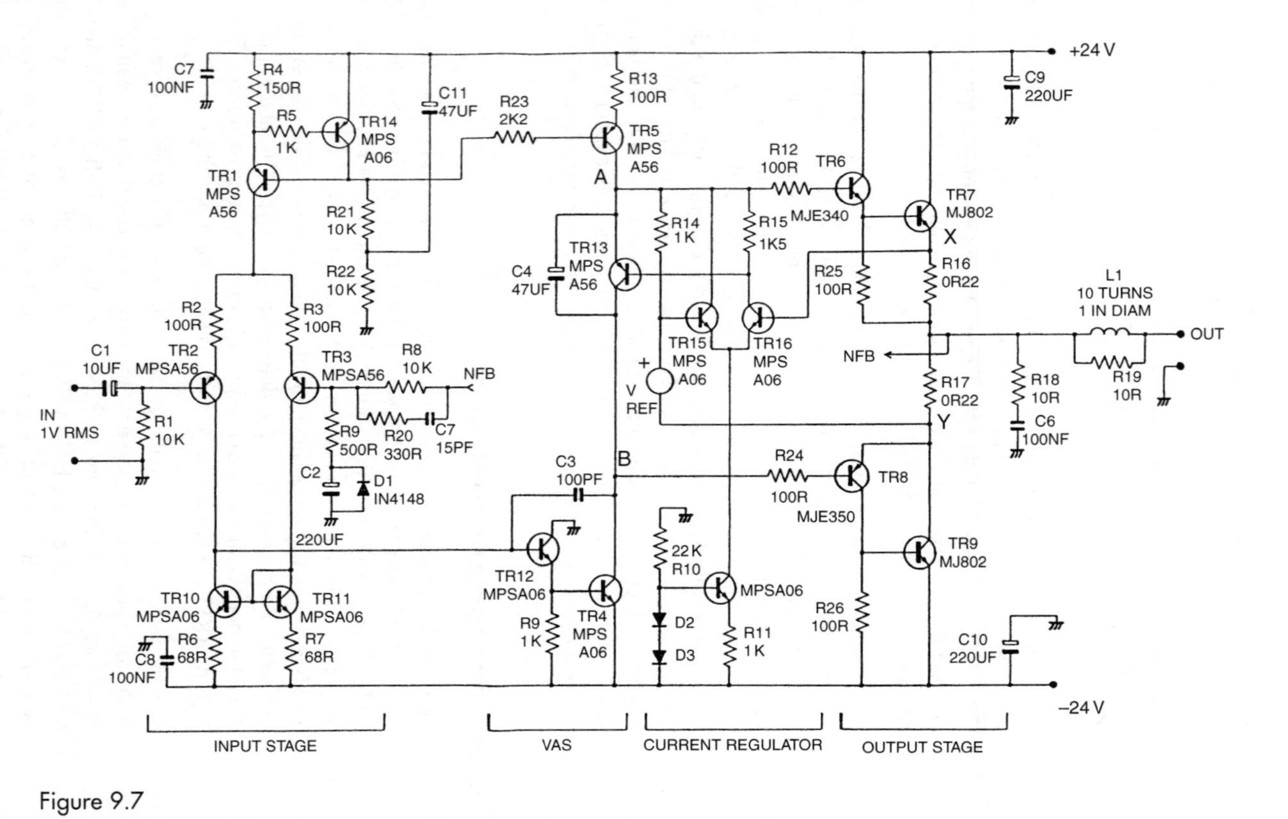

Figure 9.7
A Blameless 20 W Class-A power amplifier, using the novel current-control system

capacitor C4. Without it, the loop (rather surprisingly) works correctly, but HF oscillation at some part of the cycle is almost certain. With C4 in place the current-loop does not need to move quickly, since it is not required to transfer signal but rather to maintain a DC level.

The experimental study of Iq stability is not easy because of the inaccessibility of junction temperatures. Professional SPICE implementations like PSpice allow both the global circuit temperature and the temperature of individual devices to be manipulated; this is another aspect where simulators shine. The exact relationships of component temperatures in an amplifier is hard to predict, so I show here only the results of changing the global temperature of all devices, and changing the junction temp of TR7 alone (Figure 9.7) with different current-controllers. TR7 will be one of the hottest transistors and unlike TR9 it is not in a local NFB loop, which would greatly reduce its thermal effects.

A Class-A design

A design example of a Blameless 20 W/8 Ω Class-A power amplifier is shown in Figure 9.7. This is as close as possible in operating parameters to the previous Class-B design, to aid comparison; in particular the NFB factor remains 30 dB at 20 kHz. The front-end is as for the Class-B version, which should not be surprising as it does exactly the same job, input Distortion 1 being unaffected by output topology. As before the input pair uses a high tail current, so that R2, 3 can be introduced to linearise the transfer characteristic and set the transconductance. Distortion 2 (VAS) is dealt with as before, the beta-enhancer TR12 increasing the local feedback through Cdom. There is no need to worry about Distortion 4 (non-linear loading by output stage) as the input impedance of a Class-A output, while not constant, does not have the sharp variations shown by Class-B.

Figure 9.7 uses a standard quasi output. This may be replaced by a CFP stage without problems. In both cases the distortion is extremely low, but gratifyingly the CFP proves even better than the quasi, confirming the simulation results for output stages in isolation.

The operation of the current regulator TR13, 15, 16 has already been described. The reference used is a National LM385/1.2. Its output voltage is fixed at 1.223 V nominal; this is reduced to approximately 0.6 V by a 1 k–1 k divider (not shown). Using this band-gap reference, a 1.6 A Iq is held to within ±2 mA from a second or two after switch-on. Looking at Table 9.2, there seems no doubt that the new system is effective.

As before, a simple unregulated power supply with 10,000 μF reservoirs was used, and despite the higher prevailing ripple, no PSRR difficulties were encountered once the usual decoupling precautions were taken.

The closed-loop distortion performance (with conventional compensation) is shown in Figure 9.8 for the quasi-comp output stage, and in Figure 9.9 for a CFP output version. The THD residual is pure noise for almost all of the audio spectrum, and only above 10 kHz do small amounts of third-harmonic appear. The expected source is the input pair, but this so far remains unconfirmed.

The distortion generated by the Class-B and A design examples is summarised in Table 9.3, which shows a pleasing reduction as various measures are taken to deal with it. As a final fling two-pole compensation was applied to the most linear (CFP) of the Class-A versions, reducing distortion to a rather small 0.0012% at 20 kHz, at some cost in slew-rate. (Figure 9.10). While this may not be the fabled Distortionless Amplifier, it must be a near relation.

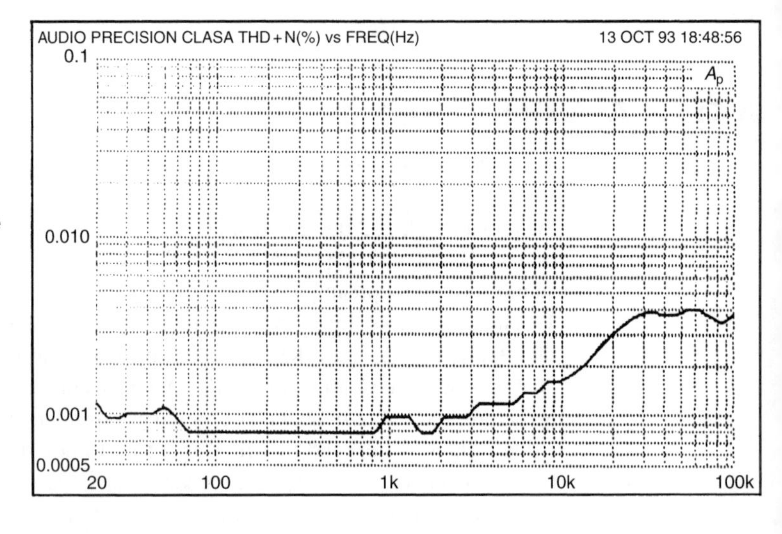

Figure 9.8
Class-A amplifier THD performance with quasi-comp output stage. The steps in the LF portion of the trace are measurement artefacts

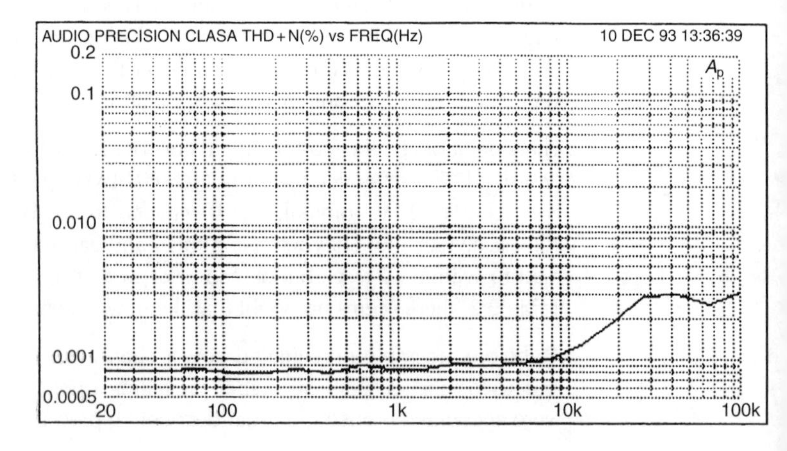

Figure 9.9
Class-A distortion performance with CFP output stage

Table 9.3

	1 kHz (%)	10 kHz (%)	20 kHz (%)	Power (W)
Class B EF	<0.0006	0.0060	0.012	50
Class B CFP	<0.0006	0.0022	0.0040	50
Class B EF 2-pole	<0.0006	0.0015	0.0026	50
Class A quasi	<0.0006	0.0017	0.0030	20
Class A CFP	<0.0006	0.0010	0.0018	20
Class A CFP 2-pole	<0.0006	0.0010	0.0012	20

(All for 8 Ω loads and 80 kHz bandwidth. Single-pole compensation unless otherwise stated)

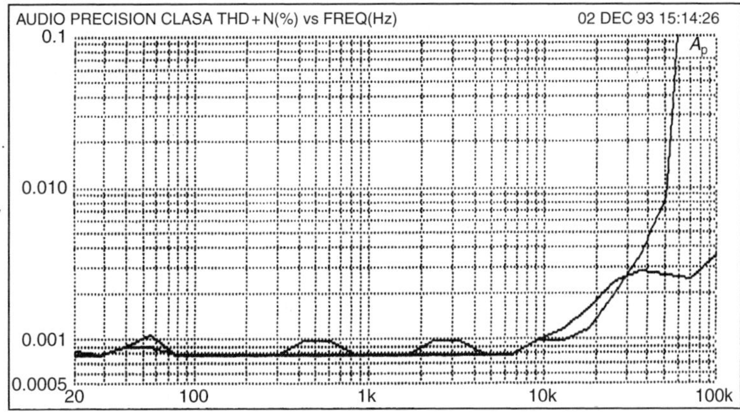

Figure 9.10 Distortion performance for CFP output stage with 2-pole compensation. The THD drops to 0.0012% at 20 kHz, but the extra VAS loading has compromised the positive-going slew capability

The trimodal amplifier

I present here my own contribution to global warming in the shape of an improved Class-A amplifier; it is believed to be unique in that it not only copes with load impedance dips by means of the most linear form of Class-AB possible, but will also operate as a Blameless Class-B engine. The power output in pure Class-A is 20 to 30 W into 8 Ω, depending on the supply-rails chosen.

This amplifier uses a Complementary-Feedback Pair (CFP) output stage for best possible linearity, and some incremental improvements have been made to noise, slew-rate and maximum DC offset. The circuit naturally bears a very close resemblance to a Blameless Class-B amplifier, and so it was decided to retain the Class-B Vbe-multiplier, and use it as a safety-circuit to prevent catastrophe if the relatively complex Class-A current-regulator failed. From this the idea arose of making the amplifier instantly switchable between Class-A and Class-B modes, which gives two kinds of amplifier for the price of one, and permits of some interesting listening tests. Now you really can do an A/B comparison.

In the Class-B mode the amplifier has the usual negligible quiescent dissipation. In Class-A the thermal dissipation is naturally considerable, as true Class-A operation is extended down to 6 Ω resistive loads for the full output voltage swing, by suitable choice of the quiescent current; with heavier loading the amplifier gracefully enters Class-AB, in which it will give full output down to 3 Ω before the Safe-Operating-Area (SOAR) limiting begins to act. Output into 2 Ω is severely curtailed, as it must be with one output pair, and this kind of load is definitely not recommended.

In short, the amplifier allows a choice between:

1 being very linear all the time (Blameless Class-B) and
2 ultra-linear most of the time (Class-A) with occasional excursions into Class-AB. The AB mode is still extremely linear by current standards, though inherently it can never be quite as good as properly handled Class-B. Since there are three classes of operation I have decided to call the design a *trimodal* power amplifier.

It is impossible to be sure that you have read all the literature; however, to the best of my knowledge this is the first ever Trimodal amplifier.

As previously said, designing a low-distortion Class-A amplifier is in general a good deal simpler than the same exercise for Class-B, as all the difficulties of arranging the best possible crossover between the output devices disappear. Because of this it is hard to define exactly what *Blameless* means for a Class-A amplifier. In Class-B the situation is quite different, and *Blameless* has a very specific meaning; when each of the eight or more distortion mechanisms has been minimised in effect, there always remains the crossover distortion inherent in Class-B, and there appears to be no way to reduce it without departing radically from what might be called the generic Lin amplifier configuration. Therefore the Blameless state appears to represent some sort of theoretical limit for Class-B, but not for Class-A.

However, Class-B considerations cannot be ignored, even in a design intended to be Class-A only, because if the amplifier does find itself driving a lower load impedance than expected, it will move into Class-AB, and then all the additional Class-B requirements are just as significant as for a Class-B design proper. Class-AB can never give distortion as low as optimally biased Class-B, but it can be made to approach it reasonably closely, if the extra distortion mechanisms are correctly handled.

In a class-A amplifier, certain sacrifices are made in the name of quality, and so it is reasonable not to be satisfied with anything less than the best possible linearity. The amplifier described here therefore uses the Complementary-Feedback Pair (CFP) type of output stage, which has the lowest distortion due to the local feedback loops wrapped around the output devices. It also has the advantage of better output efficiency than the emitter-follower (EF) version, and inherently superior quiescent current stability. It will shortly be seen that these are both important for this design.

Half-serious thought was given to labelling the Class-A mode *Distortionless* as the THD is completely unmeasurable across most of the audio band.

However, detectable distortion products do exist above 10 kHz, so this provocative idea was regretfully abandoned.

It seemed appropriate to take another look at the Class-A design, to see if it could be inched a few steps nearer perfection. The result is a slight improvement in efficiency, and a 2 dB improvement in noise performance. In addition the expected range of output DC offset has been reduced from ±50 mV to ±15 mV, still without any adjustment.

Load impedance and operating mode

The amplifier is 4 Ω capable in both A/AB and B operating modes, though it is the nature of things that the distortion performance is not quite so good. All solid-state amplifiers (without qualification, as far as I am aware) are much happier with an 8 Ω load, both in terms of linearity and efficiency; loudspeaker designers please note. With a 4 Ω load, Class-B operation gives better THD than Class-A/AB, because the latter will always be in AB mode, and therefore generating extra output stage distortion through gm-doubling. (Which should really be called gain-deficit-halving, but somehow I do not see this term catching on.) These not entirely obvious relationships are summarised in Table 9.4.

Figure 9.11 attempts to show diagrammatically just how power, load resistance, and operating mode are related. The rails have been set to ±20 V, which just allows 20 W into 8 Ω in Class-A. The curves are lines of constant power (i.e., V × I in the load), the upper horizontal line represents maximum voltage output, allowing for Vce(sat)s, and the sloping line on the right is the SOAR protection locus; the output can never move outside this area in either mode. The intersection between the load resistance lines

Table 9.4 Distortion and dissipation for different output stage classes

Load (Ω)	Mode	Distortion	Dissipation
8	A/AB	Very low	High
4	A/AB	High	High
8	B	Low	Low
4	B	Medium	Medium

(Note: *High distortion* in the context of this sort of amplifier means about 0.002% THD at 1 kHz and 0.01% at 10 kHz)

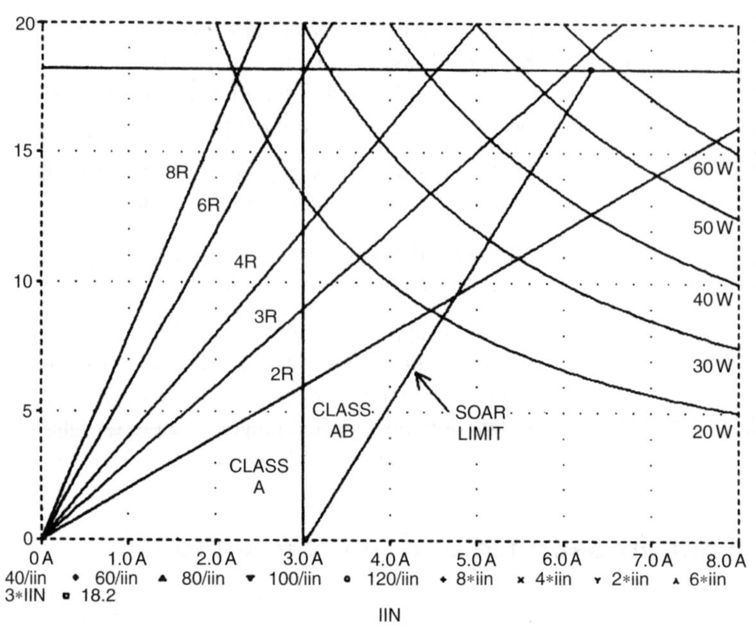

Figure 9.11
The relationships between load, mode, and power output. The intersection between the sloping load resistance lines and the ultimate limits of voltage-clipping and SOAR protection define which of the curved constant-power lines is reached. In A/AB mode, the operating point must be to the left of the vertical push-pull current-limit line for true Class-A

sloping up from the origin and the ultimate limits of voltage-clip and SOAR protection define which of the curved constant-power lines is reached.

In A/AB mode, the operating point must be left of the vertical push-pull current-limit line (at 3 A, twice the quiescent current) for Class-A. If we move to the right of this limit along one of the impedance lines, the output devices will begin turning off for part of the cycle; this is the AB operation zone. In Class-B mode, the 3 A line has no significance and the amplifier remains in optimal Class-B until clipping or SOAR limiting occurs. Note that the diagram axes represent instantaneous power in the load, but the curves show sinewave RMS power, and that is the reason for the apparent factor-of-two discrepancy between them.

Efficiency

Concern for efficiency in Class-A may seem paradoxical, but one way of looking at it is that Class-A Watts are precious things, wrought in great heat and dissipation, and so for a given quiescent power it makes sense to ensure that the amplifier approaches its limited theoretical efficiency as closely as possible. I was confirmed in this course by reading of another recent design[11] which seems to throw efficiency to the winds by using a hybrid BJT/FET cascode output stage. The voltage losses inherent in this arrangement demand ±50 V rails and six-fold output devices for a 100 W

Class-A capability; such rail voltages would give 156 W from a 100% efficient amplifier.

The voltage efficiency of a power amplifier is the fraction of the supply-rail voltage which can actually be delivered as peak-to-peak voltage swing into a specified load; efficiency is invariably less into 4 Ω due to the greater resistive voltage drops with increased current.

The Class-B amplifier I described in Chapter 6 has a voltage efficiency of 91.7% for positive swings, and 92.5% for negative, into 8 Ω. Amplifiers are not in general completely symmetrical, and so two figures need to be quoted; alternatively the lower of the two can be given as this defines the maximum undistorted sinewave. These figures above are for an emitter-follower output stage, and a CFP output does better, the positive and negative efficiencies being 94.0% and 94.7%, respectively. The EF version gives a lower output swing because it has two more Vbe drops in series to be accommodated between the supply-rails; the CFP is always more voltage-efficient, and so selecting it over the EF for the current Class-A design is the first step in maximising efficiency.

Figure 9.12 shows the basic CFP output stage, together with its two biasing elements. In Class-A the quiescent current is rigidly controlled by negative-feedback; this is possible because in Class-A the total voltage

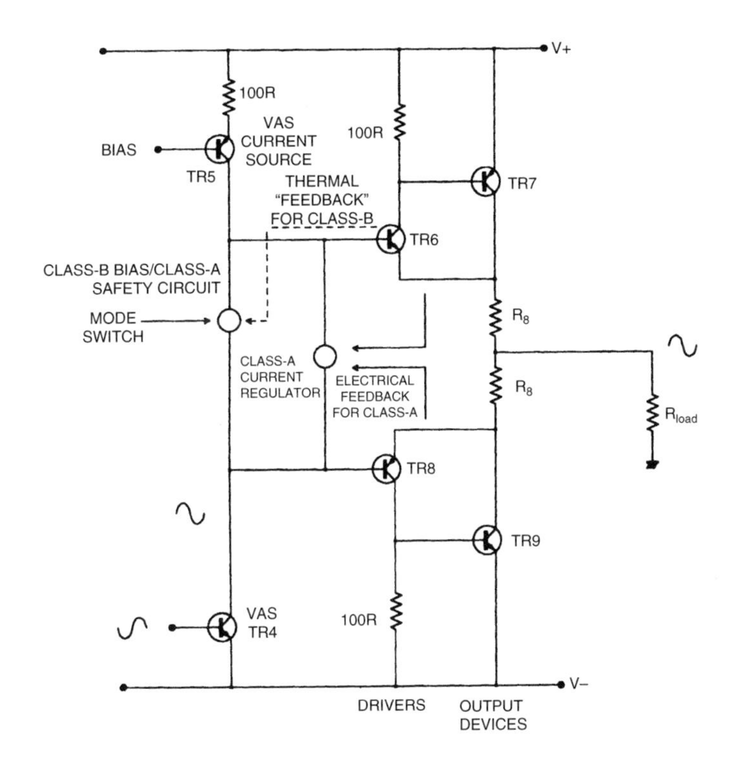

Figure 9.12
The basic CFP output stage, equally suited to operating Class B, AB and A, depending on the magnitude of Vbias. The emitter resistors Re may be from 0.1 to 0.47 Ω

across both emitter resistors Re is constant throughout the cycle. In Class-B this is not the case, and we must rely on *thermal feedback* from the output stage, though to be strictly accurate this is not *feedback* at all, but a kind of feedforward (Chapter 12). Another big advantage of the CFP configuration is that Iq depends only on driver temperature, and this is important in the Class-B mode, where true feedback control of quiescent current is not possible, especially if low-value Re's such as 0.1 Ω, are chosen, rather than the more usual 0.22 Ω; the motivation for doing this will soon become clear.

The voltage efficiency for the quasi-complementary Class-A circuit of the circuit on page 270 into 8 Ω is 89.8% positive and 92.2% negative. Converting this to the CFP output stage increases this to 92.9% positive and 93.6% negative. Note that a Class-A quiescent current (Iq) of 1.5 A is assumed throughout; this allows 31 W into 8 Ω in push-pull, if the supply-rails are adequately high. However the assumption that loudspeaker impedance never drops below 8 Ω is distinctly doubtful, to put it mildly, and so as before this design allows for full Class-A output voltage swing into loads down to 6 Ω.

So how else can we improve efficiency? The addition of extra and higher supply-rails for the small-signal section of the amplifier surprisingly does not give a significant increase in output; examination of Figure 9.13 shows why. In this region, the output device TR6 base is at a virtually constant 880 mV from the V+ rail, and as TR7 driver base rises it passes this level, and keeps going up; clipping has not yet occurred. The driver emitter follows the driver base up, until the voltage difference between this emitter

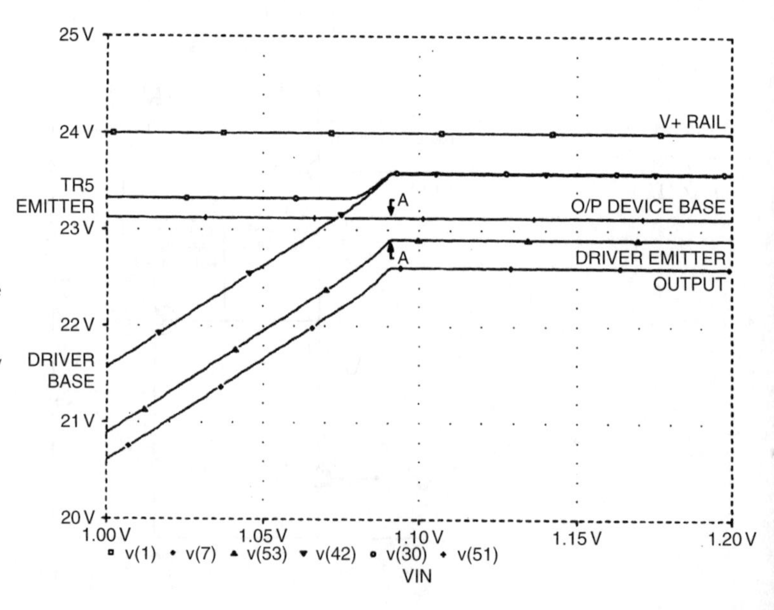

Figure 9.13 PSpice simulation showing how positive clipping occurs in the CFP output. A higher subrail for the VAS cannot increase the output swing, as the limit is set by the minimum driver Vce, and not the VAS output swing

and the output base (i.e., the driver Vce) becomes too small to allow further conduction; this choke point is indicated by the arrows A–A. At this point the driver base is forced to level off, although it is still about 500 mV below the level of V+. Note also how the voltage between V+ and TR5 emitter collapses. Thus a higher rail will give no extra voltage swing, which I must admit came as something of a surprise. Higher sub-rails for small-signal sections only come into their own in FET amplifiers, where the high Vgs for FET conduction (5 V or more) makes their use almost mandatory.

The efficiency figures given so far are all greater for negative rather than positive voltage swings. The approach to the rail for negative clipping is slightly closer because there is no equivalent to the 0.6 V bias established across R13; however this advantage is absorbed by the need to lose a little voltage in the RC filtering of the V− supply to the current-mirror and VAS. This is essential if really good ripple/hum performance is to be obtained (see Chapter 8).

In the quest for efficiency, an obvious variable is the value of the output emitter resistors Re. The performance of the current-regulator described, especially when combined with a CFP output stage, is more than good enough to allow these resistors to be reduced while retaining first-class Iq stability. I took 0.1 Ω as the lowest practicable value, and even this is comparable with PCB track resistance, so some care in the exact details of physical layout is essential; in particular the emitter resistors must be treated as four-terminal components to exclude unwanted voltage drops in the tracks leading to the resistor pads.

If Re is reduced from 0.22 Ω to 0.1 Ω then voltage efficiency improves from 92.9%/93.6%, to 94.2%/95.0%. Is this improvement worth having? Well, the voltage-limited power output into 8 Ω is increased from 31.2 to 32.2 W with ±24 V rails, at zero cost, but it would be idle to pretend that the resulting increase in SPL is highly significant; it does however provide the philosophical satisfaction that as much Class-A power as possible is being produced for a given dissipation; a delicate pleasure.

The linearity of the CFP output stage in Class-A is very slightly worse with 0.1 Ω emitter resistors, though the difference is small and only detectable open-loop; the simulated THD for 20 V pk–pk into 8 Ω is only increased from 0.0027% to 0.0029%. This is probably due simply to the slightly lower total resistance seen by the output stage.

However, at the same time, reducing the emitter resistors to 0R1 provides much lower distortion when the amplifier runs out of Class-A; it halves the size of the step gain changes inherent in Class-AB, and so effectively reduces distortion into 4 Ω loads. See Figures 9.14 and 9.15 for output linearity simulations; the measured results from a real and Blameless Trimodal

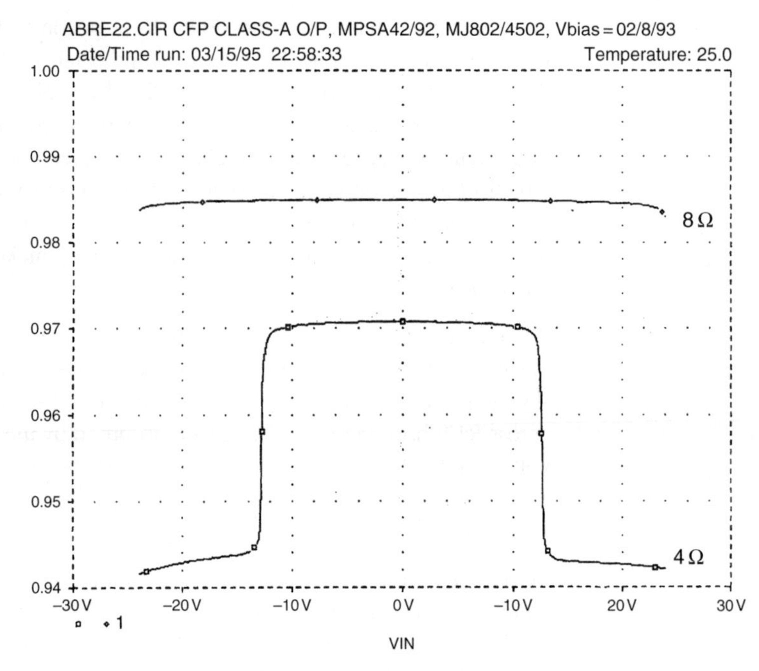

Figure 9.14
CFP output stage linearity with Re = OR22. Upper trace is Class-A into 8 Ω, lower is Class-AB operation into 4 Ω, showing step changes in gain of 0.024 units

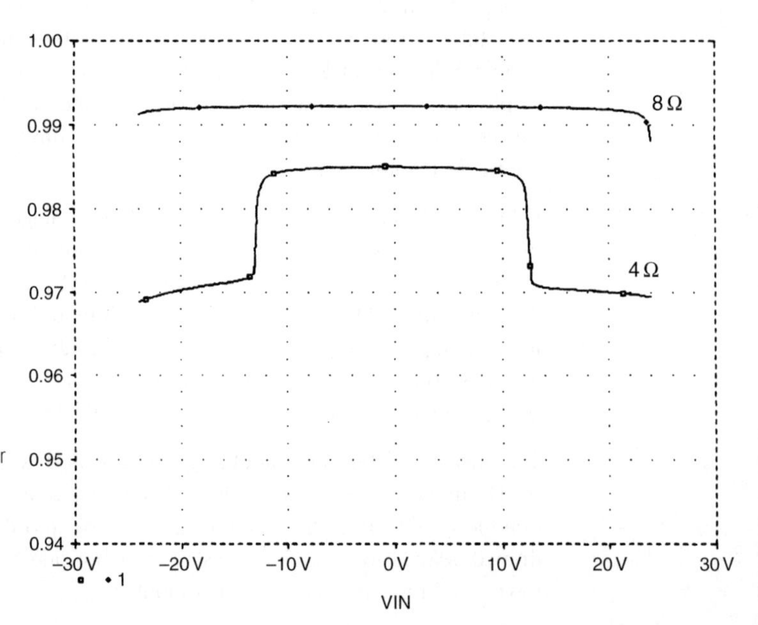

Figure 9.15
CFP output linearity with Re = OR1, re-biased to keep Iq at 1.5 A. There is slightly poorer linearity in the flat-topped Class-A region than for Re = OR22, but the 4 Ω AB steps are halved in size at 0.012 units. Note that both gains are now closer to unity; same scale as Figure 9.14

amplifier are shown in Figure 9.16, where it can be clearly seen that THD has been halved by this simple change. To the best of my knowledge this is a new result; if you must work in Class-AB, then keep the emitter resistors as low as possible, to minimise the gain changes.

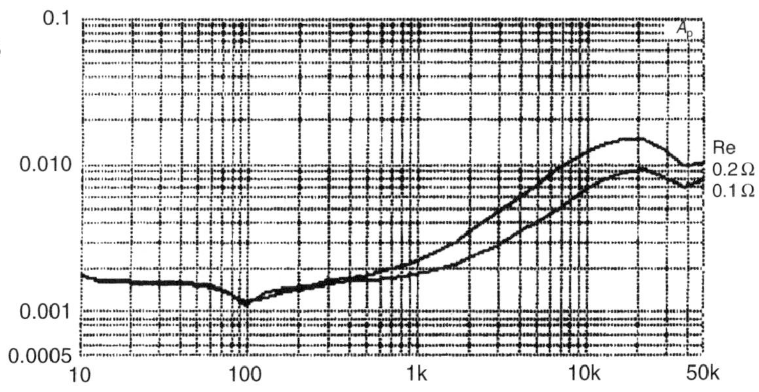

Figure 9.16 Distortion in Class-AB is reduced by lowering the value of Re

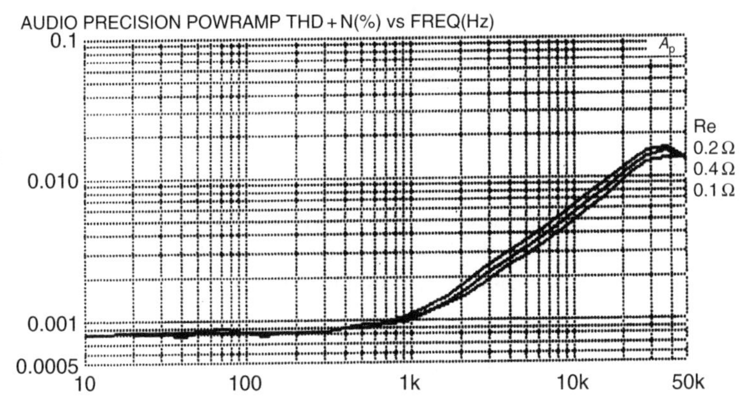

Figure 9.17 Proving that emitter resistors matter much less in Class-B. Output was 20 W in 8 Ω, with optimal bias. Interestingly, the bias does not need adjusting as the value of Re changes

Having considered the linearity of Class-A and AB, we must not neglect what effect this radical Re change has on Class-B linearity. The answer is, not very much; see Figure 9.17, where crossover distortion seems to be slightly higher with Re = 0.2 Ω than for either 0.1 or 0.4 Ω. Whether this is a consistent effect (for CFP stages anyway) remains to be seen.

The detailed mechanisms of bias control and mode-switching are described on pages 281–286.

On Trimodal biasing

Figure 9.18 shows a simplified rendering of the Trimodal biasing system; the full version appears in Figure 9.19. The voltage between points A and B is determined by one of two controller systems, only one of which can be in command at a time. Since both are basically shunt voltage regulators sitting between A and B, the result is that the lowest voltage wins. The novel Class-A current-controller introduced on page 269 is used here adapted for

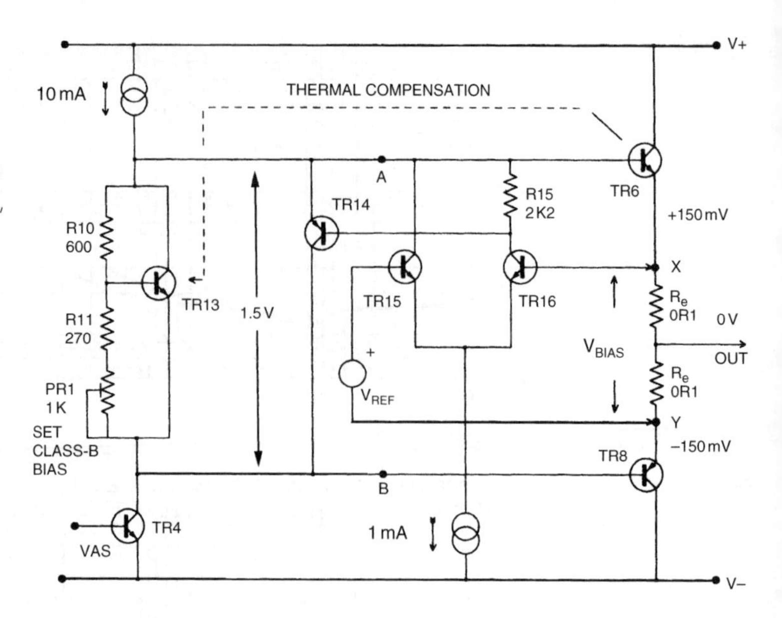

Figure 9.18
The simplified current-controller in action, showing typical DC voltages in Class-A. Points A, B, X and Y are in Figure 9.6 on page 268

0.1 Ω emitter resistors, mainly by reducing the reference voltage to 300 mV, which gives a quiescent current (Iq) of 1.5 A when established across the total emitter resistance of 0.2 Ω.

In parallel with the current-controller is the Vbe-multiplier TR13. In Class-B mode, the current-controller is disabled, and critical biasing for minimal crossover distortion is provided in the usual way by adjusting preset PR1 to set the voltage across TR13. In Class-A/AB mode, the voltage TR13 attempts to establish is increased (by shorting out PR1) to a value greater than that required for Class-A. The current-controller therefore takes charge of the voltage between X and Y, and unless it fails TR13 does not conduct. Points A, B, X, and Y are the same circuit nodes as in the simple Class-A design (see Figure 9.6c).

Class-A/AB mode

In Class-A/AB mode, the current-controller (TR14, 15, 16 in Figure 9.18) is active and TR13 is off, as TR20 has shorted out PR1. TR15, 16 form a simple differential amplifier that compares the reference voltage across R31 with the Vbias voltage across output emitter resistors R16 and R17; as explained above, in Class-A this voltage remains constant despite delivery of current into the load. If the voltage across R16, 17 tends to rise, then TR16 conducts more, turning TR14 more on and reducing the voltage between A and B. TR14, 15 and 16 all move up and down with the amplifier output, and so a tail current-source (TR17) is used.

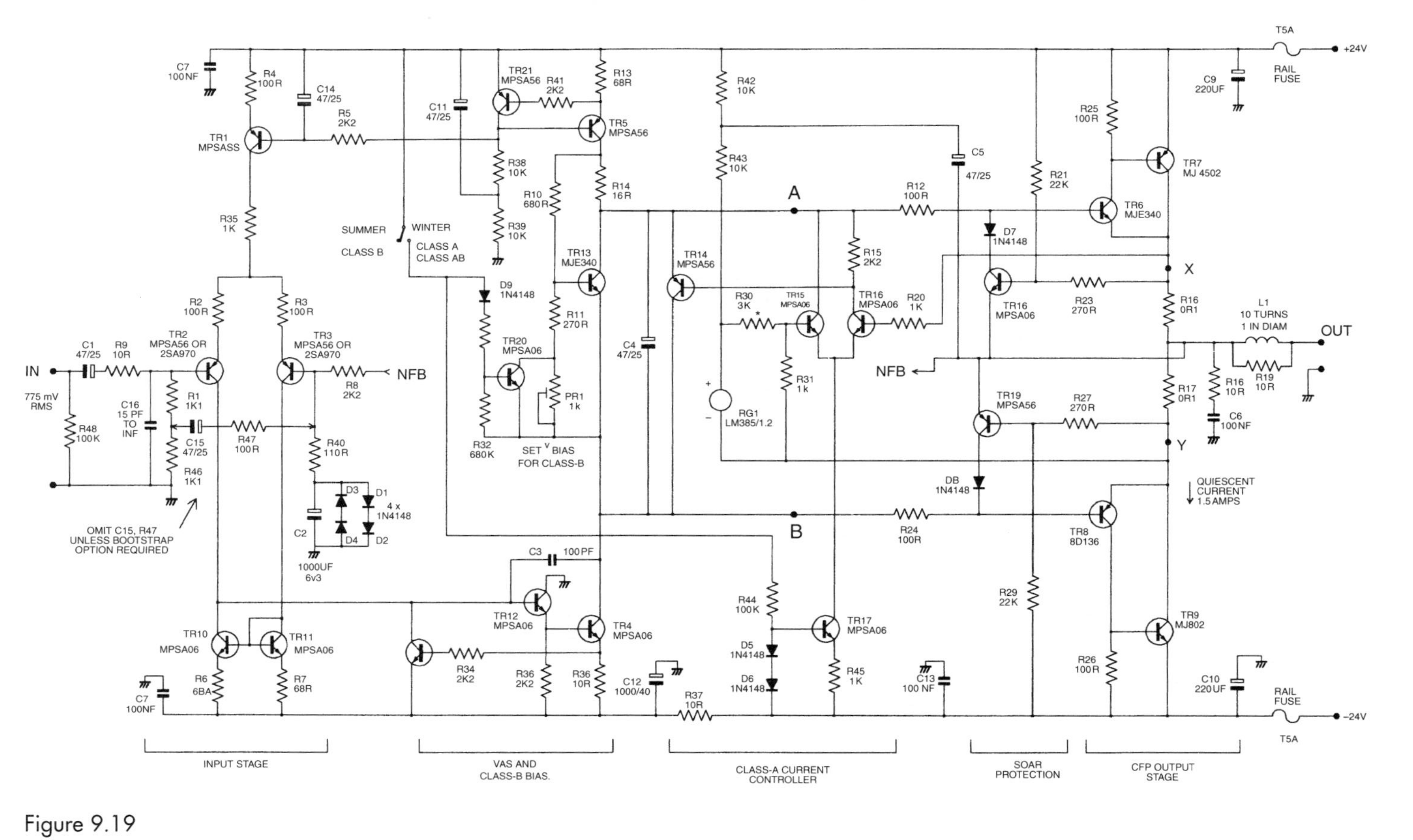

Figure 9.19
The complete circuit diagram of Trimodal amplifier, including the optional bootstrapping components, R47 and C15

I am very aware that the current-controller is more complex than the simple Vbe-multiplier used in most Class-B designs. There is an obvious risk that an assembly error could cause a massive current that would prompt the output devices to lay down their lives to save the rail fuses. The tail-source TR17 is particularly vulnerable because any fault that extinguishes the tail-current removes the drive to TR14, the controller is disabled, and the current in the output stage will be very large. In Figure 9.18 the Vbe-multiplier TR13 acts as a safety-circuit which limits Vbias to about 600 mV rather than the normal 300 mV, even if the current-controller is completely non-functional and TR14 fully off. This gives a *quiescent* of 3.0 A, and I can testify this is a survivable experience for the output devices in the short-term; however they may eventually fail from overheating if the condition is allowed to persist.

There are some important points about the current-controller. The entire tail-current for the error-amplifier, determined by TR17, is syphoned off from VAS current source TR5, and must be taken into account when ensuring that the upper output half gets enough drive current.

There must be enough tail-current available to turn on TR14, remembering that most of TR16 collector-current flows through R15, to keep the pair roughly balanced. If you feel moved to alter the VAS current, remember also that the base current for driver TR6 is higher in Class-A than Class-B, so the positive slew-rate is slightly reduced in going from Class-A to B.

The original Class-A amplifier used a National LM385/1.2, its output voltage fixed at 1.223 V nominal; this was reduced to approximately 0.6 V by a 1k–1k divider. The circuit also worked well with Vref provided by a silicon diode, 0.6 V being an appropriate Vbias drop across two 0.22 Ω output emitter resistors. This is simple, and retains the immunity of Iq to heatsink and output device temperatures, but it does sacrifice the total immunity to ambient temperature that a band-gap reference gives.

The LM385/1.2 is the lowest voltage band-gap reference commonly available; however, the voltages shown in Figure 9.18 reveal a difficulty with the new lower Vbias value and the CFP stage; points A and Y are now only 960 mV apart, which does not give the reference room to work in if it is powered from node A, as in the original circuit. The solution is to power the reference from the V+ rail, via R42 and R43. The mid-point of these two resistors is bootstrapped from the amplifier output rail by C5, keeping the voltage across R43 effectively constant. Alternatively, a current-source could be used, but this might reduce positive headroom. Since there is no longer a strict upper limit on the reference voltage, a more easily obtainable 2.56 V device could be used providing R30 is suitably increased to 5k to maintain Vref at 300 mV across R31.

In practical use, Iq stability is very good, staying within 1% for long periods. The most obvious limitation on stability is differential heating of TR15, 16 due to heat radiation from the main heatsink. TR14 should also be sited with this in mind, as heating it will increase its beta and slightly imbalance TR15, 16.

Class-B mode

In Class-B mode, the current-controller is disabled, by turning off tail-source TR17 so TR14 is firmly off, and critical biasing for minimal crossover distortion is provided as usual by Vbe-multiplier TR13. With 0.1 Ω emitter resistors Vbias (between X and Y) is approximately 10 mV. I would emphasise that in Class-B this design, if constructed correctly, will be as Blameless as a purpose-built Class-B amplifier. No compromises have been made in adding the mode-switching.

As in the previous Class-B design, the addition of R14 to the Vbe-multiplier compensates against drift of the VAS current-source TR5. To make an old but much-neglected point, the preset should always be in the bottom arm of the Vbe divider R10, 11, because when presets fail it is usually by the wiper going open; in the bottom arm this gives minimum Vbias, but in the upper it would give maximum.

In Class-B, temperature compensation for changes in driver dissipation remains vital. Thermal runaway with the CFP is most unlikely, but accurate quiescent setting is the only way to minimise cross-over distortion. TR13 is therefore mounted on the same small heatsink as driver TR6. This is often called thermal feedback, but it is no such thing as TR13 in no way controls the temperature of TR6; *thermal feedforward* would be a more accurate term.

The mode-switching system

The dual nature of the biasing system means Class-A/Class-B switching can be implemented fairly simply. A Class-A amplifier is an uneasy companion in hot weather, and so I have been unable to resist the temptation to subtitle the mode switch *Summer/Winter*, by analogy with a car air intake.

The switchover is DC-controlled, as it is not desirable to have more signal than necessary running around inside the box, possibly compromising interchannel crosstalk. In Class-A/AB mode, SW1 is closed, so TR17 is biased normally by D5, 6, and TR20 is held on via R33, shorting out preset PR1 and setting TR13 to safety mode, maintaining a maximum Vbias limit of 600 mV. For Class-B, SW1 is opened, turning off TR17 and therefore TR15, 16 and 14. TR20 also ceases to conduct, protected against reverse-bias by

D9, and reduces the voltage set by TR13 to a suitable level for Class-B. The two control pins of a stereo amplifier can be connected together, and the switching performed with a single-pole switch, without interaction or increased crosstalk.

The mode-switching affects the current flowing in the output devices, but not the output voltage, which is controlled by the global feedback loop, and so it is completely silent in operation. The mode may be freely switched while the amplifier is handling audio, which allows some interesting *A/B* listening tests.

It may be questioned why it is necessary to explicitly disable the current controller in Class-B; TR13 is establishing a lower voltage than the current controller which latter subsystem will therefore turn TR14 off as it strives futilely to increase Vbias. This is true for 8 Ω loads, but 4 Ω impedances increase the currents flowing in R16, 17 so they are transiently greater than the Class-A Iq, and the controller will therefore intermittently take control in an attempt to reduce the average current to 1.5 A. Disabling the controller by turning off TR17 via R44 prevents this.

If the Class-A controller is enabled, but the preset PR1 is left in circuit (e.g., by shorting TR20 base-emitter) we have a test mode which allows suitably cautious testing; Iq is zero with the preset fully down, as TR13 over-rides the current-controller, but increases steadily as PR1 is advanced, until it suddenly locks at the desired quiescent current. If the current-controller is faulty then Iq continues to increase to the defined maximum of 3.0 A.

Thermal design

Class-A amplifiers are hot almost by definition, and careful thermal design is needed if they are to be reliable, and not take the varnish off the Sheraton. The designer has one good card to play; since the internal dissipation of the amplifier is maximal with no signal, simply turning on the prototype and leaving it to idle for several hours will give an excellent idea of worst-case component temperatures. In Class-B the power dissipation is very program-dependent, and estimates of actual device temperatures in realistic use are notoriously hard to make.

Table 9.5 shows the output power available in the various modes, with typical transformer regulation, etc.; the output mode diagram in Figure 9.11 shows exactly how the amplifier changes mode from A to AB with decreasing load resistance. Remember that in this context *high distortion* means 0.002% at 1 kHz. This diagram was produced in the analysis section of PSpice simply by typing in equations, and without actually simulating anything at all.

Table 9.5
Power capability

	Load resistance			
	8 Ω	6 Ω	4 Ω	Distortion
Class-A	20 W	27 W	15 W	Low
Class-AB	n/a	n/a	39 W	High
Class-B	21 W	28 W	39 W	Medium

The most important thermal decision is the size of the heatsink; it is going to be expensive, so there is a powerful incentive to make it no bigger than necessary. I have ruled out fan cooling as it tends to make concern for ultra-low electrical noise look rather foolish; let us rather spend the cost of the fan on extra cooling fins and convect in ghostly silence. The exact thermal design calculations are simple but tedious, with many parameters to enter; the perfect job for a spreadsheet. The final answer is the margin between the predicted junction temperatures and the rated maximum. Once power output and impedance range are decided, the heatsink thermal resistance to ambient is the main variable to manipulate; and this is a compromise between coolness and cost, for high junction temperatures always reduce semiconductor reliability. Looking at it very roughly:

	Thermal resistance °C/W	Heat flow (W)	Temp rise (°C)	Temp (°C)
Juncn to TO3 Case	0.7	36	25	100 junction
Case to Sink	0.23	36	8	75 TO3 case
Sink to air	0.65	72	47	67 Heatsink
Total			80	20 Ambient

This shows that the transistor junctions will be 80° above ambient, i.e., at around 100°C; the rated junction maximum is 200°C, but it really is not wise to get anywhere close to this very real limit. Note the Case-Sink thermal washers were high-efficiency material, and standard versions have a slightly higher thermal resistance.

The heatsinks used in the prototype had a thermal resistance of 0.65°C/W per channel. This is a substantial chunk of metal, and since aluminium is basically congealed electricity, it's bound to be expensive.

A complete Trimodal amplifier circuit

The complete Class-A amplifier is shown in Figure 9.19, complete with optional input bootstrapping. It may look a little complex, but we have only added four low-cost transistors to realise a high-accuracy Class-A quiescent controller, and one more for mode-switching. Since the biasing system has been described above, only the remaining amplifier subsystems are dealt with here.

The input stage follows my design methodology by using a high tail current to maximise transconductance, and then linearising by adding input degeneration resistors R2, 3 to reduce the final transconductance to a suitable level. Current-mirror TR10, 11 forces the collector currents of the two input devices TR2, 3 to be equal, balancing the input stage to prevent the generation of second-harmonic distortion. The mirror is degenerated by R6, 7 to eliminate the effects of Vbe mismatches in TR10, 11. With some misgivings I added the input network R9, C15, which is definitely not intended to define the system bandwidth, unless fed from a buffer stage; with practical values the HF roll-off could vary widely with the source impedance driving the amplifier. It is intended rather to give the possibility of dealing with RF interference without having to cut tracks. R9 could be increased for bandwidth definition if the source impedance is known, fixed, and taken into account when choosing R9; bear in mind that any value over 47 Ω will measurably degrade the noise performance. The values given roll off above 150 MHz to keep out UHF.

The input-stage tail current is increased from 4 to 6 mA, and the VAS standing current from 6 to 10 mA over the original Chapter 6 circuit. This increases maximum positive and negative slew-rates from +21, −48 V/μsec to +37, −52 V/μsec; as described in Chapter 7, this amplifier architecture is bound to slew asymmetrically. One reason is feedthrough in the VAS current source; in the original circuit an unexpected slew-rate limit was set by fast edges coupling through the current-source c-b capacitance to reduce the bias voltage during positive slewing. This effect is minimised here by using the negative-feedback type of current source bias generator, with VAS collector current chosen as the controlled variable. TR21 senses the voltage across R13, and if it attempts to exceed Vbe, turns on further to pull up the bases of TR1 and TR5. C11 filters the DC supply to this circuit and prevents ripple injection from the V+ rail. R5, C14 provide decoupling to prevent TR5 from disturbing the tail-current while controlling the VAS current.

The input tail-current increase also slightly improves input-stage linearity, as it raises the basic transistor gm and allows R2, 3 to apply more local NFB.

The VAS is linearised by beta-enhancing stage TR12, which increases the amount of local NFB through Miller dominant-pole capacitor C3 (i.e., *C*dom). R36 has been increased to 2k2 to minimise power dissipation, as there seems no significant effect on linearity or slewing. Do not omit it altogether, or linearity will be affected and slewing much compromised.

As described in Chapter 8, the simplest way to prevent ripple from entering the VAS via the V– rail is old-fashioned RC decoupling, with a small R and a big C. We have some 200 mV in hand (see page 278) in the negative direction, compared with the positive, and expending this as the voltage-drop through the RC decoupling will give symmetrical clipping. R37 and C12 perform this function; the low rail voltages in this design allow the 1000 μF C12 to be a fairly compact component.

The output stage is of the Complementary-Feedback Pair (CFP) type, which as previously described, gives the best linearity and quiescent stability, due to the two local negative feedback loops around driver and output device. Quiescent stability is particularly important with R16, 17 at 0.1 Ω, and this low value might be rather dicey in a double emitter-follower (EF) output stage. The CFP voltage efficiency is also higher than the EF version. R25, 26 define a suitable quiescent collector current for the drivers TR6, 8, and pull charge carriers from the output device bases when they are turning off. The lower driver is now a BD136; this has a higher fT than the MJE350, and seems to be more immune to odd parasitics at negative clipping.

The new lower values for the output emitter resistors R16, 17 halve the distortion in Class-AB. This is equally effective when in Class-A with too low a load impedance, or in Class-B but with Iq maladjusted too high. It is now true in the latter case that too much Iq really is better than too little – but not much better, and AB still comes a poor third in linearity to Classes A and B.

SOAR (Safe Operating ARea) protection is given by the networks around TR18, TR19. This is a single-slope SOAR system that is simpler than two-slope SOAR, and therefore somewhat less efficient in terms of getting the limiting characteristic close to the true SOAR of the output transistor. In this application, with low rail voltages, maximum utilisation of the transistor SOAR is not really an issue; the important thing is to observe maximum junction temperatures in the A/AB mode.

The global negative-feedback factor is 32 dB at 20 kHz, and this should give a good margin of safety against Nyquist-type oscillation. Global NFB increases at 6 dB/octave with decreasing frequency to a plateau of around 64 dB, the corner being at a rather ill-defined 300 Hz; this is then maintained down to 10 Hz. It is fortunate that magnitude and frequency here are non-critical, as they depend on transistor beta and other doubtful parameters.

It is often stated in hi-fi magazines that semiconductor amplifiers sound better after hours or days of warm-up. If this is true (which it certainly is not in most cases) it represents truly spectacular design incompetence. This sort of accusation is applied with particular venom to Class-A designs, because it is obvious that the large heatsinks required take time to reach final temperature, so I thought it important to state that in Class-A this design stabilises its electrical operating conditions in less than a second, giving the full intended performance. No *warm-up time* beyond this is required; obviously the heatsinks take time to reach thermal equilibrium, but as described above, measures have been taken to ensure that component temperature has no significant effect on operating conditions or performance.

The power supply

A suitable unregulated power supply is that shown in Figure 8.1; a transformer secondary voltage of 20–0–20 V rms and reservoirs totalling 20,000 μF per rail will give approximately ±24 V. This supply must be designed for continuous operation at maximum current, so the bridge rectifier must be properly heat-sunk, and careful consideration given to the ripple-current ratings of the reservoirs. This is one reason why reservoir capacitance has been doubled to 20,000 μF per rail, over the 10,000 μF that was adequate for the Class-B design; the ripple voltage is halved, which improves voltage efficiency as it is the ripple troughs that determine clipping onset, but in addition the ripple current, although unchanged in total value, is now split between two components. (The capacitance was *not* increased to reduce ripple injection, which is dealt with far more efficiently and economically by making the PSRR high.) Do not omit the secondary fuses; even in these modern times rectifiers do fail, and transformers are horribly expensive.

The performance

The performance of a properly-designed Class-A amplifier challenges the ability of even the Audio Precision measurement system. To give some perspective on this, Figure 9.20 shows the distortion of the AP oscillator driving the analyser section directly for various bandwidths. There appear to be internal mode changes at 2 kHz and 20 kHz, causing step increases in oscillator distortion content; these are just visible in the THD plots for Class-A mode.

Figure 9.21 shows Class-B distortion for 20 W into 8 and 4 Ω, while Figure 9.22 shows the same in Class-A/AB. Figure 9.23 shows distortion

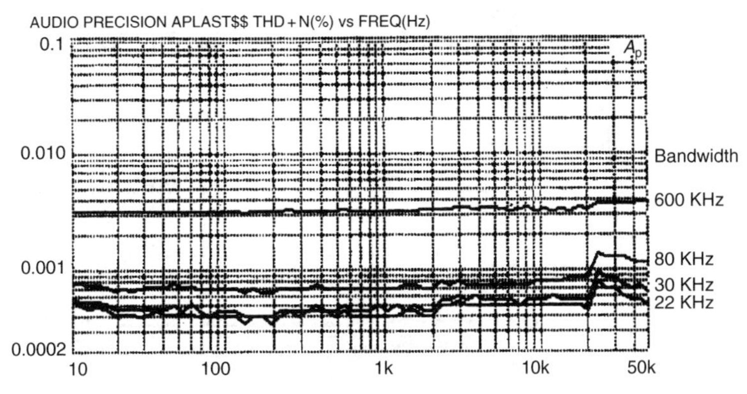

Figure 9.20
The distortion in the AP-1 system at various measurement bandwidths

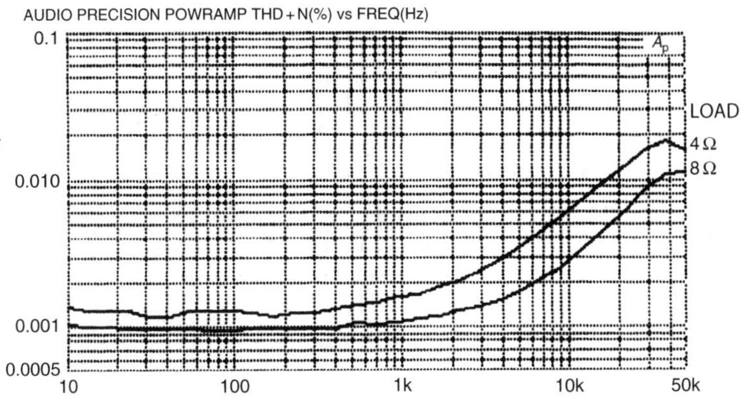

Figure 9.21
Distortion in Class-B (Summer) mode. Distortion into 4 Ω is always worse. Power was 20 W in 8 Ω and 40 W in 4 Ω, bandwidth 80 kHz

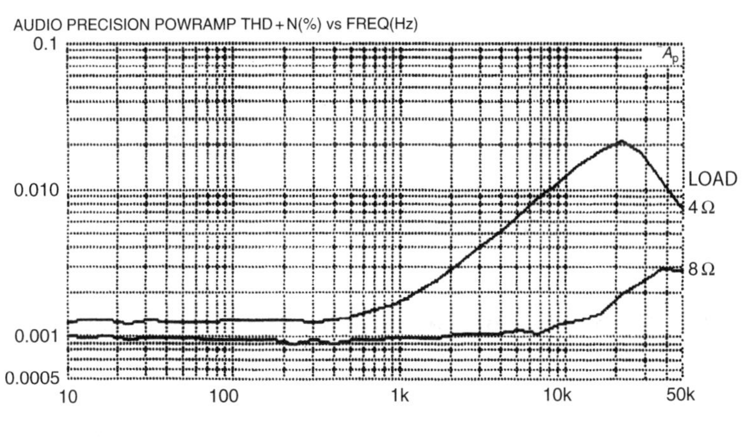

Figure 9.22
Distortion in Class-A/AB (Winter) mode, same power and bandwidth of Figure 9.21. The amplifier is in AB mode for the 4 Ω case, and so distortion is higher than for Class-B into 4 Ω. At 80 kHz bandwidth, the Class-A plot below 10 kHz merely shows the noise floor

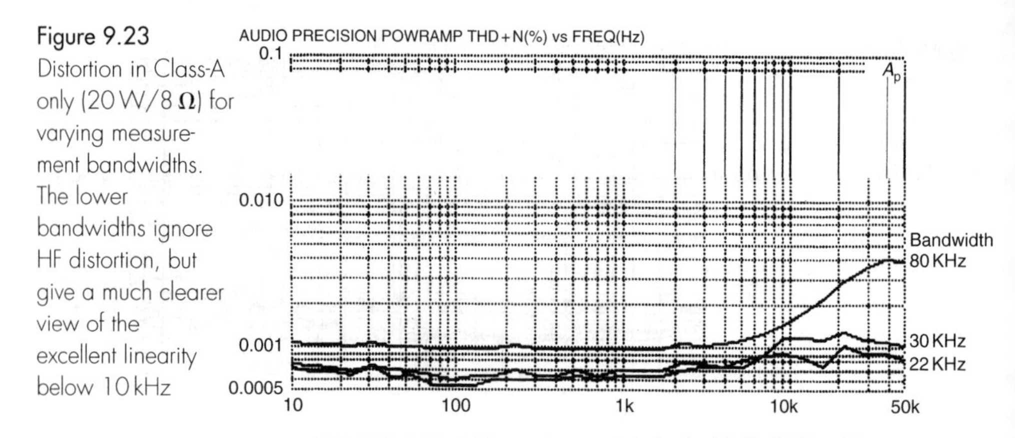

Figure 9.23 Distortion in Class-A only (20 W/8 Ω) for varying measurement bandwidths. The lower bandwidths ignore HF distortion, but give a much clearer view of the excellent linearity below 10 kHz

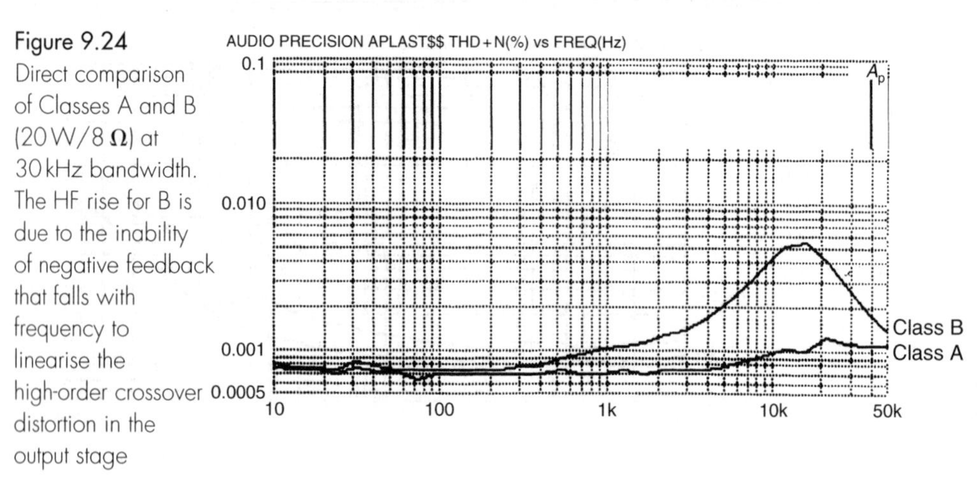

Figure 9.24 Direct comparison of Classes A and B (20 W/8 Ω) at 30 kHz bandwidth. The HF rise for B is due to the inability of negative feedback that falls with frequency to linearise the high-order crossover distortion in the output stage

in Class-A for varying measurement bandwidths. The lower bandwidths misleadingly ignore the HF distortion, but give a much clearer view of the excellent linearity below 10 kHz. Figure 9.24 gives a direct comparison of Classes A and B. The HF rise for B is due to high-order crossover distortion being poorly linearised by negative feedback that falls with frequency.

Further possibilities

One interesting extension of the ideas presented here is the Adaptive Trimodal Amplifier. This would switch into Class-B on detecting device or heatsink overtemperature, and would be a unique example of an amplifier that changed mode to suit the operating conditions. The thermal protection would need to be latching; flipping from Class-A to Class-B every few minutes would subject the output devices to unnecessary thermal cycling.

References

1. Moore, B J *An Introduction To The Psychology of Hearing* Academic Press, 1982, pp. 48–50.
2. Tanaka, S *A New Biasing Circuit for Class-B Operation* JAES, Jan/Feb 1981, p. 27.
3. Fuller, S *Private communication.*
4. Nelson Pass *Build A Class-A Amplifier* Audio, Feb 1977, p. 28 (Constant-current).
5. Linsley-Hood, *J Simple Class-A Amplifier* Wireless World, April 1969, p. 148.
6. Self, D *High-Performance Preamplifier* Wireless World, Feb 1979, p. 41.
7. Nelson-Jones, L *Ultra-Low Distortion Class-A Amplifier* Wireless World, March 1970, p. 98.
8. Giffard, T *Class-A Power Amplifier* Elektor, Nov 1991, p. 37.
9. Linsley-Hood, *J High-Quality Headphone Amp* HiFi News and RR, Jan 1979, p. 81.
10. Nelson Pass *The Pass/A40 Power Amplifier* The Audio Amateur, 1978, p. 4 (Push-pull).
11. Thagard, N *Build a 100 W Class-A Mono Amp* Audio, Jan 95, p. 43.

10

Class-G power amplifiers

Most types of audio power amplifier are less efficient than Class-B; for example, Class-AB is markedly less efficient at the low end of its power capability, while it is clear that Class-A wastes virtually all the energy put into it. Building amplifiers with higher efficiency is more difficult. Class-D, using ultrasonic pulse-width modulation, promises high efficiency and sometimes even delivers it, but it is undeniably a difficult technology. The practical efficiency of Class-D rests on details of circuit design and device characteristics. The apparently unavoidable LC output filter – second order at least – can only give a flat response into one load impedance, and its magnetics are neither cheap nor easy to design. There are likely to be some daunting EMC difficulties with emissions. Class-D is not an attractive proposition for high-quality domestic amplifiers that must work with separate speakers of unknown impedance characteristics.

There is, however, the Class-G method. Power is drawn from either high- or low-voltage rails as the signal level demands. This technology has taken a long time to come to fruition, but is now used in very-high-power amplifiers for large PA systems, where the power savings are important, and is also making its presence felt in home theatre sytems; if you have seven or eight power amplifiers instead of two their losses are rather more significant. Class-G is firmly established in powered subwoofers, and even in ADSL telephone-line drivers. It is a technology whose time has come.

The principles of Class-G

Music has a large peak-to-mean level ratio. For most of the time the power output is a long way below the peak levels, and this makes possible the improved efficiency of Class-G. Even rudimentary statistics for this ratio for various genres of music are surprisingly hard to find, but it is widely accepted that the range between 10 dB for compressed rock, and 30 dB for classical material, covers most circumstances.

If a signal spends most of its time at low power, then while this is true a low-power amplifier will be much more efficient. For most of the time lower output levels are supplied from the lowest-voltage rails, with a low voltage drop between rail and output, and correspondingly low dissipation. The most popular Class-G configurations have two or three pairs of supply rails, two being usual for hi-fi, while three is more common in high-power PA amplifiers.

When the relatively rare high-power peaks do occur they must be handled by some mechanism that can draw high power, causing high internal dissipation, but which only does so for brief periods. These infrequent peaks above the transition level are supplied from the high-voltage pair of rails. Clearly the switching between rails is the heart of the matter, and anyone who has ever done any circuit design will immediately start thinking about how easy or otherwise it will be to make this happen cleanly with a 20 kHz signal.

There are two main ways to arrange the dual-rail system: series and parallel (i.e., shunt). This chapter deals only with the series configuration, as it seems to have had the greatest application to hi-fi. The parallel version is more often used in high-power PA amplifiers.

Introducing series class-G

A series configuration Class-G output stage using two rail voltages is shown in Figure 10.1 The so-called inner devices are those that work in Class-B; those that perform the rail-switching on signal peaks are called the outer devices – by me, anyway. In this design study the EF type of output stage is chosen because of its greater robustness against local HF instability, though the CFP configuration could be used instead for inner, outer, or both sets of output devices, given suitable care. For maximum power efficiency the inner stage normally runs in Class-B, though there is absolutely no reason why it could not be run in Class-AB or even Class-A; there will be more discussion of these intriguing possibilities later. If the inner power devices are in Class-B, and the outer ones conduct for much less than 50% of a cycle, being effectively in Class-C, then according to the classification scheme I proposed[1], this should be denoted Class B + C. The plus sign indicates the series rather than shunt connection of the outer and inner power devices. This basic configuration was developed by Hitachi to reduce amplifier heat dissipation[2],[3]. Musical signals spend most of their time at low levels, having a high peak/mean ratio, and power dissipation is greatly reduced by drawing from the lower ±V1 supply rails at these times.

The inner stage TR3, 4 operates in normal Class-B. TR1, 2 are the usual drivers and R1 is their shared emitter resistor. The usual temperature compensated Vbias generator is required, shown here theoretically split in half to maintain circuit symmetry when the stage is SPICE simulated; since the

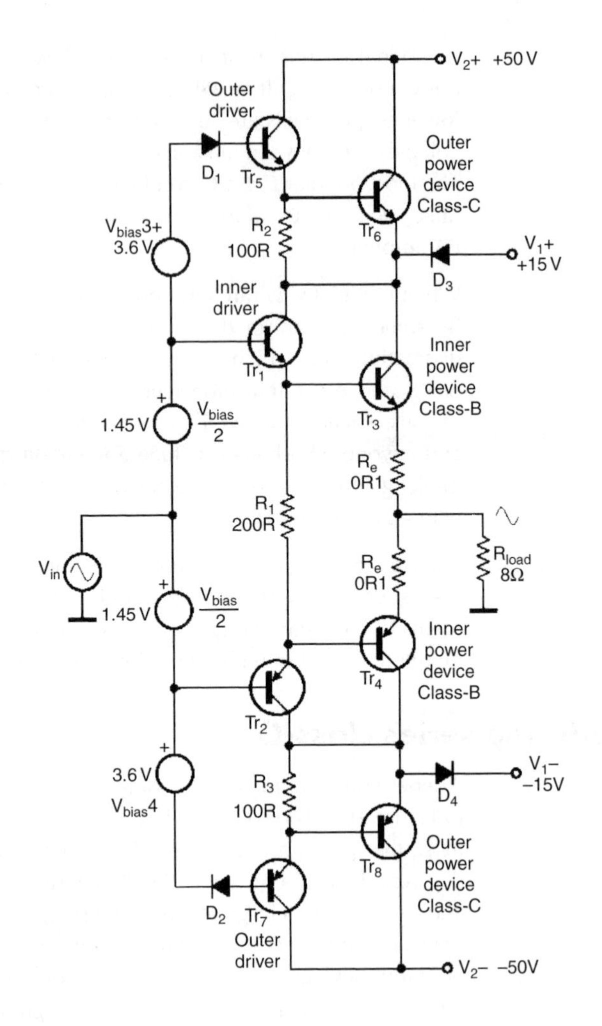

Figure 10.1
A series Class-G output stage, alternatively Class B + C. Voltages and component values are typical. The inner stage is Class-B EF. Biasing by my method

inner power devices work in Class-B it is their temperature which must be tracked to maintain quiescent conditions. Power from the lower supply is drawn through D3 and D4, often called the commutating diodes, to emphasise their rail-switching action. The word 'commutation' avoids confusion with the usual Class-B crossover at zero volts. I have called the level at which rail-switching occurs the transition level.

When a positive-going instantaneous signal exceeds low rail +V1, D1 conducts, TR5 and TR6 turn on and D3 turns off, so the entire output current is now taken from the high-voltage +V2 rail, with the voltage drop and hence power dissipation shared between TR4 and TR6. Negative-going signals are handled in exactly the same way. Figure 10.2 shows how the collector voltages of the inner power devices retreat away from the output rail as it approaches the lower supply level.

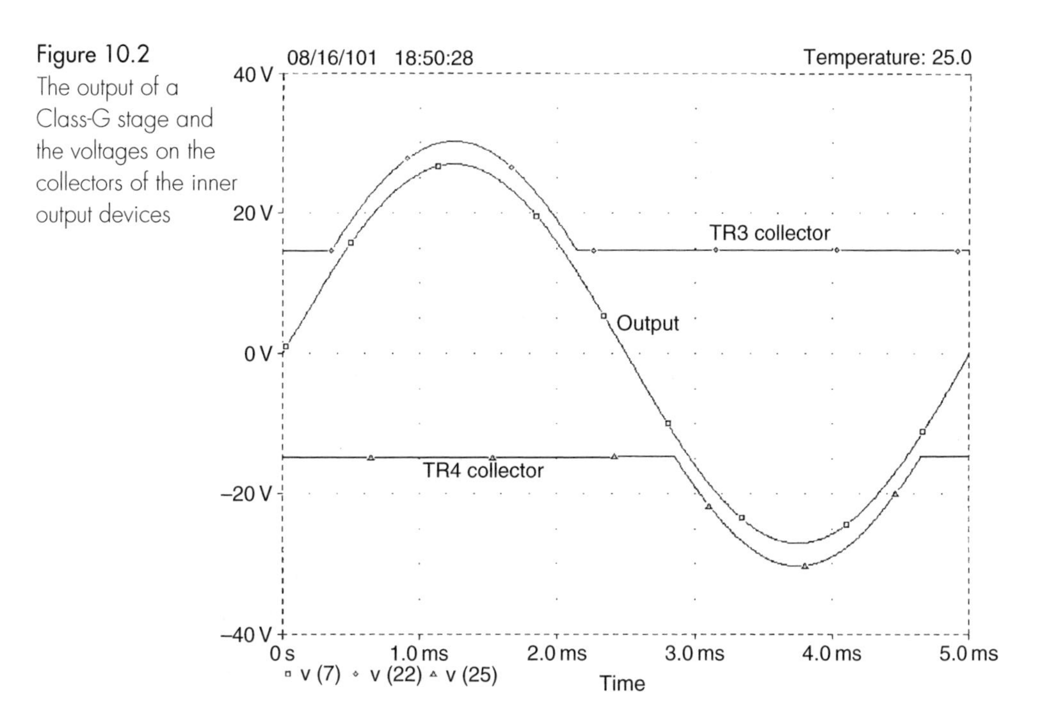

Figure 10.2 The output of a Class-G stage and the voltages on the collectors of the inner output devices

Class-G is commonly said to have worse linearity than Class-B, the blame usually being loaded onto the diodes and problems with their commutation. As usual, received wisdom is only half of the story, if that, and there are other linearity problems that are not due to sluggish diodes, as will be revealed shortly. It is inherent in the Class-G principle that if switching glitches do occur they only happen at moderate power or above, and are well displaced away from the critical crossover region where the amplifier spends most of its time. A Class-G amplifier has a low-power region of true Class-B linearity, just as a Class-AB amplifier has a low-power region of true Class-A performance.

Efficiency of Class-G

The standard mathematical derivation of Class-B efficiency with sinewave drive uses straightforward integration over a half-cycle to calculate internal dissipation against voltage fraction, i.e., the fraction of possible output voltage swing. As is well known, in Class-B the maximum heat dissipation is about 40% of maximum output power, at an output voltage fraction of 63%, which also delivers 40% of the maximum output power to the load.

The mathematics is simple because the waveforms do not vary in shape with output level. Every possible idealization is assumed, such as zero quiescent current, no emitter resistors, no Vce(sat) losses and so on. In Class-G, on the other hand, the waveforms are a strong function of output level, requiring variable limits of integration and so on, and it all gets very unwieldy.

The SPICE simulation method described by Self[4] is much simpler, if somewhat laborious, and can use any input waveform, yielding a Power Partition Diagram (PPD), which shows how the power drawn from the supply is distributed between output device dissipation and useful power in the load.

No one disputes that sinewaves are poor simulations of music for this purpose, and their main advantage is that they allow direct comparison with the purely mathematical approach. However, since the whole point of Class-G is power saving, and the waveform used has a strong effect on the results, I have concentrated here on the PPD of an amplifier with real musical signals, or at any rate, their statistical representation. The triangular Probability Distribution Function (PDF) approach is described in Self[5].

Figure 10.3 shows the triangular PDF PPD for conventional Class-B EF, while Figure 10.4 is that for Class-G with $\pm V2 = 50\,V$ and $\pm V1 = 15\,V$, i.e., with the ratio of V1/V2 set to 30%. The PPD plots power dissipated in all four output devices, the load, and the total drawn from the supply rails. It shows how the input power is partitioned between the load and the output devices. The total sums to slightly less than the input power, the remainder being accounted for as usual by losses in the drivers and Re's. Note that in Class-G power dissipation is shared, though not very equally, between the inner and outer devices, and this helps with efficient utilization of the silicon.

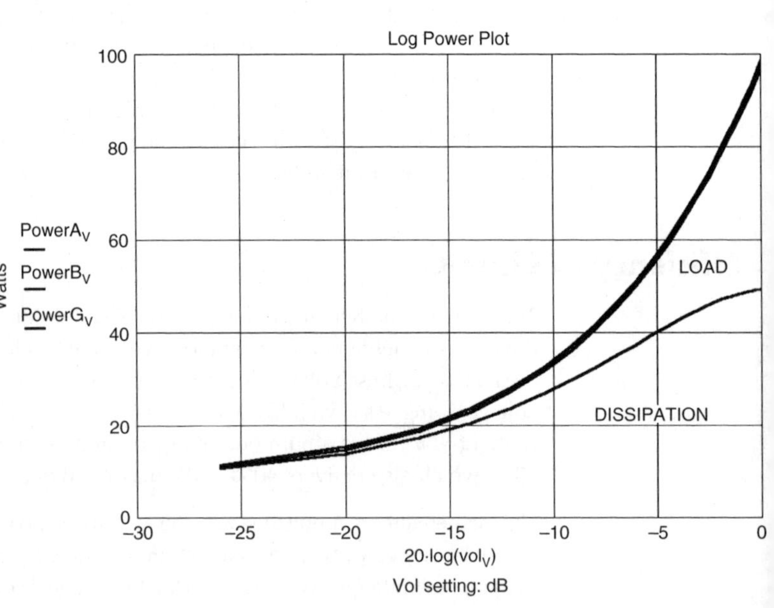

Figure 10.3
Power partition diagram for a conventional Class-B amplifier handling a typical music signal with a triangular Probability Density Function. X-axis is volume

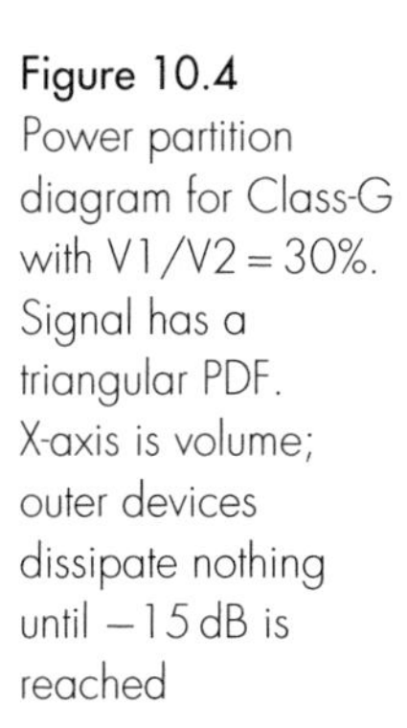

Figure 10.4 Power partition diagram for Class-G with V1/V2 = 30%. Signal has a triangular PDF. X-axis is volume; outer devices dissipate nothing until −15 dB is reached

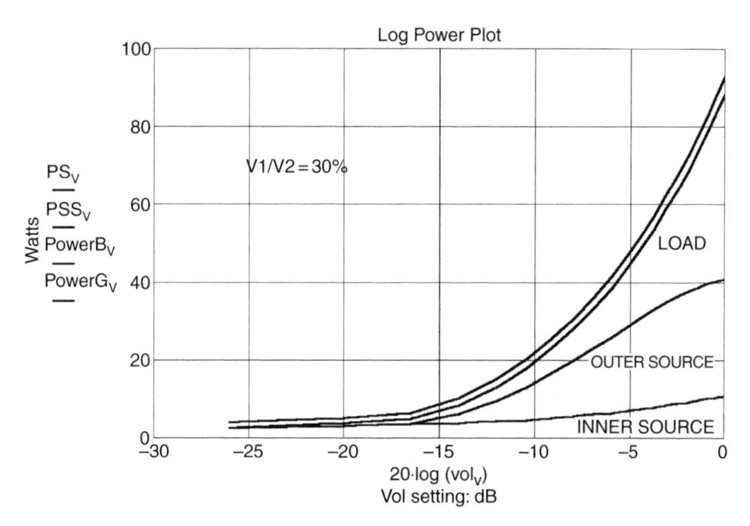

In Figure 10.4 the lower area represents the power dissipated in the inner devices and the larger area just above represents that in the outer devices; there is only one area for each because in Class-B and Class-G only one side of the amplifier conducts at a time. Outer device dissipation is zero below the rail-switching threshold at −15 dB below maximum output. The total device dissipation at full output power is reduced from 48 W in Class-B to 40 W, which may not appear at first to be a very good return for doubling the power transistors and drivers.

Figure 10.5 shows the same PPD but with ±V2 = 50 V and ±V1 = 30 V, i.e., with V1/V2 set to 60%. The low-voltage region now extends up to

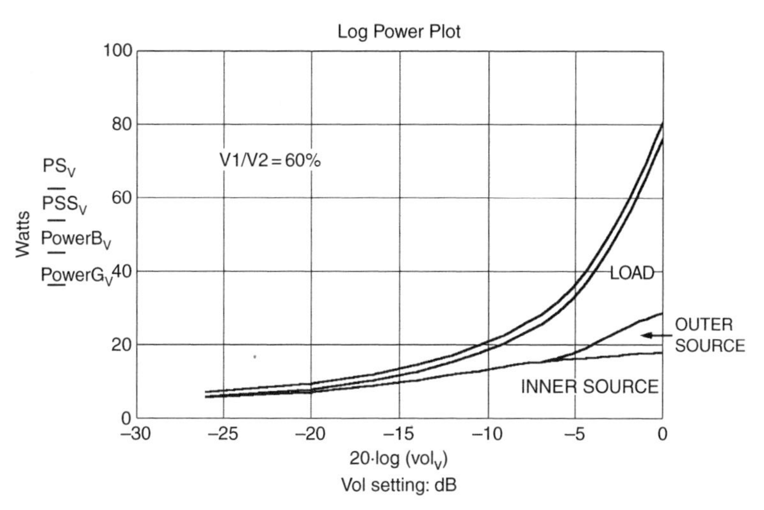

Figure 10.5 Power partition diagram for Class-G with V1/V2 = 60%. Triangular PDF. Compared with Figure 10.4, the inner devices dissipate more and the outer devices almost nothing except at maximum volume

−6 dB ref full power, but the inner device dissipation is higher due to the higher V1 rail voltages. The result is that total device power at full output is reduced from 48 W in Class-B to 34 W, which is a definite improvement. The efficiency figure is highly sensitive to the way the ratio of rail voltages compares with the signal characteristics. Domestic hi-fi amplifiers are not operated at full volume all the time, and in real life the lower option for the V1 voltage is likely to give lower general dissipation. I do not suggest that V1/V2 = 30% is the optimum lower-rail voltage for all situations, but it looks about right for most domestic hi-fi.

Practicalities

In my time I have wrestled with many 'new and improved' output stages that proved to be anything but. When faced with a new and intriguing possibility, I believe the first thing to do is sketch out a plausible circuit such as Figure 10.1 and see if it works in SPICE simulation. It duly did.

The next stage is to build it, power it from low supply rails to minimise any resulting explosions, and see if it works for real at 1 kHz. This is a bigger step than it looks.

SPICE simulation is incredibly useful but it is not a substitute for testing a real prototype. It is easy to design clever and complex output stages that work beautifully in simulation but in reality prove impossible to stabilise at high frequencies. Some of the most interesting output-triple configurations seem to suffer from this.

The final step – and again it is a bigger one than it appears – is to prove real operation at 20 kHz and above. Again it is perfectly possible to come up with a circuit configuration that either just does not work at 20 kHz, due to limitations on power transistor speeds, or is provoked into oscillation or other misbehaviour that is not set off by a 1 kHz testing.

Only when these vital questions are resolved is it time to start considering circuit details, and assessing just how good the amplifier performance is likely to be.

The biasing requirements

The output stage bias requirements are more complex than for Class-B. Two extra bias generators Vbias3, Vbias4 are required to make TR6 turn on before TR3 runs out of collector voltage. These extra bias voltages are not critical, but must not fall too low, or become much too high. Should these bias voltages be set too low, so the outer devices turn on too late, then the Vce across TR3 becomes too low, and its current sourcing capability is reduced. When evaluating this issue bear in mind the lowest impedance load the amplifier is planned to drive, and the currents this will draw from the output devices. Fixed Zener diodes of normal commercial tolerance are quite accurate and stable enough for setting Vbias3 and Vbias4.

Alternatively, if the bias voltage is set too low, then the outer transistors will turn on too early, and the heat dissipation in the inner power devices becomes greater than it need be for correct operation. The latter case is rather less of a problem so if in doubt this bias should be chosen to be on the high side rather than the low.

The original Hitachi circuit[1] put Zeners in series with the signal path to the inner drivers to set the output quiescent bias, their voltage being subtracted from the main bias generator which was set at 10 V or so, a much higher voltage than usual (see Figure 10.6). SPICE simulation showed me that the presence of Zener diodes in the forward path to the inner power devices gave poor linearity, which is not exactly a surprise. There is also the problem that the quiescent conditions will be affected by changes in the

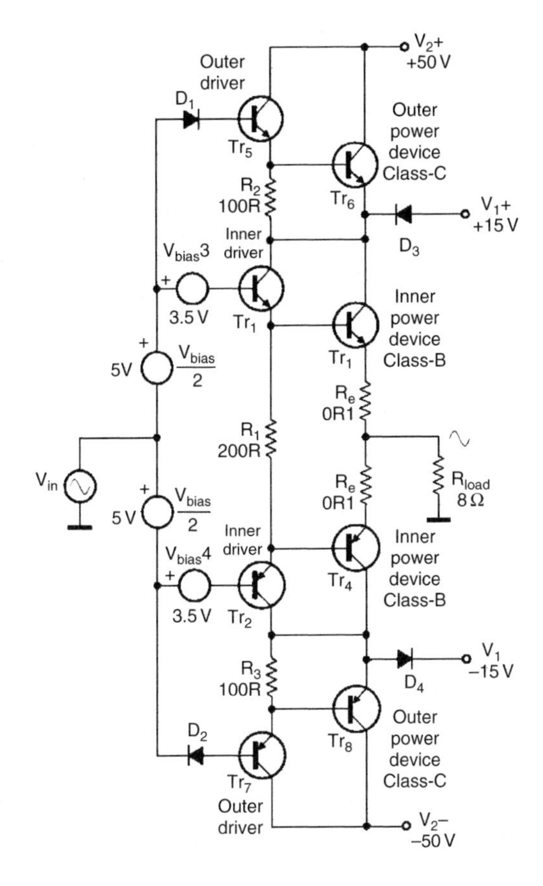

Figure 10.6
The original Hitachi Class-G biasing system, with inner device bias derived by subtracting Vbias3, 4 from the main bias generator

Zener voltage. The 10 V bias generator, if it is the usual Vbe multiplier, will have much too high a temperature coefficient for proper thermal tracking.

I therefore rearrange the biasing as in Figure 10.1. The amplifier forward path now goes directly to the inner devices, and the two extra bias voltages are in the path to the outer devices; since these do not control the output directly, the linearity of this path is of little importance. The Zeners are out of the forward path and the bias generator can be the standard sort. It must be thermally coupled to the inner power devices; the outer ones have no effect on the quiescent conditions.

The linearity issues of series Class-G

Series Class-G has often had its linearity called into question because of difficulties with supply-rail commutation. Diodes D3, D4 must be power devices capable of handling a dozen amps or more, and conventional silicon rectifier diodes that can handle such currents take a long time to turn off due to their stored charge carriers. This has the following unhappy effect: when the voltage on the cathode of D3 rises above V1, the diode tries to turn off abruptly, but its charge carriers sustain a brief but large reverse current as they are swept from its junction. This current is supplied by TR6, attempting as an emitter-follower to keep its emitter up to the right voltage. So far all is well.

However, when the diode current ceases, TR6 is still conducting heavily, due to its own charge-carrier storage. The extra current it turned on to feed D3 in reverse now goes through TR3 collector, which accepts it because of TR3's low Vce, and passes it onto the load via TR3 emitter and Re.

This process is readily demonstrated by a SPICE commutation transient simulation; see Figures 10.7 and 10.8. Note there are only two of these events per cycle – not four, as they only occur when the diodes turn off. In the original Hitachi design this problem was reportedly tackled by using fast transistors and relatively fast gold-doped diodes, but according to Sampei et al.[2] this was only partially successful.

It is now simple to eradicate this problem. Schottky power diodes are readily available, as they were not in 1976, and are much faster due to their lack of minority carriers and charge storage. They have the added advantage of a low forward voltage drop at large currents of 10 A or more. The main snag is a relatively low reverse withstand voltage, but fortunately in Class-G usage the commutating diodes are only exposed at worst to the difference between V2 and V1, and this only when the amplifier is in its low power domain of operation. Another good point about Schottky power diodes is that they do appear to be robust; I have subjected 50 amp Motorola devices to 60 amps-plus repeatedly without a single failure. This is a good sign. The spikes disappear completely from the SPICE plot if the

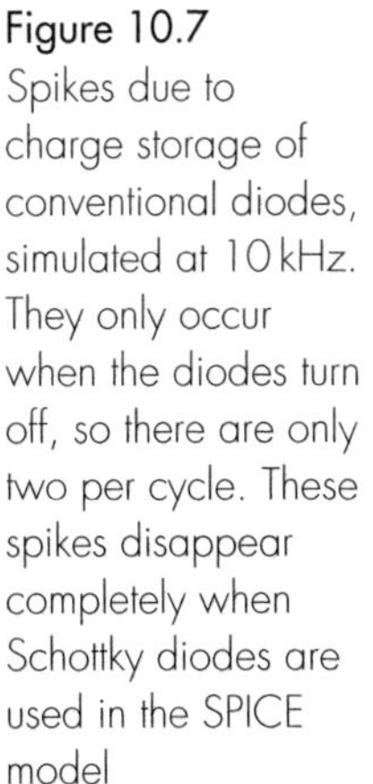

Figure 10.7
Spikes due to charge storage of conventional diodes, simulated at 10 kHz. They only occur when the diodes turn off, so there are only two per cycle. These spikes disappear completely when Schottky diodes are used in the SPICE model

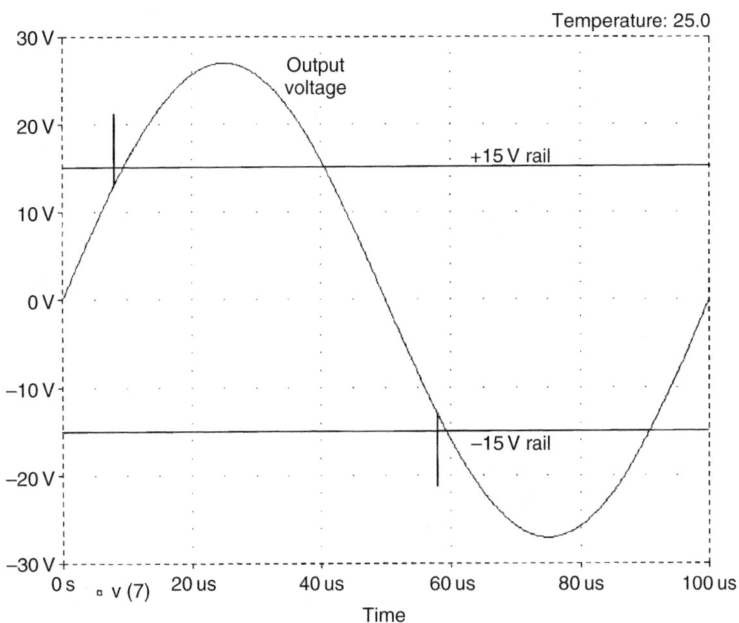

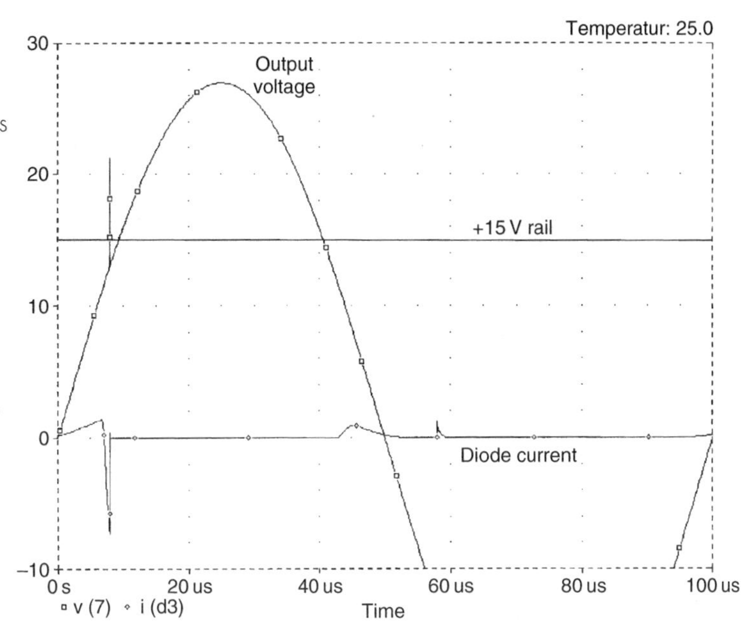

Figure 10.8
A close-up of the diode transient. Diode current rises as output moves away from zero, then reverses abruptly as charge carriers are swept out by reverse-biasing. The spike on the output voltage is aligned with the sudden stop of the diode reverse current

commutating diodes are Schottky rectifiers. Motorola MBR5025L diodes capable of 50 A and 25 PIV were used in simulation.

The static linearity

SPICE simulation shows in Figure 10.9 that the static linearity (i.e., that ignoring dynamic effects like diode charge-storage) is distinctly poorer than for Class-B. There is the usual Class-B gain-wobble around the crossover region, exactly the same size and shape as for conventional Class-B, but also there are now gain-steps at ±16 V. The result with the inner devices biased into push-pull Class-A is also shown, and proves that the gain-steps are not in any way connected with crossover distortion. Since this is a DC analysis the gain-steps cannot be due to diode switching-speed or other dynamic phenomena, and Early Effect was immediately suspected. (Early Effect is the increase in collector current when the collector voltage increases, even though the Vbe remains constant.) When unexpected distortion appears in a SPICE simulation of this kind, and effects due to finite transistor beta and associated base currents seem unlikely, a most useful diagnostic technique is to switch off the simulation of Early Effect for each transistor in turn. In SPICE transistor models the Early Effect can be totally disabled by setting the parameter VAF to a much higher value than the default of 100, such as 50,000. This experiment demonstrated in

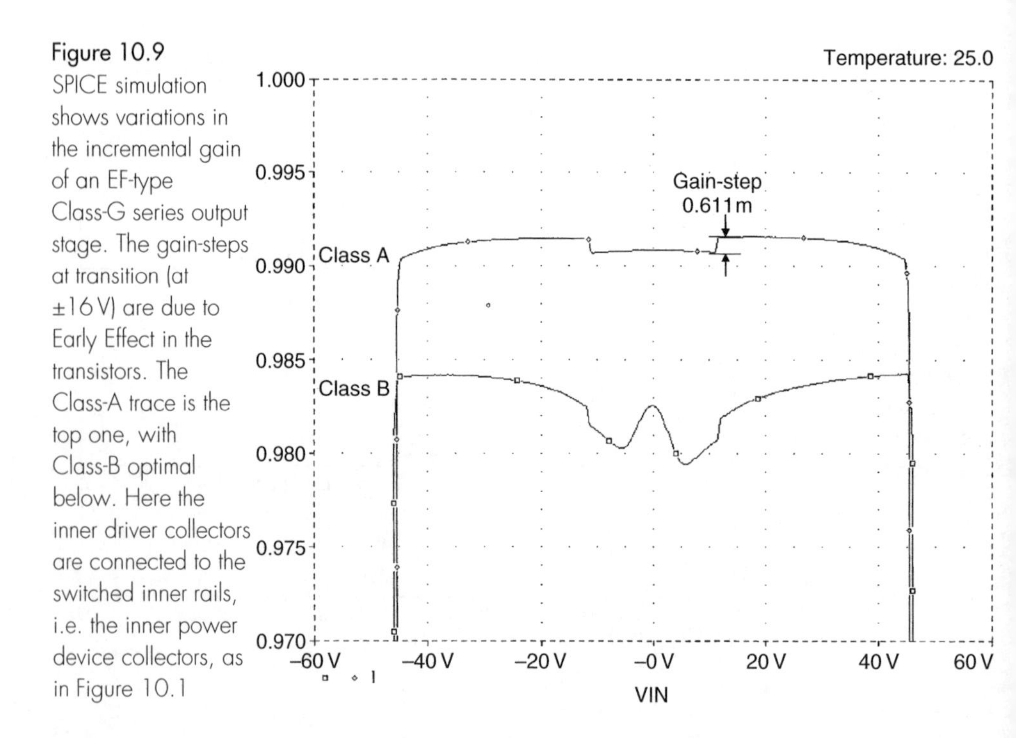

Figure 10.9 SPICE simulation shows variations in the incremental gain of an EF-type Class-G series output stage. The gain-steps at transition (at ±16 V) are due to Early Effect in the transistors. The Class-A trace is the top one, with Class-B optimal below. Here the inner driver collectors are connected to the switched inner rails, i.e. the inner power device collectors, as in Figure 10.1

short order that the gain-steps were caused wholly by Early Effect acting on both inner drivers and inner output devices. The gain-steps are completely abolished. When TR6 begins to act, TR3 Vce is no longer decreasing as the output moves positive, but substantially constant as the emitter of Q6 moves upwards at the same rate as the emitter of Q3. This has the effect of a sudden change in gain, which naturally degrades the linearity.

This effect appears to occur in drivers and output devices to the same extent. It can be easily eliminated in the drivers by powering them from the outer rather than the inner supply rails. This prevents the sudden changes in the rate in which driver Vce varies. The improvement in linearity is seen in Figure 10.10, where the gain-steps have been halved in size. The resulting circuit is shown in Figure 10.11. Driver power dissipation is naturally increased by the increased driver Vce, but this is such a small fraction of the power consumed that the overall efficiency is not significantly reduced. It is obviously not practical to apply the same method to the output devices, because then the low-voltage rail would never be used and the amplifier is no longer working in Class-G. The small-signal stages naturally have to work from the outer rails to be able to generate the full voltage swing to drive the output stage.

We have now eliminated the commutating diode glitches, and halved the size of the unwanted gain-steps in the output stage. With these

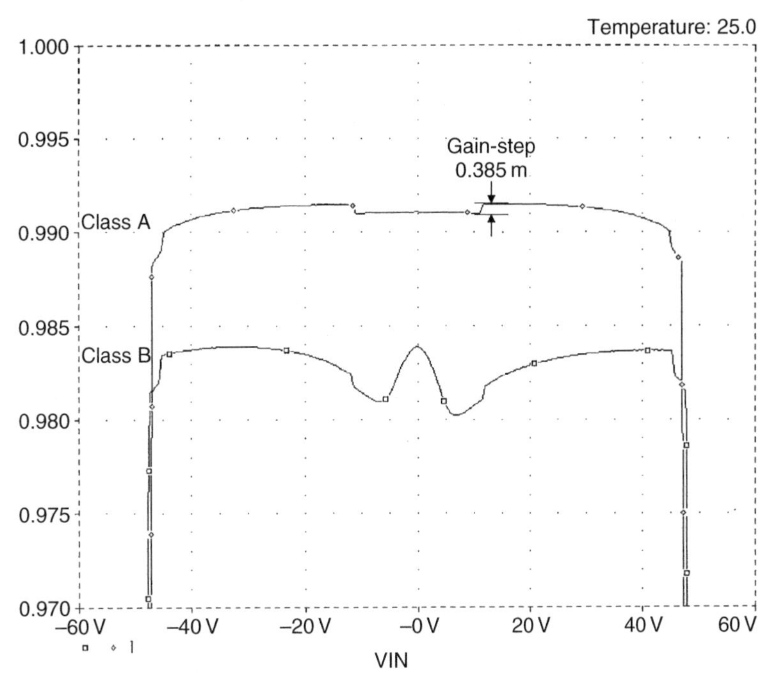

Figure 10.10 Connecting the inner driver collectors to the outer V2 rails reduces Early Effect non-linearities in them, and halves the transition gain-steps

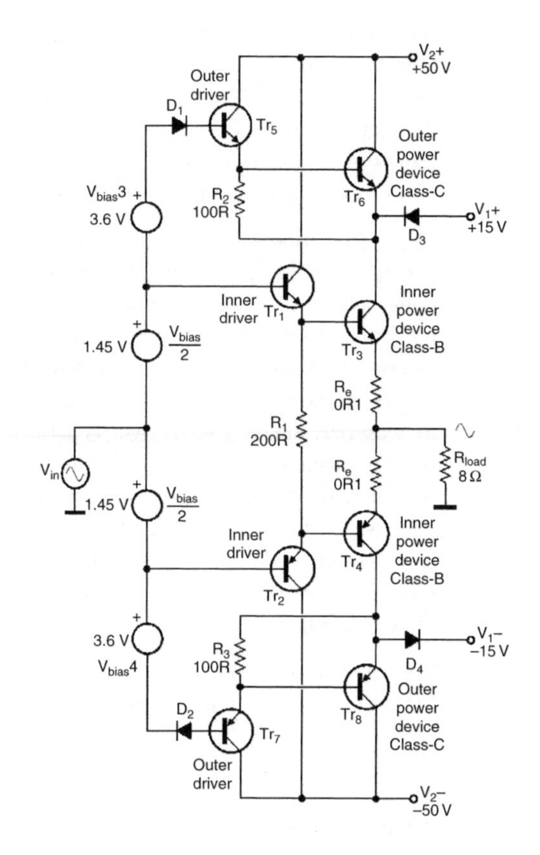

Figure 10.11
A Class-G output stage with the drivers powered from the outer supply rails

improvements made it is practical to proceed with the design of a Class-G amplifier with midband THD below 0.002%.

Practical Class-G design

The Class-G amplifier design expounded here uses very similar small-signal circuitry to the Blameless Class-B power amplifier, as it is known to generate very little distortion of its own. If the specified supply voltages of ±50 and ±15 V are used, the maximum power output is about 120 W into 8 Ω, and the rail-switching transition occurs at 28 W.

This design incorporates various techniques described in this book, and closely follows the Blameless Class-B amp described on page 179, though some features derive from the Trimodal (page 283) and Load Invariant (page 138) amplifiers. A notable example is the low-noise feedback network, complete with its option of input bootstrapping to give a high impedance when required. Single-slope VI limiting is incorporated for overload protection; this is implemented by Q12, 13. Figure 10.12 shows the circuit.

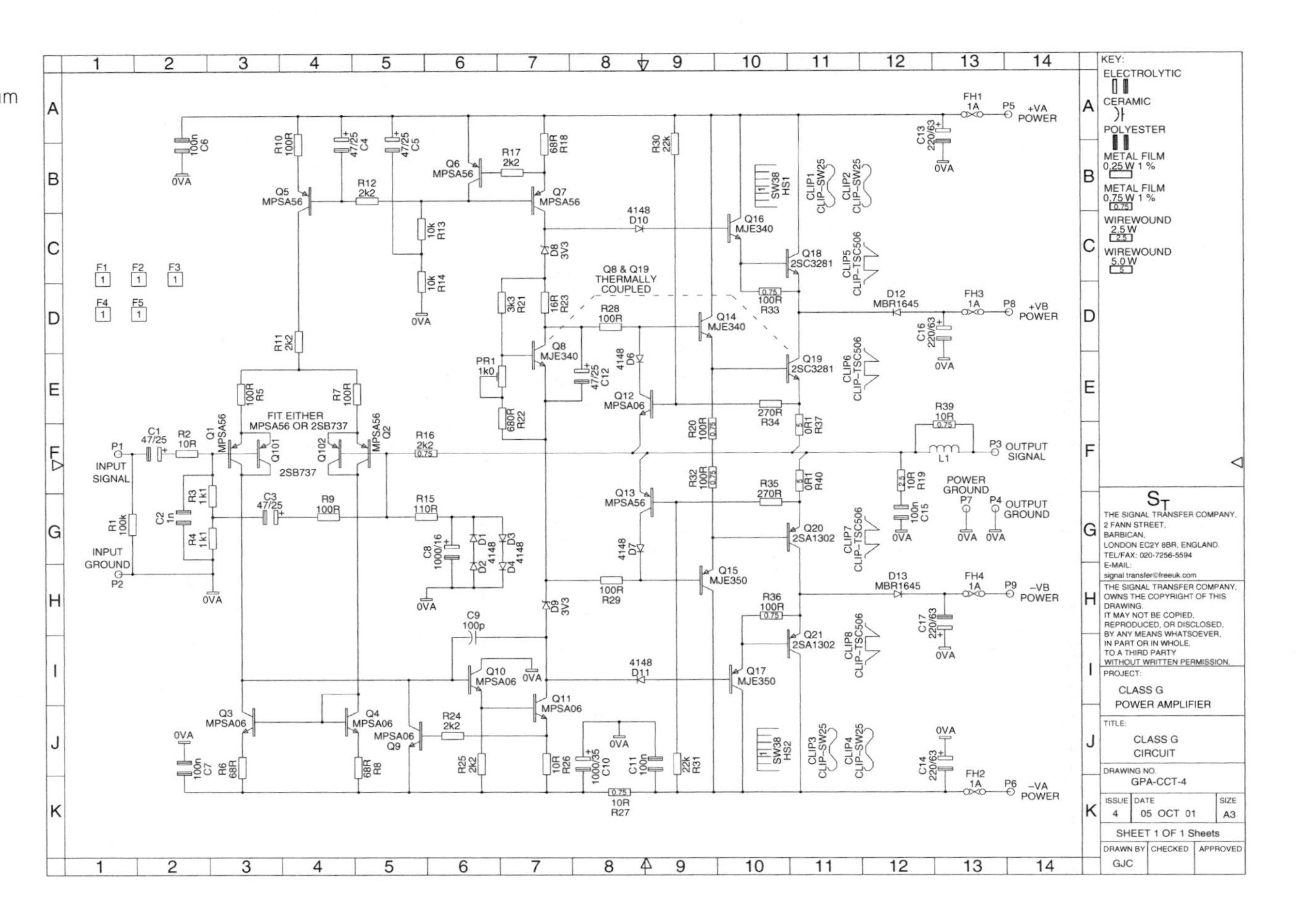

Figure 10.12 The circuit diagram of the Class-G amplifier

As usual in my amplifiers the global NFB factor is a modest 30 dB at 20 kHz.

Controlling small-signal distortion

The distortion from the small-signal stages is kept low by the same methods as for the other amplifier designs in this book, and so this is only dealt with briefly here. The input stage differential pair Q1, 2 is given local feedback by R5 and R7 to delay the onset of third-harmonic Distortion 1. Internal re variations in these devices are minimised by using an unusually high tail current of 6 mA. Q3, 4 are a degenerated current-mirror that enforces accurate balance of the Q1, 2 collector currents, preventing the production of second-harmonic distortion. The input resistance (R3 + R4) and feedback resistance R16 are made equal and made unusually low, so that base current mismatches stemming from input device beta variations give a minimal DC offset. Vbe mismatches in Q1 and Q2 remain, but these are much smaller than the effects of Ib. Even if Q1 and Q2 are high-voltage types with relatively low beta, the DC offset voltage at the output should be kept to less than ±50 mV. This is adequate for all but the most demanding applications. This low-impedance technique eliminates the need for balance presets or DC servo systems, which is most convenient.

A lower value for R16 implies a proportionally lower value for R15 to keep the gain the same, and this reduction in the total impedance seen by Q2 improves noise performance markedly. However, the low value of R3 plus R4 at 2k2 gives an input impedance which is not high enough for many applications.

There is no problem if the amplifier is to have an additional input stage, such as a balanced line receiver. Proper choice of op-amp will allow the stage to drive a 2k2 load impedance without generating additional distortion. Be aware that adding such a stage – even if it is properly designed and the best available op-amps are used – will degrade the signal-to-noise ratio significantly. This is because the noise generated by the power amplifier itself is so very low – equivalent to the Johnson noise of a resistor of a few hundred ohms – that almost anything you do upstream will degrade it seriously.

If there is no separate input stage then other steps must be taken. What we need at the input of the power amplifier is a low DC resistance, but a high AC resistance; in other words we need either a 50 Henry choke or recourse to some form of bootstrapping. There is to my mind no doubt about the way to go here, so bootstrapping it is. The signal at Q2 base is almost exactly the same as the input, so if the mid-point of R3 and R4 is driven by C3, so far as input signals are concerned R3 has a high AC impedance. When I first used this arrangement I had doubts about its high-frequency stability, and so added resistor R9 to give some isolation between the bases of Q1 and Q2. In the event I have had no trouble with instability, and no reports of any from the many constructors of the Trimodal and Load-Invariant designs, which incorporate this option.

The presence of R9 limits the bootstrapping factor, as the signal at R3–R4 junction is thereby a little smaller than at Q2 base, but it is adequate. With R9 set to 100R, the AC input impedance is raised to 13 k, which should be high enough for almost all purposes. Higher values than this mean that an input buffer stage is required.

The value of C8 shown (1000 μF) gives an LF roll-off in conjunction with R15 that is −3 dB at 1.4 Hz. The purpose is not impossibly extended sub-bass, but the avoidance of a low-frequency rise in distortion due to non-linearity effects in C8. If a 100 μF capacitor is used here the THD at 10 Hz worsens from <0.0006% to 0.0011%, and I regard this as unacceptable aesthetically – if not perhaps audibly. This is not the place to define the low-frequency bandwidth of the system – this must be done earlier in the signal chain, where it can be properly implemented with more accurate non-electrolytic capacitors. The protection diodes D1 to D4 prevent damage to C2 if the amplifier suffers a fault that makes it saturate in either direction; it looks like an extremely dubious place to put diodes but since they normally have no AC or DC voltage across them no measurable or detectable distortion is generated.

The Voltage-Amplifier-Stage (VAS) Q11 is enhanced by emitter-follower Q10 inside the Miller-compensation loop, so that the local negative feedback that linearises the VAS is increased. This effectively eliminates VAS non-linearity. Thus increasing the local feedback also reduces the VAS collector impedance, so a VAS-buffer to prevent Distortion 4 (loading of VAS collector by the non-linear input impedance of the output stage) is not required. Miller capacitor Cdom is relatively big at 100 pF, to swamp transistor internal capacitances and circuit strays, and make the design predictable. The slew rate calculates as 40 V/μsec use in each direction. VAS collector-load Q7 is a standard current source.

Almost all the THD from a Blameless amplifier derives from crossover distortion, so keeping the quiescent conditions optimal to minimise this is essential. The bias generator for an EF output stage, whether in Class-B or Class-G, is required to cancel out the Vbe variations of four junctions in series; those of the two drivers and the two output devices. This sounds difficult, because the dissipation in the two types of devices is quite different, but the problem is easier than it looks. In the EF type of output stage the driver dissipation is almost constant as power output varies, and so the problem is reduced to tracking the two output device junctions. The bias generator Q8 is a standard Vbe-multiplier, with R23 chosen to minimise variations in the quiescent conditions when the supply rails change. The bias generator should be in contact with the top of one of the inner output devices, and not the heatsink itself. This position gives much faster and less attenuated thermal feedback to Q8. The VAS collector circuit incorporates not only bias generator Q8 but also the two Zeners D8, D9 which determine how early rail-switching occurs as the inner device emitters approach the inner (lower) voltage rails.

The output stage was selected as an EF (emitter-follower) type as this is known to be less prone to parasitic or local oscillations than the CFP configuration, and since this design was to the same extent heading into the unknown it seemed wise to be cautious where possible. R32 is the usual shared emitter resistor for the inner drivers. The outer drivers Q16 and Q17 have their own emitter resistors R33 and R36, which have their usual role of establishing a reasonable current in the drivers as they turn on, to increase driver transconductance, and also in speeding up turn-off of the outer output devices by providing a route for charge-carriers to leave the output device bases.

As explained above, the inner driver collectors are connected to the outer rails to minimise the gain-steps caused by the abrupt change in collector voltage when rail transition occurs.

Deciding the size of heatsink required for this amplifier is not easy, mainly because the heat dissipated by a Class-G amplifier depends very much on the rail voltages chosen and the signal statistics. A Class-B design giving 120 W into 8R would need a heatsink with thermal resistance of the order of 1°C/W (per channel); a good starting point for a Class-G version giving the same power would be about half the size, i.e. 2°C/W. The Schottky commutating diodes do not require much heatsinking, as they conduct only intermittently and have a low forward voltage drop. It is usually convenient to mount them on the main heatsink, even if this does mean that most of the time they are being heated rather than cooled.

C15 and R38 make up the usual Zobel network. The coil L1, damped by R39, isolates the amplifier from load capacitance. A component with 15 to 20 turns at 1 inch diameter should work well; the value of inductance for stability is not all that critical.

The performance

Figure 10.13 shows the THD at 20 W/50 W (into 8 Ω) and I think this demonstrates at once that the design is a practical competitor for Class-B amplifiers. Compare these results with the upper trace of Figure 10.14, taken from a Blameless Class-B amplifier at 50 W, 8 Ω. Note the lower trace of Figure 10.14 is for 30 kHz bandwidth, used to demonstrate the lack of distortion below 1 kHz; the THD data above 30 kHz is in this case

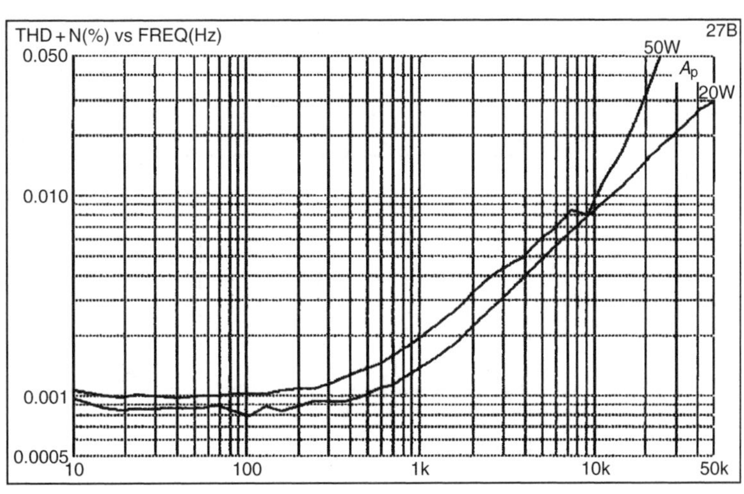

Figure 10.13 THD versus frequency, at 20 W (below transition) and 50 W into an 8 Ω load. The joggle around 8 kHz is due to a cancellation of harmonics from crossover and transition. 80 kHz bandwidth

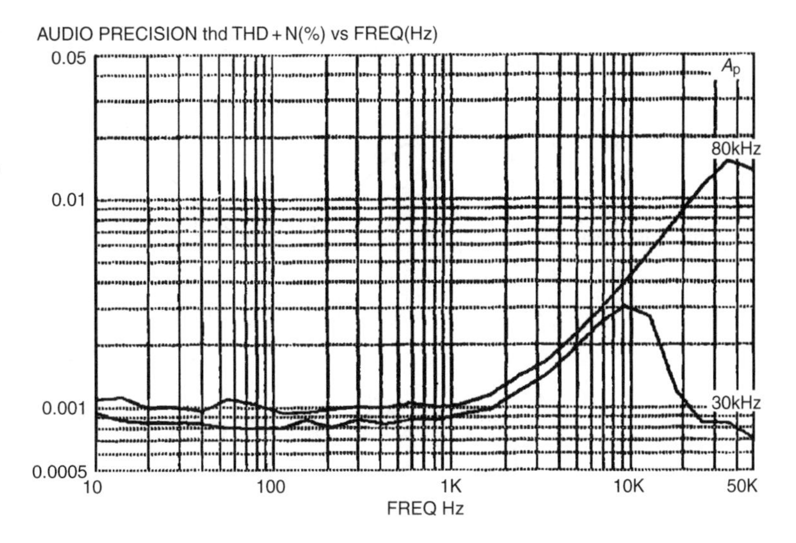

Figure 10.14 THD versus frequency for a Blameless Class-B amplifier at 50 W, 8 Ω

meaningless as all the harmonics are filtered out. All the Class-G plots here are taken at 80 kHz to make sure any high-order glitching is properly measured.

Figure 10.15 shows the actual THD residual at 50 W output power. The glitches from the gain-steps are more jagged than the crossover disturbances, as would be expected from the output stage gain plot in Figures 10.9, 10.11. Figure 10.16 confirms that at 20 W, below transition, the residual is indistinguishable from that of a Blameless Class-B amplifier; in this region, where the amplifier is likely to spend most of its time, there are no compromises on quality.

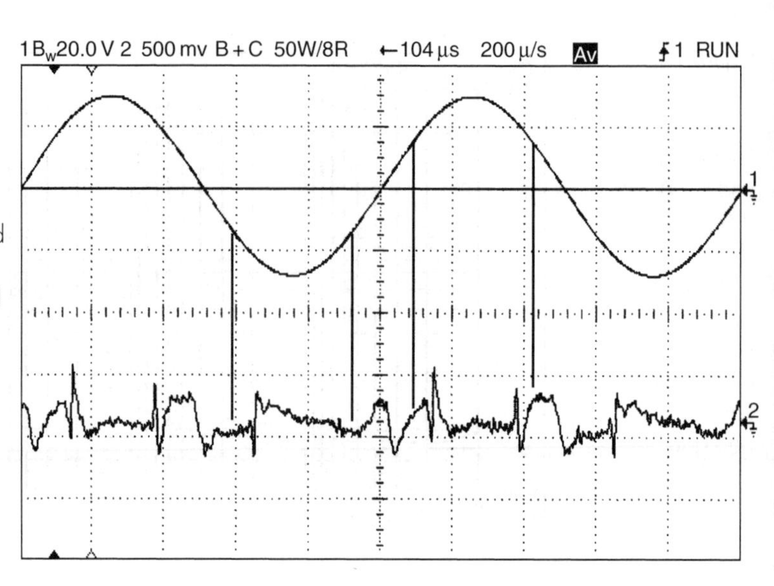

Figure 10.15
The THD residual waveform at 50 W into 8 Ω. This residual may look rough, but in fact it had to be averaged eight times to dig the glitches and crossover out of the noise; THD is only 0.0012%. The vertical lines show where transition occurs

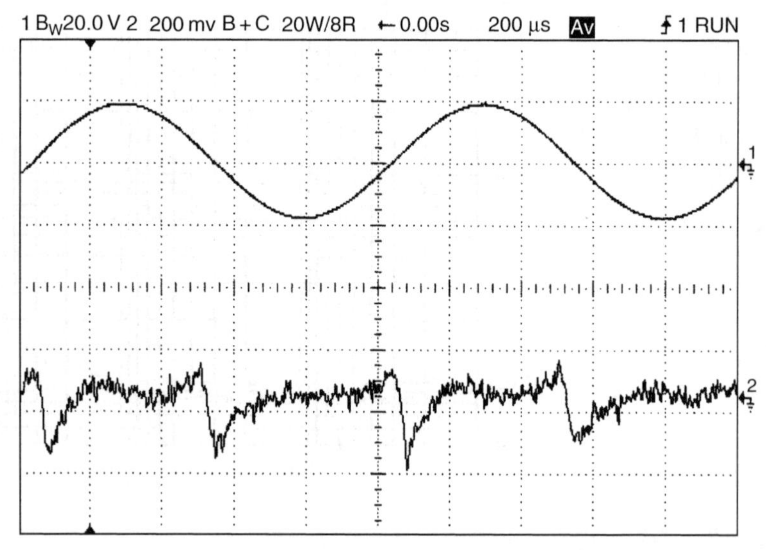

Figure 10.16
The THD residual waveform at 20 W into 8 Ω, below transition. Only crossover artefacts are visible as there is no rail-switching

Figure 10.17 shows THD versus level, demonstrating how THD increases around 28 W as transition begins. The steps at about 10 W are nothing to do with the amplifier – they are artefacts due to internal range-switching in the measuring system.

Figure 10.18 shows for real the benefits of powering the inner drivers from the outer supply rails. In SPICE simulation (see above) the gain-steps were roughly halved in size by this modification, and Figure 10.18 duly confirms that the THD is halved in the HF region, the only area where it is sufficiently clear of the noise floor to be measured with any confidence.

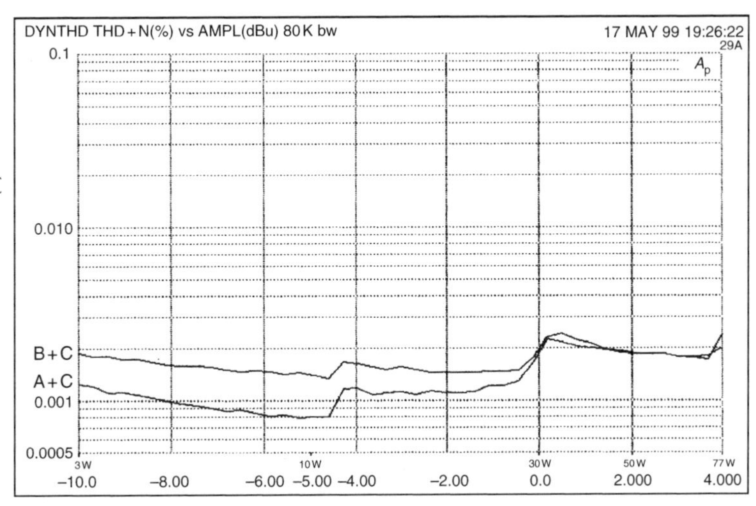

Figure 10.17 THD versus level, showing how THD increases around 28 W as transition begins. Class-A + C is the lower and Class-B + C the upper trace

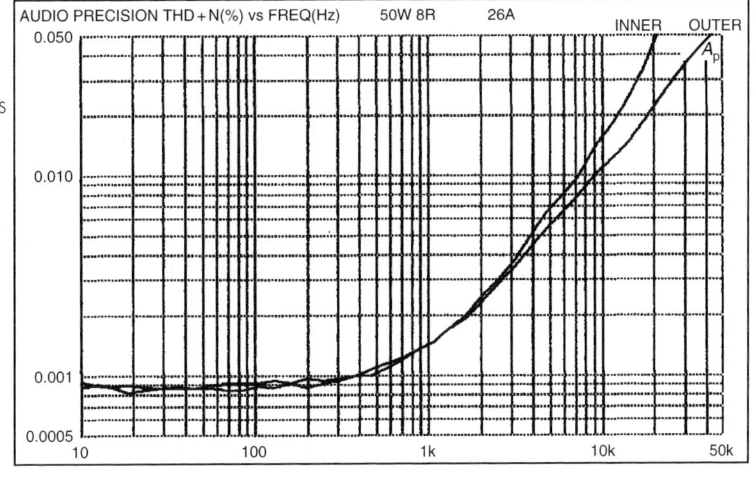

Figure 10.18 THD plot of real amplifier driving 50 W into 8 Ω. Rails were ±40 and ±25 V. Distortion at HF is halved by connecting the inner drivers to the outer supply rails rather than the inner rails

Deriving a new kind of amplifier: Class-A + C

A conventional Class-B power amplifier can be almost instantly converted to push-pull Class-A simply by increasing the bias voltage to make the required quiescent current flow. This is the only real circuit change, though naturally major increases in heatsinking and power-supply capability are required for practical use. Exactly the same principle applies to the Class-G amplifier. Recently I suggested a new and much more flexible system for classifying amplifier types[6] and here it comes in very handy. Describing Class-G operation as Class-B + C immediately indicates that only a bias increase is required to transform it into Class-A + C, and a new type of amplifier is born. This amplifier configuration combines the superb linearity of classic Class-A up to the transition level, with only minor distortion artefacts occurring at higher levels, as demonstrated for Class-B + C above. Using Class-A means that the simple Vbe-multiplier bias generator can be replaced with precise negative feedback control of the quiescent current, as implemented in the Trimodal amplifier in this book. There is no reason why an amplifier could not be configured as a Class-G Trimodal, i.e., manually switchable between Classes A and B. That would indeed be an interesting machine.

In Figure 10.19 is shown the THD plot for such an A + C amplifier working at 20 W and 30 W into 8 Ω. At 20 W the distortion is very low indeed, no higher than a pure Class-A amplifier. At 30 W the transition gain-steps appear, but the THD remains very well controlled, and no higher than a Blameless Class-B design. Note that as in Class-B, when the THD does start to rise it only does so at 6 dB/octave. The quiescent current was set to 1.5 A.

Figure 10.20 reveals the THD residual during A + C operation. There are absolutely no crossover artefacts, and the small disturbances that do occur happen at such a high signal level that I really do think it is safe to assume they could never be audible. Figure 10.21 shows the complete absence of artefacts on the residual when this new type of amplifier is working below transition; it gives pure Class-A linearity. Finally, Figure 10.22 gives the THD when the amplifier is driving the full 50 W into 8 Ω; as before the A + C THD plot is hard to distinguish from Class-B, but there is the immense advantage that there is no crossover distortion at low levels, and no critical bias settings.

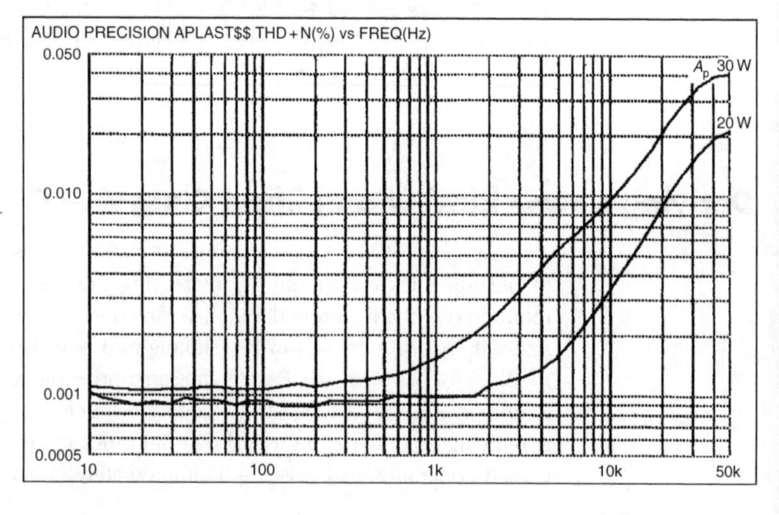

Figure 10.19 The THD plot of the Class-A + C amplifier (30 W and 20 W into 8 Ω). Inner drivers powered from outer rails

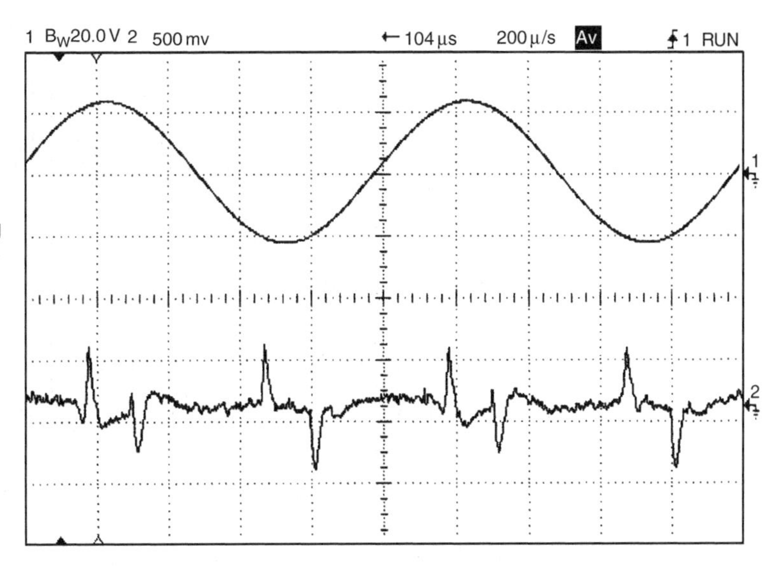

Figure 10.20
The THD residual waveform of the Class-A + C amplifier above transition, at 30 W into 8 Ω. Switching artifacts are visible but not crossover distortion

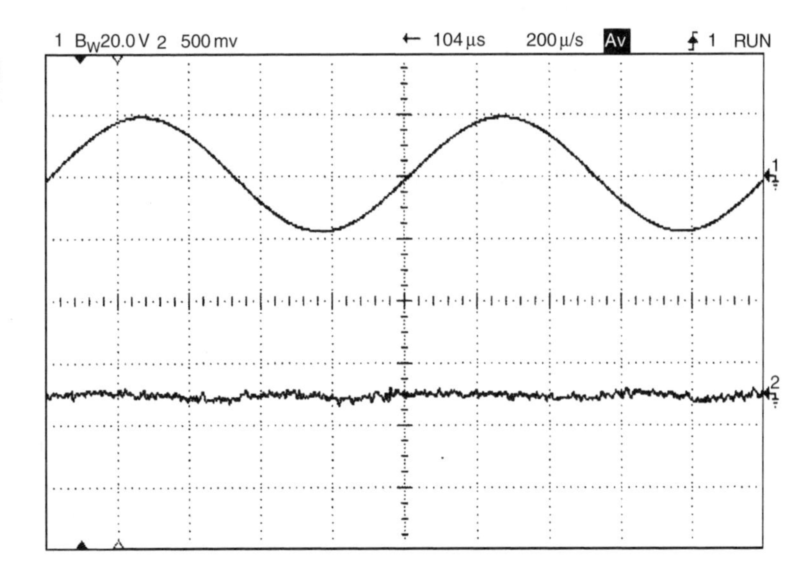

Figure 10.21
The THD residual waveform plot of the Class-A + C amplifier (20 W into 8 Ω)

Adding two-pole compensation

I have previously shown elsewhere in this book that amplifier distortion can be very simply reduced by changes to the compensation; which means a scheme more sophisticated than the near-universal dominant pole method. It must be borne in mind that any departure from the conventional 6 dB/octave-all-the-way compensation scheme is likely to be a move away from unconditional stability. (I am using this phrase in its proper meaning;

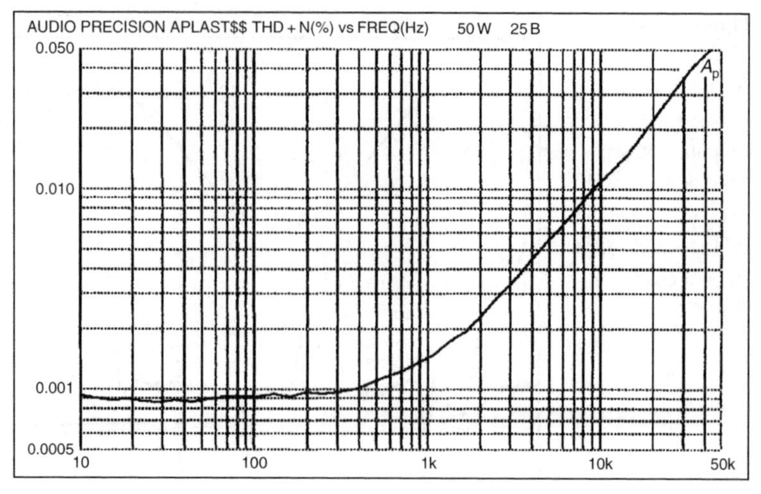

Figure 10.22 The THD plot of the Class-A + C amplifier (50 W into 8 Ω). Inner drivers powered from outer rails

in Control Theory unconditional stability means that increasing open-loop gain above a threshold causes instability, but the system is stable for all lower values. Conditional Stability means that lower open-loop gains can also be unstable.)

A conditionally stable amplifier may well be docile and stable into any conceivable reactive load when in normal operation, but shows the cloven hoof of oscillation at power-up and power-down, or when clipping. This is because under these conditions the effective open-loop gain is reduced.

Class-G distortion artefacts are reduced by normal dominant-pole feedback in much the same way as crossover non-linearities, i.e., not all that effectively, because the artefacts take up a very small part of the cycle and are therefore composed of high-order harmonics. Therefore a compensation system that increases the feedback factor at high audio frequencies will be effective on switching artefacts, in the same way that it is for crossover distortion. The simplest way to implement two-pole circuit compensation is shown in Figure 10.23. Further details are given on page 191.

The results of two-pole compensation for B + C are shown in Figure 10.24; comparing it with Figure 10.13 (the normally compensated B + C amplifier) the above-transition (30 W) THD at 10 kHz has dropped from 0.008% to 0.005%; the sub-transition (20 W) THD at 10 kHz has fallen from 0.007% to 0.003%. Comparisons have to be done at 10 kHz or thereabouts to ensure there is enough to measure.

Now comparing the two-pole B + C amplifier with Figure 10.19 (the A + C amplifier) the above-transition (30 W) THD at 10 kHz of the former is lower at 0.005% compared with 0.008%. As I have demonstrated before, proper

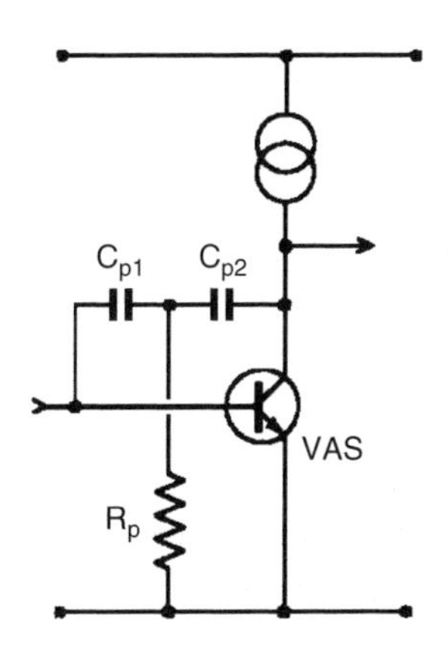

Figure 10.23 The circuit modification for two-pole compensation

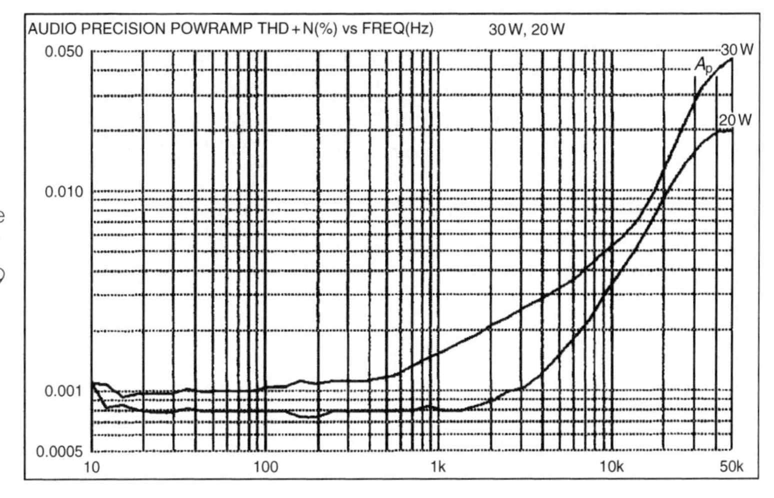

Figure 10.24 The THD plot for B + C operation with two-pole compensation (20 W and 30 W into 8 Ω). Compare with Figures 10.13 (B + C) and 10.19 (A + C)

use of two-pole compensation can give you a Class-B amplifier that is hard to distinguish from Class-A – at least until you put your hand on the heatsink.

Further variations on Class-G

This by no means exhausts the possible variations that can be played on Class-G. For example, it is not necessary for the outer devices to operate synchronously with the inner devices. So long as they turn on in time, they can turn off much later without penalty except in terms of increased dissipation. In so-called syllabic Class-G, the outer devices turn on fast but then typically remain on for 100 msec or so to prevent glitching; see Funada and Akiya[7] for one version. Given the good results obtained with straight Class-G, this no longer seems a promising route to explore.

With the unstoppable advance of multichannel amplifier and powered sub-woofers, Class-G is at last coming into its own. It has recently even appeared in a Texas ADSL driver IC. I hope I have shown how to make it work, and then how to make it work better. From the results of a Web search done today, I would modestly suggest that this might be the lowest distortion Class-G amplifier so far.

References

1. Self, D *Self On Audio* pub. Newnes, ISBN 0–7506–4765–5, p. 347.
2. Sampei et al. *Highest Efficiency and Super Quality Audio Amplifier Using MOS Power FETs in Class-G Operation* IEEE Trans on Consumer Electronics, Vol CE–24, #3 Aug 1978, p. 300.
3. Feldman *Class-G High Efficiency Hi-Fi Amplifier Radio Electronics,* Aug 1976, p. 47.
4. Self, D *Self On Audio* pub. Newnes, ISBN 0–7506–4765–5, p. 369.
5. Self, D *Self On Audio* pub. Newnes, ISBN 0–7506–4765–5, p. 386.
6. Self, D *Self On Audio* pub. Newnes, ISBN 0–7506–4765–5, p. 293.
7. Funada and Akiya *A Study of High-Efficiency Audio Power Amplifiers Using a Voltage Switching Method* JAES Vol 32 #10, Oct 1984, p. 755.

11

Class-D amplifiers

Since the first edition of this book, Class-D amplifiers have increased enormously in popularity. This is because Class-D gives the highest efficiency of any of the amplifier classes, although the performance, particularly in terms of linearity, is not so good. The rapid rate of innovation means that this section of the book is much more of a snapshot of a fast-moving scene than the rest of the material. I do not want to keep repeating 'At the time of writing' as each example is introduced, so I hope you will take that as read.

The fields of application for Class-D amplifiers can be broadly divided into two areas; low and high power outputs. The low power field reaches from a few milliwatts (for digital hearing aids) to around 5 W, while the high-power applications go from 80 W to 1400 W. At present there seems to be something of a gap in the middle, for reasons that will emerge.

The low-power area includes applications such as mobile phones, personal stereos, and laptop computer audio. These products are portable, and battery driven, so power economy is very important. A major application of Class-D is the production of useful amounts of audio power from a single low-voltage supply rail. A good example is the National Semiconductor LM4671, a single-channel amplifier IC that gives 2.1 W into a 4 Ω speaker from a 5 V supply rail, using a 300 kHz switching frequency. This is a very low voltage by conventional power amplifier standards, and requires an H-bridge output structure, of which more later.

The high-power applications include PA amplifiers, home theatre systems, and big sub-woofers. These are all energised from the mains supply, so power economy is not such a high priority. Here Class-D is used because it keeps dissipation and therefore power supply and heatsink size to a minimum, leading to a smaller and neater product. High-power Class-D amplifiers are also used in car audio systems, with power capabilities of 1000 W or more into 2 Ω; here minimising the power drain is of rather greater importance, as the capabilities of the engine-driven alternator that provides the 12 V supply are finite.

There is a middle ground between these two areas, where an amplifier is powered from the mains but of no great output power – say a stereo unit with an output of 30 W into 8 Ω per channel. The heatsinks will be small, and eliminating them altogether will not be cost saving. The power supply will almost certainly be a conventional toroid-and-bridge-rectifier arrangement, and the cost savings on reducing the size of this component by using Class-D will not be large. In this area the advantage gained by accepting the limitations of Class-D are not at present enough to justify it.

Class-D amplifiers normally come as single ICs or as chip sets with separate output stages. Since the circuitry inside these ICs is complex, and not disclosed in detail, they are not very instructive to those planning to design their own discrete Class-D amplifier.

History

The history of the Class-D amplifier goes back, as is so often the case with technology, further back than you might think. The principle is generally regarded as having surfaced in the 1950s, but the combination of high switching frequencies and valve output transformers probably did not appear enticing. The first public appearance of Class-D in the UK was the Sinclair X-10, which claimed an output of 10 W. This was followed by the X20, alleging a more ambitious 20 W. I resurrected one of the latter in 1976, when my example proved to yield about 3 W into 8 Ω. The THD was about 5% and the rudimentary output filter did very little to keep the low switching frequency out of the load. The biggest problem of the technology at that time was that bipolar transistors of suitable power-handling capacity were too slow for the switching frequencies required; this caused serious losses that undermined the whole point of Class-D, and also produced unappealingly high levels of distortion. It was not until power FETs, with their very fast switching times, appeared that Class-D began to become a really practical proposition.

Basic principles

Amplifiers working in Class-D differ radically from the more familiar Classes of A, B and G. In Class-D there are no output devices operating in the linear mode. Instead they are switched on and off at an ultrasonic frequency, the output being connected alternately to each supply rail. When the mark-space ratio of the input signal is varied, the average output voltage varies with it, the averaging being done by a low-pass output filter, or by the loudspeaker inductance alone. Note that the output is also directly proportional to the supply voltage; there is no inherent supply rejection at all with this sort of output stage, unlike the Class-B output stage. The use of negative feedback helps with this. The switching frequencies used range from 50 kHz to 1 MHz. A higher frequency makes the output filter simpler and smaller, but tends to increase switching losses and distortion.

The classic method of generating the drive signal is to use a differential comparator. One input is driven by the incoming audio signal, and the other by a sawtooth waveform at the required switching frequency. A basic Class-D amplifier is shown in Figure 11.1, and the PWM process is illustrated in Figure 11.2.

Clearly the sawtooth needs to be linear (i.e., with constant slope) to prevent distortion being introduced at this stage. There are other ways to create the required waveform, such as a sigma-delta modulator.

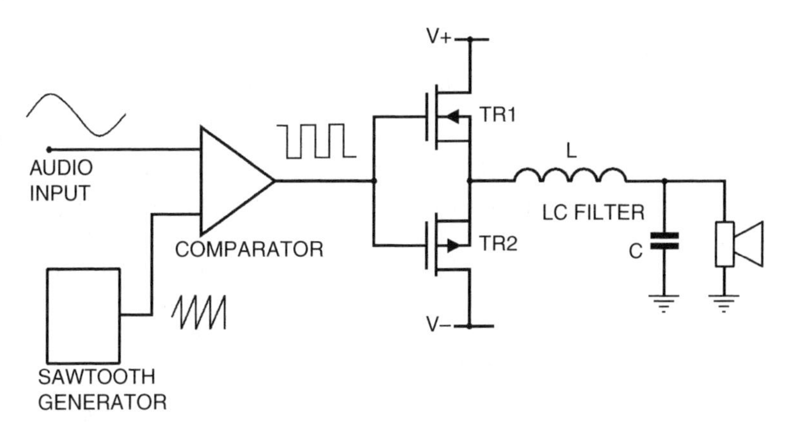

Figure 11.1
A basic Class-D amplifier with PWM comparator, FET output stage, and second-order LC output filter

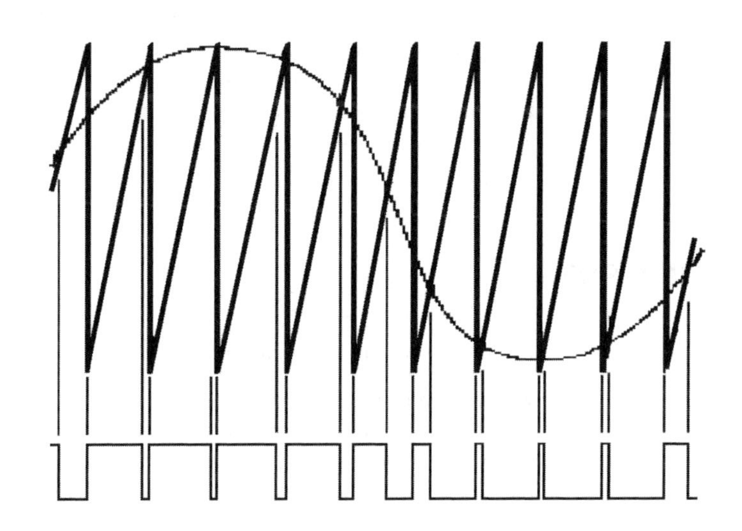

Figure 11.2
The PWM process as performed by a differential comparator

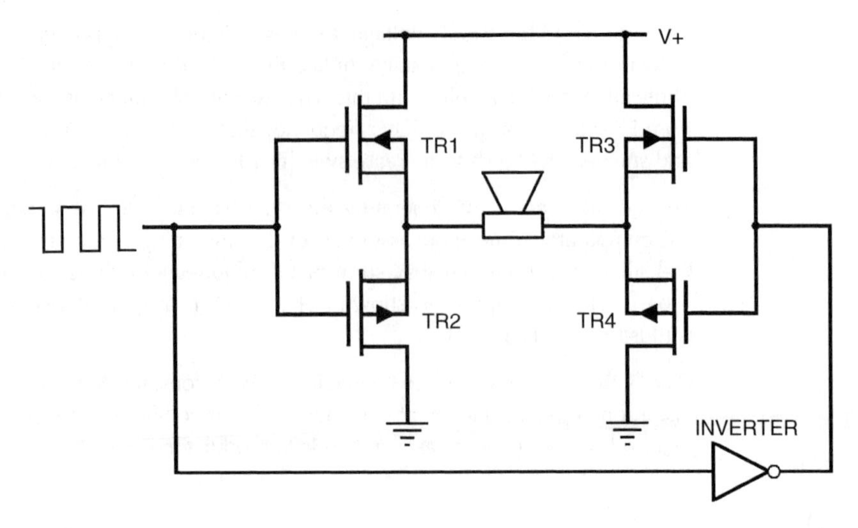

Figure 11.3
The H-bridge output configuration. The output filter is not shown

When the aim is to produce as much audio power as possible from a low voltage supply such as 5 V, the H-bridge configuration is employed, as shown in Figure 11.3. It allows twice the voltage-swing across the load, and therefore theoretically four times the output power, and also permits the amplifier to run from one supply rail without the need for bulky output capacitors of doubtful linearity. This method is also called the Bridge-Tied Load, or BTL.

The use of two amplifier outputs requires a somewhat more complex output filter. If the simple 2-pole filter of Figure 11.4a is used, the switching frequency is kept out of the loudspeaker, but the wiring to it will carry a large common-mode signal from OUT. A balanced filter is therefore commonly used, in either the Figure 11.4b or Figure 11.4c versions. Figure 11.4d illustrates a four-pole output filter – note that you can save a capacitor. This is only used in quality applications because inductors are never cheap.

Technology

The theory of Class-D has a elegant simplicity about it; but in real life complications quickly begin to intrude.

While power FETs have a near-infinite input resistance at the gate, they require substantial current to drive them at high frequencies, because of the large device capacitances, and the gate drive circuitry is a non-trivial part of the amplifier. Power FETs, unlike bipolars, require several volts on

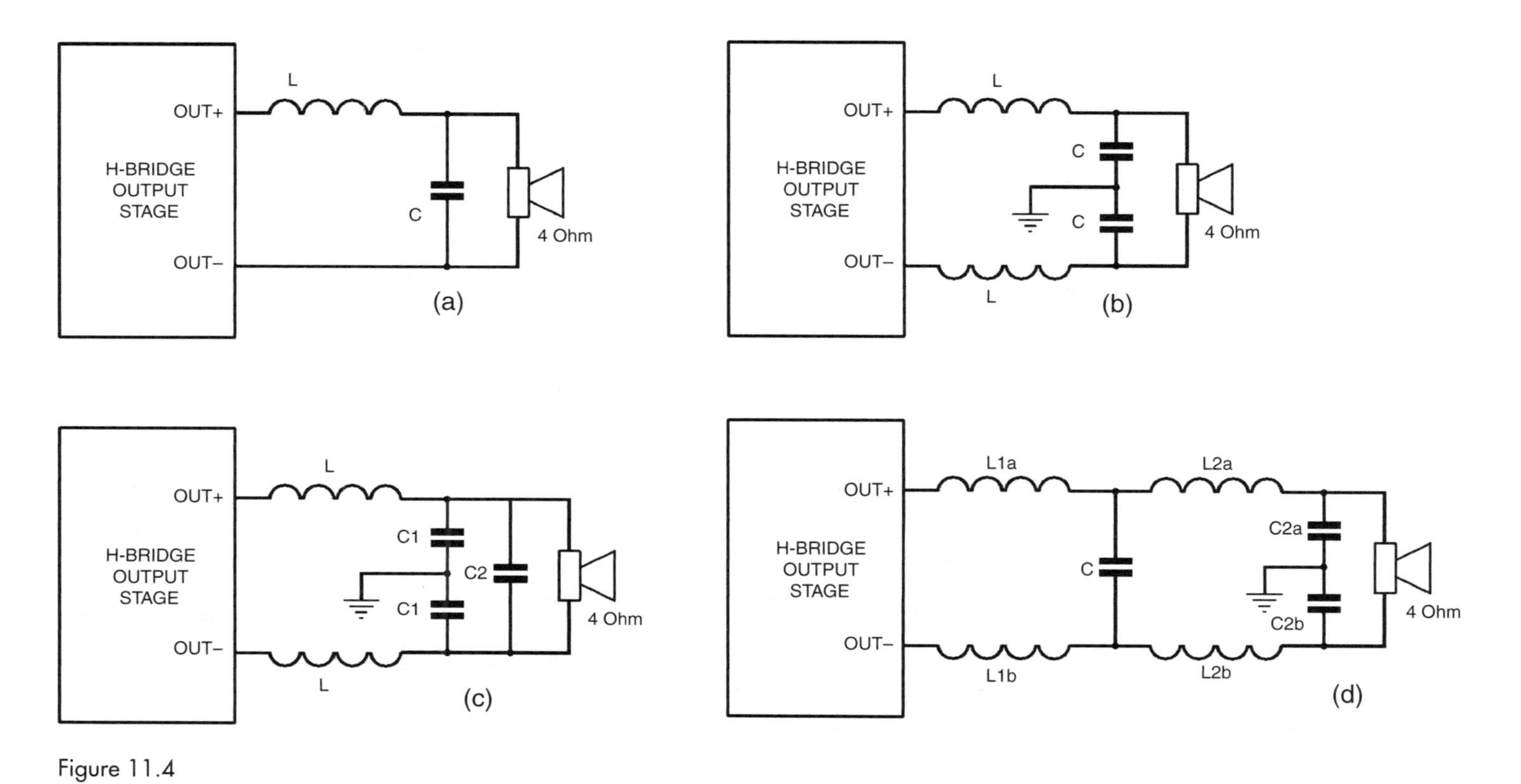

Figure 11.4
Filter arrangements for the H-bridge output. 4a is the simplest but allows a common-mode signal on the speaker cabling. 4b and 4c are the most usual versions. 4d is a 4-pcle filter

the gate to turn them on. This means that the gate drive voltage needed for the high-side FET TR1 in Figure 11.1 is actually above the positive voltage rail. In many designs a bootstrap supply driven from the output is used to power the gate-drive circuitry. Since this supply will not be available until the high-side FET is working, special arrangements are needed at start-up.

The more powerful amplifiers usually have external Schottky diodes connected from output to the supply rails for clamping flyback pulses generated by the inductive load. These are not merely to protect the output stage from damage, but to improve efficiency; as described in the section below.

The application of negative feedback to reduce distortion and improve supply rail rejection is complicated by the switched nature of the output waveform. Feedback can be taken from after the output filter, or alternatively taken from before it and passed through an op-amp active filter to remove the switching frequency. In either case the filtering adds phase shift and limits the amount of negative feedback that can be applied while still retaining Nyquist stability.

Other enhancements that are common are selectable input gain, and facilities for synchronising the switching clocks of multiple amplifiers to avoid audible heterodyne tones. Figure 11.5 shows a Class-D amplifier including these features.

Protection

All the implementations of Class-D on the market have internal protection systems to prevent excessive output currents and device temperatures.

In the published circuitry DC offset protection is conspicuous by its absence. It is understandable that there is little enthusiasm for adding output relays to personal stereos – they might consume more power than the amplifier. However it is surprising that they also appear to be absent from 500 W designs where relay size and power consumption are minor issues. Are such amplifiers really that reliable?

Most Class-D systems also have undervoltage protection. If the supply voltage falls too low then there may not be enough gate drive voltage to turn the output FETs fully on, and they will dissipate excessive power. A lockout circuit prevents operation below a certain voltage. A shutdown facility is almost always provided; this inhibits any switching in the output stage and allows power consumption to be very low indeed in the standby mode.

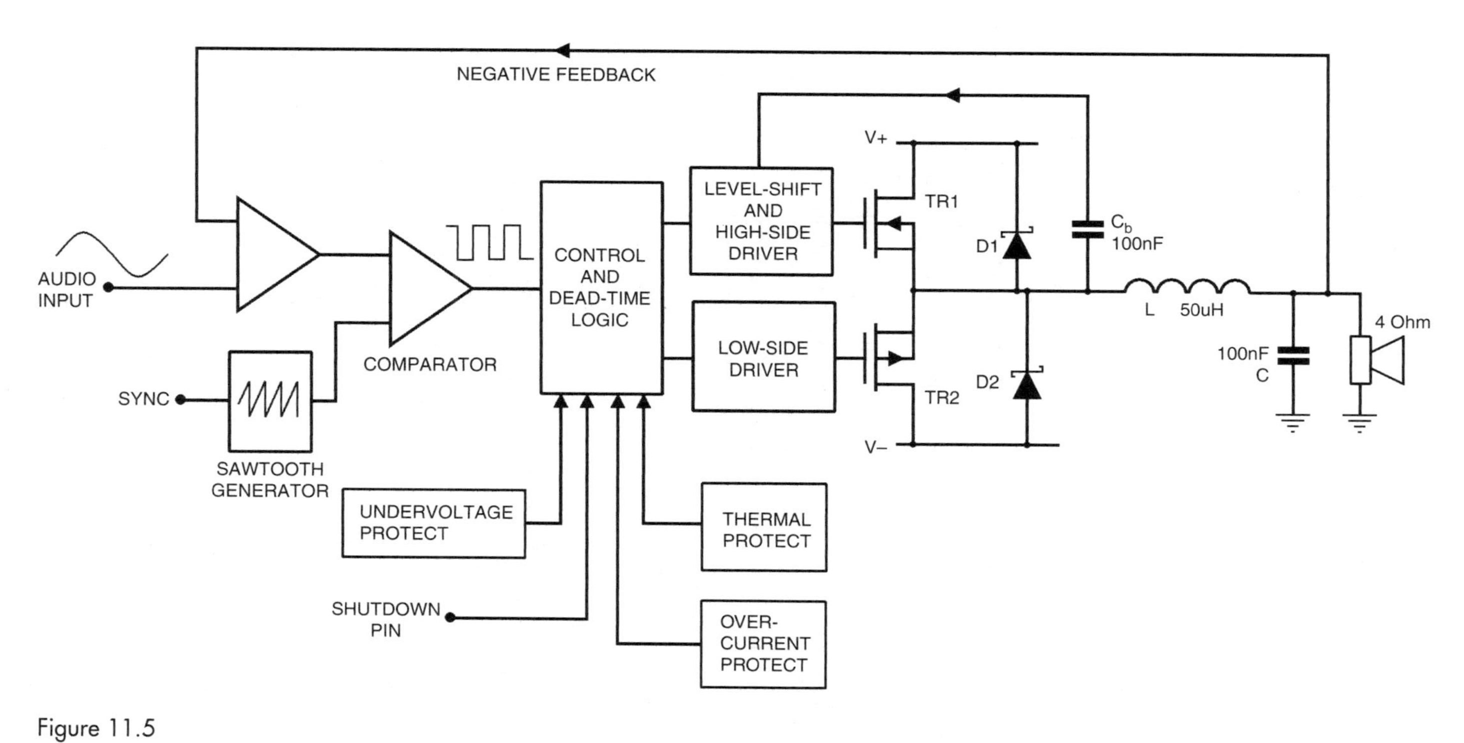

Figure 11.5
Showing the main features of a practical Class-D amplifier including Schottky clamp diodes, bootstrap supply, and one form of negative feedback

Output filters

The purpose of the output filter is to prevent radiation of switching frequencies for amplifiers that have external speaker cables, and also to improve efficiency. The inductance of a loudspeaker coil alone will in general be low enough to allow some of the switching frequency energy to pass through it to ground, causing significant losses. While some low-power integrated applications have no output filter at all, most Class-D amplifiers have a second-order LC filter between the amplifier output and the loudspeaker. In some cases a fourth-order filter is used, as in Figure 11.4d. The Butterworth alignment is usually chosen to give maximal flatness of frequency response.

As described in the chapter on real speaker loads, a loudspeaker, even a single-element one, is a long way from being a resistive load. It is therefore rather surprising that at least one manufacturer provides filter design equations that assume just that. When a Class-D amplifier is to be used with separate loudspeakers of unknown impedance characteristic, the filter design can only proceed on the basis of plausible assumptions, and there are bound to be some variations in frequency response.

The inductor values required are typically in the region 10 μH–50 μH, which is much larger than the 1 μH–2 μH air-cored coils used to ensure stability with capacitative loads in Class-B amplifiers. It is therefore necessary to use ferrite-cored inductors, and care must be taken that they do not saturate at maximum output.

Efficiency

At the most elementary level of theory, the efficiency of a Class-D amplifier is always 100%, at all output levels. In practice, of course, the mathematical idealisations do not hold, and the real-life efficiency of most implementations is between 80% and 90% over most of the power output range. At very low powers the efficiency falls off steeply, as there are fixed losses that continue to dissipate power in the amplifier when there is no audio output at all (See Figure 11.6).

The losses in the output stage are due to several mechanisms. The most important are:

First, the output FETs have a non-zero resistance even when they are turned hard on. This is typically in the range 100 to 200 mΩ, and can double as the device temperature increases from 0 to 150°C, the latter being the usual maximum operating temperature. This resistance causes I^2R losses.

Second, the output devices have non-zero times for switching on and off. In the period when the FET is turning on or turning off, it has an intermediate

Figure 11.6
A typical efficiency curve for a Class-D amplifier driving a 4-Ω load at 1 kHz

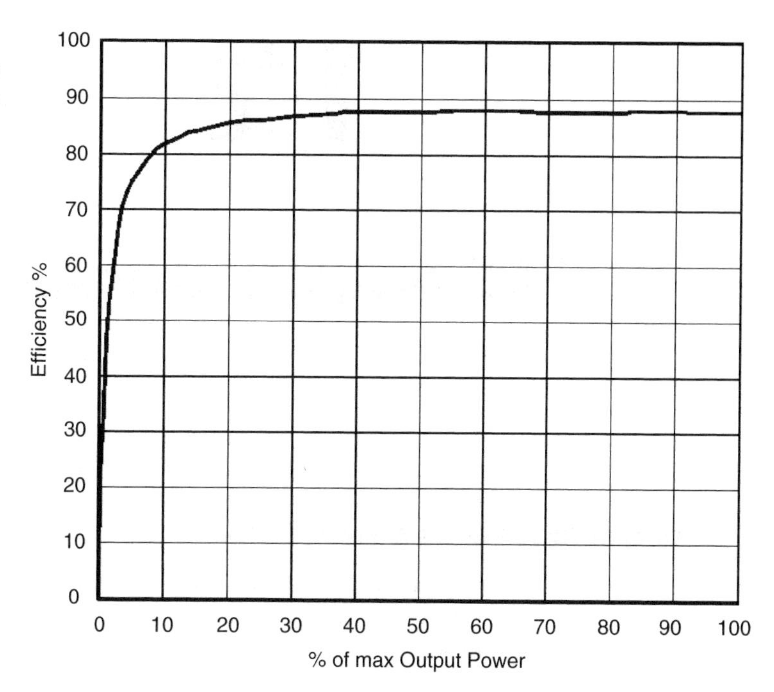

value of resistance which again causes I^2R losses. It is essential to minimise the stray inductance in the drain and source circuits as this not only extends the switching times but also causes voltage transients at turn-off that can overstress the FETs.

Third, flyback pulses generated by an inductive load can cause conduction of the parasitic diodes that are part of the FET construction. These diodes have relatively long reverse recovery times and more current will flow than is necessary. To prevent this many Class-D designs have Schottky clamp diodes connected between the output line and supply rails as in Figure 11.5. These turn on at a lower voltage than the parasitic FET diodes and deal with the flyback pulses. They also have much faster reverse recovery times.

Last, and perhaps most dangerous, is the phenomenon known as 'shoot-through'. This somewhat opaque term refers to the situation when one FET has not stopped conducting before the other starts. This gives rise to an almost direct short between the supply rails, although very briefly, and large amounts of unwanted heating can occur. To prevent this the gate drive to the FET that is about to be turned on is slightly delayed, by a 'dead-time' circuit. The introduction of dead-time increases distortion, so only the minimum is applied; a 40 nsec delay is sufficient to create more than 2% THD in a 1 kHz sinewave.

12

FET output stages

The characteristics of power FETS

An FET is essentially a voltage-controlled device. So are BJTs, despite the conventional wisdom that persists in regarding them as current-controlled. They are not, even if BJT base currents are non-negligible.

The power FETs normally used are enhancement devices – in other words, with no voltage between gate and source they remain off. In contrast, the junction FETs found in small-signal circuitry are depletion devices, requiring the gate to be taken negative of the source (for the most common N-channel devices) to reduce the drain current to usable proportions. (Please note that the standard information on FET operation is in many textbooks and will not be repeated here.)

Power FETs have large internal capacitances, both from gate to drain, and from gate to source. The gate-source capacitance is effectively bootstrapped by the source-follower configuration, but the gate-drain capacitance, which can easily total 2000 pF, remains to be driven by the previous stage. There is an obvious danger that this will compromise the amplifier slew-rate if the VAS is not designed to cope.

FETs tend to have much larger bandwidths than BJT output devices. My own experience is that this tends to manifest itself as a greater propensity for parasitic oscillation rather than anything useful, but the tempting prospect of higher global NFB factors due to a higher output stage pole remains. The current state of knowledge does not yet permit a definitive judgement on this.

A great deal has been said on the thermal coefficients of the Vbias voltage. It is certainly true that the temp coefficient at high drain currents is negative – in other words drain current falls with increasing temperature – but on the other hand the coefficient reverses sign at low drain currents, and this implies that precise quiescent-current setting will be very difficult. A negative-temperature coefficient provides good protection against thermal runaway, but this should never be a problem anyway.

FET versus BJT output stages

On beginning any power amplifier design, one of the first decisions that must be made is whether to use BJTs or FETs in the output stage. This decision may of course already have been taken for you by the marketing department, as the general mood of the marketplace is that if FETs are more expensive, they must be better. If however, you are lucky enough to have this crucial decision left to you, then FETs normally disqualify themselves on the same grounds of price. If the extra cost is not translated into either better performance and/or a higher sustainable price for the product, then it appears to be foolish to choose anything other than BJTs.

Power MOSFETS are often hailed as the solution to all amplifier problems, but they have their own drawbacks, not the least being low transconductance, poor linearity, and a high ON-resistance that makes output efficiency mediocre. The high-frequency response may be better, implying that the second pole P2 of the amplifier response will be higher, allowing the dominant pole P1 be raised with the same stability margin, and so in turn giving more NFB to reduce distortion. However, we would need this extra feedback (if it proves available in practice) to correct the worse open-loop distortion, and even then the overall linearity would almost certainly be worse. To complicate matters, the compensation cannot necessarily be lighter because the higher output-resistance makes more likely the lowering of the output pole by capacitative loading.

The extended FET frequency response is, like so many electronic swords, two-edged if not worse, and the HF capabilities mean that rigorous care must be taken to prevent parasitic oscillation, as this is often promptly followed by an explosion of disconcerting violence. FETs should at least give freedom from switchoff troubles (Distortion 3c) as they do not suffer from BJT charge-storage effects.

Advantages of FETs

1. For a simple complementary FET output stage, drivers are not required. This is somewhat negated by the need for gate-protection zener diodes.
2. There is no second-breakdown failure mechanism. This may simplify the design of overload protection systems, especially when arranging for them to cope with highly reactive loads.
3. There are no charge-storage effects to cause switchoff distortion.

Disadvantages of FETs

1 Linearity is very poor by comparison with a BJT degenerated to give the same transconductance. The Class-B conduction characteristics do not cross over smoothly, and there is no equivalent to the optimal Class-B bias condition that is very obvious with a BJT output stage.
2 The Vgs required for conduction is usually of the order of 4–6 V, which is much greater than the 0.6–0.8 V required by a BJT for base drive. This greatly reduces the voltage efficiency of the output stage unless the preceding small-signal stages are run from separate and higher-voltage supply rails.
3 The minimum channel resistance of the FET, known as Rds(on), is high and gives a further reduction in efficiency compared with BJT outputs.
4 Power FETs are liable to parasitic oscillation. In severe cases a plastic-package device will literally explode. This is normally controllable in the simple complementary FET output stage by adding gate-stopper resistors, but is a serious disincentive to trying radical experiments in output stage circuit design.
5 Some commentators claim that FET parameters are predictable; I find this hard to understand as they are notorious for being anything but. From one manufacturer's data (Harris), the Vgs for the IRF240 FET varies between 2.0 and 4.0 V for an Id of 250 μA; this is a range of two to one. In contrast the Vbe/Ic relation in bipolars is fixed by a mathematical equation for a given transistor type, and is much more reliable. Nobody uses FETs in log converters.
6 Since the Vgs spreads are high, this will complicate placing devices in parallel for greater power capability. Paralleled BJT stages rarely require current-sharing resistors of greater than 0.1 Ω, but for the FET case they may need to be a good deal larger, reducing efficiency further.
7 At the time of writing, there is a significant economic penalty in using FETs. Taking an amplifier of given power output, the cost of the output semiconductors is increased by between 1.5 and 2 times with FETs.

IGBTs

Insulated-Gate Bipolar Transistors (IGBTs) represent a relatively new option for the amplifier designer. They have been held up as combining the best features of FETs and BJTs. In my view, this is a dubious proposition as I find the advantages of FETs for audio to be heavily outweighed by the drawbacks, and if IGBTs have any special advantages they have not so far emerged. According to the Toshiba application notes[1], IGBTs consist of an FET controlling a bipolar power transistor; I have no information on the linearity of these devices, but the combination does not sound promising.

The most discouraging aspect of IGBTs is the presence of a parasitic BJT that turns the device hard on above a critical current threshold. This inbuilt self-destruct mechanism will at the very least make overload protection an extremely critical matter; it seems unlikely that IGBTs will prove popular for audio amplification.

Power FET output stages

Three types of FET output stage are shown in Figure 12.1, and Figures 12.2–12.5 show SPICE gain plots, using 2SK135/2SJ50 devices. Most FET amplifiers use the simple source-follower configuration in Figure 12.1a; the large-signal gain plot at Figure 12.2 shows that the gain for a given load is lower (0.83 rather than 0.97 for bipolar, at 8 Ω) because of low *gm*, and this, with the high on-resistance, reduces output efficiency seriously. Open-loop distortion is markedly higher; however LSN does not increase with heavier loading, there being no equivalent of *Bipolar Gain-Droop*. The crossover region has sharper and larger gain deviations than a bipolar stage, and generally looks pretty nasty; Figure 12.3 shows the impossibility of finding a *correct* Vbias setting.

Figure 12.1b shows a hybrid (i.e., bipolar/FET) quasi-complementary output stage, first described in Self[2]. This topology is intended to maximise economy rather than performance, once the decision has been made (presumably for marketing reasons) to use FETs, by making both output devices cheap N-channel devices; complementary MOSFET pairs remain relatively rare and expensive. The basic configuration is badly asymmetrical, the hybrid lower half having a higher and more constant gain than the source-follower upper half. Increasing the value of Re2 gives a reasonable match between the gains of the two halves, but leaves a daunting crossover discontinuity.

The hybrid full-complementary stage in Figure 12.1c was conceived[3] to maximise FET performance by linearising the output devices with local feedback and reducing Iq variations due to the low power dissipation of the bipolar drivers. It is very linear, with no gain-droop at heavier loadings (Figure 12.4), and promises freedom from switchoff distortions; however, as shown, it is rather inefficient in voltage-swing. The crossover region in Figure 12.5 still has some unpleasant sharp corners, but the total crossover gain deviation (0.96–0.97 at 8 Ω) is much smaller than for the quasi-hybrid (0.78–0.90) and so less high-order harmonic energy is generated.

Table 12.1 summarizes the SPICE curves for 4 and 8 Ω loadings. Each was subjected to Fourier analysis to calculate THD% results for a ±40 V input. The BJT results from Chapter 5 are included for comparison.

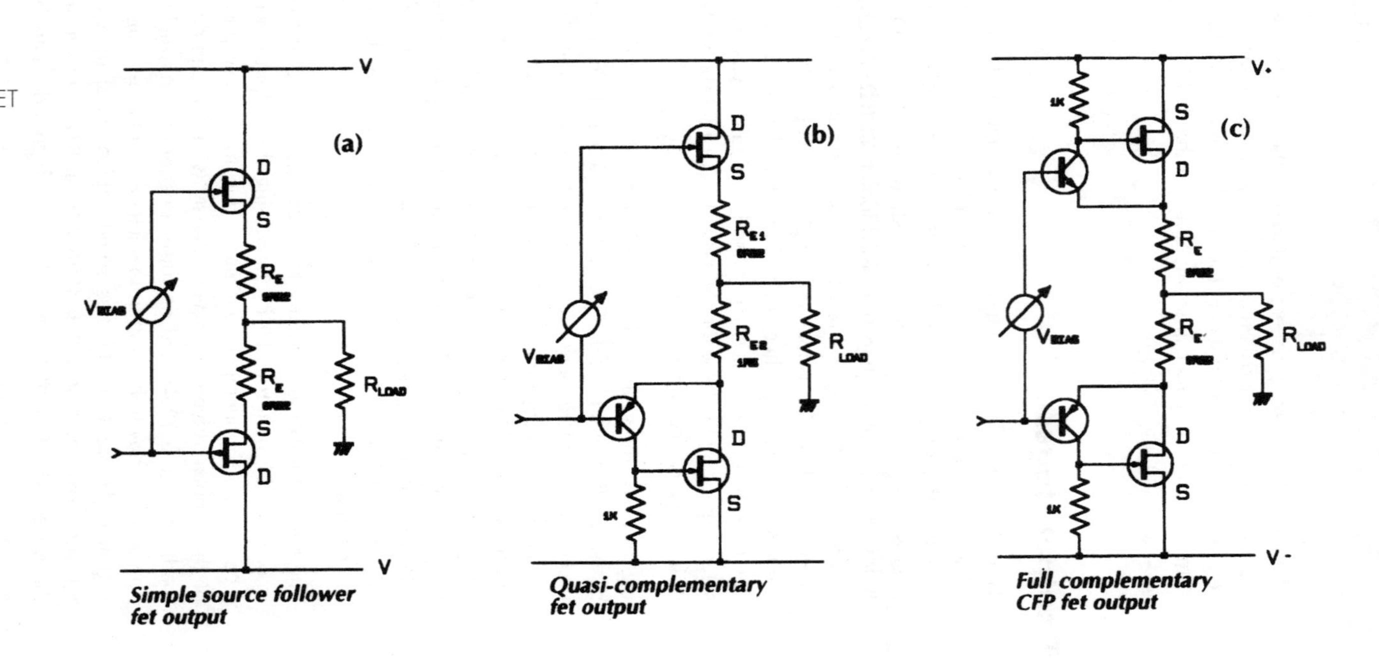

Figure 12.1 Three MOSFET output architectures

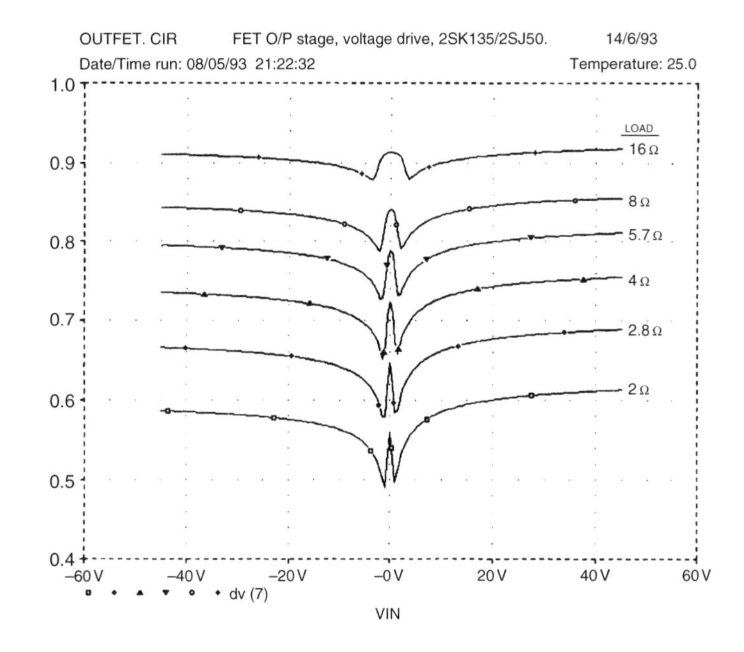

Figure 12.2
Source-Follower FET large-signal gain versus output

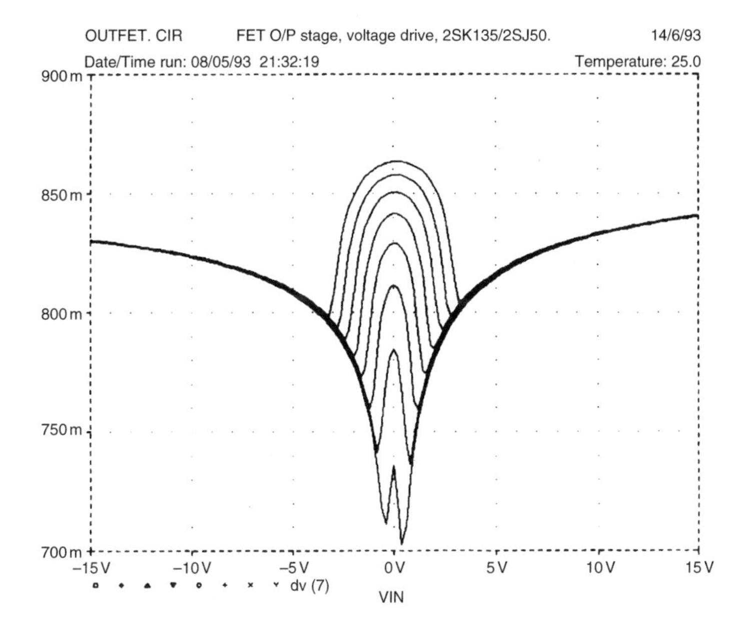

Figure 12.3
Source-Follower FET crossover region ±15 V range

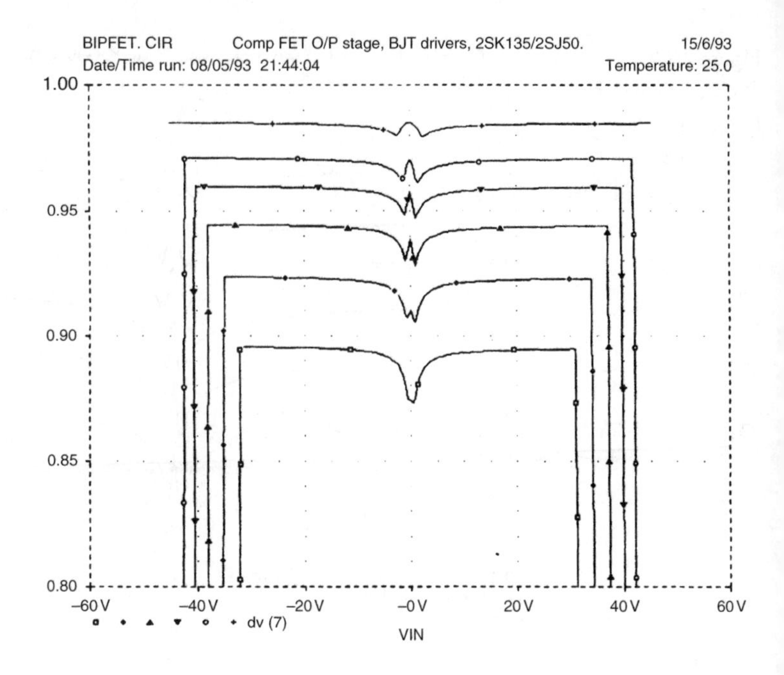

Figure 12.4
Complementary Bipolar-FET gain versus output

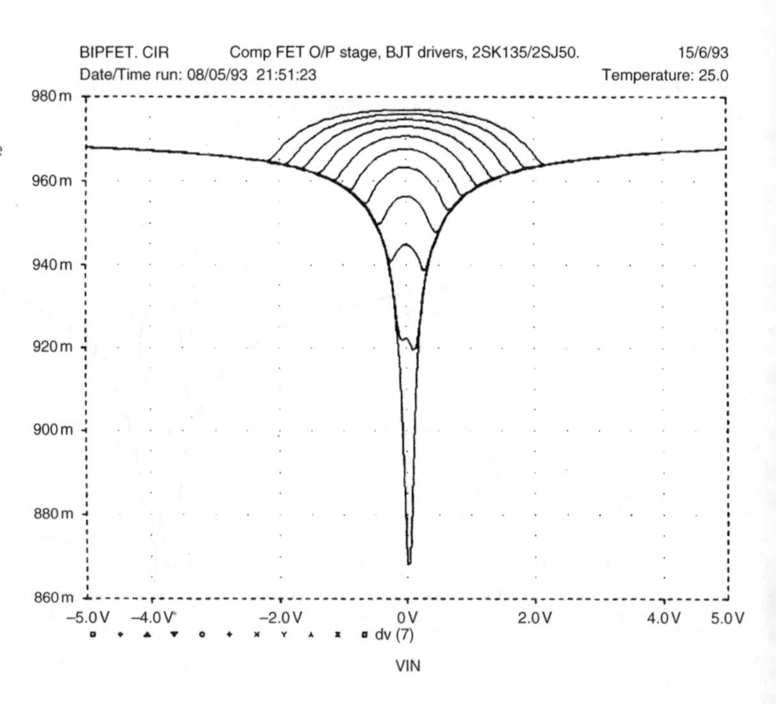

Figure 12.5
Complementary BJT–FET crossover region ±15 V range

Table 12.1 THD percentages and average gains for various types of output stage, for 8 and 4 ohm loading

	Emitter Follower	CFP	Quasi Simple	Quasi Bax	Triple Type 1	Simple MOSFET	Quasi MOSFET	Hybrid MOSFET
8 Ω (%)	0.031	0.014	0.069	0.050	0.13	0.47	0.44	0.052
Gain	0.97	0.97	0.97	0.96	0.97	0.83	0.84	0.97
4 Ω (%)	0.042	0.030	0.079	0.083	0.60	0.84	0.072	0.072
Gain	0.94	0.94	0.94	0.94	0.92	0.72	0.73	0.94

Power FETs and bipolars: the linearity comparison

There has been much debate as to whether power FETs or bipolar junction transistors (BJTs) are superior in power amplifier output stages, e.g., Hawtin[4]. As the debate rages, or at any rate flickers, it has often been flatly stated that power FETs are more linear than BJTs, usually in tones that suggest that only the truly benighted are unaware of this.

In audio electronics it is a good rule of thumb that if an apparent fact is repeated times without number, but also without any supporting data, it needs to be looked at very carefully indeed. I therefore present my own view of the situation here.

I suggest that it is now well-established that power FETs when used in conventional Class-B output stages are a good deal less linear than BJTs. The gain-deviations around the crossover region are far more severe for FETs than the relatively modest wobbles of correctly biased BJTs, and the shape of the FET gain-plot is inherently jagged, due to the way in which two square-law devices overlap. The incremental gain range of a simple FET output stage is 0.84–0.79 (range 0.05) and this is actually much greater than for the Bipolar stages examined in Chapter 5; the EF stage gives 0.965–0.972 into 8 Ω (range 0.007) and the CFP gives 0.967–0.970 (range 0.003). The smaller ranges of gain-variation are reflected in the much lower THD figures when PSpice data is subjected to Fourier analysis.

However, the most important difference may be that the bipolar gain variations are gentle wobbles, while all FET plots seem to have abrupt changes that are much harder to linearise with NFB that must decline with rising frequency. The basically exponential Ic/Vbe characteristics of two BJTs approach much more closely the ideal of conjugate (i.e., always adding up to 1) mathematical functions, and this is the root cause of the much lower crossover distortion.

A close-up examination of the way in which the two types of device begin conducting as their input voltages increase shows that FETs move abruptly into the square-law part of their characteristic, while the exponential behaviour of bipolars actually gives a much slower and smoother start to conduction.

Similarly, recent work[5] shows that less conventional approaches, such as the CC-CE configuration of Mr Bengt Olsson[6], also suffer from the non-conjugate nature of FETs, and show sharp changes in gain. Gevel[7] shows that this holds for both versions of the stage proposed by Olsson, using both N- and P-channel drivers. There are always sharp gain-changes.

FETs in Class-A stages

It occurred to me that the idea that FETs are more linear was based not on Class-B power-amplifier applications, but on the behaviour of a single device in Class-A. It might be argued that the roughly square-law nature of a FET's Id/Vgs law is intuitively more *linear* than the exponential Ic/Vbe law of a BJT, but it is a bit difficult to know quite how to define *linear* in this context. Certainly a square-law device will generate predominantly low-order harmonics, but this says nothing about the relative amounts produced.

In truth the BJT/FET contest is a comparison between apples and aardvarks, the main problem being that the raw transconductance (gm) of a BJT is far higher than for any power FET. Figure 12.6 illustrates the conceptual test circuit; both a TO3 BJT (MJ802) and a power-FET (IRF240) have an increasing DC voltage Vin applied to their base/gate, and the resulting collector and drain currents from PSpice simulation are plotted in Figure 12.7. Voffset is used to increase the voltage applied to FET M1 by 3.0 V because nothing much happens below Vgs = 4 V, and it is helpful to have the curves on roughly the same axis. Curve A, for the BJT, goes almost vertically skywards, as a result of its far higher *gm*. To make the comparison meaningful, a small amount of local negative feedback is added to Q1 by Re, and as this emitter degeneration is increased from 0.01 Ω to 0.1 Ω, the Ic curves become closer in slope to the Id curve.

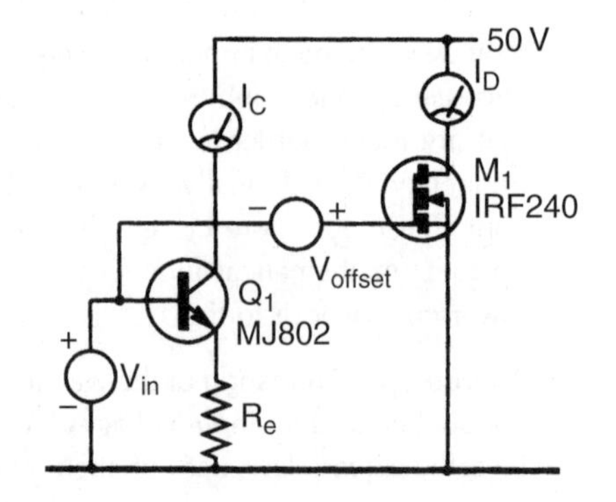

Figure 12.6
The linearity test circuit. Voffset adds 3 V to the DC level applied to the FET gate, purely to keep the current curves helpfully adjacent on a graph

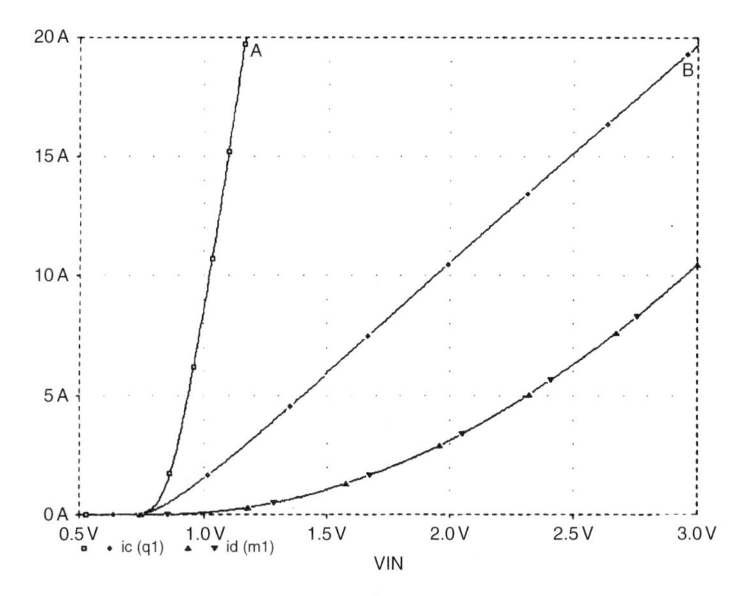

Figure 12.7
Graph of Ic and Id for the BJT and the FET. Curve A shows Ic for the BJT alone, while Curve B shows the result for Re = 0.1 Ω. The curved line is the Id result for a power FET without any degeneration

Because of the curved nature of the FET Id plot, it is not possible to pick an Re value that allows very close *gm* equivalence; Re = 0.1 Ω was chosen as a reasonable approximation; see Curve B. However, the important point is that I think no-one could argue that the FET Id characteristic is more linear than Curve B.

This is made clearer by Figure 12.8, which directly plots transconductance against input voltage. There is no question that FET transconductance increases in a beautifully linear manner – but this 'linearity' is what results in a square-law Id increase. The near-constant *gm* lines for the BJT are a much more promising basis for the design of a linear amplifier.

To forestall any objections that this comparison is all nonsense because a BJT is a current-operated device, I add here a small reminder that this is untrue. The BJT is a voltage operated device, and the base current that flows is merely an inconvenient side-effect of the collector current induced by said base voltage. This is why beta varies more than most BJT parameters; the base current is an unavoidable error rather than the basis of transistor operation.

The PSpice simulation shown here was checked against manufacturers' curves for the devices, and the agreement was very good – almost unnervingly so. It therefore seems reasonable to rely on simulator output for these kind of studies – it is certainly infinitely quicker than doing the real measurements, and the comprehensive power FET component libraries that are part of PSpice allow the testing to be generalised over a huge number of component types without actually buying them.

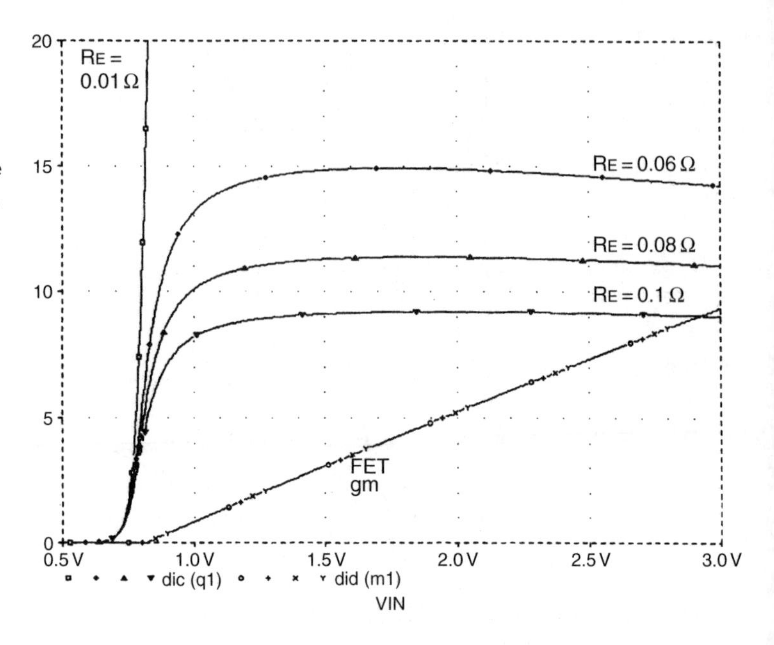

Figure 12.8 Graph of transconductance versus input voltage for BJT and FET. The near-horizontal lines are BJT gm for various Re values

To conclude, I think it is probably irrelevant to simply compare a naked BJT with a naked FET. Perhaps the vital point is that a bipolar device has much more raw transconductance gain to begin with, and this can be handily converted into better linearity by local feedback, i.e., adding a little emitter degeneration. If the transconductance is thus brought down roughly to FET levels, the bipolar has far superior large-signal linearity. I must admit to a sneaking feeling that if practical power BJTs had come along after FETs, they would have been seized upon with glee as a major step forward in power amplification.

References

1. Langdon, S *Audio amplifier designs using IGBT's, MOSFETs, and BJTs* Toshiba Application Note X3504, Vol 1, Mar 1991.
2. Self, D *Sound MOSFET Design* Electronics and Wireless World, Sept 1990, p. 760.
3. Self, D *MOSFET Audio Output* Letter, Electronics and Wireless World, May 1989, p. 524 (see also reference 2 above).
4. Hawtin, V *Letters, Electronics World*, Dec 1994, p. 1037.
5. Self, D *Two-Stage Amplifiers and the Olsson Output Stage* Electronics World, Sept 1995, p. 762.
6. Olsson, B *Better Audio From Non-Complements?* Electronics World, Dec 1994, p. 988.
7. Gevel, M *Private communication*, Jan 1995.

13

Thermal compensation and thermal dynamics

Why quiescent conditions are critical

In earlier sections of this book we looked closely at the distortion produced by amplifier output stages, and it emerged that a well-designed Class-B amplifier with proper precautions taken against the easily-fixed sources of non-linearity, but using basically conventional circuitry, can produce startlingly low levels of THD. The distortion that actually is generated is mainly due to the difficulty of reducing high-order crossover non-linearities with a global negative-feedback factor that declines with frequency; for 8 Ω loads this is the major source of distortion, and unfortunately crossover distortion is generally regarded as the most pernicious of non-linearities. For convenience, I have chosen to call such an amplifier, with its small signal stages freed from unnecessary distortions, but still producing the crossover distortion inherent in Class-B, a Blameless amplifier (see Chapter 3).

Page 145 suggests that the amount of crossover distortion produced by the output stage is largely fixed for a given configuration and devices, so the best we can do is ensure the output stage runs at optimal quiescent conditions to minimise distortion.

Since it is our only option, it is therefore particularly important to minimise the output-stage gain irregularities around the crossover point by holding the quiescent conditions at their optimal value. This conclusion is reinforced by the finding that for a Blameless amplifier increasing quiescent current to move into Class-AB makes the distortion worse, not better, as gm-doubling artefacts are generated. In other words the quiescent setting will only be correct over a relatively narrow band, and THD measurements show that too much quiescent current is as bad (or at any rate very little better) than too little.

The initial quiescent setting is simple, given a THD analyser to get a good view of the residual distortion; simply increase the bias setting from minimum until the sharp crossover spikes on the residual merge into the noise. Advancing the preset further produces edges on the residual that move apart from the crossover point as bias increases; this is gm-doubling at work, and is a sign that the bias must be reduced again.

It is easy to attain this optimal setting, but keeping it under varying operating conditions is a much greater problem because quiescent current (Iq) depends on the maintenance of an accurate voltage-drop *V*q across emitter resistors Re of tiny value, by means of hot transistors with varying Vbe drops. It's surprising it works as well as it does.

Some kinds of amplifier (e.g. Class-A or current-dumping types) manage to evade the problem altogether, but in general the solution is some form of thermal compensation, the output-stage bias voltage being set by a temperature-sensor (usually a Vbe-multiplier transistor) coupled as closely as possible to the power devices.

There are inherent inaccuracies and thermal lags in this sort of arrangement, leading to program-dependency of Iq. A sudden period of high power dissipation will begin with the Iq increasing above the optimum, as the junctions will heat up very quickly. Eventually the thermal mass of the heatsink will respond, and the bias voltage will be reduced. When the power dissipation falls again, the bias voltage will now be too low to match the cooling junctions and the amplifier will be under-biased, producing crossover spikes that may persist for some minutes. This is very well illustrated in an important paper by Sato et al[1].

Accuracy required of thermal compensation

Quiescent stability depends on two main factors. The first is the stability of the Vbias generator in the face of external perturbations, such as supply voltage variations. The second and more important is the effect of temperature changes in the drivers and output devices, and the accuracy with which Vbias can cancel them out.

Vbias must cancel out temperature-induced changes in the voltage across the transistor base-emitter junctions, so that Vq remains constant. From the limited viewpoint of thermal compensation (and given a fixed Re) this is very much the same as the traditional criterion that the quiescent current must remain constant, and no relaxation in exactitude of setting is permissible.

I have reached some conclusions on how accurate the Vbias setting must be to attain minimal distortion. The two major types of output stage, the EF and the CFP, are quite different in their behaviour and bias requirements, and this complicates matters considerably. The results are approximate, depending partly on visual assessment of a noisy residual signal, and may change slightly with transistor type, etc. Nonetheless, Table 13.1 gives a much-needed starting point for the study of thermal compensation.

From these results, we can take the permissible error band for the EF stage as about ±100 mV, and for the CFP as about ±10 mV. This goes some way to explaining why the EF stage can give satisfactory quiescent stability despite its dependence on the Vbe of hot power transistors.

Returning to the PSpice simulator, and taking Re = OR1, a quick check on how the various transistor junction temperatures affect Vq yields:

- The EF output stage has a Vq of 42 mV, with a Vq sensitivity of −2 mV/°C to driver temperature, and −2 mV/°C to output junction temperature. No surprises here.
- The CFP stage has a much smaller Vq (3.1 mV) Vq sensitivity is −2 mV/°C to driver temperature, and only −0.1 mV/°C to output device temperature. This confirms that local NFB in the stage makes Vq relatively independent of output device temperature, which is just as well as Table 13.1 shows it needs to be about ten times more accurate.

The CFP output devices are about 20 times less sensitive to junction temperature, but the Vq across Re is something like 10 times less; hence the actual relationship between output junction temperature and crossover distortion is not so very different for the two configurations, indicating that as regards temperature stability the CFP may only be twice as good as the EF, and not vastly better, which is perhaps the common assumption. In fact, as will be described, the CFP may show poorer thermal performance in practice.

In real life, with a continuously varying power output, the situation is complicated by the different dissipation characteristics of the drivers as

Table 13.1
Vbias tolerance for 8 Ω

		EF output	CFP output
Crossover spikes obvious	Underbias	2.25 V	1.242 V
Spikes just visible	Underbias	2.29	1.258
Optimal residual	Optimal	2.38	1.283
gm-doubling just visible	Overbias	2.50	1.291
gm-doubling obvious	Overbias	2.76	1.330

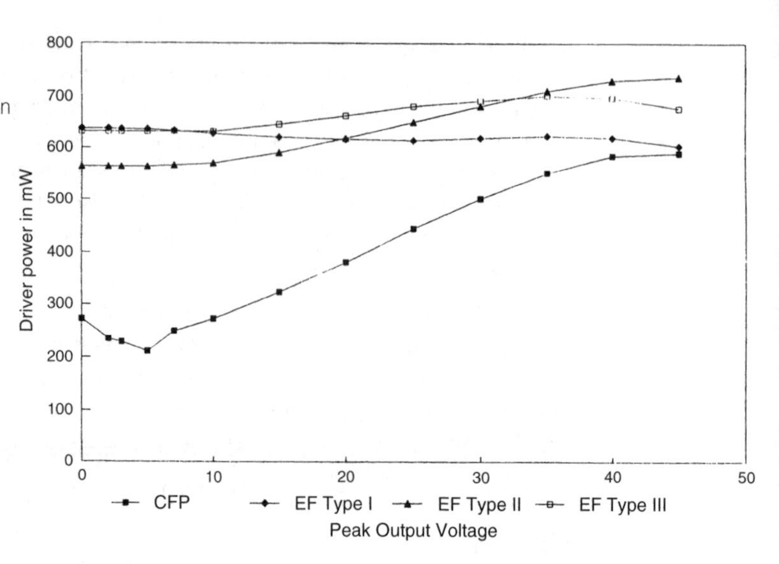

Figure 13.1
Driver dissipation versus output level. In all variations on the EF configuration, power dissipation varies little with output; CFP driver power however varies by a factor of two or more

output varies. See Figure 13.1, which shows that the CFP driver dissipation is more variable with output, but on average runs cooler. For both configurations driver temperature is equally important, but the EF driver dissipation does not vary much with output power, though the initial drift at switch-on is greater as the standing dissipation is higher. This, combined with the two-times-greater sensitivity to output device temperature and the greater self-heating of the EF output devices, may be the real reason why most designers have a general feeling that the EF version has inferior quiescent stability. The truth as to which type of stage is more thermally stable is much more complex, and depends on several design choices and assumptions.

Having assimilated this, we can speculate on the ideal thermal compensation system for the two output configurations. The EF stage has Vq set by the subtraction of four dissimilar base-emitter junctions from Vbias, all having an equal say, and so all four junction temperatures ought to be factored into the final result. This would certainly be comprehensive, but four temperature-sensors per channel is perhaps overdoing it. For the CFP stage, we can ignore the output device temperatures and only sense the drivers, which simplifies things and works well in practice.

If we can assume that the drivers and outputs come in complementary pairs with similar Vbe behaviour, then symmetry prevails and we need only consider one half of the output stage, so long as Vbias is halved to suit. This assumes the audio signal is symmetrical over timescales of seconds to minutes, so that equal dissipations and temperature rises occur in the top and bottom halves of the output stage. This seems a pretty safe bet, but the unaccompanied human voice has positive and negative peak values that may differ by up to 8 dB, so prolonged acapella performances have at least the potential to mislead any compensator that assumes symmetry. One amplifier that does use separate sensors for the upper and lower output sections is the Adcom GFA-565.

For the EF configuration, both drivers and outputs have an equal influence on the quiescent Vq, but the output devices normally get much hotter than the drivers, and their dissipation varies much more with output level. In this case the sensor goes on or near one of the output devices, thermally close to the output junction. It has been shown experimentally that the top of the TO3 can is the best place to put it, see page 350. Recent experiments have confirmed that this holds true also for the TO3P package (a large flat plastic package like an overgrown TO220, and nothing like TO3) which can easily get 20° hotter on its upper plastic surface than does the underlying heatsink.

In the CFP the drivers have most effect and the output devices, although still hot, have only one-twentieth the influence. Driver dissipation is also much more variable, so now the correct place to put the thermal sensor is as near to the driver junction as you can get it.

Schemes for the direct servo control of quiescent current have been mooted[2], but all suffer from the difficulty that the quantity we wish to control is not directly available for measurement, as except in the complete absence of signal it is swamped by Class-B output currents. In contrast the quiescent current of a Class-A amplifier is easily measured, allowing very precise feedback control; ironically its value is not critical for distortion performance.

So, how accurately must quiescent current be held? This is not easy to answer, not least because it is the wrong question. Page 151 established that the crucial parameter is not quiescent current (hereafter Iq) as such, but rather the quiescent voltage-drop Vq across the two emitter resistors Re. This takes a little swallowing – after all people have been worrying about quiescent current for 30 years or more – but it is actually good news, as the value of Re does not complicate the picture. The voltage across the output stage inputs (Vbias) is no less critical, for once Re is chosen Vq and Iq vary proportionally. The two main types of output stage, the Emitter-Follower (EF) and the Complementary-Feedback Pair (CFP) are shown in Figure 13.2. Their Vq tolerances are quite different.

From the measurements on page 341 above the permissible error band for Vq in the EF stage is ±100 mV, and for the CFP is ±10 mV. These tolerances are not defined for all time; I only claim that they are realistic

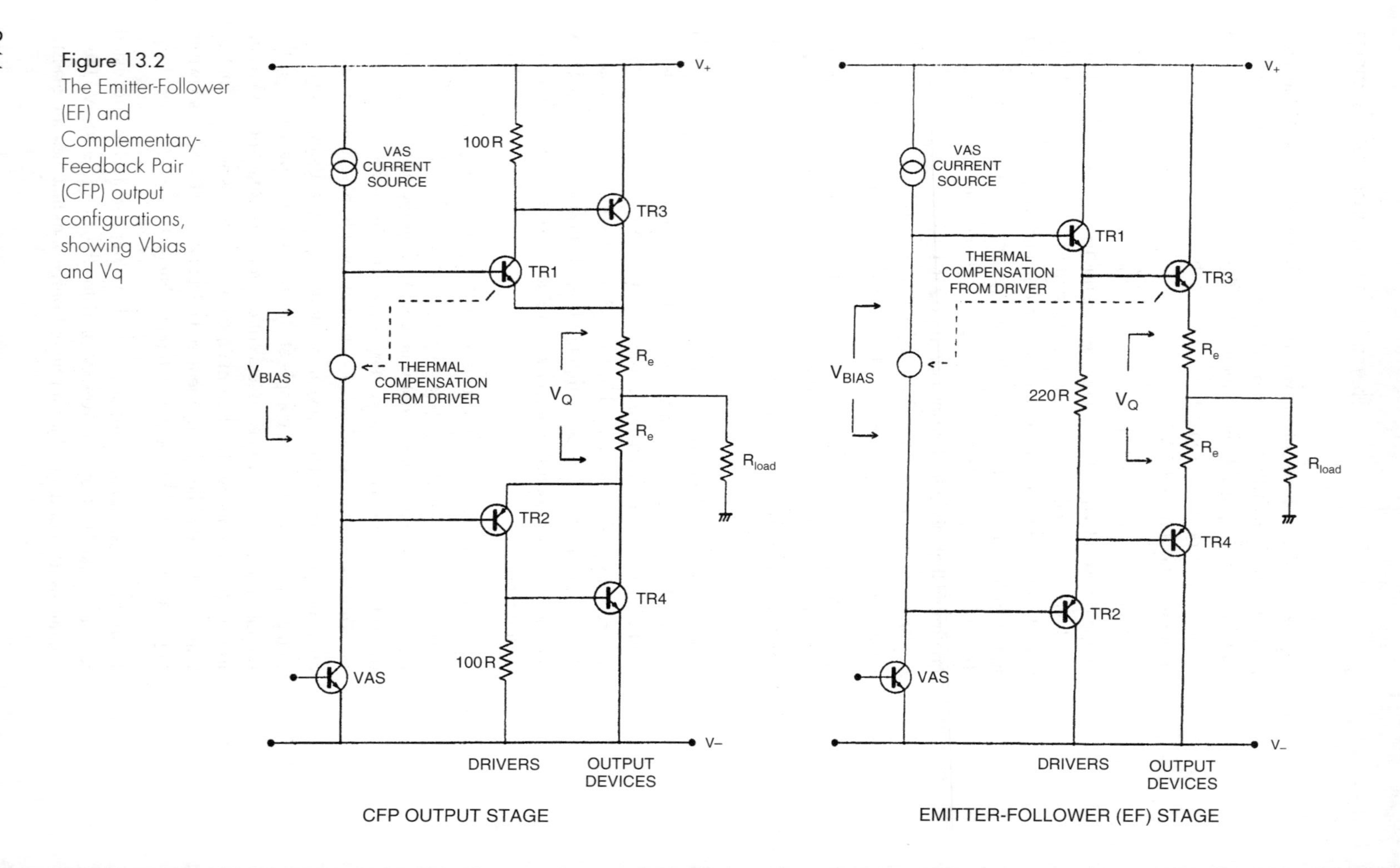

Figure 13.2 The Emitter-Follower (EF) and Complementary-Feedback Pair (CFP) output configurations, showing Vbias and Vq

and reasonable. In terms of total Vbias, the EF needs 2.93 V ±100 mV, and the CFP 1.30 V ±10 mV. Vbias must be higher in the EF as four Vbe's are subtracted from it to get Vq, while in the CFP only two driver Vbe's are subtracted.

The CFP stage appears to be more demanding of Vbias compensation than EF, needing 1% rather than 3.5% accuracy, but things are not so simple. Vq stability in the EF stage depends primarily on the hot output devices, as EF driver dissipation varies only slightly with power output. Vq in the CFP depends almost entirely on driver junction temperature, as the effect of output device temperature is reduced by the local negative-feedback; however CFP driver dissipation varies strongly with power output so the superiority of this configuration cannot be taken for granted.

Driver heatsinks are much smaller than those for output devices, so the CFP Vq time-constants promise to be some ten times shorter.

Basic thermal compensation

In Class B, the usual method for reducing quiescent variations is *thermal feedback*. Vbias is generated by a thermal sensor with a negative temperature-coefficient, usually a Vbe-multiplier transistor mounted on the main heatsink. This system has proved entirely workable over the last 30-odd years, and usually prevents any possibility of thermal runaway. However, it suffers from thermal losses and delays between output devices and temperature sensor that make maintenance of optimal bias rather questionable, and in practice quiescent conditions are a function of recent signal and thermal history. Thus the crossover linearity of most power amplifiers is intimately bound up with their thermal dynamics, and it is surprising this area has not been examined more closely; Sato et al.[1] is one of the few serious papers on the subject, though the conclusions it reaches appear to be unworkable, depending on calculating power dissipation from amplifier output voltage without considering load impedance.

As is almost routine in audio design, things are not as they appear. So-called *thermal feedback* is not feedback at all – this implies the thermal sensor is in some way controlling the output stage temperature; it is not. It is really a form of approximate *feedforward* compensation, as shown in Figure 13.3. The quiescent current (Iq) of a Class-B design causes a very small dissipation compared with the signal, and so there is no meaningful feedback path returning from Iq to the left of the diagram. (This might be less true of Class-AB, where quiescent dissipation may be significant.) Instead this system aspires to make the sensor junction temperature mimic the driver or output junction temperature, though it can never do this promptly or exactly

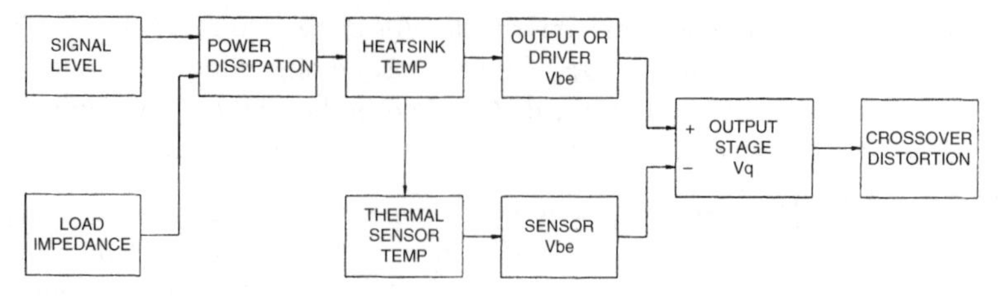

Figure 13.3
Thermal signal flow of a typical power amplifier, showing that there is no thermal feedback to the bias generator. There is instead feedforward of driver junction temperature, so that the sensor Vbe will hopefully match the driver Vbe

because of the thermal resistances and thermal capacities that lie between driver and sensor temperatures in Figure 13.3. It does not place either junction temperature or quiescent current under direct feedback control, but merely aims to cancel out the errors. Hereafter I simply call this *thermal compensation*.

Assessing the bias errors

The temperature error must be converted to mV error in Vq, for comparison with the tolerance bands suggested above. In the CFP stage this is straightforward; both driver Vbe and the halved Vbias voltage decrease by 2 mV per °C, so temperature error converts to voltage error by multiplying by 0.002 Only half of each output stage will be modelled, exploiting symmetry, so most of this chapter deals in half-Vq errors, etc. To minimise confusion this use of half-amplifiers is adhered to throughout, except at the final stage when the calculated Vq error is doubled before comparison with the tolerance bands quoted above.

The EF error conversion is more subtle. The EF Vbias generator must establish four times Vbe plus Vq, so the Vbe of the temperature-sensing transistor is multiplied by about 4.5 times, and so decreases at 9 mV/°C. The CFP Vbias generator only multiplies 2.1 times, decreasing at 4 mV/°C. The corresponding values for a half-amplifier are 4.5 and 2 mV/°C.

However, the EF drivers are at near-constant temperature, so after two driver Vbe's have been subtracted from Vbias, the remaining voltage decreases faster with temperature than does output device Vbe. This runs counter to the tendency to under-compensation caused by thermal attenuation between output junctions and thermal sensor; in effect the compensator has thermal *gain*, and this has the potential to reduce long-term Vq errors. I suspect this is the real reason why the EF stage, despite looking unpromising, can in practice give acceptable quiescent stability.

Thermal simulation

Designing an output stage requires some appreciation of how effective the thermal compensation will be, in terms of how much delay and attenuation the *thermal signal* suffers between the critical junctions and the Vbias generator.

We need to predict the thermal behaviour of a heatsink assembly over time, allowing for things like metals of dissimilar thermal conductivity, and the very slow propagation of heat through a mass compared with near-instant changes in electrical dissipation. Practical measurements are very time-consuming, requiring special equipment such as multi-point thermocouple recorders. A theoretical approach would be very useful.

For very simple models, such as heat flow down a uniform rod, we can derive analytical solutions to the partial differential equations that describe the situation; the answer is an equation directly relating temperature to position along-the-rod and time. However, even slight complications (such as a non-uniform rod) involve rapidly increasing mathematical complexities, and anyone who is not already deterred should consult Carslaw and Jaeger[3]; this will deter them.

To avoid direct confrontation with higher mathematics, finite-element and relaxation methods were developed; the snag is that Finite-Element-Analysis is a rather specialised taste, and so commercial FEA software is expensive.

I therefore cast about for another method, and found I already had the wherewithall to solve problems of thermal dynamics; the use of electrical analogues is the key. If the thermal problem can be stated in terms of lumped electrical elements, then a circuit simulator of the SPICE type can handle it, and as a bonus has extensive capabilities for graphical display of the output. The work here was done with PSpice. A more common use of electrical analogues is in the electro-mechanical domain of loudspeakers; see Murphy[4] for a virtuoso example.

The simulation approach treats temperature as voltage, and thermal energy as electric charge, making thermal resistance analogous to electrical resistance, and thermal capacity to electrical capacitance. Thermal capacity is a measure of how much heat is required to raise the temperature of a mass by 1°C. (And if anyone can work out what the thermal equivalent of an inductor is, I would be interested to know.) With the right choice of units the simulator output will be in Volts, with a one-to-one correspondence with degrees Celsius, and Amps, similarly representing Watts of heat flow; see Table 13.2. It is then simple to produce graphs of temperature against time.

Table 13.2

	Reality	Simulation
Temperature	°C	Volts
Heat quantity	Joules (Watt-seconds)	Coulombs (Amp-seconds)
Heat flowrate	Watts	Amps
Thermal resistance	°C/Watt	ohms
Thermal capacity	°C/Joule	Farads
Heat source	Dissipative element, e.g., transistor	Current source
Ambient	Medium-sized planet	Voltage source

Since heat flow is represented by current, the inputs to the simulated system are current sources. A voltage source would force large chunks of metal to change temperature instantly, which is clearly wrong. The ambient is modelled by a voltage source, as it can absorb any amount of heat without changing temperature.

Modelling the EF output stage

The major characteristic of Emitter-Follower (EF) output stages is that the output device junction temperatures are directly involved in setting Iq. This junction temperature is not accessible to a thermal compensation system, and measuring the heatsink temperature instead provides a poor approximation, attenuated by the thermal resistance from junction to heatsink mass, and heavily time-averaged by heatsink thermal inertia. This can cause serious production problems in initial setting up; any drift of Iq will be very slow as a lot of metal must warm up.

For EF outputs, the bias generator must attempt to establish an output bias voltage that is a summation of four driver and output Vbe's. These do not vary in the same way. It seems at first a bit of a mystery how the EF stage, which still seems to be the most popular output topology, works as well as it does. The probable answer is Figure 13.1, which shows how driver dissipation (averaged over a cycle) varies with peak output level for the three kinds of EF output described on page 116, and for the CFP configuration. The SPICE simulations used to generate this graph used a triangle waveform, to give a slightly closer approximation to the peak–average ratio of real waveforms. The rails were ±50 V, and the load 8 Ω.

It is clear that the driver dissipation for the EF types is relatively constant with power output, while the CFP driver dissipation, although generally lower, varies strongly. This is a consequence of the different operation of these two kinds of output. In general, the drivers of an EF output remain conducting to some degree for most or all of a cycle, although the output devices are certainly off half the time. In the CFP, however, the drivers turn off almost in synchrony with the outputs, dissipating an amount of power that varies much more with output. This implies that EF drivers will work at roughly the same temperature, and can be neglected in arranging thermal compensation; the temperature-dependent element is usually attached to the main heatsink, in an attempt to compensate for the junction temperature of the outputs alone. The Type I EF output keeps its drivers at the most constant temperature; this may (or may not) have something to do with why it is the most popular of the EF types.

(The above does not apply to integrated Darlington outputs, with drivers and assorted emitter resistors combined in one ill-conceived package, as the driver sections are directly heated by the output junctions. This would seem to work directly against quiescent stability, and why these compound devices are ever used in audio amplifiers remains a mystery to me.)

The drawback with most EF thermal compensation schemes is the slow response of the heatsink mass to thermal transients, and the obvious solution is to find some way of getting the sensor closer to one of the output junctions (symmetry of dissipation is assumed). If TO3 devices are used, then the flange on which the actual transistor is mounted is as close as we can get without a hacksaw. This is however clamped to the heatsink, and almost inaccessible, though it might be possible to hold a sensor under one of the mounting bolts. A simpler solution is to mount the sensor on the top of the TO3 can. This is probably not as accurate an estimate of junction temperature as the flange would give, but measurement shows the top gets much hotter much faster than the heatsink mass, so while it may appear unconventional, it is probably the best sensor position for an EF output stage. Figure 13.4 shows the results of an experiment designed to test this. A TO3 device was mounted on a thick aluminium L-section thermal coupler in turn clamped to a heatsink; this construction is representative of

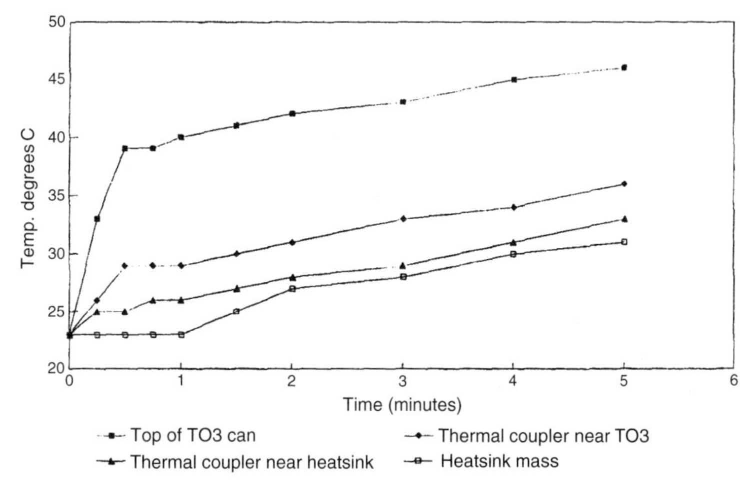

Figure 13.4 Thermal response of a TO3 device on a large heatsink when power is suddenly applied. The top of the TO3 can responds most rapidly

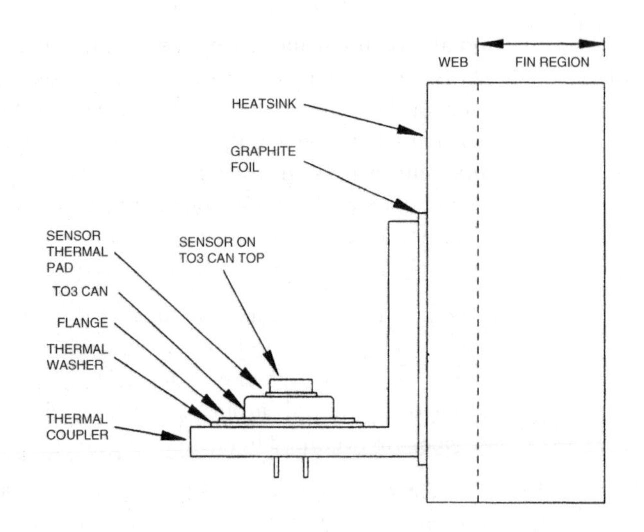

Figure 13.5
A TO3 power transistor attached to a heatsink by a thermal coupler. Thermal sensor is shown on can top; more usual position would be on thermal coupler

many designs. Dissipation equivalent to 100 W/8 Ω was suddenly initiated, and the temperature of the various parts monitored with thermocouples. The graph clearly shows that the top of the TO3 responds much faster, and with a larger temperature change, though after the first two minutes the temperatures are all increasing at the same rate. The whole assembly took more than an hour to asymptote to thermal equilibrium.

Figure 13.5 shows a TO3 output device mounted on a thermal coupling bar, with a silicone thermal washer giving electrical isolation. The coupler is linked to the heatsink proper via a second conformal material; this need not be electrically insulating so highly efficient materials like graphite foil can be used. This is representative of many amplifier designs, though a good number have the power devices mounted directly on the heatsink; the results hardly differ. A simple thermal-analogue model of Figure 13.5 is shown in Figure 13.6; the situation is radically simplified by treating each

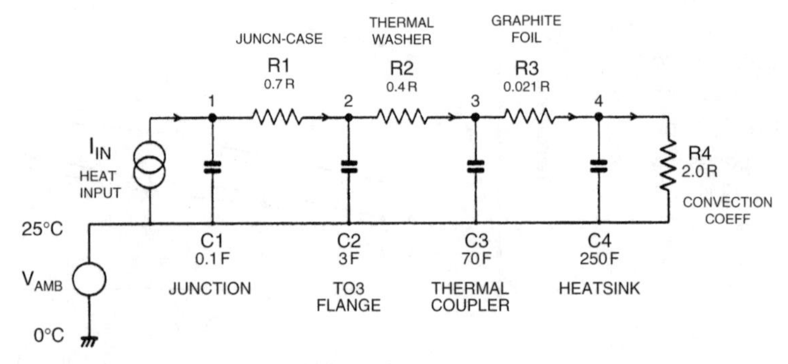

Figure 13.6
A thermal/electrical model of Figure 13.5, for half of one channel only. Node 1 is junction temperature, node 2 flange temperature, and so on. Vamb sets the baseline to 25°C. Arrows show heat flow

mass in the system as being at a uniform temperature, i.e., isothermal, and therefore representable by one capacity each. The boundaries between parts of the system are modelled, but the thermal capacity of each mass is concentrated at a notional point. In assuming this we give capacity elements zero thermal resistance; e.g., both sides of the thermal coupler will always be at the same temperature. Similarly, elements such as the thermal washer are assumed to have zero heat capacity, because they are very thin and have negligible mass compared with other elements in the system. Thus the parts of the thermal system can be conveniently divided into two categories; pure thermal resistances and pure thermal capacities. Often this gives adequate results; if not, more sub-division will be needed. Heat losses from parts other than the heatsink are neglected.

Real output stages have at least two power transistors; the simplifying assumption is made that power dissipation will be symmetrical over anything but the extreme short-term, and so one device can be studied by slicing the output stage, heatsink, etc., in half.

It is convenient to read off the results directly in °C, rather than temperature rise above ambient, so Figure 13.6 represents ambient temperature with a voltage source Vamb that offsets the baseline (node 10) 25°C from simulator ground, which is inherently at 0°C (0 V).

Values of the notional components in Figure 13.6 have to be filled in with a mixture of calculation and manufacturer's data. The thermal resistance R1 from junction to case comes straight from the data book, as does the resistance R2 of the TO3 thermal washer; also R4, the convection coefficient of the heatsink itself, otherwise known as its thermal resistance to ambient. This is always assumed to be constant with temperature, which it very nearly is. Here R4 is 1°C/W, so this is doubled to 2 as we cut the stage in half to exploit symmetry.

R3 is the thermal resistance of the graphite foil; this is cut to size from a sheet and the only data is the bulk thermal resistance of 3.85 W/mK, so R3 must be calculated. Thickness is 0.2 mm, and the rectangle area in this example was 38 × 65 mm. We must be careful to convert all lengths to metres;

$$\text{Heat flow/°C} = \frac{3.85 \times \text{Area}}{\text{Thickness}} = \frac{3.83 \times (0.038 \times 0.065)}{0.0002} = 47.3\,\text{W/°C} \qquad \text{Equation 13.1}$$

$$\text{So thermal resistance} = \frac{1}{47.3} = 0.021\,\text{°C/W}$$

Thermal resistance is the reciprocal of heat flow per degree, so R3 is 0.021°C/W, which just goes to show how efficient thermal washers can be if they do not have to be electrical insulators as well.

In general all the thermal capacities will have to be calculated, sometimes from rather inadequate data, thus:

$$\text{Thermal capacity} = \text{Density} \times \text{Volume} \times \text{Specific heat}$$

A power transistor has its own internal structure, and its own internal thermal model (Figure 13.7). This represents the silicon die itself, the solder that fixes it to the copper header, and part of the steel flange the header is welded to. I am indebted to Motorola for the parameters, from an MJ15023 TO3 device[5]. The time-constants are all extremely short compared with heatsinks, and it is unnecessary to simulate in such detail here.

The thermal model of the TO3 junction is therefore reduced to lumped component C1, estimated at 0.1 J/°C; with a heat input of 1 W and no losses its temperature would increase linearly by 10°C/sec. The capacity C2 for the transistor package was calculated from the volume of the TO3 flange (representing most of the mass) using the specific heat of mild steel. The thermal coupler is known to be aluminium alloy (not pure aluminium, which is too soft to be useful) and the calculated capacity of 70J/°C should be reliable. A similar calculation gives 250 J/°C for the larger mass of the aluminium heatsink. Our simplifying assumptions are rather sweeping here, because we are dealing with a substantial chunk of finned metal which will never be truly isothermal.

The derived parameters for both output TO3's and TO-225 AA drivers are summarised in Table 13.3. The drivers are assumed to be mounted onto

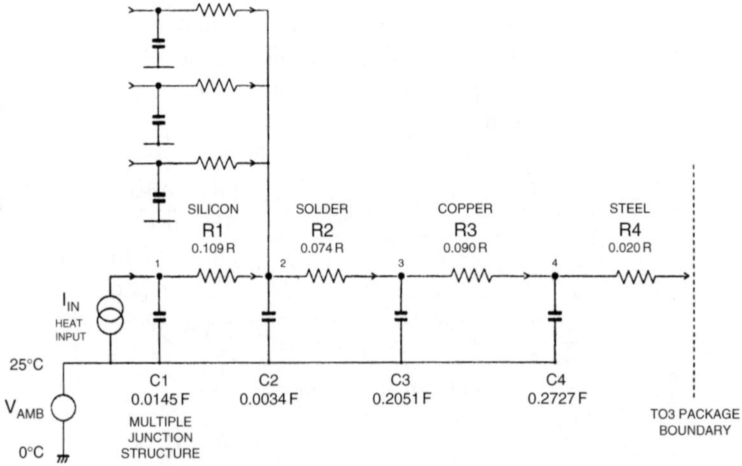

Figure 13.7 Internal thermal model for a TO3 transistor. All the heat is liberated in the junction structure, shown as N multiples of C1 to represent a typical interdigitated power transistor structure

Table 13.3

			Output device	Driver
C1	Junction capacity	J/°C	0.1	0.05
R1	Junction-case resistance	°C/W	0.7	6.25
C2	Transistor package capacity	3.0	0.077	
R2	Thermal washer res		0.4	6.9
C3	Coupler capacity		70	–
R3	Coupler-heatsink res		0.021	–
C4	Heatsink capacity		250	20.6
R4	Heatsink convective res		2.0	10.0

small individual heatsinks with an isolating thermal washer; the data is for the popular Redpoint SW38-1 vertical heatsink.

Figures 13.8 and 13.9 show the result of a step-function in heat generation in the output transistor; 20 W dissipation is initiated, corresponding approximately to a sudden demand for full sinewave power from a quiescent 100 W amplifier. The junction temperature V(1) takes off near-vertically, due to its small mass and the substantial thermal resistance between it and the TO3 flange; the flange temperature V(2) shows a similar but smaller step as R2 is also significant. In contrast the thermal coupler, which is so efficiently bonded to the heatsink by graphite foil that they might almost

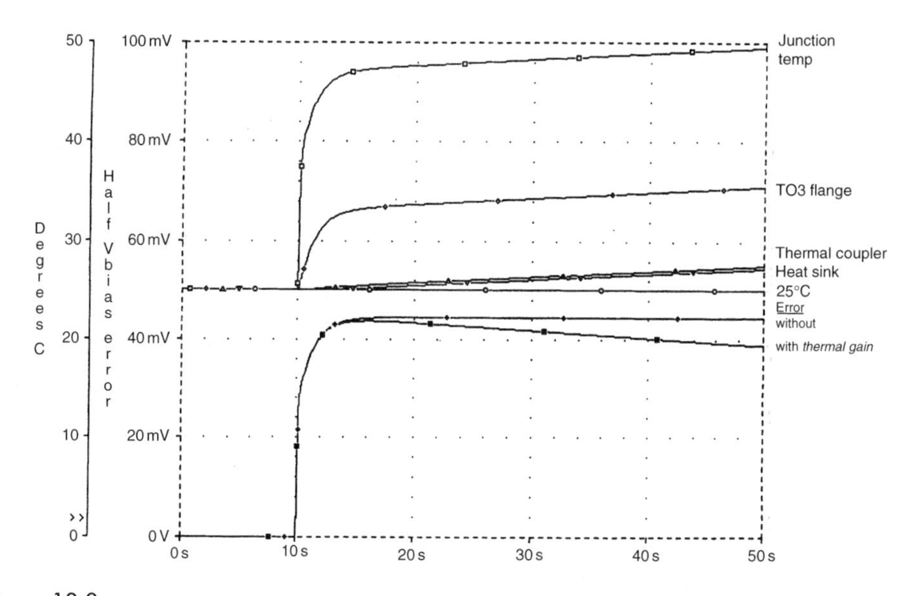

Figure 13.8
Results for Figure 13.6, with step heat input of 20 W to junction initiated at Time = 10 sec. Upper plot shows temperatures, lower the Vbias error for half of output stage

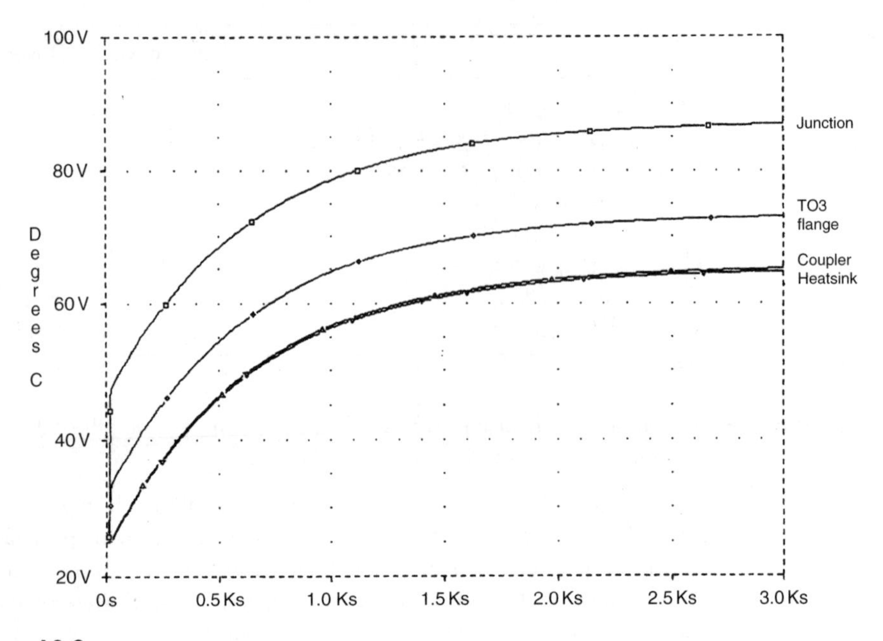

Figure 13.9
The long-term version of Figure 13.8, showing that it takes over 40 minutes for the heatsink to get within 1° of final temperature

be one piece of metal, begins a slow exponential rise that will take a very long time to asymptote. Since after the effect of C1 and C2 have died away the junction temp is offset by a constant amount from the temp of C3 and C4, V(1) also shows a slow rise. Note the X-axis of Figure 13.9 must be in kilo-seconds, because of the relatively enormous thermal capacity of the heatsink.

This shows that a temperature sensor mounted on the main heatsink can never give accurate bias compensation for junction temperature, even if it is assumed to be isothermal with the heatsink; in practice there will be some sensor cooling which will make the sensor temperature slightly under-read the heatsink temperature V(4). Initially the temperature error V(1)–V(4) increases rapidly as the TO3 junction heats, reaching 13° in about 200 ms. The error then increases much more slowly, taking 6 sec to reach the effective final value of 22°. If we ignore the *thermal-gain* effect mentioned above, the long-term Vq error is +44 mV, i.e., Vq is too high. When this is doubled to allow for both halves of the output stage we get +88 mV, which uses up nearly all of the ±100 mV error band, without any other inaccuracies. (Hereafter all Vbias/Vq error figures quoted have been doubled and so apply to a complete output stage.) Including the thermal gain actually makes little difference over a 10-sec timescale; the lower

Vq-error trace in Figure 13.8 slowly decays as the main heatsink warms up, but the effect is too slow to be useful.

The amplifier Vq and Iq will therefore rise under power, as the hot output device Vbe voltages fall, but the cooler bias generator on the main heatsink reduces its voltage by an insufficient amount to compensate.

Figure 13.9 shows the long-term response of the system. At least 2500 sec pass before the heatsink is within a degree of final temperature.

In the past I have recommended that EF output stages should have the thermal sensor mounted on the top of the TO3 can, despite the mechanical difficulties. This is not easy to simulate as no data is available for the thermal resistance between junction and can top. There must be an additional thermal path from junction to can, as the top very definitely gets *hotter* than the flange measured at the very base of the can. In view of the relatively low temperatures, this path is probably due to internal convection rather than radiation.

A similar situation arises with TO3P packages (a large plastic package, twice the size of TO220) for the top plastic surface can get at least 20° hotter than the heatsink just under the device.

Using the real thermocouple data from page 349, I have estimated the parameters of the thermal paths to the TO3 top. This gives Figure 13.10, where the values of elements R20, R21, C5 should be treated with

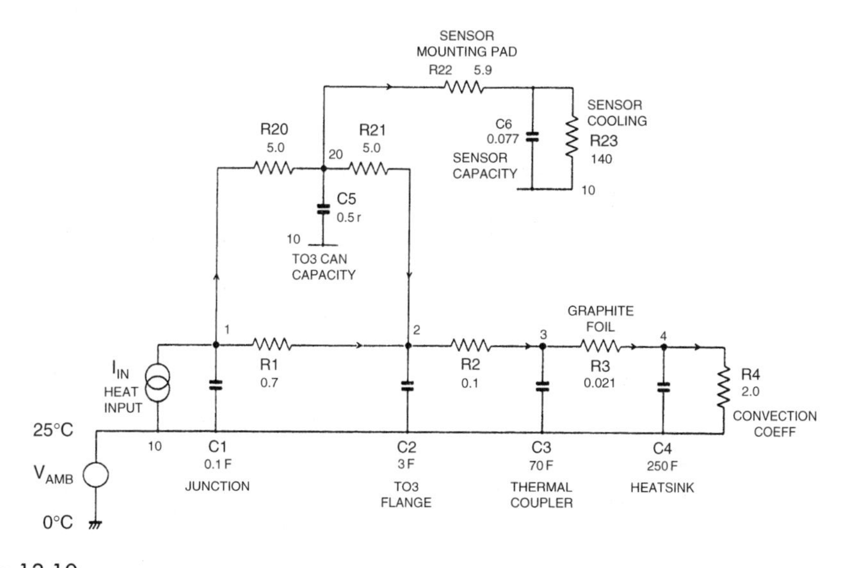

Figure 13.10
Model of EF output stage with thermal paths to TO3 can top modelled by R20, R21. C5 simulates can capacity. R23 models sensor convection cooling; node 21 is sensor temperature

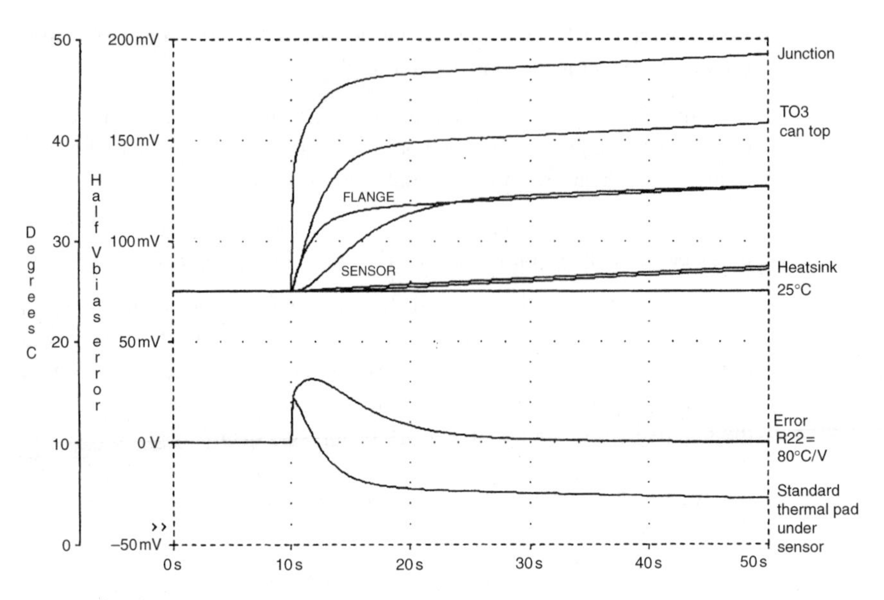

Figure 13.11
The simulation results for Figure 13.10; lower plot shows Vbias errors for normal thermal pad under sensor, and 80°C/W semi-insulator. The latter has near-zero long-term error

considerable caution, though the temperature results in Figure 13.11 match reality fairly well; the can top (V20) gets hotter faster than any other accessible point. R20 simulates the heating path from the junction to the TO3 can and R21 the can-to-flange cooling path, C5 being can thermal capacity.

Figure 13.10 includes approximate representation of the cooling of the sensor transistor, which now matters. R22 is the thermal pad between the TO3 top and the sensor, C6 the sensor thermal capacity, and R23 is the convective cooling of the sensor, its value being taken as twice the datasheet free-air thermal resistance as only one face is exposed.

Placing the sensor on top of the TO3 would be expected to reduce the steady-state bias error dramatically. In fact, it overdoes it, as after factoring in the thermal-gain of a Vbe-multiplier in an EF stage, the bottom-most trace of Figure 13.11 shows that the bias is over-compensated; after the initial positive transient error, Vbias falls too low giving an error of −30 mV, slowly worsening as the main heatsink warms up. If thermal-gain had been ignored, the simulated error would have apparently fallen from +44 (Figure 13.8) to +27 mV; apparently a useful improvement, but actually illusory.

Since the new sensor position over-compensates for thermal errors, there should be an intermediate arrangement giving near-zero long-term error. I found this condition occurs if R22 is increased to 80°C/W, requiring some sort of semi-insulating material rather than a thermal pad, and gives the upper error trace in the lower half of Figure 13.11. This peaks at +30 mV after 2 sec, and then decays to nothing over the next twenty. This is much superior to the persistent error in Figure 13.8, so I suggest this new technique may be useful.

Modelling the CFP output stage

In the CFP configuration, the output devices are inside a local feedback loop, and play no significant part in setting Vq, which is dominated by thermal changes in the driver Vbe's. Such stages are virtually immune to thermal runaway; I have found that assaulting the output devices with a powerful heat gun induces only very small Iq changes. Thermal compensation is mechanically simpler as the Vbe-multiplier transistor is usually mounted on one of the driver heatsinks, where it aspires to mimic the driver junction temperature.

It is now practical to make the bias transistor of the same type as the drivers, which should give the best matching of Vbe[6], though how important this is in practice is uncertain. This also avoids the difficulty of trying to attach a small-signal (probably TO92) transistor package to a heatsink.

Since it is the driver junctions that count, output device temperatures are here neglected. The thermal parameters for a TO225AA driver (e.g., MJE340/350) on the SW38-1 vertical heatsink are shown in Table 13.3; the drivers are on individual heatsinks so their thermal resistance is used directly, without doubling.

In the simulation circuit (Figure 13.12) V(3) is the heatsink temperature; the sensor transistor (also MJE340) is mounted on this sink with thermal washer R4, and has thermal capacity C4. R5 is convective cooling of the sensor. In this case the resulting differences in Figure 13.13 between sink V(3) and sensor V(4) are very small.

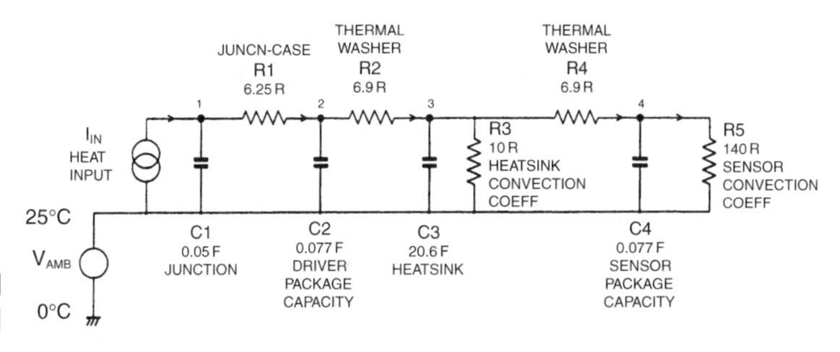

Figure 13.12 Model of a CFP stage. Driver transistor is mounted on a small heatsink, with sensor transistor on the other side. Sensor dynamics and cooling are modelled by R4, C4 and R5

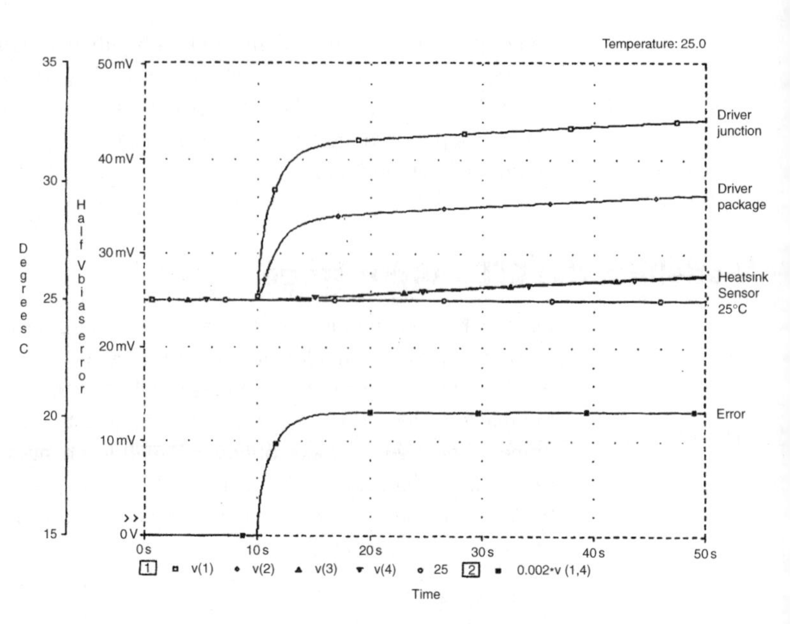

Figure 13.13 Simulation results for CFP stage, with step heat input of 0.5 W. Heatsink and sensor are virtually isothermal, but there is a persistent error as driver is always hotter than heatsink due to R1, R2

We might expect the CFP delay errors to be much shorter than in the EF; however, simulation with a heat step-input suitably scaled down to 0.5 W (Figure 13.13) shows changes in temperature error V(1)–V(4) that appear rather paradoxical; the error reaches 5° in 1.8 seconds, levelling out at 6.5° after about 6 sec. This is markedly slower than the EF case, and gives a total bias error of +13 mV, which after doubling to +26 mV is well outside the CFP error band of ±10 mV.

The initial transients are slowed down by the much smaller step heat input, which takes longer to warm things up. The *final* temperature however, is reached in 500 rather than 3000 sec, and the timescale is now in hundreds rather than thousands of seconds. The heat input is smaller, but the driver heatsink capacity is also smaller, and the overall time-constant is less.

It is notable that both timescales are much longer than musical dynamics.

The integrated absolute error criterion

Since the thermal sensor is more or less remote from the junction whose gyrations in temperature will hopefully be cancelled out, heat losses and thermal resistances cause the temperature change reaching the sensor to be generally too little and too late for complete compensation.

As in the previous section, all the voltages and errors here are for one-half of an output stage, using symmetry to reduce the work involved. These *half-amplifiers* are used throughout this chapter, for consistency, and the error voltages are only doubled to represent reality (a complete output stage) when they are compared against the tolerance bands previously quoted.

We are faced with errors that vary not only in magnitude, but also in their persistence over time; judgement is required as to whether a prolonged small error is better than a large error which quickly fades away.

The same issue faces most servomechanisms, and I borrow from Control Theory the concept of an *Error Criterion* which combines magnitude and time into one number[7],[8]. The most popular criterion is the Integrated Absolute Error (IAE) which is computed by integrating the absolute-value of the error over a specified period after giving the system a suitably provocative stimulus; the absolute-value prevents positive and negative errors cancelling over time. Another common criterion is the Integrated Square Error (ISE) which solves the polarity problem by squaring the error before integration – this also penalises large errors much more than small ones. It is not immediately obvious which of these is most applicable to bias-control and the psychoacoustics of crossover distortion that changes with time, so I have chosen the popular IAE.

One difficulty is that the IAE error criterion for bias voltage tends to accumulate over time, due to the integration process, so any constant bias error quickly comes to dominate the IAE result. In this case, the IAE is little more than a counter-intuitive way of stating the constant error, and must be quoted over a specified integration time to mean anything at all. This is why the IAE concept was not introduced earlier in this chapter.

Much more useful results are obtained when the IAE is applied to a situation where the error decays to a very small value after the initial transient, and stays there. This can sometimes be arranged in amplifiers, as I hope to show. In an ideal system where the error decayed to zero without overshoot, the IAE would asymptote to a constant value after the initial transient. In real life, residual errors make the IAE vary slightly with time, so for consistency all the IAE values given here are for 30 sec after the step-input.

Improved thermal compensation: the emitter-follower stage

It was shown above that the basic emitter-follower (EF) stage with the sensor on the main heatsink has significant thermal attenuation error and therefore under-compensates temperature changes. (The Vq error is +44 mV, the positive sign showing it is too high. If the sensor is on the TO3 can top it over-compensates instead) (Vq error −30 mV).

If an intermediate configuration is contrived by putting a layer of controlled thermal resistance (80°C/W) between the TO3 top and the sensor, then the 50-sec timescale component of the error can be reduced to near-zero. This is the top error trace in bottom half of Figure 13.14; the lower trace shows the wholly misleading result if sensor heat losses are neglected in this configuration.

Despite this medium-term accuracy, if the heat input stimulus remains constant over the very long-term (several kilo-seconds) there still remains a very slow drift towards over-compensation due to the slow heating of the main heatsink (Figure 13.15).

This long-term drift is a result of the large thermal inertia of the main heatsink and since it takes 1500 sec (25 minutes) to go from zero to −32 mV is of doubtful relevance to the time-scales of music and signal level changes. On doubling to −64 mV, it remains within the EF Vq tolerance of ±100 mV. On the shorter 50-sec timescale, the half-amplifier error remains within a ±1 mV window from 5 sec to 60 sec after the step-input.

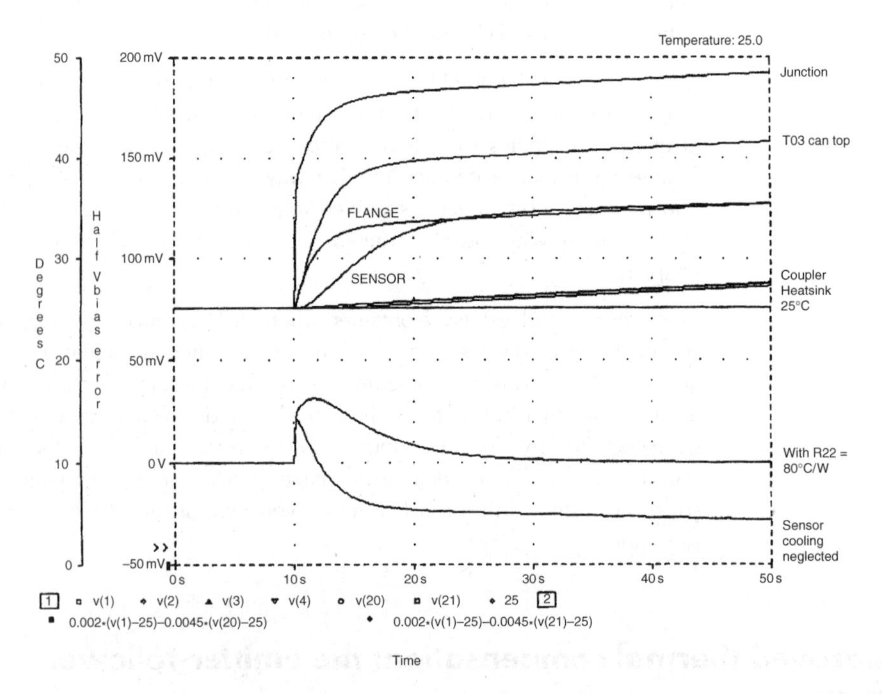

Figure 13.14
EF behaviour with semi-insulating pad under sensor on TO3 can top. The sensor in the upper temperature plot rises more slowly than the flange, but much faster than the main heatsink or coupler. In lower Vq-error section, upper trace is for a 80°C/W thermal resistance under the sensor, giving near-zero error. Bottom trace shows serious effect of ignoring sensor-cooling in TO3-top version

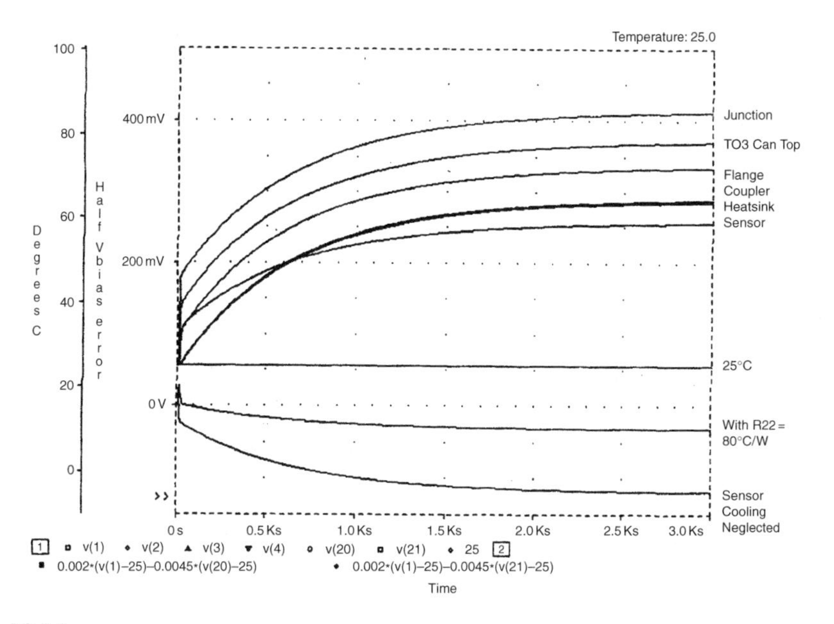

Figure 13.15
Over a long timescale, the lower plot shows that the Vq error, although almost zero in Figure 13.14, slowly drifts into over-compensation as the heatsink temperature (upper plot) reaches asymptote

For the EF stage, a very-long-term drift component will always exist so long as the output device junction temperature is kept down by means of a main heatsink that is essentially a weighty chunk of finned metal.

The EF system stimulus is a 20 W step as before, being roughly worst-case for a 100 W amplifier. Using the 80°C/W thermal semi-insulator described above gives the upper error trace in Figure 13.16, and an IAE of 254 mV-sec after 30 sec. This is relatively large because of the extra time-delay caused by the combination of an increased R22 with the unchanged sensor thermal capacity C6. Once more, this figure is for a half-amplifier, as are all IAEs in this chapter.

Up to now I have assumed that the temperature coefficient of a Vbe-multiplier bias generator is rigidly fixed at −2 mV/°C times the Vbe-multiplication factor, which is about 4.5× for EF and 2× for CFP. The reason for the extra *thermal gain* displayed by the EF was set out on page 348.

The above figures are for both halves of the output stage, so the half-amplifier value for EF is −4.5 mV/°C, and for CFP −2 mV/°C. However, if we boldly assume that the Vbias generator can have its thermal coefficient varied at will, the insulator and its aggravated time-lag can be eliminated.

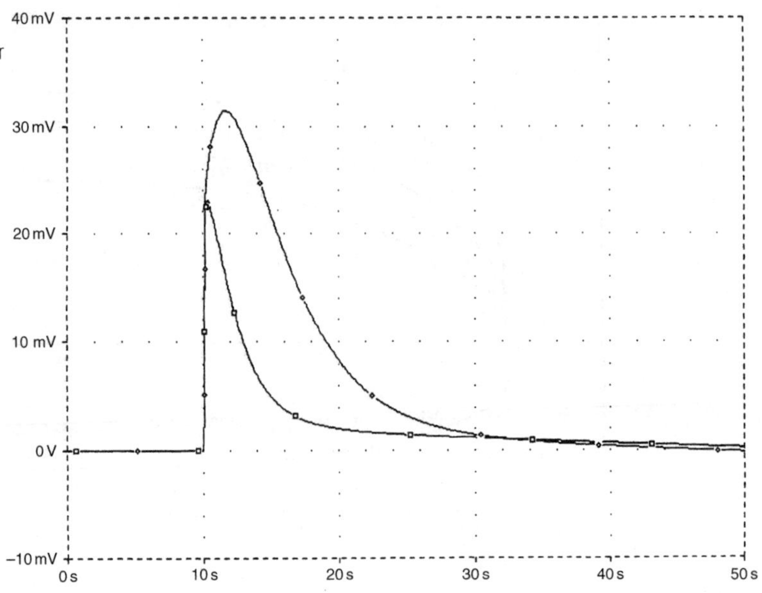

Figure 13.16 The transient error for the semi-insulating pad and the low-tempco version. The latter responds much faster, with a lower peak error, and gives less than half the Integrated Absolute Error (IAE)

If a thermal pad of standard material is once more used between the sensor and the TO3 top, the optimal Vbias coefficient for minimum error over the first 40 sec proves −2.8 mV/°C, which is usefully less than than −4.5. The resulting 30-sec IAE is 102 mV-sec, more than a two times improvement; see the lower trace in Figure 13.16, for comparison with the semi-insulator method described above.

In view of the fixed time-constants, dependant upon a certain weight of metal being required for heat dissipation, it appears that the only way this performance could be significantly improved upon might be to introduce a new kind of output transistor with an integral diode that would sense the actual junction temperature, being built into the main transistor junction structure. Although it would be of immense help to amplifier makers, no one seems to be keen to do this.

From here on I am assuming that a variable-temperature-coefficient (tempco) bias generator can be made when required; the details of how to do it are not given here. It is an extremely useful device, as thermal attenuation can then be countered by increasing the *thermal gain*; it does not however help with the problem of thermal delay.

In the second EF example above, the desired tempco is −2.8 mV/°C, while an EF output stage plus has an actual tempco of −4.5 mV/°C. (This inherent thermal gain in the EF was explained on page 348.) In this case we need a bias generator that has a *smaller* tempco than the standard circuit. The conventional EF with its temp sensor on the relatively cool main heatsink would require a *larger* tempco than standard.

A potential complication is that amplifiers should also be reasonably immune to changes in ambient temperature, quite apart from changes due to dissipation in the power devices. The standard tempco gives a close approach to this automatically, as the Vbe-multiplication factor is naturally almost the same as the number of junctions being biased. However, this will no longer be true if the tempco is significantly different from standard, so it is necessary to think about a bias generator that has one tempco for power-device temperature changes, and another for ambient changes. This sounds rather daunting, but is actually fairly simple.

Improved compensation for the CFP output stage

As revealed on page 343, the Complementary-Feedback-Pair (CFP) output stage has a much smaller bias tolerance of $\pm 10\,mV$ for a whole amplifier, and surprisingly long time-constants. A standard CFP stage therefore has larger relative errors than the conventional Emitter-Follower (EF) stage with thermal sensor on the main heatsink; this is the opposite of conventional wisdom. Moving the sensor to the top of the TO3 can was shown to improve the EF performance markedly, so we shall attempt an analogous improvement with driver compensation.

The standard CFP thermal compensation arrangements have the sensor mounted on the driver heatsink, so that it senses the heatsink temperature rather than that of the driver itself. (See Figure 13.17a for mechanical arrangement, and Figure 13.18 for thermal model.) As in the EF, this gives a constant long-term error due to the sustained temperature difference between the driver junction and heatsink mass; see the upper traces in Figure 13.20, plotted for different bias tempcos. The CFP stimulus is a 0.5 W step, as before. This constant error cannot be properly dealt with by choosing a tempco that gives a bias error passing through a zero in the first fifty seconds, as was done for the EF case with a TO3-top sensor, as the heatsink thermal inertia causes it to pass through zero very quickly and head rapidly South in the direction of ever-increasing negative error. This is because it has allowed for thermal attenuation but has not decreased thermal delay. It is therefore pointless to compute an IAE for this configuration.

A better sensor position

By analogy with the TO3 and TO3P transistor packages examined earlier, it will be found that driver packages such as TO225AA on a heatsink get hotter faster on their exposed plastic face than any other accessible point.

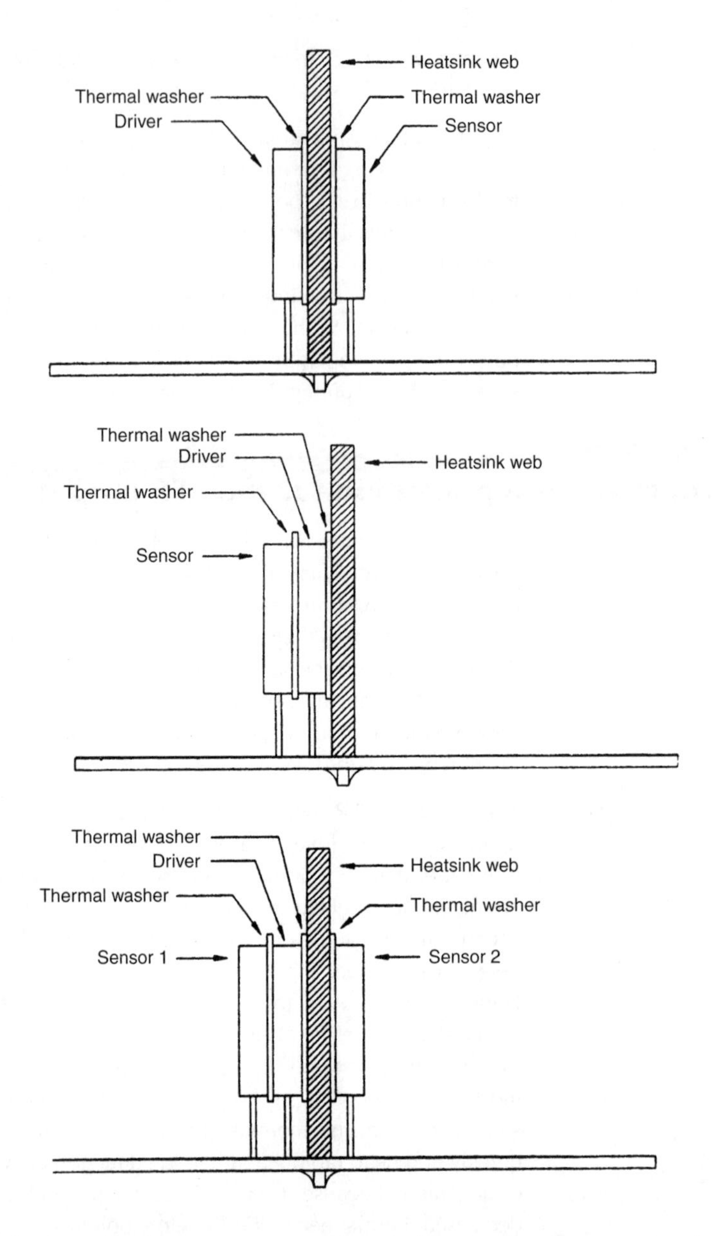

Figure 13.17
(a) The sensor transistor on the driver heatsink
(b) An improved version, with the sensor mounted on top of the driver itself, is more accurate
(c) Using two sensors to construct a junction-estimator

It looks as if a faster response will result from putting the sensor on top of the driver rather than on the other side of the sink as usual. With the Redpoint SW 38-1 heatsink this is fairly easy as the spring-clips used to secure one plastic package will hold a stack of two TO225AA's with only a little physical persuasion. A standard thermal pad is used between the top of the driver and the metal face of the sensor, giving the sandwich shown in Figure 13.17b. The thermal model is Figure 13.19. This scheme greatly reduces both thermal attenuation and thermal delay (lower traces in

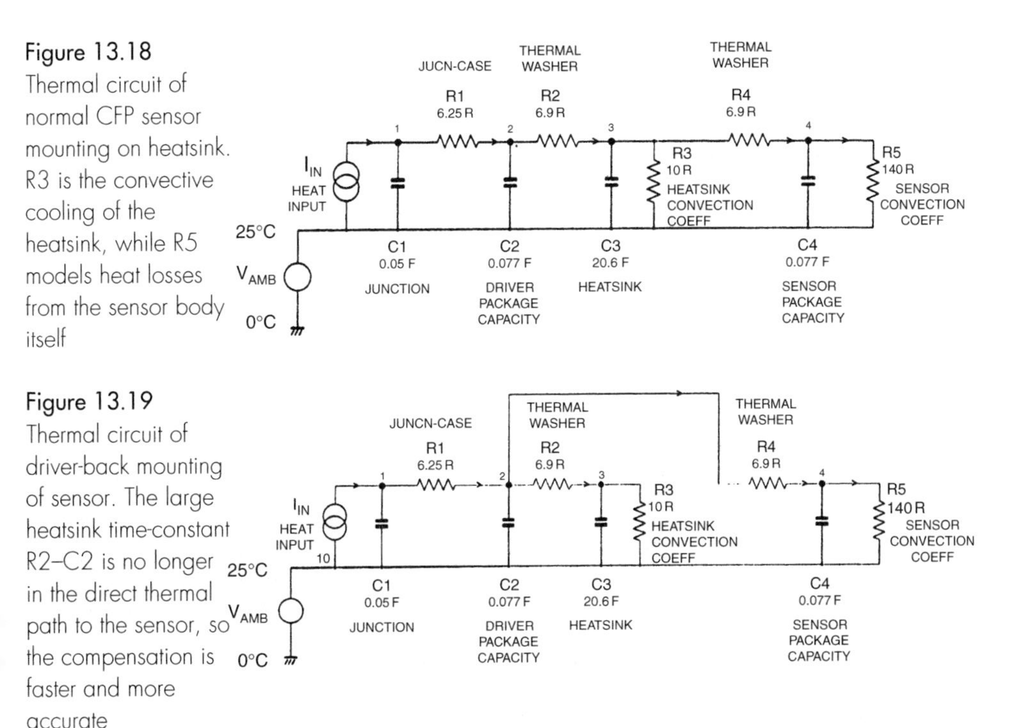

Figure 13.18 Thermal circuit of normal CFP sensor mounting on heatsink. R3 is the convective cooling of the heatsink, while R5 models heat losses from the sensor body itself

Figure 13.19 Thermal circuit of driver-back mounting of sensor. The large heatsink time-constant R2–C2 is no longer in the direct thermal path to the sensor, so the compensation is faster and more accurate

Figure 13.20) giving an error that falls within a ±1 mV window after about 15.5 seconds, when the tempco is set to −3.8 mV/°C. The IAE computes to 52 mV, as shown in Figure 13.21, which demonstrates how the IAE criterion tends to grow without limit unless the error subsides to zero. This value is a distinct improvement on the 112 mV IAE which is the best that could be got from the EF output.

The effective delay is much less because the long heatsink time-constant is now partly decoupled from the bias compensation system.

A junction-temperature estimator

It appears that we have reached the limit in what can be done, as it is hard to get one transistor closer to another than they are in Figure 13.17b. It is however possible to get better performance, not by moving the sensor position, but by using more of the available information to make a better estimate of the true driver junction temperature. Such *estimator* subsystems are widely used in servo control systems where some vital variable is inaccessible, or only knowable after such a time-delay as to render the data useless[9]. It is often almost as useful to have a *model system*, usually just an abstract set of gains and time-constants, which all give an estimate of what the current value of the unknown variable must be, or ought to be.

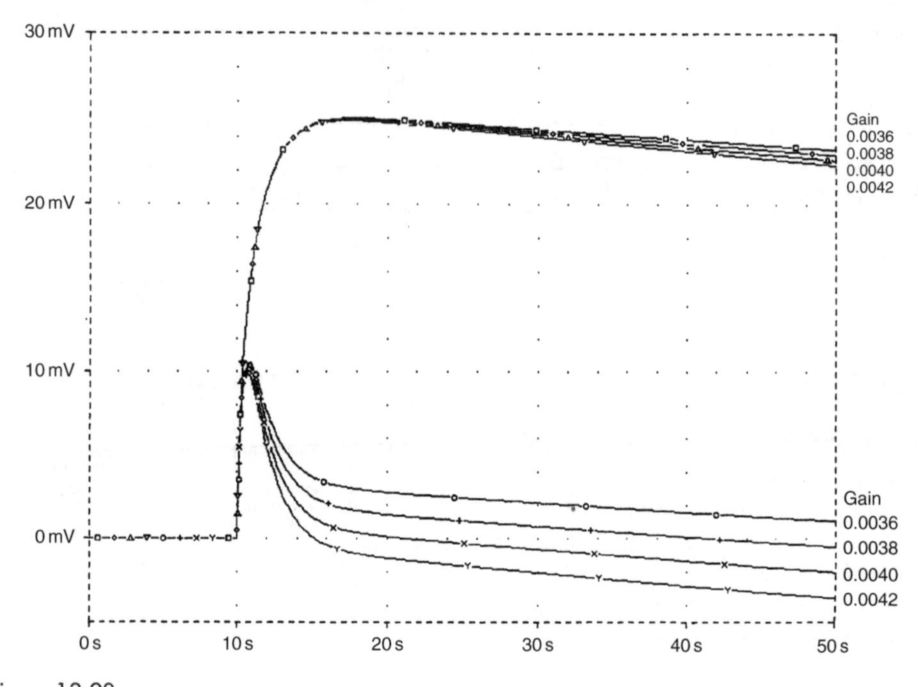

Figure 13.20
The Vq errors for normal and improved sensor mounting, with various tempcos. The improved method can have its tempco adjusted to give near-zero error over this timescale. Not so for the usual method

The situation here is similar, and the first approach makes a better guess at the junction temperature V(1) by using the known temperature drop between the package and the heatsink. The inherent assumption is made that the driver package is isothermal, as it is modelled by one temperature value V(2).

If two sensors are used, one placed on the heatsink as usual, and the other on top of the driver package, as described above (Figure 13.17c), then things get interesting. Looking at Figure 13.19, it can be seen that the difference between the driver junction temperature and the heatsink is due to R1 and R2; the value of R1 is known, but not the heat flow through it. Neglecting small incidental losses, the temperature drop through R1 is proportional to the drop through R2. Since C2 is much smaller than C3, this should remain reasonably true even if there are large thermal transients. Thus, measuring the difference between V(2) and V(3) allows a reasonable estimate of the difference between V(1) and V(2); when this difference is added to the known V(2), we get a rather good estimation of the inaccessible V(1). This system is shown conceptually in Figure 13.22, which gives only the basic method of operation; the details of

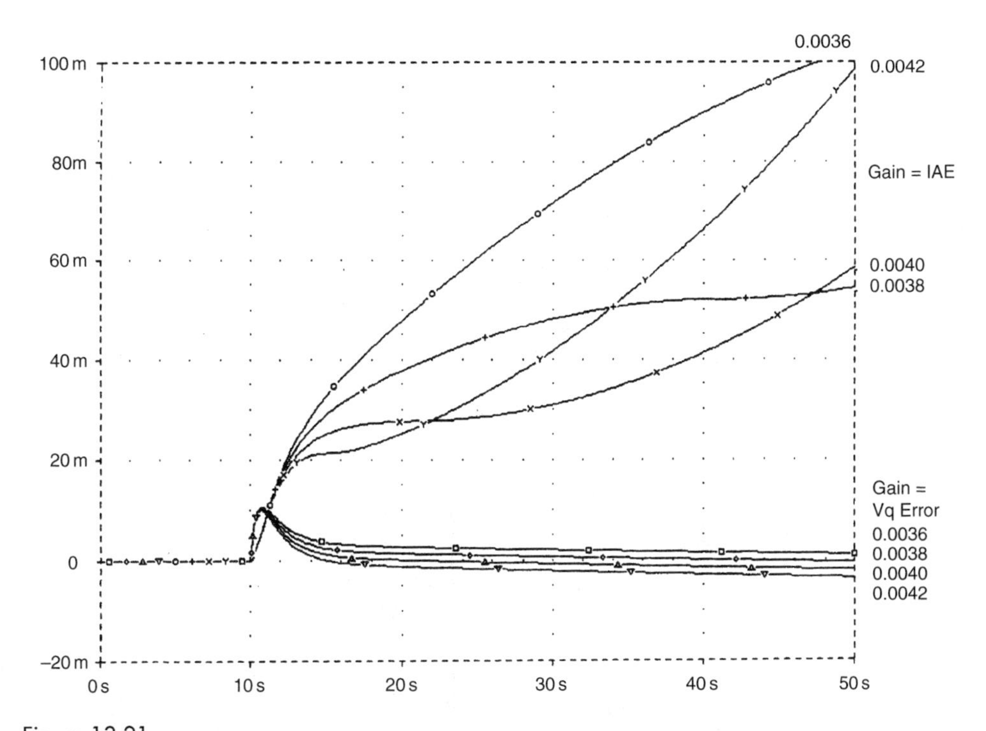

Figure 13.21
The Vq error and IAE for the improved sensor mounting method on driver back. Error is much smaller, due both to lower thermal attenuation and to less delay. Best IAE is 52 mV-sec (with gain = 0.0038); twice as good as the best EF version

the real circuitry must wait until we have decided exactly what we want it to do.

We can only measure V(2) and V(3) by applying thermal sensors to them, as in Figure 13.17c, so we actually have as data the sensor temperatures V(4) and V(5). These are converted to bias voltage and subtracted, thus estimating the temperature drop across R1. The computation is done by Voltage-Controlled-Voltage-Source E1, which in PSpice can have any equation assigned to define its behaviour. Such definable VCVSs are very handy as little *analogue computers* that do calculations as part of the simulation model. The result is then multiplied by a scaling factor called *estgain* which is incorporated into the defining equation for E1, and is adjusted to give the minimum error; in other words the variable-tempco bias approach is used to allow for the difference in resistance between R1 and R2.

The results are shown in Figure 13.23, where an *estgain* of 1.10 gives the minimum IAE of 25 mV-sec. The transient error falls within a ±1 mV window after about 5 sec. This is a major improvement, at what promises to be little cost.

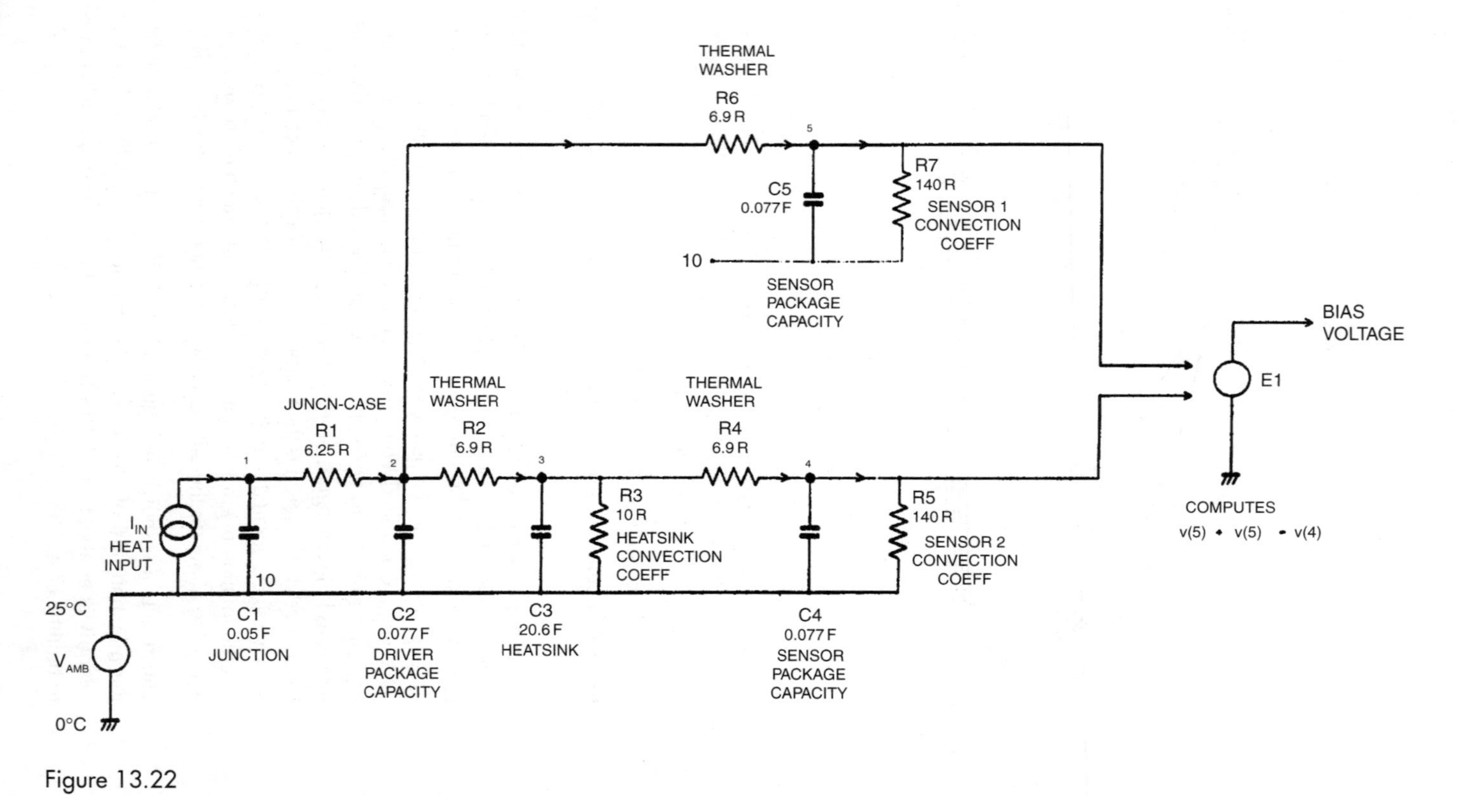

Figure 13.22

Conceptual diagram of the junction-estimator. Controlled-voltage-source E1 acts as an analogue computer performing the scaling and subtraction of the two sensor temperatures V(4) and V(5), to derive the bias voltage

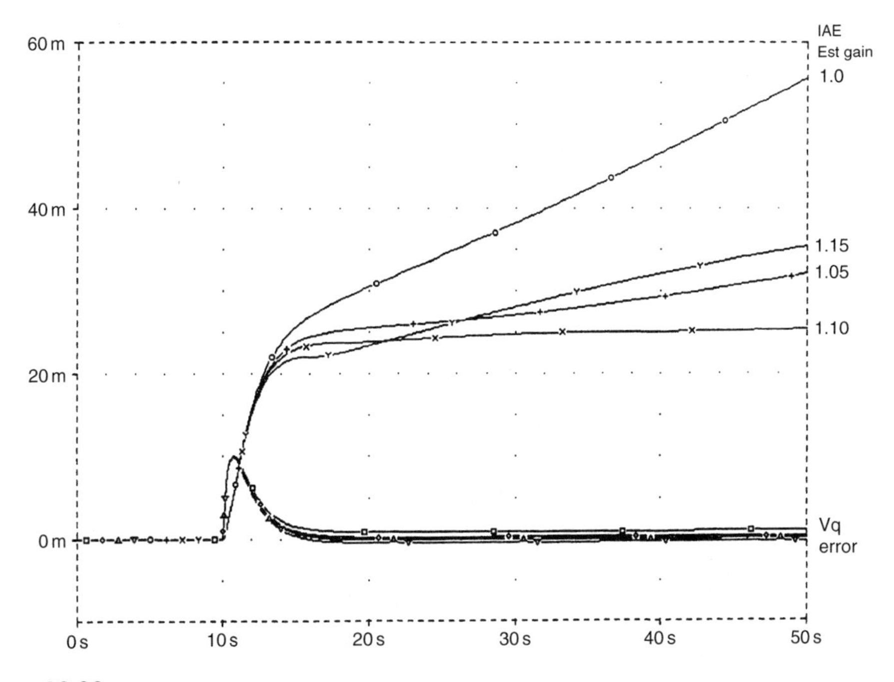

Figure 13.23
Simulation results for the junction-estimator, for various values of estgain. The optimal IAE is halved to 25 mV-sec; compare with Figure 13.21

A junction estimator with dynamics

The remaining problem with the junction-estimator scheme is still its relatively slow initial response; nothing can happen before heat flows through R6 into C5, in Figure 13.22. It will take even longer for C4 to respond, due to the inertia of C3, so we must find a way to speed up the dynamics of the junction-estimator.

The first obvious possibility is the addition of phase-advance to the forward bias-compensation path. This effectively gives a high gain initially, to get things moving, which decays back over a carefully set time to the original gain value that gave near-zero error over the 50-sec timescale. The conceptual circuit in Figure 13.24 shows the phase-advance *circuitry* added to the compensation path; the signal is attenuated 100× by R50 and R51, and then scaled back up to the same level by VCVS E2, which is defined to give a gain of 110 times incorporating estimated gain = 1.10. C causes fast changes to bypass the attenuation, and its value in conjunction with R50, R51 sets the degree of phase-advance or lead. The slow behaviour of the circuit is thus unchanged, but transients pass through C and are greatly amplified by comparison with steady-state signals.

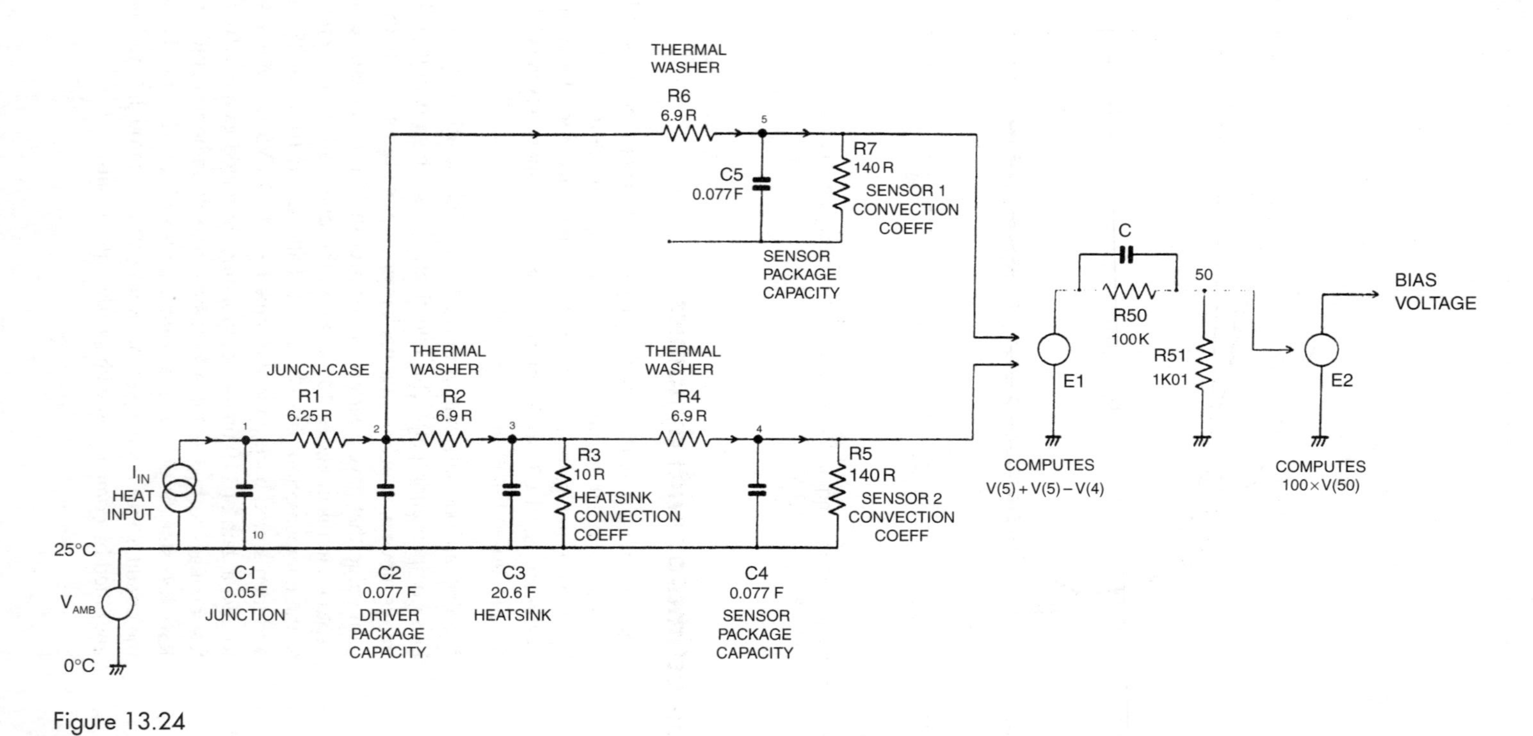

Figure 13.24
The conceptual circuit of a junction-estimator with dynamics. C gives higher gain for fast thermal transients and greatly reduces the effects of delay

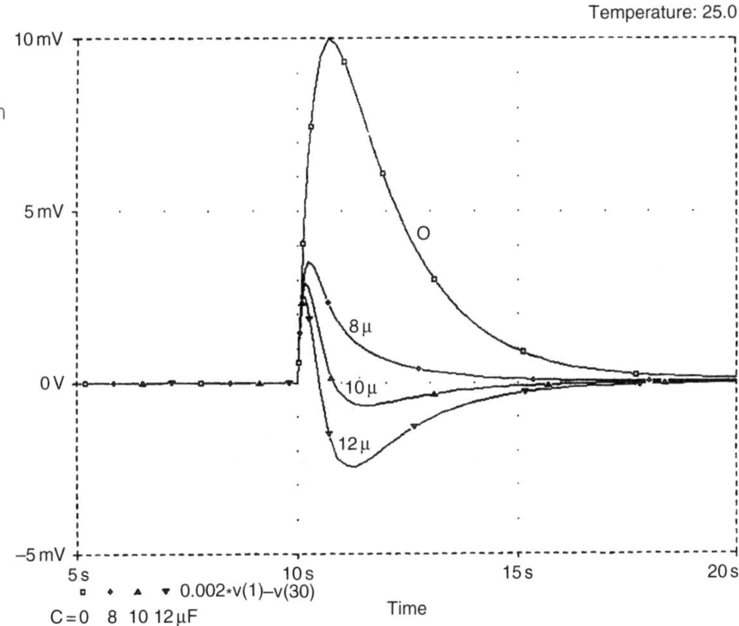

Figure 13.25
The initial transient errors for different values of C. Too high a value causes undershoot

The result on the initial error transient of varying C around its optimal value can be seen in the expanded view of Figure 13.25. The initial rise in Vq error is pulled down to less than a third of its value if C is made 10 μF; with a lower C value the initial peak is still larger than it need be, while a higher value introduces some serious undershoot that causes the IAE to rise again, as seen in the upper traces in Figure 13.26. The big difference between no phase-advance, and a situation where it is even approximately correct, is very clear.

With C set to 10 μF, the transient error falls a ±1 mV window after only 0.6 sec, which is more than which is more than twenty times faster than the first improved CFP version (sensor put on driver) and gives a nicely reduced IAE of 7.3 mV-sec at 50 sec. The real-life circuitry to do this has not been designed in detail, but presents no obvious difficulties. The result should be the most accurately bias-compensated Class-B amplifier ever conceived.

Conclusion

Some of the results of these simulations and tests were rather unexpected. I thought that the CFP would show relatively smaller bias errors than the EF, but it is the EF that stays within its much wider tolerance bands, with either heatsink or TO3-top mounted sensors. The thermal-gain effect in the EF stage seems to be the root cause of this, and this in turn is a consequence of the near-constant driver dissipation in the EF configuration.

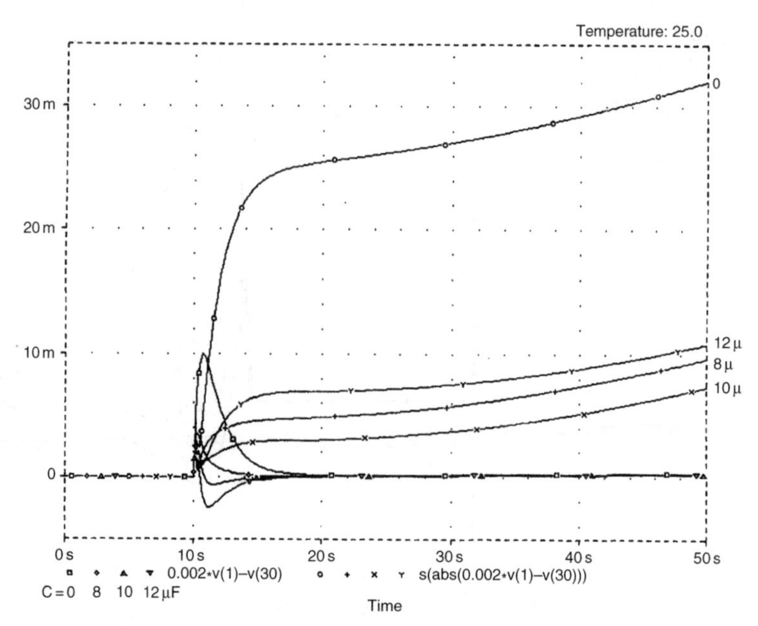

Figure 13.26
The IAE for different values of C. 10 μF is clearly best for minimum integrated error (IAE = 7.3 mV–sec) but even a rough value is a great improvement

However, the cumulative bias errors of the EF stage can only be reduced to a certain extent, as the system is never free from the influence of the main heatsink with its substantial thermal inertia. In contrast the CFP stage gives much more freedom for sensor placement and gives scope for more sophisticated approaches that reduce the errors considerably.

Hopefully it is clear that it is no longer necessary to accept *Vbe-multiplier on the heatsink* as the only option for the crucial task of Vbias compensation. The alternatives presented promise greatly superior compensation accuracy.

Variable-tempco bias generators

The standard Vbe-multiplier bias generator has a temperature coefficient that is fixed by the multiplication factor used, and so ultimately by the value of Vbias required. At many points in this chapter it has been assumed that it is possible to make a bias generator with an arbitrary temperature coefficient. This section shows how to do it.

Figure 13.27 shows two versions of the usual Vbe-multiplier bias generator. Here the lower rails are shown as grounded to simplify the results. The first version in Figure 13.27a is designed for an EF (Emitter-Follower) output stage, where the voltage Vbias to be generated is $(4 \times \text{Vbe}) + \text{Vq}$, which totals +2.93 V. Recall that Vq is the small quiescent voltage across the emitter-resistors Re; it is *this* quantity we are aiming to keep constant, rather

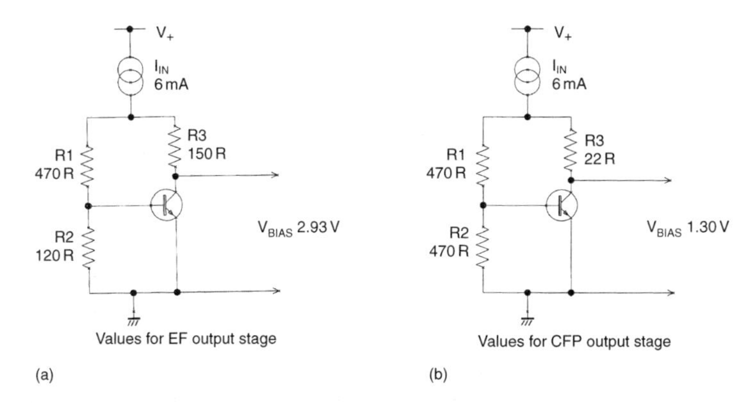

Figure 13.27
The classical Vbe-multiplier bias generator. Two versions are shown: for biasing EF (a) and (b) CFP output stages. The EF requires more than twice the bias voltage for optimal crossover performance

than the quiescent current, as is usually assumed. The optimal Vq for an EF stage is in the region of 50 mV.

The second bias generator in Figure 13.27b is intended for a CFP (Complementary-Feedback Pair) output stage, for which the required Vbias is less at (2 × Vbe) + Vq, or approximately 1.30 V in total. Note that the optimal Vq is also much smaller for the CFP type of output stage, being about 5 mV.

It is assumed that Vbias is trimmed by varying R2, which will in practice be a preset with a series end-stop resistor to limit the maximum Vbias setting. It is important that this is the case, because a preset normally fails by the wiper becoming disconnected, and if it is in the R2 position the bias will default to minimum. In the R1 position an open-circuit preset will give maximum bias, which may blow fuses or damage the output stage. The adjustment range provided should be no greater than that required to take up production tolerances; it is, however, hard to predict just how big that will be, so the range is normally made wide for pre-production manufacture, and then tightened in the light of experience.

The EF version of the bias generator has a higher Vbias, so there is a larger Vbe-multiplication factor to generate it. This is reflected in the higher temperature coefficient (hereafter shortened to 'tempco'). See Table 13.4.

Creating a higher tempco

A higher (i.e., more negative) tempco than normal may be useful to compensate for the inability to sense the actual output junction temperatures. Often the thermal losses to the temperature sensor are the major source

Table 13.4

	Vbias volts	R1 Ω	R2 Ω	R3 Ω	Tempco mV/°C
EF	2.93	120R	470R	22R	−9.3
CFP	1.30	470R	470R	150R	−3.6

of steady-state Vbias error, and to reduce this a tempco is required that is larger than the standard value given by: 'Vbe-multiplication factor times −2 mV/°C'. Many approaches are possible, but the problem is complicated because in the CFP case the bias generator has to work within two rails only 1.3 V apart. Additional circuitry outside this voltage band can be accommodated by bootstrapping, as in the Trimodal amplifier biasing system in Chapter 9, but this does add to the component count.

A simple new idea is shown in Figure 13.28. The aim is to increase the multiplication factor (and hence the negative tempco) required to give the same Vbias. The diagram shows a voltage source V1 inserted in the R2 arm. To keep Vbias the same, R2 is reduced. Since the multiplication factor (R1 + R2)/R2 is increased, the tempco is similarly increased. In Table 13.5, a CFP bias circuit has its tempco varied by increasing V1 in 100 mV steps; in each case the value of R2 is then reduced to bring Vbias back to the desired value, and the tempco is increased.

A practical circuit is shown in Figure 13.29, using a 2.56 V bandgap reference to generate the extra voltage across R4. This reference has to work outside the bias generator rails, so its power-feed resistors R7, R8 are bootstrapped by C from the amplifier output, as in the Trimodal amplifier design.

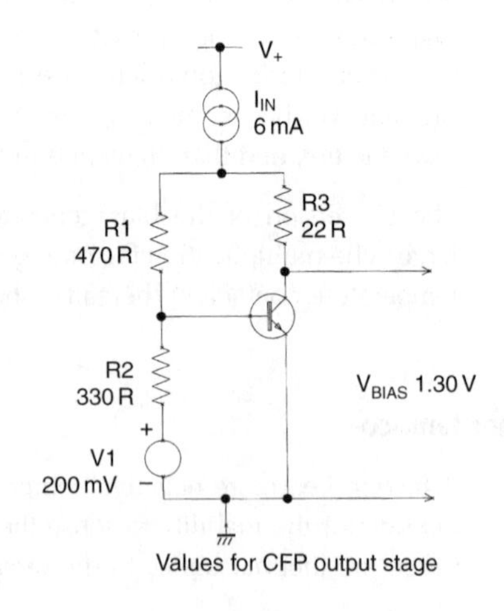

Figure 13.28
Principle of a Vbe multiplier with increased tempco. Adding voltage source V1 means the voltage-multiplication factor must be increased to get the same Vbias. The tempco is therefore also increased, here to −4.4 mV/°C

Table 13.5

V1 mV	Vbias V	R2 Ω	Tempco mV/C
0	1.287	470	−3.6
100	1.304	390	−4.0
200	1.287	330	−4.4
300	1.286	260	−5.0
400	1.285	190	−6.9

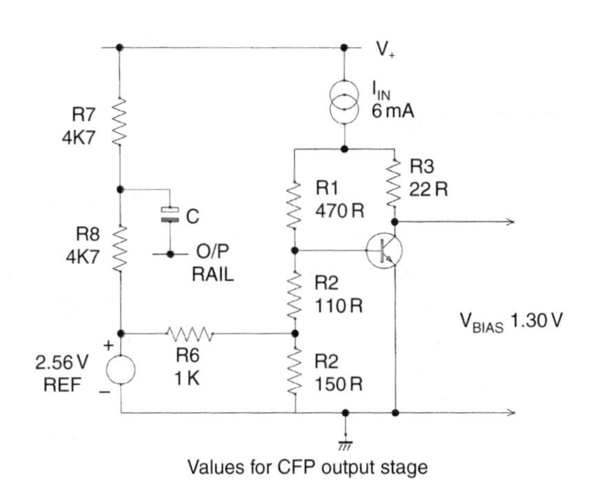

Figure 13.29 Shows a practical version of a Vbe multiplier with increased tempco. The extra voltage source is derived from the bandgap reference by R6, R4. Tempco is increased to −5.3 mV/°C

Ambient temperature changes

Power amplifiers must be reasonably immune to ambient temperature changes, as well as changes due to dissipation in power devices. The standard compensation system deals with this pretty well, as the Vbe-multiplication factor is inherently almost the same as the number of junctions being biased. This is no longer true if the tempco is significantly modified. Ideally we require a bias generator that has one increased tempco for power-device temperature changes only, and another standard tempco for ambient changes affecting all components. One approach to this is Figure 13.30, where V1 is derived via R6, R4 from a silicon diode rather than a bandgap reference, giving a voltage reducing with temperature. The tempco for temperature changes to Q1 only is −4.0 mV/°C, while the tempco for global temperature changes to *both* Q1 and D1 is lower at −3.3 mV/°C. Ambient temperatures vary much less than output device junction temperatures, which may easily range over 100°C.

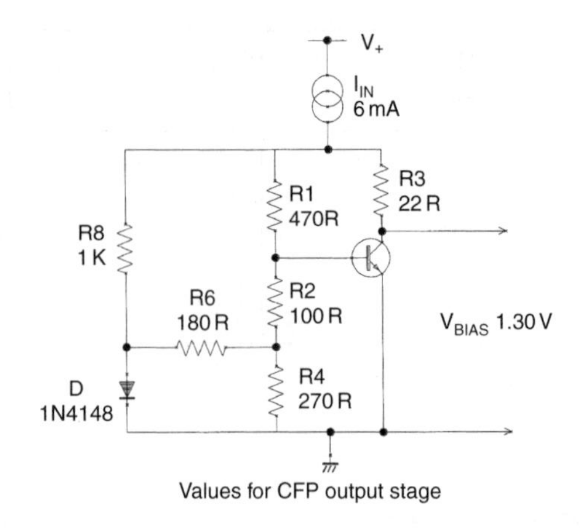

Figure 13.30 Practical Vbe multiplier with increased tempco, and also improved correction for ambient temperature changes, by using diode D to derive the extra voltage

Creating a lower tempco

Earlier in this chapter I showed that an EF output stage has 'thermal gain' in that the thermal changes in Vq make it appear that the tempco of the Vbias generator is higher than it really is. This is because the bias generator is set up to compensate for four base-emitter junctions, but in the EF output configuration the drivers have a roughly constant power dissipation with changing output power, and therefore do not change much in junction temperature. The full effect of the higher tempco is thus felt by the output junctions, and if the sensor is placed on the power device itself rather than the main heatsink, to reduce thermal delay, then the amplifier can be seriously over compensated for temperature. In other words, after a burst of power Vq will become too low rather than too high, and crossover distortion will appear. We now need a Vbias generator with a *lower* tempco than the standard circuit.

The principle is exactly analogous to the method of increasing the tempco. In Figure 13.31, a voltage source is inserted in the upper leg of potential divider R1, R2; the required Vbe-multiplication factor for the same Vbias is reduced, and so therefore is the tempco.

Table 13.6 shows how this works as V1 is increased in 100 mV steps. R1 has been varied to keep Vbias constant, in order to demonstrate the symmetry of resistor values with Table 13.5; in reality R2 would be the variable element, for the safety reasons described above.

Current compensation

Both bias generators in Figure 13.27 are fitted with a current-compensation resistor R3. The Vbe multiplier is a very simple shunt regulator, with low loop gain, and hence shows a significant series resistance. In other words, the Vbias generated shows unwanted variations in voltage with changes in

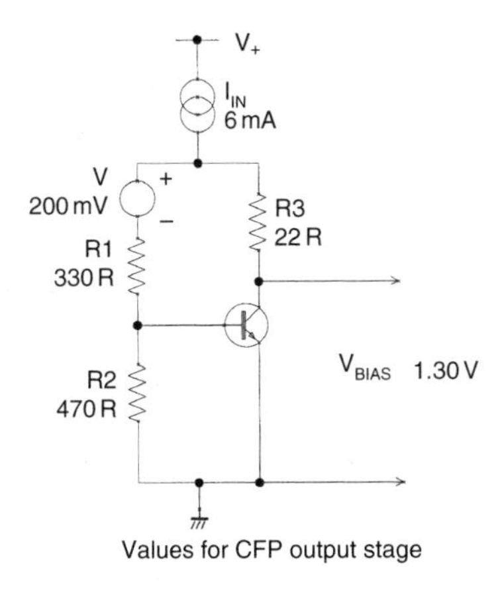

Figure 13.31
The principle of a Vbe multiplier with reduced tempco. The values shown give –3.1 mV/°C

Table 13.6

V1 mV	Vbias V	R1 Ω	Tempco mV/C
0	1.287	470	–3.6
100	1.304	390	–3.3
200	1.287	330	–3.1
300	1.286	260	–2.8
400	1.285	190	–2.5

the standing current through it. R3 is added to give first-order cancellation of Vbias variations caused by these current changes. It subtracts a correction voltage proportional to this current. Rather than complete cancellation, this gives a peaking of the output voltage at a specified current, so that current changes around this peak value cause only minor voltage variations. This peaking philosophy is widely used in IC bias circuitry.

R3 should never be omitted, as without it mains voltage fluctuations can seriously affect Vq. Table 13.4 shows that the optimal value for peaking at 6 mA depends strongly on the Vbe multiplication factor.

Figure 6.14 demonstrates the application of this method to the Class-B amplifier. The graph shows the variation of Vbias with current for different values of R3. The slope of the uncompensated (R3 = 0) curve at 6 mA is approximately 20 Ω, and this linear term is cancelled by setting R3 to 18 Ω in Figure 6.13.

The current through the bias generator will vary because the VAS current source is not a perfect circuit element. Biasing this current source with the usual pair of silicon diodes does not make it wholly immune to supply-rail variations. I measured a generic amplifier (essentially the original Class-B Blameless design) and varied the incoming mains from 212 V to 263 V, a range of 20%. This in these uncertain times is perfectly plausible for a power amplifier travelling around Europe. The VAS current-source output varied from 9.38 mA to 10.12 mA, which is a 7.3% range. Thanks to the current-compensating resistor in the bias generator, the resulting change in quiescent voltage Vq across the two Re's is only from 1.1 mV (264 V mains) to 1.5 mV (212 V mains). This is a very small absolute change of 0.4 mV, and within the Vq tolerance bands. The ratio of change is greater, because Vbias has had a large fixed quantity (the device Vbe's) subtracted from it, so the residue varies much more. Vq variation could be further suppressed by making the VAS current source more stable against supply variations.

The finite ability of even the current-compensated bias generator to cope with changing standing current makes a bootstrapped VAS collector load much less attractive than the current-source version; from the above data, it appears that Vq variations will be at least three times greater.

A quite different approach reduces Vbias variations by increasing the loop gain in the Vbe multiplier. Figure 13.32 shows the circuit of a two-transistor version that reduces the basic resistance slope from 20 to 1.7 Ω. The first transistor is the sensor. An advantage is that Vbias variations will be smaller for all values of VAS current, and no optimisation of a resistor value is required. A drawback is slightly greater complexity in an area where reliability is vital. Figure 13.33 compares the two-transistor configuration with the standard version (without R3). Multi-transistor feedback loops raise the possibility of instability and oscillation, and this must be carefully guarded against, as it is unlikely to improve amplifier reliability.

This section of the Thermal Dynamics chapter describes simple Vbias generators with tempcos ranging from −2.5 to −6.9 mV/°C. It is hoped that this, in combination with the techniques described earlier, will enable the

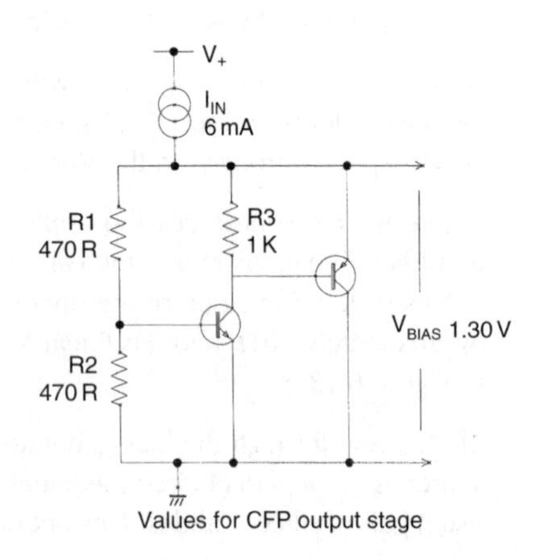

Figure 13.32 Circuit of a two-transistor Vbe multiplier. The increased loop gain holds Vbias more constant against current changes

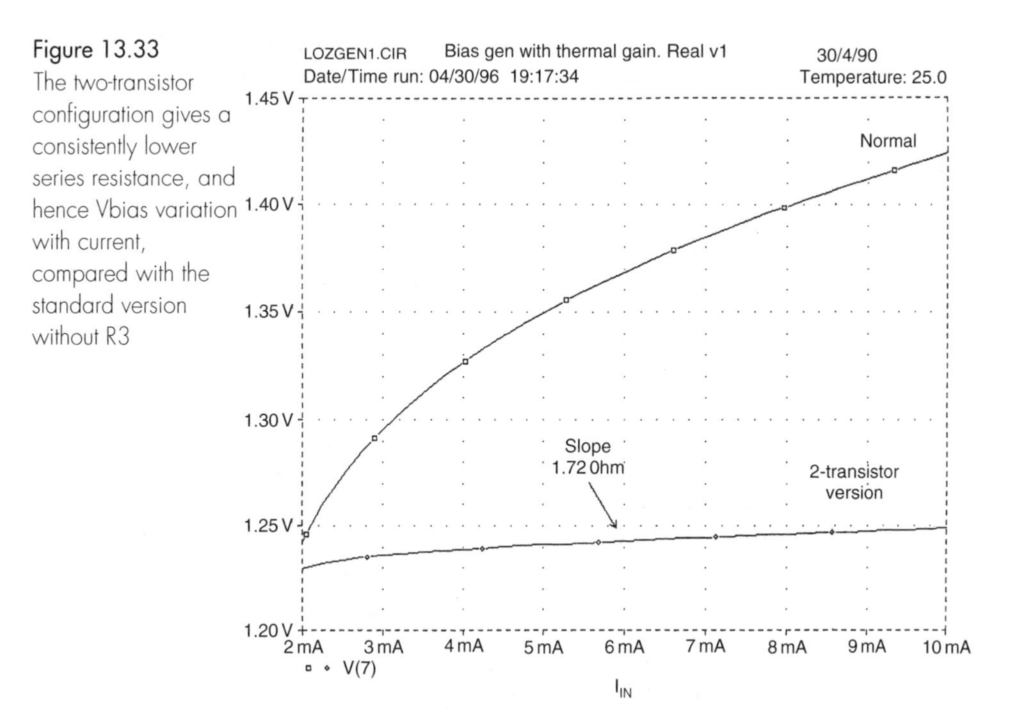

Figure 13.33
The two-transistor configuration gives a consistently lower series resistance, and hence Vbias variation with current, compared with the standard version without R3

design of Class-B amplifiers with greater bias accuracy, and therefore less afflicted by crossover distortion.

Thermal dynamics in reality

One of the main difficulties in the study of amplifier thermal dynamics is that some of the crucial quantities, such as transistor junction temperatures, are not directly measurable. The fact that bias conditions are altering is usually recognised from changes in the THD residual as viewed on a scope. However, these temperatures are only a means to an end – low distortion. What really matters is the crossover distortion produced by the output stage. Measuring this gets to the heart of the matter. The method is as follows. The amplifier under study is deliberately underbiased by a modest amount. I chose a bias setting that gave about 0.02% THD with a peak responding measurement mode. This is to create crossover spikes that are clear of the rest of the THD residual, to ensure the analyser is reading these spikes and ignoring noise and other distortions at a lower level. The AP System-1 has a mode that plots a quantity against time (it has to be said that the way to do this is not at all obvious from the AP screen menus – essentially 'time' is treated as an external stimulus – but it is in the manual) and this effectively gives that most desirable of plots – crossover conditions against time. In both cases below the amplifier was turned on with the input signal already present, so that dissipation conditions stabilised within a second or so.

Results

The first test amplifier examined has a standard EF output stage. The drivers have their own small heatsinks and have no thermal coupling with the main output device heatsink. The most important feature is that the bias sensor transistor is not mounted on the main heatsink, as is usual, but on the back of one of the output devices, as I recommended above. This puts the bias sensor much closer thermally to the output device junction. A significant feature of this test amplifier is its relatively high supply rails. This means that even under no load, there is a drift in the bias conditions due to the drivers heating up to their working temperature. This drift can be reduced by increasing the size of the driver heatsinks, but not eliminated. Figure 13.34 shows the THD plot taken over 10 minutes, starting from cold and initiating some serious power dissipation at $t = 0$. The crossover distortion drops at once; Figure 13.1 shows that driver dissipation is not much affected by output level, so this must be due to the output device junctions heating up and increasing Vq. There is then a slower reduction until the THS reading stabilises at about 3 minutes.

The second amplifier structure examined is more complex. It is a triple-EF design with drivers and output devices mounted on a large heatsink with considerable thermal inertia. The pre-drivers are TO220 devices mounted separately without heatsinks. It may seem perverse to mount the drivers on

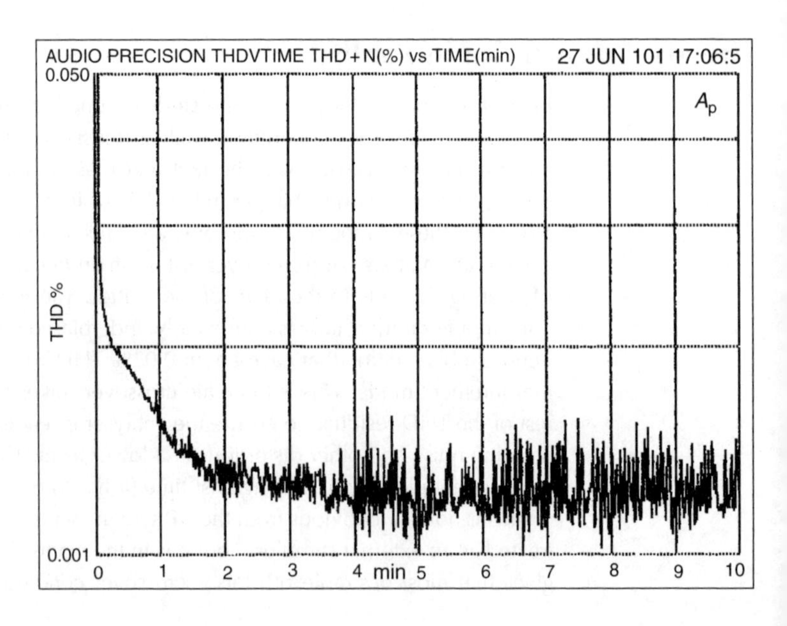

Figure 13.34
Peak THD vs time over 10 minutes

the same heatsink as the outputs, because some of the time they are being heated up rather than cooled down, which is exactly the opposite of what is required to minimise Vbe changes. However, they need a heatsink of some sort, and given the mechanical complications of providing a separate thermally isolated heatsink just for the drivers, they usually end up on the main heatsink. All that can be done (as in this case) is to put them in the heatsink position that stays coolest in operation. Once more the bias sensor transistor is not mounted on the main heatsink, but on the back of one of the output devices. See Figure 13.35 for the electrical circuit and thermal coupling paths.

The results are quite different. Figure 13.36 shows at A the THD plot taken over 10 minutes, again starting from cold and initiating dissipation at $t = 0$. Initially THD falls rapidly, as before, as the output device junctions heat. It then commences a slow rise over 2 minutes, indicative of falling bias, and this represents the timelag in heating the sensor transistor. After this there is a much slower drift downwards, at about the same rate as the main heatsink is warming up. There are clearly at least three mechanisms operating with very different time-constants. The final time-constant is very long, and the immediate suspicion is that it must be related to the slow warming of the main heatsink. Nothing else appears to be changing over this sort of timescale. In fact this long-term increase in bias is caused by cooling of the bias sensor compared with the output device it is mounted on. This effect was theoretically predicted above, and it is pleasing to see that it really exists, although it does nothing but further complicate the quest for optimal Class-B operation. As the main heatsink gets hotter, the heat losses from the sensor become more significant, and its temperature is lower than it should be. Therefore the bias voltage generated is too high, and this effect grows over time as the heatsink warms up.

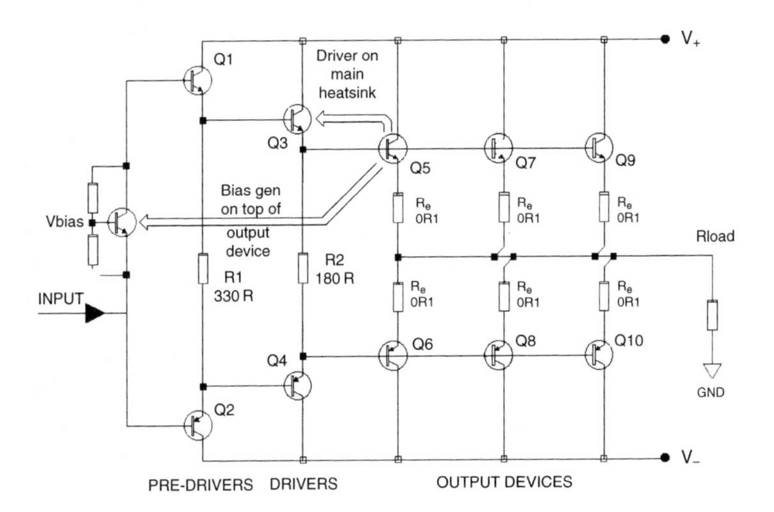

Figure 13.35 Circuit and thermal paths of the triple-EF output stage

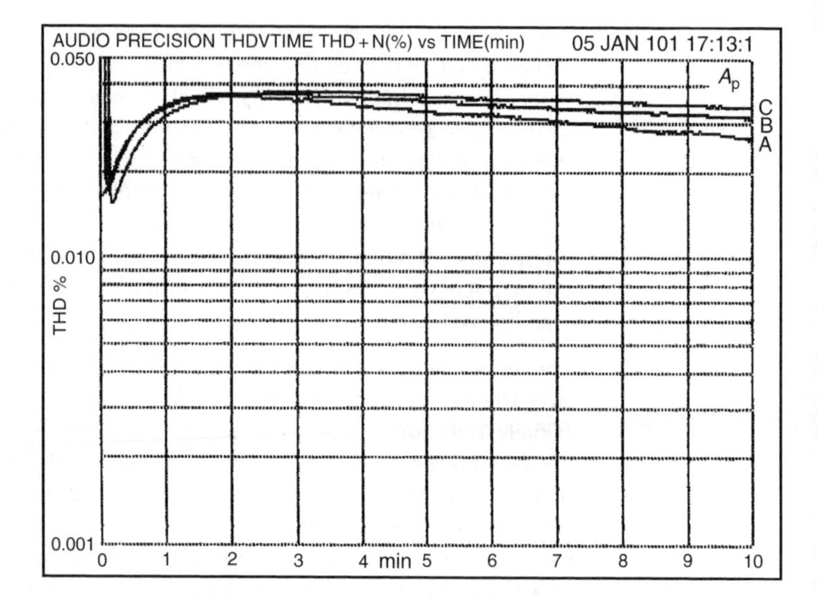

Figure 13.36
Peak THD versus time over 10 sec

Knowledge of how the long-term drift occurs leads at once to a strategy for reducing it. Adding thermal insulation to cover the sensor transistor, in the form of a simple pad of plastic foam, gives plot B, with the long-term variation reduced. Plot C reduces it still further by more elaborate insulation; a rectangular block of foam with a cutout for the sensor transistor. This is about as far as it is possible to go with sensor insulation; the long-term variation is reduced to about 40% of what it was. While this technique certainly appears to improve bias control, bear in mind that it is being tested with a steady sinewave. Music is noted for not being at the same level all the time, and its variations are much faster than the slow effect we are examining. It is very doubtful if elaborate efforts to reduce sensor cooling are worthwhile. I must admit this is the first time I have applied thermal lagging to an amplifier output stage.

Early effect in output stages

There is another factor that affects the accuracy with which quiescent conditions can be maintained. If you take a typical power amplifier and power it from a variable-voltage transformer, you are very likely to find that Vq varies with the mains voltage applied. This at first seems to indicate that the apparently straightforward business of compensating the bias generator for changes in standing current has fallen somewhat short of success (see page 181). However, even if this appears to be correct, and the constant-current source feeding the bias generator and VAS is made absolutely stable, the quiescent conditions are still likely to vary. At first this seems utterly mysterious, but the true reason is that the transistors in the output stage are reacting directly to the change in their collector–emitter voltage (Vce). As Vce increases, so does the Vq and the quiescent current. This is called Early Effect. It is a narrowing of the base-collector region as Vce increases, which will cause an increase in the collector current Ic even if Vbe and Ib are held constant. In a practical EF output stage the result is a significant variation in quiescent conditions when the supply voltage is varied over a range such as ±10%.

Table 13.7 shows the effect as demonstrated by SPICE simulation, using MJE340/50 for drivers and MJ15022/23 as output devices, with fixed bias voltage of 2.550 V, which gave optimal crossover in this case. It is immediately obvious that (as usual) things are more complicated than they at first appear. The Vq increases with rail voltage, which matches reality. However, the way in which this occurs is rather unexpected. The Vbe's of the drivers Q1 and Q2 reduce with increasing Vce as expected. However, the output devices Q3 and Q4 show a Vbe that increases – but by a lesser amount, so that after subtracting all the Vbe drops from the fixed bias voltage the aggregate effect is that Vq, and hence quiescent current Iq, both increase. Note that the various voltages have been summed as a check that they really do add up to 2.550 V in each case.

Table 13.8 has the results of real Vbe measurements. These are not easy to do, because any increase in Iq increases the heating in the various

Table 13.7 SPICE Vbe changes with supply rail voltage (MJE340/50 and MJ15022/3). All devices held at 25°C

±rail V	Vq mV	Q1 Vbe mV	Q3 Vbe mV	Q2 Vbe mV	Q4 Vbe mV	Sum V
10	7.8	609	633	654	646	2.550
20	13	602	640	647	648	2.550
30	18	597	643	641	649	2.550
40	23	593	647	637	650	2.550
50	28	589	649	634	650	2.550

Table 13.8 Real Vbe changes with supply rail voltage (2SC4382, 2SA1668 drivers and 2SC2922, 2SA1216 output)

±rail V	Vq mV	Q1 Vbe mV	Q3 Vbe mV	Q2 Vbe mV	Q4 Vbe mV	Sum V
40	1.0	554	568	541	537	2.201
45	1.0	544	556	533	542	2.176
50	1.0	534	563	538	536	2.172
55	1.0	533	549	538	540	2.161
60	1.0	527	552	536	535	2.151
65	1.0	525	540	536	539	2.141
70	1.0	517	539	537	539	2.133

transistors, which will cause their Vbe's to drift. This happens to such an extent that sensible measurements are impossible. The measurement technique was therefore slightly altered. The amplifier was powered up on the minimum rail voltage, with its Vq set to 1.0 mV only. This is far too low for good linearity, but minimises heating while at the same time ensuring that the output devices are actually conducting. The various voltages were measured, the rail voltage increased by 5 V, and then the bias control turned down as quickly as possible to get Vq back to 1.0 mV, and the process is repeated. The results are inevitably less tidy as the real Vbe's are prone to wander around by a millivolt or so, but it is clear that in reality, as in SPICE, most of the Early Effect is in the drivers, and there is a general reduction in aggregate Vbe as rail voltage increases. The sum of Vbe's is no longer constant as Vq has been constrained to be constant instead.

It may seem at this point as if the whole business of quiescent control is just too hopelessly complicated. Not so. The cure for the Early Effect problem is to overcompensate for standing current changes, by making the value of resistor R3 above larger than usual. The best and probably the only practical way to find the right value is the empirical method. Wind the HT up and down on the prototype design and adjust the value of R3 until the Vq change is at a minimum. (Unfortunately this interacts with the bias setting, so there is a bit of twiddling to do – however, for a given design you only need to find the optimal value for R3 once.) This assumes that the supply-rail rejection of the VAS current source is predictable and stable; with the circuits normally used this seems to be the case, but some further study in this area is required.

References

1. Sato et al *Amplifier Transient Crossover Distortion Resulting from Temperature Change of Output Power Transistors* AES Preprint. AES Preprint 1896 for 72nd Convention, Oct 1982.
2. Brown, I *Opto-Bias Basis for Better Power Amps* Electronics World, Feb 1992, p. 107.
3. Carslaw and Jaeger *Conduction of Heat in Solids* Oxford Univ. Press 1959, ISBN 0-19-853368-3.
4. Murphy, D *Axisymmetric Model of a Moving-Coil Loudspeaker* JAES, Sept 1993, p. 679.
5. Motorola, Toulouse *Private communication.*
6. Evans, J *Audio Amplifier Bias Current Letters* Electronics & Wireless World, Jan 1991, p. 53.
7. Chen, C-T *Analog & Digital Control System Design* Saunders-HBJ 1993, p. 346.
8. Harriot, P *Process Control* McGraw-Hill 1964, pp. 100–102.
9. Liptak, B, ed. *Instrument Engineer's Handbook-Process Control* Butterworth-Heinemann 1995, p. 66.

14

The design of DC servos

In the section of this book dealing with input stages I have gone to some lengths to demonstrate that a plain unassisted amplifier – if designed with care – can provide DC offset voltages at the output which are low enough for most practical purposes, without needing either an offset-nulling preset or a DC servo system. For example, the Trimodal amplifier can be expected not to exceed $\pm15\,\text{mV}$ at the output. However, there may be premium applications where this is not good enough. In this case the choice is between manual adjustment and DC servo technology.

DC offset trimming

Preset adjustment to null the offset voltage has the advantage that it is simple in principle and most unlikely to cause any degradation of audio performance. In servicing the offset should not need renulling unless one of relatively few components are changed; the input devices have the most effect, because the new parts are unlikely to have exactly the same beta, but the feedback resistors also have some influence as the input stage base currents flow through them.

The disadvantages are that an extra adjustment is required in production, and since this is a set-and-forget preset, it can have no effect on DC offsets that may accumulate due to input stage thermal drift or component ageing.

Figure 14.1 shows one simple way to add a DC trim control to an amplifier, by injecting a small current of whatever polarity is required into the feedback point. Since the trim circuit is powered from the main HT rails, which are assumed to be unregulated, careful precautions against the injection of noise, ripple and DC fluctuations must be taken. The diodes D1,D2 set up a stable voltage across the potentiometer. They do of course have a thermal coefficient, but this is not likely to be significant over the normal temperature range. R3 and C1 form a low-pass filter to reduce noise and

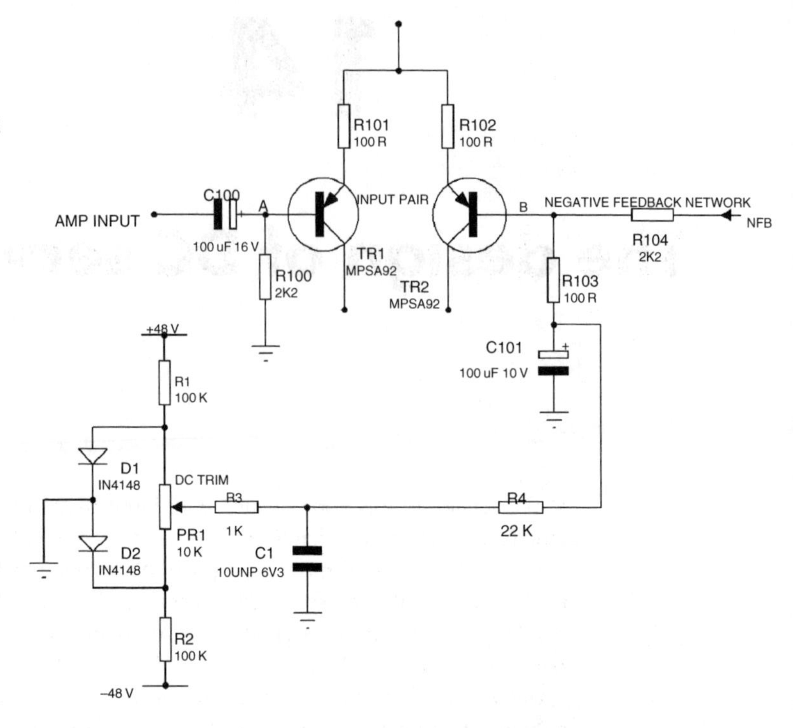

Figure 14.1
DC offset trim with injection into the negative feedback network

ripple, and the trimming current is injected through R4. This resistor has a relatively high value to minimise its effect on the closed-loop gain, and to give a powerful filtering action in conjunction with the large value of C101, to remove any remaining noise and ripple. Note that the trim current is injected at the bottom of R103, and not into the actual feedback point at B, as this would feed any disturbances on C1 directly into the amplifier path. From the point of view of the amplifier, R4 is simply a resistance to ground in parallel with R101, so its effect on the gain can be easily taken into account if required. This DC trim circuit should not degrade the noise performance of the amplifier when it is added, even though the amplifier itself is unusually quiet due to the low impedance of the feedback network.

So long as the input is properly AC-coupled (DC-blocked) the trim current can also be fed into the input at point A, but the possible effect on the noise and hum performance is less predictable as the impedance feeding the amplifier input is not known.

DC offset control by servo-loop

A DC servo system (presumably so-called to emphasise that it does not get directly involved in the main feedback loop) provides continuous active nulling of the amplifier offset by creating another feedback path that has a high gain at DC and very low frequencies, but limited control of the output DC level. This second path uses an op-amp, usually configured as an integrator, to perform the feedback subtraction in which the output DC level is compared with ground. It is straightforward to select an op-amp whose input offset specification is much better than the discrete input stage, because DC precision is where op-amp technology can really excel. For example, both the Analog Devices AD711JN and OPA134 offer a maximum offset of ±2 mV at 25°C, rising to 3 mV over the full commercial temperature range. Performance an order of magnitude better than this is available – e.g., the OPA627, but the price goes up by an order of magnitude too. FET input op-amps are normally used to avoid bias current offsets with high-value resistors.

An unwelcome complication is the need to provide ±15 V (or thereabouts) supply rails for the op-amp, if it does not already exist. It is absolutely essential that this supply is not liable to dropout if the main amplifier reproduces a huge transient that pulls down the main supply rails. If it does dropout sufficiently to disrupt the operation of the servo, disturbances will be fed into the main amplifier, possibly causing VLF oscillation. This may not damage the amplifier, but is likely to have devastating results for the loudspeakers connected to it.

The advantages of DC servos

1. The output op-amp DC offset of the amplifier can be made almost as low as desired. The technology of DC precision is mature and well-understood.
2. The correction process is continuous and automatic, unlike the DC trimming approach. Thermal drift and component ageing are dealt with, and there is only one part on which the accuracy of offset nulling depends-the servo op-amp, which should not significantly change its characteristics over time.
3. The low-frequency roll-off of the amplifier can be made very low without using huge capacitors. It can also be made more accurate, as the frequency is now set by a non-electrolytic capacitor.
4. The use of electrolytics in the signal path can be avoided, and this will impress some people.
5. The noise performance of the power amplifier can be improved because lower value resistances can be used in the feedback network, yielding a very quiet amplifier indeed.

Points 3, 4, and 5 are all closely related, so they are dealt with at greater length below.

Basic servo configurations

Figure 14.2a shows a conventional feedback network, as used in the Load-Invariant amplifier in this book. The usual large capacitor C is present at the bottom of the feedback network; its function is to improve offset accuracy by reducing the closed-loop gain to unity at DC. Figure 14.2b shows a power amplifier with a DC servo added, in the form of a long-time-constant integrator feeding into the feedback point. C is no longer required, as the servo does all the work. It had better be said at once that if the integrator constant is suitably long, a negligible amount of the audio signal passes through it, and the noise and distortion of the main amplifier should not be degraded in any way; more on this later.

As with manual trimming, there are many ways to implement a DC servo. This method works very well, and I have used it many times. One important

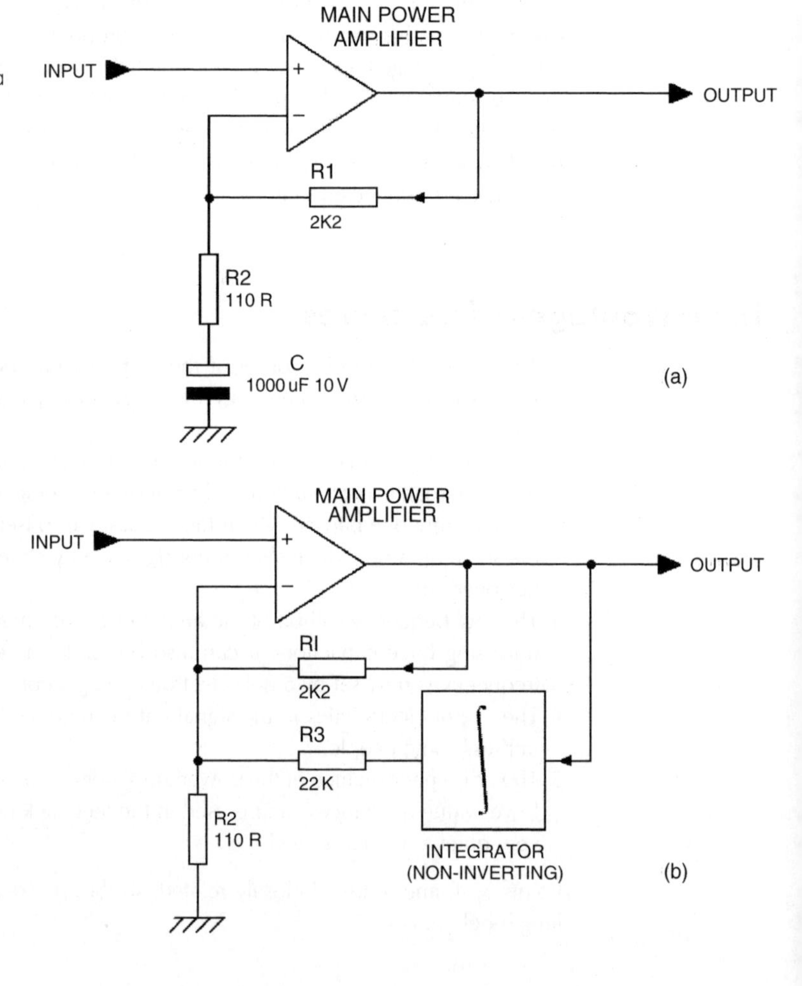

Figure 14.2
Power amplifiers without and with a DC servo in the feedback path

point is that the integrator block must be non-inverting for the servo feedback to be in the correct phase. The standard shunt-feedback integrator is of course inverting, so something needs to be done about that. Several non-inverting integrators are examined below.

Injecting the servo signal into the input is possible, and in this case a standard inverting integrator can be used. However, as for manual trimming, using the input gives a greater degree of uncertainty in the operating conditions as the source impedance is unknown. If there is no DC blocking on the input, the DC servo will probably not work correctly as the input voltage will be controlled by the low impedance of a preamp output. If there is DC blocking then the blocking capacitor may introduce an extra pole into the servo response, which if nothing else complicates things considerably.

Injection of the servo correction into the amplifier forward path is not a good idea as the amplifier has its own priorities – in particular keeping the input pair exactly balanced. If, for example, you feed the servo output into the current-mirror at the bottom of the input pair, the main amplifier can only accommodate its control demands by unbalancing the input pair collector currents, and this will have dire effects on the high-frequency distortion performance.

Noise, component values, and the roll-off

When you design an amplifier feedback network, there is a big incentive to keep the Johnson noise down by making the resistor values as low as possible. In the simple feedback network shown in Figure 14.2a, the source impedance seen by the input stage of the amplifier is effectively that of R2; if the rest of the amplifier has been thoughtfully designed then this will be a significant contributor to the overall noise level. Since the Johnson noise voltage varies as the square root of the resistance, minor changes (such as allowing for the fact that R1 is effectively in parallel in R2) are irrelevant. Because of the low value of R2, the feedback capacitor C tends to be large as its RC time-constant with R2 (not R1 + R2) is what sets the LF roll-off. If R2 is low then C is big, and practical values of C put a limit on how far R2 can be reduced. Hence there is a tradeoff between low-frequency response and noise performance, controlled by the physical size of C.

When a DC servo is fitted, it is usual to let it do all the work, by removing capacitor C from the bottom arm of the negative-feedback network. The components defining the LF roll-off are now transferred to the servo, which will use high-value resistors and small non-electrolytic capacitors. The value of R2 is no longer directly involved in setting the LF roll-off and there is the possibility that its resistance can be further reduced to minimise its noise contribution, while at the same time the LF response is extended to whatever frequency is thought desirable. The limit of this approach to noise reduction is set by how much power it is desirable to dissipate in R1.

There is a temptation to fall for the techno-fallacy that if it can be done, it should be done. A greatly extended LF range (say below 0.5 Hz) exposes the amplifier to some interesting new problems of DC drift. A design with its lower point set at 0.1 Hz is likely to have its output wavering up and down by tens on milliVolts, as a result of air currents differentially cooling the input pair, introducing variations that are slow but still too fast for the servo to correct. Whether these perturbations are likely to cause subtle intermodulations in speaker units is a moot point; it is certain that it does not look good on an oscilloscope, and could cause reviewers to raise their eyebrows. Note that unsteady air currents can exist even in a closed box due to convection from internal heating.

A cascode input stage reduces this problem by greatly lowering the voltage drop across the input transistors, and hence their dissipation, package temperature, and vulnerability to air currents. While it has been speculated that an enormously extended LF range benefits reproduction by reducing phase distortion at the bottom of the audio spectrum, there seems to be no hard evidence for this, and in practical terms there is no real incentive to extend the LF bandwidth greatly beyond what is actually necessary.

Non-inverting integrators

The obvious way to build a non-inverting integrator is to use a standard inverting integrator followed by an inverter. The first op-amp must have good DC accuracy as it is here the the amplifier DC level is compared with 0 V. The second op-amp is wholly inside the servo loop so its DC accuracy is not important. This arrangement is shown in Figure 14.3. It is not a popular approach because it is perfectly possible to make a non-inverting integrator with one op-amp. It does however have the advantage of being conceptually simple; it is very easy to calculate. The frequency response of the integrator is needed to calculate the low-frequency response of the whole system.

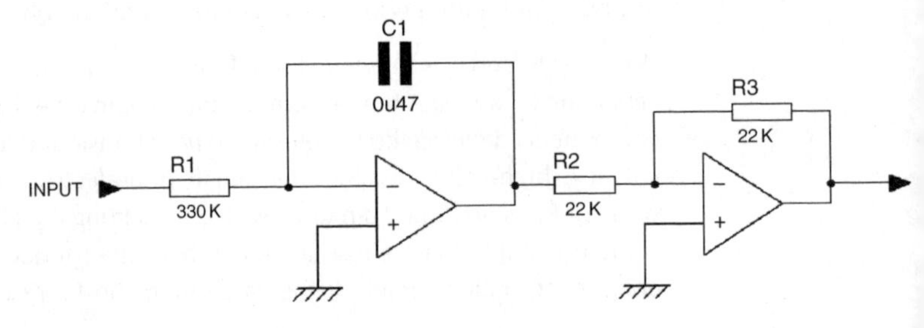

Figure 14.3
A conventional inverting integrator followed by an inverter

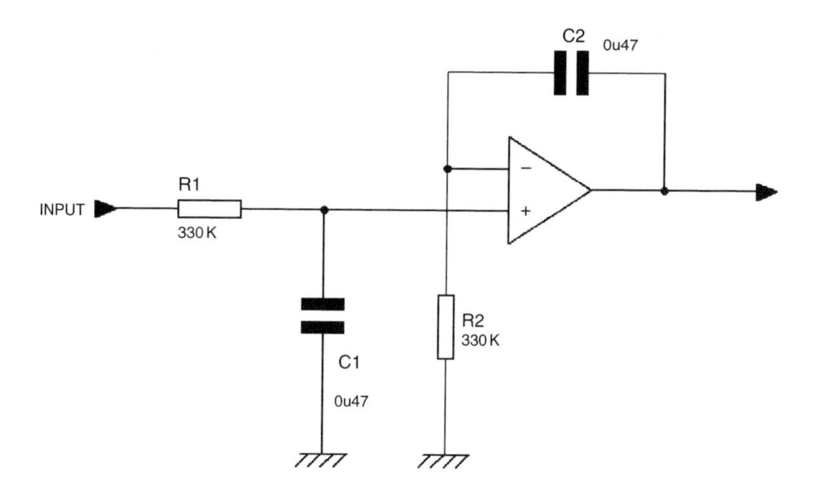

Figure 14.4
A non-inverting integrator that requires only one op-amp

The component values shown in Figure 14.3 give unity gain at 1 Hz.

Figure 14.4 shows a non-inverting integrator that has often been used in DC servo applications, having the great advantage of requiring one op-amp. It does however use two capacitors; if you are aiming for a really low roll these can become quite large for non-electrolytics and will be correspondingly expensive. Despite the presence of two RC time-constants, this circuit is still a simple integrator with a standard −6 dB/octave frequency response.

At the input is a simple RC lag, with the usual exponential time response to step changes; its deviation from being an integrator is compensated for by the RC lead network in the feedback network. A good question is what happens if the two RC time-constants are not identical; does the circuit go haywire? Fortunately not. A mismatch only causes gain errors at very low frequencies, and these are unlikely to be large enough to be a problem. An RC mismatch of ±10% leads to an error of ±0.3 dB at 1.0 Hz, and this error has almost reached its asymptote of ±0.8 dB at 0.1 Hz.

The frequency domain response of Figure 14.4 is:

$$A = \frac{1}{j\omega RC} \qquad \text{Equation 14.1}$$

where $\omega = 2\pi f$ exactly as for the simple integrator of Figure 14.3. The values shown give unity gain at 1 Hz.

Figure 14.5 displays a rather superior non-inverting integrator circuit that requires only one op-amp and one capacitor. How it works is by no means immediately obvious, but work it does. R1 and C1 form a simple lag circuit at the input. By itself, this naturally does not give the desired integrator response of a steadily rising or falling capacitor voltage as a result of a step input; instead it gives the familiar exponential response, because as the

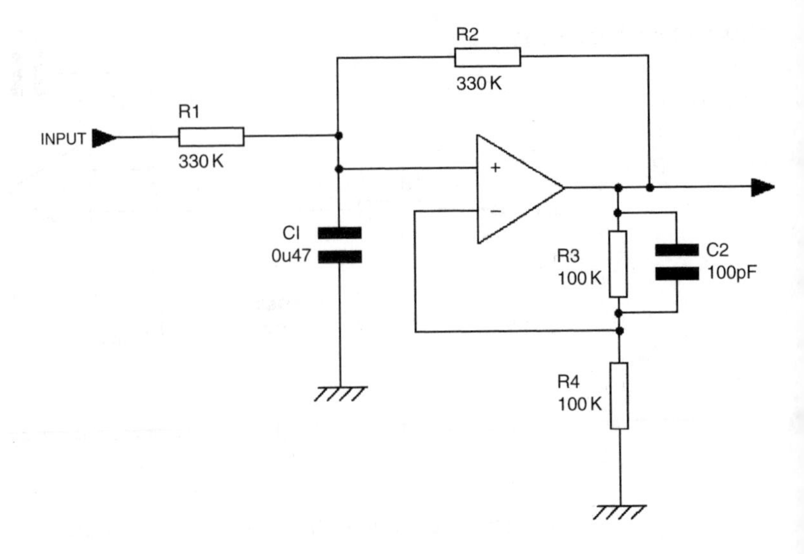

Figure 14.5
A non-inverting integrator that requires only one op-amp and one capacitor

capacitor voltage rises the voltage across R1 falls, and the rate of capacitor charging is reduced. In this circuit however, as the capacitor voltage rises the output of the op-amp rises at twice the rate, due to the gain set up by R3 and R4, and so the increasing current flowing into C1 through R2 exactly compensates for the decreasing current flowing through R1, and the voltage on C1 rises linearly, as though it were being charged from a constant current source. This is in fact the case, because the circuit can be viewed as equivalent to a Howland current source driving into a capacitor.

As for the previous circuit, doubts may be entertained as to what happens when the compensation is less than perfect. For example, here it depends on R1 and R2 being the same value, and also the equality of R3 and R4, to set a gain of exactly two. Normal circuit tolerances do not cause problems with its use as an amplifier servo.

Note that R3 and R4 can be high value resistors. Stray capacitances are dealt with by the addition of C2, which in most cases will be found to be essential for the HF stability of this configuration.

The frequency domain response is now different

$$A = \frac{1}{j\omega \frac{R}{2} C} \qquad \text{Equation 14.2}$$

$$R = R1 = R2$$

The R/2 term appears because C1 is now being charged through two equal resistors R1 and R2. The values shown therefore give unity gain at 2 Hz.

Choice of op-amps

All of these integrator circuits use high resistor values to keep the size of the capacitors down. It is essential to use FET-input op-amps, with their near-zero bias and offset currents. Bipolar op-amps have many fine properties, but they are not useful here. You will need a reasonably high quality FET op-amp to beat the non-servo amplifiers whose offset does not exceed ±15 mV offset at the output.

Here are some prime candidates, giving the maximum ± offset voltages, and the relative cost at the time of writing

Op-amp specs compared

Type	Offset at 25°C (mV)	Offset over −40 to +85°C (mV)	Relative cost
TL051	1.5	2.5	1.00
OPA134	2	3	1.34
AD711JN	2	3	1.48
OPA627AP	0.28	0.5	16.0

Note that the TL051 looks like quite a bargain, and going for a serious improvement on this with the OPA627AP will cost you deep in the purse.

Servo authority

The phrase "servo authority" refers to the amount of control that the DC servo system has over the output DC level of the amplifier. It is, I hope, clear that the correct approach is to design a good input stage that gives a reasonably small DC offset unaided, and then add the servo system to correct the last few dozen millivolts, rather than to throw together something that needs to be hauled into correct operation by brute-force servo action.

In the latter case, the servo must have high authority in order to do its job, and if the servo op-amp dies and its output hits one of its rails, the amplifier will follow suit. The DC offset protection should come into action to prevent disaster, but it is still an unhappy situation.

However, if the input stage is well-designed, so the servo is only called upon to make fine adjustments, it is possible to limit the servo authority, by proportioning the circuit values so that R3 in Figure 14.2 is relatively high. Then, even if the op-amp fails, the amplifier offset will be modest. In many cases it is possible for the amplifier to continue to function without any ill-effects on the loudspeakers. This might be valuable in sound reinforcement applications and the like.

Calculating the effects of op-amp failure in the circuit of Figure 14.2 is straightforward. The system appears as a shunt-feedback amplifier where

R3 is the input resistance and R1 is the feedback resistance. Thus if the op-amp is working from ±15 rails, then, ignoring saturation effects, the main amplifier output will be displaced by ±1.5 V.

When limiting servo authority, it is of course essential to allow enough adjustment to deal with any combination of component tolerances that may happen along. Do not limit it too much.

Design of LF roll-off point

Calculating the frequency response of the servo-controlled system is surprisingly easy. The −3 dB point will occur where the feedback through the normal network and the integrating servo path are equal in amplitude; it is −3 dB rather than −6 dB because the two signals are displaced in phase by 90°. This is exactly the same as the −3 dB point obtained with a RC circuit, which happens at the frequency where the impedance of the R and C are equal in magnitude, though displaced in phase by 90°.

As a first step, decide what overall gain is required; this sets the ratio of R1 and R2. Next determine how low R1 can conveniently be made, to minimise the noise contribution of R2. This establishes the actual values of R1 and R2. It is important to remember that the servo injection resistor R3, being connected to an effective AC ground at the servo op-amp output, is effectively in parallel with R2 and has a small influence on the main amplifier gain. Third, decide how low a −3 dB point you require for the overall system, and what servo authority you are prepared to allow. I shall take 0.2 Hz as an example, to demonstrate how a servo system makes such a low value easy to attain. Using the values shown in Figure 14.2, the section above demonstrates that the servo authority is more than enough to deal with any possible offset errors, while not being capable of igniting the loudspeakers if the worst happens. R3 is therefore 22 K, which is ten times R2, so at the −3 dB point the integrator output must be ten times the main amplifier output; in other words it must have a gain of ten at 0.2 Hz.

The next step is to choose the integrator type; the one op-amp, one capacitor version of Figure 14.5 is clearly the most economical so we will use that. The frequency response equation given above is then used to set suitable values for R1, R2 and C1 in Figure 14.5. Non-electrolytic capacitors of 470 nF are reasonably priced and this gives a value for R1, R2 of 338 kΩ; the preferred value of 330 K is quite near enough.

The final step is to check that the integrator will not be overdriven by the audio-frequency signals at the amplifier output, bearing in mind that the op-amp will be running off lower supply rails that are half or less of the main amplifier rail voltages; here I will assume the amplifier rails are ±45 V, i.e., three times the op-amp rails. Hence the integrator will clip with full amplifier output at the frequency where its gain is 1/3. The integrator we have just designed has a gain of ten times at 0.2 Hz and a slope of −6 dB per octave, so its gain will have fallen to unity at 2 Hz, and to 1/3 at 6 Hz. Hence the integrator can handle any amplifier output down to maximum power at 6 Hz, which is somewhat below the realm of audio, and all should be well.

One problem with servo designs of this type is that they are difficult to test; frequencies of 0.2 Hz and below are well outside the capabilities of normal audio test equipment. It is not too hard to find a function generator that will produce the range 0.1 to 1.0 Hz, but measuring levels to find the −3 dB point is difficult. A storage oscilloscope will give approximate results if you have one; the accuracy is not usually high.

One possibility is the time-honoured method of measuring the tilt on a low-frequency square wave. Accuracy is still limited, but you can use an ordinary oscilloscope. Even very low frequency roll-offs put an easily visible tilt on a 20 Hz square wave, and this should be fast enough to give reasonable synchronisation on a non-storage oscilloscope. Here is a rough guide:

Tilt on 20 Hz square wave with different LF roll-off frequencies

−3 dB point in Hz	Tilt %
0.15	2.5
0.23	3.5
0.32	5.0
0.50	7.4
0.70	10.5
1.0	15.2
1.4	20
2.1	28

Note that the tilt is expressed as a percentage of the zero to peak voltage, not peak-to-peak.

Performance issues

The advantages of using a DC servo have been listed above, without mentioning any disadvantages, apart from the obvious one that more parts are required. It could easily be imagined that another and serious drawback is that the presence of an op-amp in the negative feedback network of an amplifier could degrade both the noise and distortion performance. However this is not the case. When the system in Figure 14.2b is tested with a load-invariant amplifier, and an OPA134 op-amp as a servo there is no measurable effect on either quantity.

The distortion performance is unaffected because the servo integrator passes very little signal at audio frequencies. The noise performance is preserved because the integrators are very quiet due to their falling frequency response, and with the long integration constants used here they are working at a noise-gain of unity at audio frequencies. Both parameters benefit from the fact that the servo feedback path via R3 has one-tenth of the gain of the main feedback path through R1.

Multipole servos

All the servos shown above use an integrator and therefore have a single pole. It is possible to use servos that have more than one pole, and they have been used in some designs, though the motivation for doing it is somewhat unclear. The usual arrangement has a single-op-amp non-inverting integrator followed by a simple RC lag network that feeds into the feedback point. Naturally, once you have more than one pole in a system there is the possibility of an under-damped response and gain peaking, so this approach demands careful design, not least because measuring gain peaking at 0.1 Hz is not that easy.

15

Amplifier and loudspeaker protection

Categories of amplifier protection

The protection of solid-state amplifiers against overload is largely a matter of safeguarding them from load impedances that are too low and endanger the output devices; the most common and most severe condition being a short across the output. This must be distinguished from the casual use of the word *overload* to mean excessive signal that causes clipping and audible distortion.

Overload protection is not the only safety precaution required. An equally vital requirement is DC-offset protection – though here it is the loudspeaker load that is being protected from the amplifier, rather than the other way round.

Similarly, thermal protection is also required for a fully equipped amplifier. Since a well-designed product will not overheat in normal operation, this is required to deal with two abnormal conditions:

1 The amplifier heatsinking is designed to be adequate for the reproduction of speech and music (which has a high peak-to-volume ratio, and therefore brings about relatively small dissipation) but cannot sustain long-term sinewave testing into the minimum specified load impedance without excessive junction temperatures. Heatsinking forms a large part of the cost of any amplifier, and so economics makes this a common state of affairs.

   Similar considerations apply to the rating of amplifier mains transformers, which are often designed to indefinitely supply only 70% of the current required for extended sinewave operation. Some form of thermal cut-out in the transformer itself then becomes essential (see Chapter 8).

2 The amplifier is designed to withstand indefinite sinewave testing, but is vulnerable to having ventilation slots, etc. blocked, interfering either with natural convection or fan operation.

Finally, all amplifiers require internal fusing to minimise the consequences of a component failure – i.e., protecting the amplifier from itself – and to maintain safety in the event of a mains wiring fault.

Semiconductor failure modes

Solid-state output devices have several main failure modes, including excess current, excess power dissipation, and excess voltage. These are specified in manufacturer's data sheets as Absolute Maximum Ratings, usually defined by some form of words such as *exceeding these ratings even momentarily may cause degradation of performance and/or reduction in operating lifetime*. For semiconductor power devices ratings are usually plotted as a Safe Operating Area (SOA) which encloses all the permissible combinations of voltage and current. Sometimes there are extra little areas, notably those associated with second-breakdown in BJTs, with time limits (usually in microseconds) on how long you can linger there before something awful happens.

It is of course also possible to damage the base-emitter junction of a BJT by exceeding its current or reverse voltage ratings, but this is unlikely in power amplifier applications. In contrast the insulated gate of an FET is more vulnerable and zener clamping of gate to source is usually considered mandatory, especially since FET amplifiers often have separate higher supply-rails for their small-signal sections.

BJTs have an additional important failure mode known as second breakdown, which basically appears as a reduction in permissible power dissipation at high voltages, due to local instability in current flow. The details of this mechanism may be found in any textbook on transistor physics.

Excessive current usually causes failure when the I^2R heating in the bond wires becomes too great and they fuse. This places a maximum on the current-handling of the device no matter how low the voltage across it, and hence the power dissipation. In a TO3 package only the emitter bond wire is vulnerable, as the collector connection is made through the transistor substrate and flange. If this wire fails with high excess current then on some occasions the jet of vaporised metal will drill a neat hole through the top of the TO3 can – an event which can prove utterly mystifying to those not in the know.

Any solid-state device will fail from excess dissipation, as the internal heating will raise the junction temperatures to levels where permanent degradation occurs.

Excess emitter-collector or source-drain voltage will also cause failure. This failure mode does not usually require protection as such, because designing against it should be fairly easy. With a resistive load the maximum voltage is defined by the power supply-rails, and when the amplifier output is hard against one rail the voltage across the device that is turned off will be the sum of the two rails, assuming a DC-coupled design. If devices with a Vce(max.) greater than this is selected there should be no possibility of failure. However, practical amplifiers will be faced with reactive load impedances, and this can double the Vce seen by the output devices. It is therefore necessary to select a device that can withstand at least twice the sum of the HT rail voltages, and allow for a further safety margin on top of this. Even greater voltages may be generated by abrupt current changes in inductive loads, and these may go outside the supply-rail range causing device failure by reverse biasing. This possibility is usually dealt with by the addition of *catching* diodes to the circuit (see below) and does not in itself affect the output device specification required.

Power semiconductors have another failure mode initiated by repeated severe temperature changes. This is usually known as *thermal cycling* and results from stresses set up in the silicon by the differing expansion coefficients of the device chip and the header it is bonded to. This constitutes the only real wearout mechanism that semiconductors are subject to. The average lifetimes of a device subjected to temperature variations ΔT can be approximately predicted by

$$N = 10^7 \cdot e^{-0.05 \cdot \Delta T} \qquad \text{Equation 15.1}$$

where N = cycles to failure, and ΔT is the temperature change.

This shows clearly that the only way open to the designer to minimise the risk of failure is to reduce the temperature range or the number of temperature cycles. Reducing the junction temperature range requires increasing heatsink size or improving the thermal coupling to it. Thermal coupling can be quickly improved by using high-efficiency thermal washers, assuming their increased fragility is acceptable in production, and this is much more cost-effective than increasing the weight of heatsink. The number of cycles can only be minimised by leaving equipment (such as Class-A amplifiers) powered long-term, which has distinct disadvantages in terms of energy consumption and possibly safety.

Overload protection

Solid-state output devices are much less tolerant to overload conditions than valves, and often fail virtually instantaneously. Some failure modes (such as overheating) take place slowly enough for human intervention, but this can never be relied upon. Overload protection is therefore always an important issue, except for specialised applications such as amplifiers built into powered loudspeakers, where there are no external connections and no possibility of inadvertent short-circuits.

Driven by necessity, workable protection systems were devised relatively early in the history of solid-state amplifiers; see Bailey[1], Becker[2] and Motorola[3]. Part of the problem is defining what constitutes adequate current delivery into a load. Otala[4] has shown that a complex impedance, i.e., containing energy-storage elements, can be made to draw surprisingly large currents if specially optimised pulse waveforms are used that catch the load at the worst part of the cycle; however it seems to be the general view that such waveforms rarely if ever occur in real life.

Verifying that overload protection works as intended over the wide range of voltages, currents, and load impedances possible is not a light task. Peter Baxandall introduced a most ingenious method of causing an amplifier to plot its own limiting lines[5].

Overload protection by fuses

The use of fuses in series with the output line for overload protection is no longer considered acceptable, as it is virtually impossible to design a fuse that will blow fast enough to protect a semiconductor device, and yet be sufficiently resistant to transients and turn-on surges. There are also the obvious objections that the fuse must be replaced every time the protection is brought into action, and there is every chance it will be replaced by a higher value fuse which will leave the amplifier completely vulnerable. Fuses can react only to the current flowing through them, and are unable to take account of other important factors such as the voltage drop across the device protected.

Series output fuses are sometimes advocated as a cheap means of DC offset protection, but they are not dependable in this role.

Placing a fuse in series with the output will cause low-frequency distortion due to cyclic thermal changes in the fuse resistance. The distortion problem can, in theory at least, be side-stepped by placing the fuse inside the global feedback loop; however what will the amplifier do when its feedback is abruptly removed when the fuse blows? (See also page 411 on DC offset protection below.)

One way of so enclosing fuses that I have seen advocated is to use them instead of output emitter-resistors Re; I have no personal experience of this technique, but since it appears to add extra time-dependent thermal uncertainties (due to the exact fuse resistance being dependant upon its immediate thermal history) to a part of the amplifier where they already cause major difficulties, I do not see this as a promising path to take. There is the major difficulty that the failure of only one fuse will generate a maximal DC offset, so we may have dealt with the overload, but there is now a major DC offset to protect the loudspeaker from. The other fuse may blow as a consequence of the large DC current flow, but sizing a fuse to protect properly against both overload and DC offset may prove impossible.

Amplifier circuitry should always include fuses in each HT line. These are not intended to protect the output devices, but to minimise the damage when the output devices have already failed. They can and should therefore be of the slow-blow type, and rated with a good safety margin, so that they are entirely reliable; a fuse operated anywhere near its nominal fusing current has a short life-time, due to heating and oxidation of the fuse wire. HT fuses cannot save the output devices, but they do protect the HT wiring and the bridge rectifier, and prevent fire. There should be separate DC fuses for each channel, as this gives better protection than one fuse of twice the size, and allows one channel to keep working in an emergency.

Similarly, the mains transformer secondaries should also be fused. If this is omitted, a failure of the rectifier will inevitably cause the mains transformer to burn out, and this could produce a safety hazard. The secondary fuses can be very conservatively rated to make them reliable, as the mains transformer should be able to withstand a very large fault current for a short time. The fuses must be of the slow-blow type to withstand the current surge into the reservoir capacitors at switch-on.

The final fuse to consider is the mains fuse. The two functions of this are to disconnect the live line if it becomes shorted to chassis, and to protect against gross faults such as a short between live and neutral. This fuse must also be of the slow-blow type, to cope with the transformer turn-on current surge as well as charging the reservoirs. In the UK, there will be an additional fuse in the moulded mains plug. This does not apply to mains connectors in other countries and so a mains fuse built into the amplifier itself is absolutely essential.

Electronic overload protection

There are various approaches possible to overload protection. The commonest form (called electronic protection here to distinguish it from fuse methods) uses transistors to detect the current and voltage conditions in the output devices, and shunts away the base drive from the latter when the conditions become excessive. This is cheap and easy to implement (at least in principle) and since it is essentially a clamping method requires no resetting. Normal output is resumed as soon as the fault conditions are removed. The disadvantage is that a protection scheme that makes good use of the device SOA may allow substantial dissipation for as long as the fault persists undetected, and while this should not cause short-term failure if the protection has been correctly designed, the high temperatures generated may impair long-term reliability.

An alternative approach drops out the DC protection relay when overload is detected. The relay may either be opened for a few seconds delay, after which it resets, or stay latched open until the protection circuit is reset. This is normally done by cycling the mains power on and off, to avoid the expense of a reset button that would rarely be used.

If the equipment is essentially operated unattended, so that an overload condition may persist for some time, the self-resetting system will subject the output semiconductors to severe temperature changes, which may shorten their operational lifetime.

Plotting the protection locus

The standard method of representing the conditions experienced by output devices, of whatever technology, is to draw loadlines onto a diagram of the component's SOA, to determine where they cross the limits of the area. This is shown in Figure 15.1, for an amplifier with ±40 V HT rails, which would give 100 W into 8 Ω and 200 W into 4 Ω, ignoring losses; the power transistor is a Motorola MJ15024. You do not need to fix the HT voltage before drawing most of the diagram; the position of the SOA limits is fixed by the device characteristics. The line AB represents the maximum

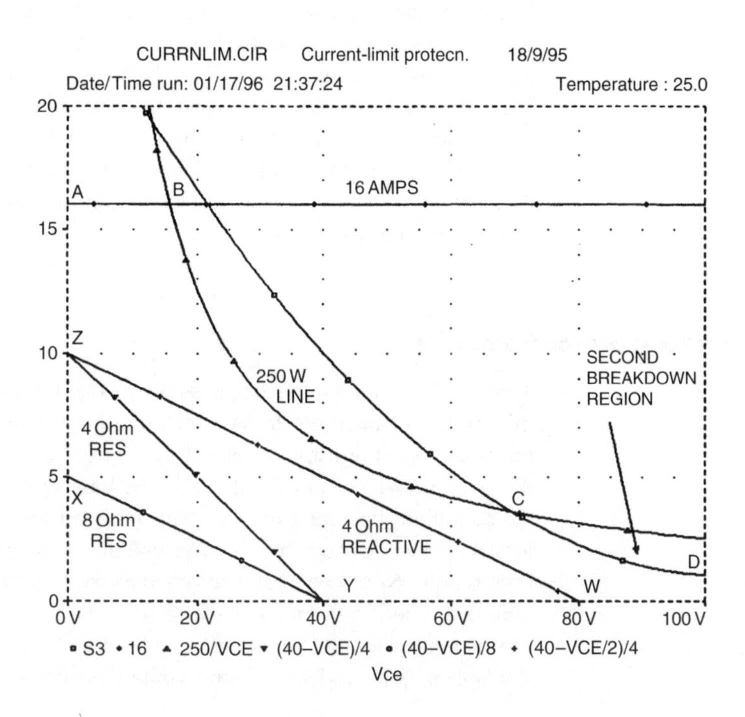

Figure 15.1 The Safe Operating Area (SOA) of a typical TO3 high-power transistor, in this case the Motorola MJ15024

current rating of 16 A, and the reciprocal curve BC the maximum power dissipation of 250 W. The maximum Vce is 250 V, and is far off the diagram to the right. Line CD defines the second-breakdown region, effectively an extra area removed from the high-voltage end of the power-limited region. Second-breakdown is an instability phenomenon that takes a little time to develop, so manufacturer's data often allows brief excursions into the region between the second-breakdown line and the power limit. The nearer these excursions go towards the power limit, the briefer they must be if the device is to survive, and trying to exploit this latitude in amplifiers is living dangerously, because the permitted times are very short (usually tens of microseconds) compared with the duration of audio waveforms.

The resistive loadline XY represents an 8Ω load, and as a point moves along it, the co-ordinates show the instantaneous voltage across the output device and the current through it. At point X, the current is maximal at 5.0 A with zero voltage across the device, as Vce(sats) and the like can be ignored without significant error. The power dissipated in the device is zero, and what matters is that point X is well below the current-limit line AB. This represents conditions at clipping.

At the other end, at Y, the loadline has hit the X-axis and so the device current is zero, with one rail voltage (40 V) across it. This represents the normal quiescent state of an amplifier, with zero volts at its output, and zero device dissipation once more. So long as Y is well to the left of the maximum-voltage line all is well. Note that while you do not need to decide the HT voltage when drawing the SOA for the device, you must do so before the loadlines are drawn, as all lines for purely resistive loads intersect the X-axis at a voltage representing one of the HT rails.

Intermediate points along XY represent instantaneous output voltages between 0 V and clipping; voltage and current co-exist and so there is significant device dissipation. If the line cuts the maximum-power rating curve BC, the dissipation is too great and the device will fail.

Different load resistances are represented by lines of differing slope; ZY is for a 4 Ω load. The point Y must be common to both lines, for the current is zero and the rail voltage unchanged no matter what load is connected to a quiescent amplifier. Point Z is however at twice the current, and there is clearly a greater chance of this low-resistance line intersecting the power limit BC. Resistive loads cannot reach the second-breakdown region with these rails.

Unwelcome complications are presented by reactive loading. Maximum current no longer coincides with the maximum voltage, and vice-versa. A typical reactive load turns the line XY into an ellipse, which gets much nearer to the SOA limit. The width (actually the minor axis, to be mathematical) of the ellipse is determined by the amount of reactance involved, and since this is another independent variable, the diagram could soon become over-complex. The solution is to take the worst-case for all possible reactive loads of the form R + jX, and instead of trying to draw hundreds of ellipses, to simply show the envelope made up of all their closest approaches to the SOA limit. This is another straight line, drawn from the same maximum current point Z to a point W at twice the rail voltage. There is clearly a much greater chance that the ZW line will hit the power-limit or second-breakdown lines than the 4 Ω resistive line ZY, and the power devices must have an SOA large enough to give a clear safety margin between its boundary and the reactive envelope line for the lowest rated load impedance. The protection locus must fit into this gap, so it must be large enough to allow for circuit tolerances.

The final step is plot the protection locus on the diagram. This locus, which may be a straight line, a series of lines, or an arbitrary curve, represents the maximum possible combinations of current and voltage that the protection circuitry permits to exist in the output device. Most amplifiers use some form of VI limiting, in which the permitted current reduces as the voltage across the device increases, putting a rough limit on device power dissipation. When this relationship between current and voltage is plotted, it forms the protection locus.

This locus must always be above and to the right of the reactive envelope line for the lowest rated load, or the power output will be restricted by the protection circuitry operating prematurely. It must also always be to the left and below the SOA limit, or it will allow forbidden combinations of voltage and current that will cause device failure.

Simple current-limiting

The simplest form of overload protection is shown in Figure 15.2, with both upper and lower sections shown. For positive output excursions, R1 samples the voltage drop across emitter-resistor Re1, and when it exceeds the Vbe of approximately 0.6 V, TR1 conducts and shunts current away from TR2 base. The component values in Figure 15.2 give a 5.5 A constant-current regime as shown in Figure 15.3, which was simulated using a model like Figure 15.8 below. The loadlines shown represent 8 Ω and 4 Ω resistive, and 4 Ω worst-case reactive (ZW). The current-limit line is exactly horizontal, though it would probably show a slight slope if the simulation was extended to include more of the real amplifier, such as real current sources, etc.

The value of Re1 is usually determined by the requirements of efficiency or quiescent stability, and so the threshold of current-limiting is set by R1 and R2. This circuit can only operate at a finite speed, and so R1 must be large enough to limit TR1 base current to a safe value. 100 Ω seems

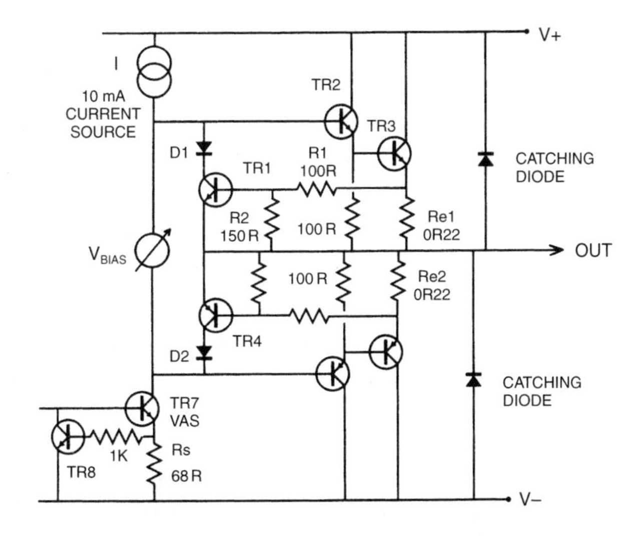

Figure 15.2
Simple current-limit circuit

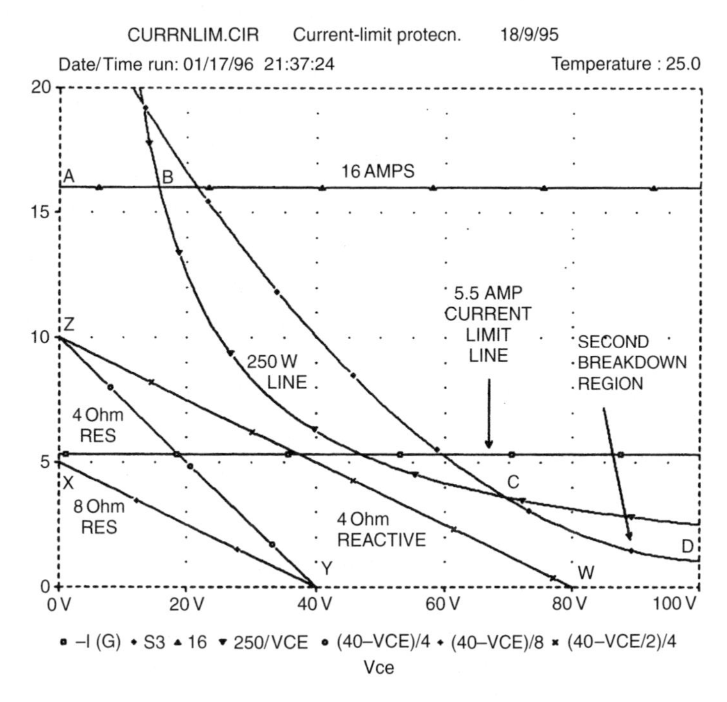

Figure 15.3
Current-limiting with ±40 V HT rails

sufficient in practice. Re1 is usually the output emitter resistor, as well as current sensor, and so does double duty.

The current drawn by TR1 in shunting away TR2 base drive is inherently limited by I, the constant-current load of the VAS. There is no such limit on TR4, which can draw large and indeterminate currents through VAS transistor TR7. If this is a TO-92 device it will probably fail. It is therefore essential to limit the VAS current in some way, and a common approach is shown in Figure 15.2. There is now a secondary layer of current-limiting, with TR8 protecting TR7 in the same way that TR1 protects TR2, 3. The addition of Rs to sense the VAS current does not significantly affect VAS operation, and does not constitute local negative feedback. This is because the input to TR7 is a current from the input stage, and not a voltage; the development of a voltage across Rs does not affect the value of this current, as it is effectively being supplied from a constant-current source.

It has to be faced that this arrangement often shows signs of HF instability when current limiting, and this can prove difficult or impossible to eradicate completely. (This applies to single and double-slope VI limiting also.) The basic cause appears to be that under limiting conditions there are two feedback systems active, each opposing the other. The global voltage feedback is attempting to bring the output to the demanded voltage level, while the overload protection must be able to override this to safeguard the output devices. HF oscillation is often a danger to BJT output devices, but in this case it does not seem to adversely affect survivability. Extensive tests have shown that in a conventional BJT output stage, the oscillation seems to reduce rather than increase the average current through the output devices, and it is arguable that it does more good than harm. It has to be said, however, that the exact oscillation mechanism remains obscure despite several investigations, and the state of our knowledge in this area is far from complete.

The diodes D1, D2 in the collectors of TR1, TR4 prevent them conducting in the wrong half cycle if the Re voltage drops are large enough to make the collector voltage go negative. Under some circumstances you may be able to omit them, but the cost saving is negligible.

The *loadline* for an output short-circuit on the SOA plot is a vertical line, starting upwards from Y, the HT rail voltage on the X-axis, and representing that current increases indefinitely without any reduction of the voltage drop across the output devices. An example is shown in Figure 15.3 for ±40 V rails. When the short-circuit line is prolonged upwards it hits the 5.5 A limiting locus at 40 V and 5.5 A; at 220 W this is just inside the power-limit section of the SOA. The devices are therefore safe against short-circuits; however the 4 Ω resistive loadline also intersects the 5.5 A line, at Vce = 18 V and Ic = 5.5 A, limiting the 4 Ω output capability to 12 V peak. This gives 18 W rather than 200 W in the load, despite the fact that full 4 Ω output would in fact be perfectly safe. The full 8 Ω output of 100 W is possible as the whole of XY lies below 5.5 A.

With 4 Ω reactive loads the situation is worse. The line ZW cuts the 5.5 A line at 38 V, leaving only 2 V for output, and limiting the power to a feeble 0.5 W.

The other drawback of constant current protection is that if the HT rails were increased only slightly, to ±46 V, the intersection of a vertical line from Y the X-axis centre would hit the power-limit line, and the amplifier would no longer be short-circuit proof unless the current limit was reduced.

Single-slope VI limiting

Simple current-limiting makes very poor use of the device SOA; single-slope VI limiting is greatly superior because it uses more of the available information to determine if the output devices are endangered. The Vce as well as the current is taken into account. The most popular circuit arrangement is seen in Figure 15.4, where R3 has been added to reduce the current-limit threshold as Vce increases. This simple summation of voltage and current seems crude at first sight, but Figure 15.5 shows it to be an enormous improvement over simply limiting the current.

The protection locus has now a variable slope, making it much easier to fit between reactive load lines and the SOA boundary; the slope is set by R3. In Figure 15.5, Locus 1 is for R3 = 15 k, and Locus 2 for 10 k. If Locus 2 is chosen the short-circuit current is reduced to 2 A, while still allowing the full 4 Ω resistive output.

Current capability at Vce = 20 V is increased from 5.5 A to 7.5 A.

Dual-slope VI limiting

The motivation for more complex forms of protection than single-slope VI limiting is usually the saving of money, by exploiting more of the output device SOA. In a typical amplifier required to give 165 W into 8 Ω and 250 W into 4 Ω (assuming realistic losses) the number of device pairs in the output stage can be reduced from three to two by the use of dual-slope protection, and the cost saving is significant. The single-slope limiting line

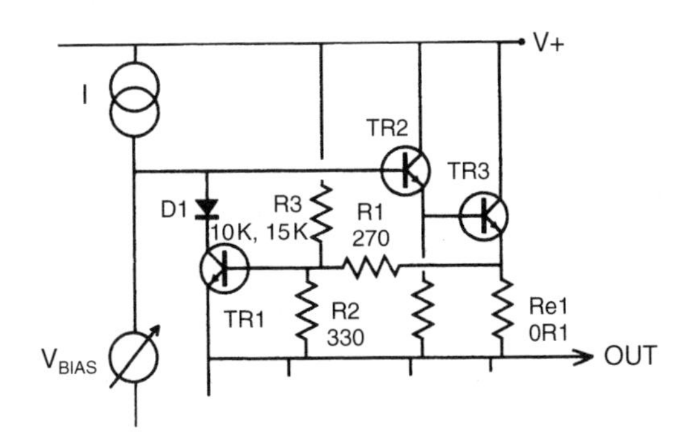

Figure 15.4
Single-slope VI limiter circuit

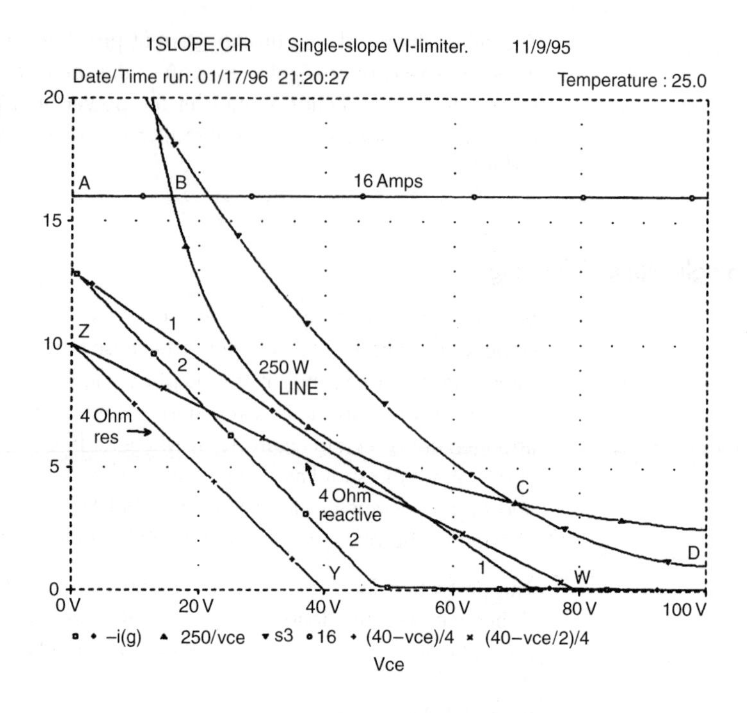

Figure 15.5
Single-slope locus plotted on MJ15024 SOA

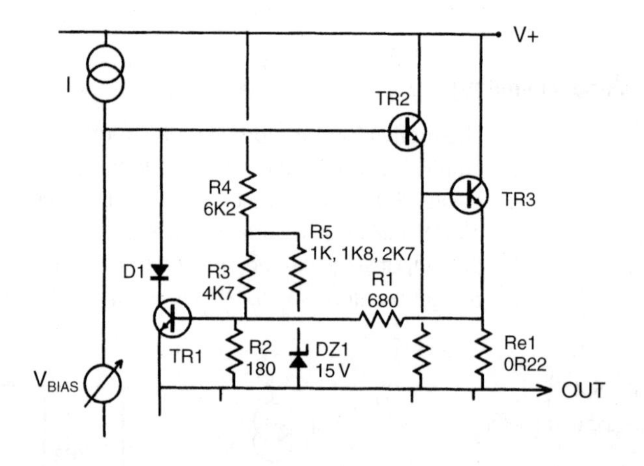

Figure 15.6
Dual-slope VI limiter circuit

is made dual-slope by introducing a breakpoint in the locus so it is made of two straight-line sections as in Figure 15.7, allowing it to be moved closer to the curved SOA limit; the current delivery possible at low device voltages is further increased.

A dual-slope system is shown in Figure 15.6. The action of the Vce component on sensing transistor TR1 is reduced when Vce is high enough

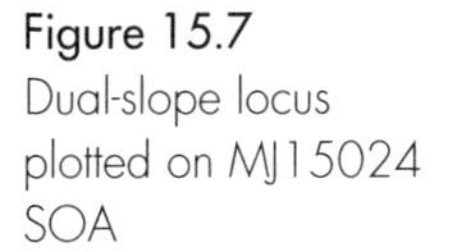

Figure 15.7
Dual-slope locus plotted on MJ15024 SOA

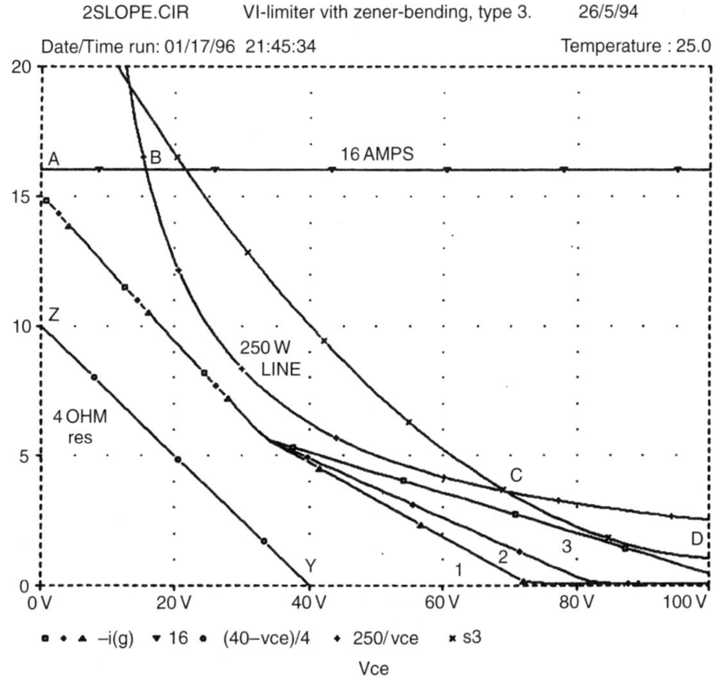

for Zener diode DZ1 to conduct. The series combination of R4 and R1 is chosen to give the required initial slope with low Vce (i.e., the left-hand slope) but as the voltage increases the Zener conducts and diverts current through R5, whose value controls the right-hand slope of the protection locus. Locii 1, 2 and 3 are for R5 = 2 k7, 1 k8 and 1 k, respectively.

Current capability at Vce = 20 V is further increased from 7.5 A to 9.5 A.

Simulating overload protection systems

The calculations for protection circuitry can be time-consuming. Simulation is quicker; Figure 15.8, shows a conceptual model of a dual-slope VI limiter, which allows the simulated protection locus to be directly compared with the loadline and the SOA. The amplifier output stage is reduced to one half (the positive or upper half) by assuming symmetry, and the combination of the actual output device and the load represented by voltage-controlled current-source G. The output current from controlled-source G is the same as the output device current in reality, and passes through current-sense resistor Re1.

The 6 mA current-source I models the current from the previous stage that TR1 must shunt away from the output device. Usually this is an accurate model because the VAS collector load will indeed be a current-source.

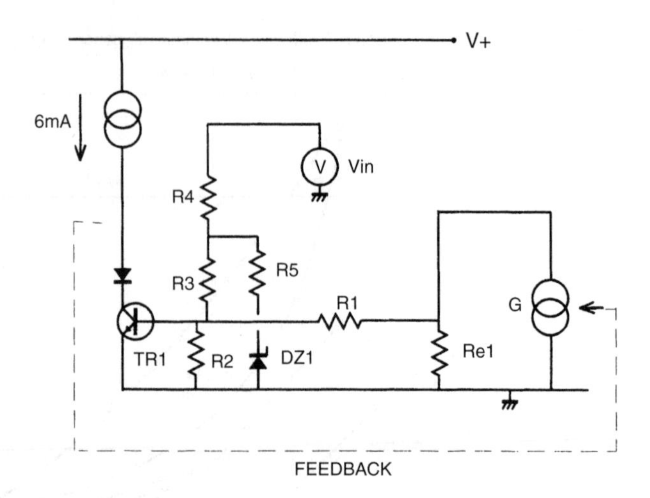

Figure 15.8
A conceptual model of an overload protection circuit that implements dual-slope limiting

The feedback loop is closed by making the voltage at the collector of TR1 control the current flowing through G and hence Re1.

In this version of VI-protection the device voltage is sensed by R4 and the current thus engendered is added to that from R1 at the base of TR1. This may seem a crude way of approximating a constant power curve, and indeed it is, but it provides very effective protection for low and medium-powered amplifiers.

Vin models the positive supply-rail, and exercises the simulation through the possible output voltage range. In reality the emitter of TR1 and Re1 would be connected to the amplifier output, which would be move up and down to vary the voltage across the output devices, and hence the voltage applied across R1, R2. Here it is easier to alter the voltage source V, as the only part of the circuit connected to HT+. V+ is fixed at a suitable HT voltage, e.g., +50 V.

The simulation only produces the protection locus, and the other lines making up the SOA plot are added at the display stage. Ic(max.) is drawn by plotting a constant to give a horizontal line at 16 A. P(max.) is drawn as a line of a constant power, by using the equation 250/Vce to give a 250 W line. In PSpice there seems to be no way to draw a strictly vertical line to represent Vce(max.), but in the case of the MJ15024 this is 250 V, and is for most practical purposes off the right-hand end of the graph anyway. The second-breakdown region is more difficult to show, for in the manufacturer's data the region is shown as bounded by a non-linear curve. The voltage/current co-ordinates of the boundary were read from manufacturer's data, and approximately modelled by fitting a second order polynomial. In this case it is:

$$I = 24.96 - 0.463 \,.\, Vce + 0.00224 \,.\, Vce^2 \qquad \text{Equation 15.2}$$

This is only valid for the portion that extends below the 250 W constant-power line, at the bottom right of the diagram.

Catching diodes

These are reverse-biased power diodes connected between the supply-rails and the output of the amplifier, to allow it to absorb transients generated by fast current-changes into an inductive load. All moving-coil loudspeakers present an inductive impedance over some frequencies.

When an amplifier attempts to rapidly change the current flowing in an inductive load, the inductance can generate voltage spikes that drive the amplifier output outside its HT rail voltages; in other words, if the HT voltage is ±50 V, then the output might be forced by the inductive back-EMF to 80 V or more, with the likelihood of failure of the reverse-biased output devices. Catching diodes prevent this by conducting and clamping the output so it cannot move more than about 1 V outside the HT rails. These diodes are presumably so-called because they *catch* the output line if it attempts to move outside the rails.

Diode current rating should be not less than 2 A, and the PIV 200 V or greater, and at least twice the sum of the HT rails. I usually specify 400 PIV 3 A diodes, and they never seem to fail.

DC-offset protection

In some respects, any DC-coupled power amplifier is an accident waiting to happen. If the amplifier suffers a fault that causes its output to sit at a significant distance away from ground, then a large current is likely to flow through the loudspeaker system. This may cause damage either by driving the loudspeaker cones beyond their mechanical limits or by causing excessive thermal dissipation in the voice-coils, the latter probably being the most likely. In either case the financial loss is likely to be serious. There is also a safety issue, in that overheating of voice-coils or crossover components could presumably cause a fire.

Since most power amplifiers consist of one global feedback loop, there are many possible component failures that could produce a DC offset at the output, and in most cases this will result in the output sitting at one of the HT rail voltages. The only way to save the loudspeaker system from damage is to remove this DC output as quickly as possible. The DC protection system must be functionally quite separate from the power amplifier itself or the same fault may disable both.

There are several possible ways to provide DC protection:

1 By fusing in the output line, the assumption being that a DC fault will give a sustained current flow that will blow the fuse when music-type current demands will not.
2 By means of a relay in the output line, which opens when a DC offset is detected.
3 By triggering a crowbar that shunts the output line to ground, and blows the HT fuses. The crowbar device is usually a triac, as the direction of offset to be dealt with is unpredictable.
4 By shutting down the power supply when a DC fault is detected. This can be done simply by an inhibit input if a switched-mode PSU is used. Conventional supplies are less easy.

DC protection by fuses

Fuses in series with the output line are sometimes recommended for DC offset protection, but their only merit is cheapness. It may be true that they have a slightly better chance of saving expensive loudspeakers than the HT fuses, but there are at least three snags:

- Selection of the correct fuse size is not at all easy. If the fuse rating is small and fast enough to provide some real loudspeaker protection, then it is likely to be liable to nuisance blowing on large bass transients. A good visual warning is given by behaviour of the fuse wire; if this can be seen sagging on transients, then it is going to fail sooner rather than later. At least one writer on DIY Class-A amplifiers gave up on the problem, and coolly left the tricky business of fuse selection to the constructor!
- Fuses running within sight of their nominal rated current generate distortion at LF due to cyclic changes in their resistance caused by I^2R heating; the THD would be expected to rise rapidly as frequency falls, and Greiner[6] states that harmonic and intermodulation distortion near the burn-out point can reach 4%. It should be possible to eradicate this by including the fuse inside the global feedback network, for the distortion will be generated at low frequencies where the feedback factor is at its greatest, but there are problems with amplifier behaviour after the fuse has blown.

In my tests, the distortion generated was fairly pure third harmonic. Figure 15.9 shows the THD measured before and after a T1A (slow-blow) fuse in series with an 8 Ω load at 25 W. Below 100 Hz the distortion completely swamps that produced by the amplifier, reaching 0.007% at 20 Hz. The distortion rises at rather less than 6 dB/octave as frequency falls. The fuse in this test is running close to its rating, as increasing the power to 30 W caused it to blow.

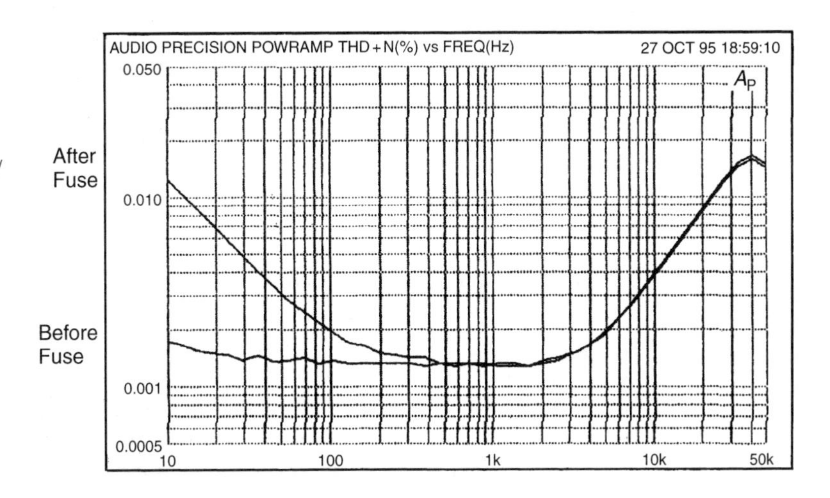

Figure 15.9 Fuse distortion. THD measured before and after the fuse at 25 W into 8 Ω

- Fuses obviously have significant resistance (otherwise they would not blow) so putting one in series with the output will degrade the theoretical damping factor. However, whether this is of any audible significance is very doubtful.

Note that the HT rail fuses, as opposed to fuses in the output line, are intended only to minimise amplifier damage in the event of output device failure. They must *not* be relied upon for speaker protection against DC offset faults. Often when one HT fuse is caused to blow the other also does so, but this cannot be relied upon, and obviously asymmetrical HT fuse blowing will in itself give rise to a large DC offset.

Relay protection and muting control

Relay protection against DC offsets has the merit that, given careful relay selection and control-circuitry design, it is virtually foolproof. The relay should be of the normally-open type so that if the protection fails it will be to a safe condition.

The first problem is to detect the fault condition as soon as possible. This is usually done by low-pass filtering the audio output, to remove all signal frequencies, before the resulting DC level is passed to a comparator that trips when a set threshold is exceeded. This is commonly in the range of 1–2 V, well outside any possible DC-offsets associated with normal operation; these will almost certainly be below 100 mV. Any low-pass filter must introduce some delay between the appearance of the DC fault and the comparator tripping, but with sensible design this will be too brief to endanger normal loudspeakers. There are other ways of tackling the fault-detection problem, for example by detecting when the global negative feedback has failed, but the filtering approach appears to be the simplest method and is generally satisfactory. First-order filtering seems to be quite adequate, though at first sight a second-order active filter would give a faster response time for the same discrimination against false-triggering on bass transients. In general there is much to be said for keeping protection circuitry as simple and reliable as possible.

Let us now examine DC offset detection circuitry in more detail. The problem falls neatly into two halves – distinguishing between acceptable large AC signals of up to 30 V rms or more, and DC offsets which may only be a volt or so before stern action is desired, and applying the result to a circuit which can detect both positive and negative transgressions. To perform the first task, relatively straightforward lowpass filtering is often adequate, but the bidirectional detection can tackled in many ways, and sometimes presents a few unexpected problems.

At this point we might consider how quickly the DC offset protection must operate to be effective. Clearly there will always be some delay, as we are discriminating against normal high-amplitude bass information, but otherwise the quicker the better if the loudspeaker is to be saved. My experience of deliberately setting fire to loudspeaker elements is limited (and I hasten to point out that I have so far never set fire to one accidentally) but here is one test I can report.

I once had the entertaining task of determining just how long a speaker element – the LF unit, obviously, as the tweeter was protected by the crossover from any DC – could sustain an amplifier DC fault. The tests, which were conducted outdoors to avoid triggering the fire alarms, showed that a well-designed and conservatively rated loudspeaker could be turned into smouldering potential landfill in less than a second. The loudspeaker unit in question was a high-quality LF unit with the relatively small diameter of 5 in., made by a respected manufacturer. The test involved applying +40 V to it, as if its accompanying amplifier had failed. The cone and voice-coil assembly shot out of the magnetic gap as if propelled by explosives, and then burst into flames in less than a second. All we could really conclude as the smoke cleared was that a second was way too long a reaction time for a protection system.

Filtering for DC Protection

A good DC protection filter is that which discriminates best between powerful low frequency signals and a genuine amplifier problem. It is easy to make the filter time-constant so long that it will never be false-triggered by a thumping great bass note, but then its time-domain response will be so slow that your precious loudspeakers will be history before the amplifier reacts to protect them.

The simplest possible filter is a single-pole circuit that requires only one RC time constant; in many cases this is quite good enough, but some more sophisticated approaches are also described here.

The Single RC Filter

The time-constant needs to be long enough to filter out the lowest frequency anticipated, at the full voltage output of the amplifier. The ability to sustain 10 Hz at the onset of clipping is usually adequate for audio, but if you are designing subsonic amplifiers to drive vibration tables, you will need to go a bit lower. Figure 15.10a shows the single-pole filter with typical values of 47 K and 47 µF that give a −3 dB point at 0.07 Hz. This is appropriate for low to medium amplifier powers, when feeding a later bidirectional detector that will trigger on an offset of the order of a Volt. The value of R1 is set by the current demands of this later stage – these can be significant, as we will see in the next section. The value of C1 is then determined by

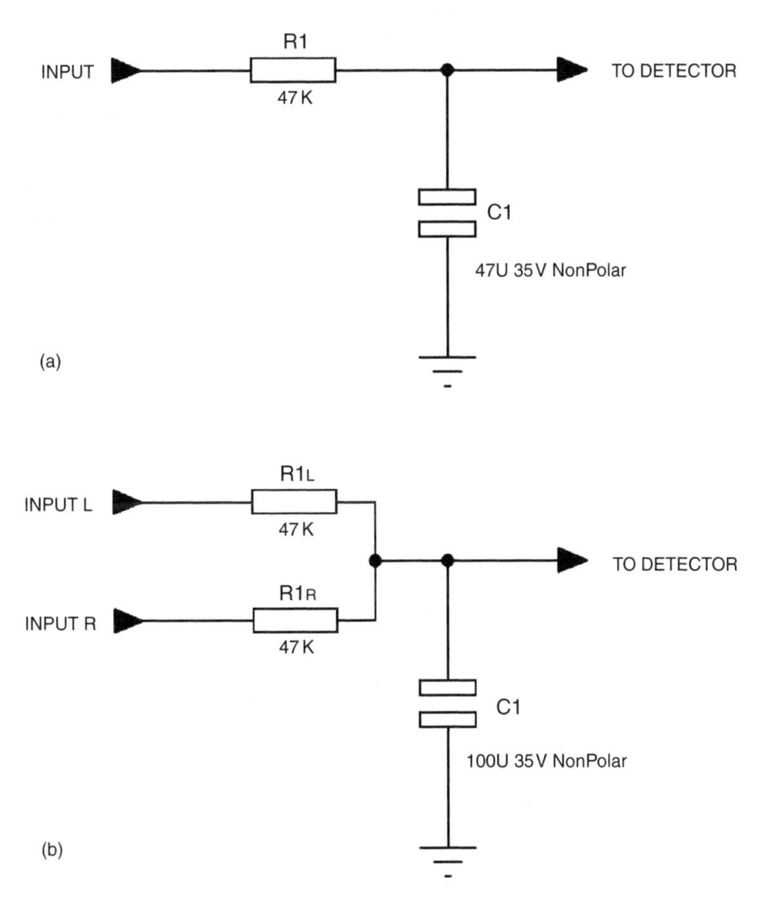

Figure 15.10
Mono and stereo single-pole filters for offset protection. The −3 dB point is 0.07 Hz

the required −3 dB frequency, and this means that it will be an electrolytic. It is important to remember at this point that DC offsets may arrive with either polarity, and may persist for long periods before someone notices there is a problem, so C1 needs to be either a non-polar electrolytic or constructed from two ordinary electrolytics connected back-to-back in the time-honoured fashion. Both methods are effective so it comes down to the fine details of the economics of component sourcing. Some amplifiers remove the supply from the power amplifier sections, so the offset does not persist, and this precaution may seem unnecessary; however there is no point in trying to save fractions of a penny by possibly compromising the reliability of something as important as the DC offset protection. C1 should have a voltage rating at least equal to the supply rails of the amplifier concerned.

The single-pole filter in Figure 15.10a is −3 dB at 0.07 Hz. To evaluate it, it was fed from a power amplifier giving 55 V peak, and the filter output connected to a bidirectional detector that had trip points at ±2.0 V. This setup triggered at 2.0 Hz when a 55 V peak signal starting at 50 Hz was slowly reduced in frequency. This corresponds to a filter attenuation of −28.8 dB at 2.0 Hz, and this frequency was used as the criterion for bass rejection thereafter. When a fault was simulated so the input to the filter shot up to +55 V, and stayed there, the detector gave a DC offset indication after 78 msec.

This circuit is easily adapted to stereo usage by having two resistors feeding into it, as in Figure 15.10b. If the resistors remain the same value, then the resistance seen by C is halved, and its capacitance must be doubled to maintain the same roll-off frequency. The incoming DC offset is also halved, so the detector sensitivity must be doubled if it is to trigger from the same level of offset on one of the stereo amplifier outputs. You could also object that a positive offset on one channel might be cancelled out by a negative offset on the other; this seems laughably unlikely until you recall that bridged amplifiers are driven with input signals that are in anti-phase, so a DC error in the drive circuitry could present just this situation. More sophisticated circuits provide two independent inputs that do not interact, avoiding this problem. More on this later, in the section on detectors.

The dual RC filter

The thinking behind the use of more complicated filtering is that a faster response roll-off will give better discrimination against high-amplitude bass events, so a higher −3 dB frequency can be used with (hopefully) a quicker response in the time domain.

The simplest method is to cascade two single-pole RC filters, as shown in Figure 15.11. This obviously gives a rather soggy roll-off, but has the merit of not introducing any more semiconductors that might fail. The

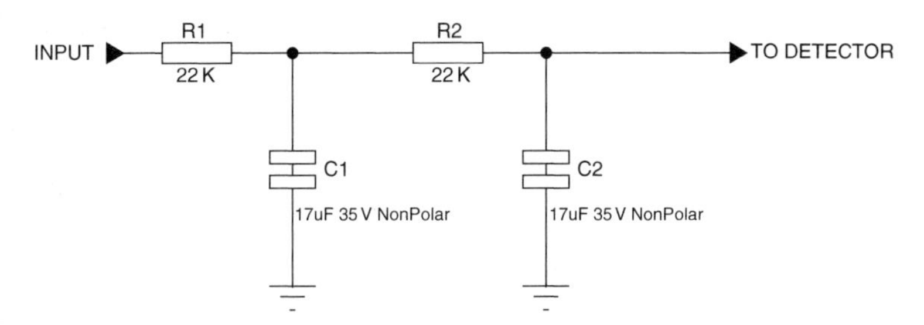

Figure 15.11
A dual RC filter for offset protection

non-standard capacitor values shown give the same attenuation of −28.8 dB at 2.0 Hz as the previous circuit. The only real snag to this scheme is that it does not work. The time to react was 114 msec, half as long again as the simple filter above. However I have seen it used in several designs, so you might come across it.

The second-order active filter

Some amplifier designs use an active filter to separate the bass from the breakdowns. This obviously allows a nice sharp roll-off, and gives the freedom to set the filter damping factors and so on. But does it deliver? I tested the circuit of Figure 15.12, a Sallen-and-Key configuration which with the values shown gives a second-order Butterworth (maximal flatness) characteristic, with a −3 dB point at 0.23 Hz; due to the increased filter slope the attenuation is once more −28.8 dB at 2.0 Hz. The reaction time is 109 msec, which is better than the dual RC filter but yet somewhat inferior to the single-pole filter of Figure 15.10a. Most disappointing. The Bessel filter characteristic is noted for a better response in the time domain, at the expense of a sharp roll-off, so I tried that. The component values in Figure 15.3 are now R1 = R2 = 35 K, C1 = 13.3 μF, and C2 = 10 μF. The reaction time is actually worse, at 131 msec, which was rather a surprise.

Building active filters usually means using op-amps. Putting an op-amp into the system creates a need for low voltage supplies within the power amplifier, which is highly inconvenient if they do not exist already. Most protection designs use discrete transistors throughout, and one of the advantages

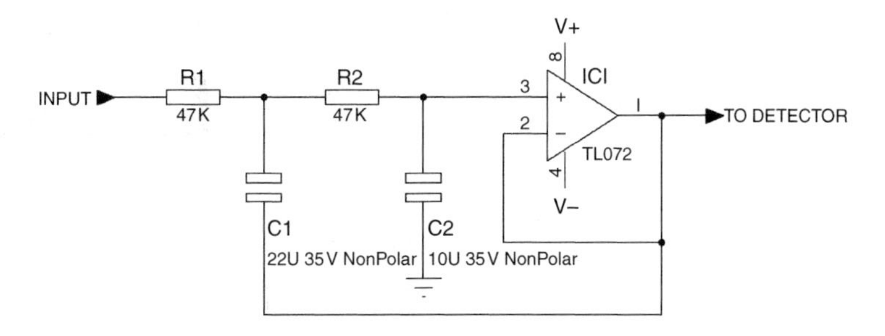

Figure 15.12
A second-order Sallen-and-Key filter for offset protection

of the Sallen-and-Key configuration is that it can be realised using a simple emitter-follower.

An important consideration is that op-amps have a limited common-mode voltage capability, and they will not appreciate having the full power amplifier supply rail applied to them directly. It will be necessary to scale down the incoming voltages and allow for this when setting the detector thresholds.

The conclusion seems inescapable that for once, the simplest circuit is the best; the single-pole filter is the way to go.

Bidirectional DC detection

There are many, many ways to construct circuits that will respond to both positive and negative signals of a defined level, and here some of the more common and more useful ones are examined.

The conventional two-transistor circuit

The circuit in Figure 15.13 is probably the most common approach to bidirectional detection. When the input exceeds ±0.6 V, Q1 turns on and the output voltage falls while Q2 stays off. When the input goes negative, Q2 operates in common-base mode, and conducts, Q1 remaining off as its base-emitter junction is reverse-biased. In either case currentis drawn through R10 and the output voltage drops to signal an offset. There is a certain elegance in the way that the conducting base-emitter junction protects its neighbour from excess reverse bias, but this circuit has one great disadvantage. Since Q2 operates in common-base mode, it has near unity current gain, as opposed to Q1, which is in common-emitter mode and therefore has current gain equal to the device beta.

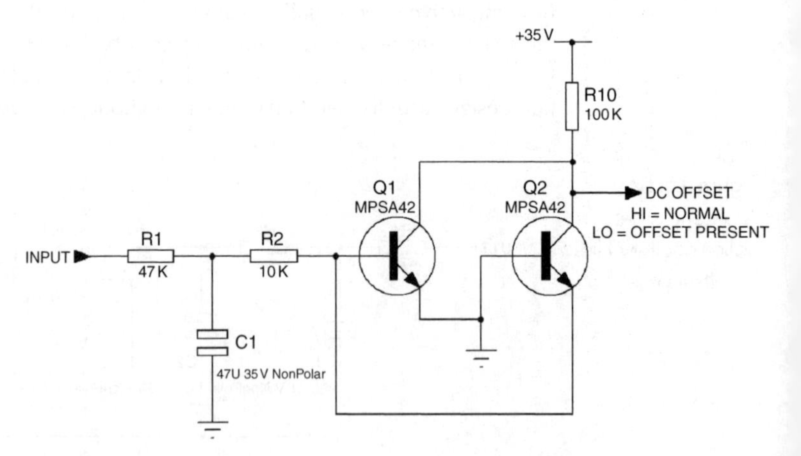

Figure 15.13
A common bidirectional detect circuit, giving very different thresholds for positive and negative inputs

This makes the two thresholds very asymmetrical. When the detector is driven from a single-pole RC filter with R1 = 47 K, the positive threshold is +1.05 V but the negative threshold is −5.5 V. To reduce this asymmetry R1 needs to be kept low, which leads to inconveniently large values of C1.

The one-transistor version

Figure 15.14 shows a variation on this theme, saving a transistor by adding diodes and resistors. With current component pricing the economic benefit is trivial, but it is still a circuit that has seen a great deal of use in Japanese amplifier designs. For positive inputs, D1, Q1 and D2 conduct. For negative inputs, D3 and Q1 conduct, the latter getting its base current through R2. As for the previous circuit, the current-gain differences between common-base and common-emitter modes of transistor operation gives asymmetrical thresholds; slightly less so because of the effect of D2 during positive inputs.

The differential detector

The interesting circuit of Figure 15.15, which has also seen use in Japanese hi-fi equipment, is based on a differential pair. This removes the objection to all the other circuits here, which is that it takes 0.6 V on the base to turn on a transistor directly, and so the detection thresholds will be that or more, due to extra diodes and so on. In this circuit the differential pair Q1,Q2 cancels out the 0.6 V Vbe-drop, and sensitivity can be much higher; under what conditions this is actually necessary is a moot point. There is no low pass filter as such; instead the same effect is achieved by high-pass filtering the signal, to remove DC and

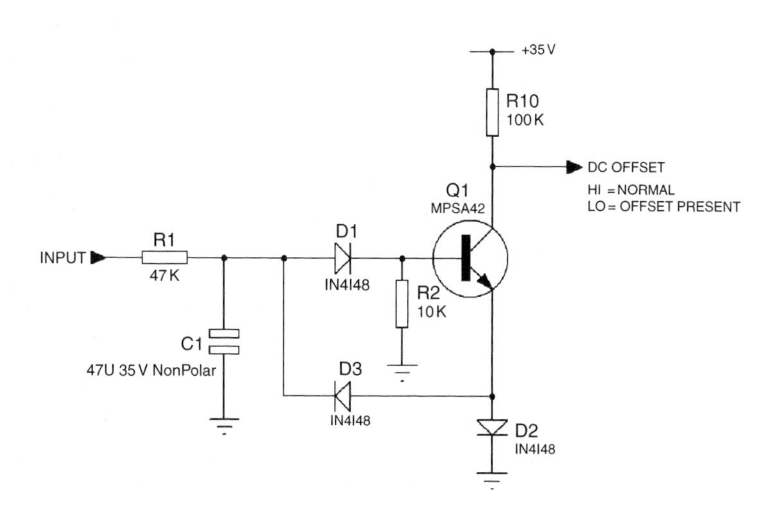

Figure 15.14 Another implementation of the same principle, saving a transistor but retaining the problem of asymmetrical thresholds

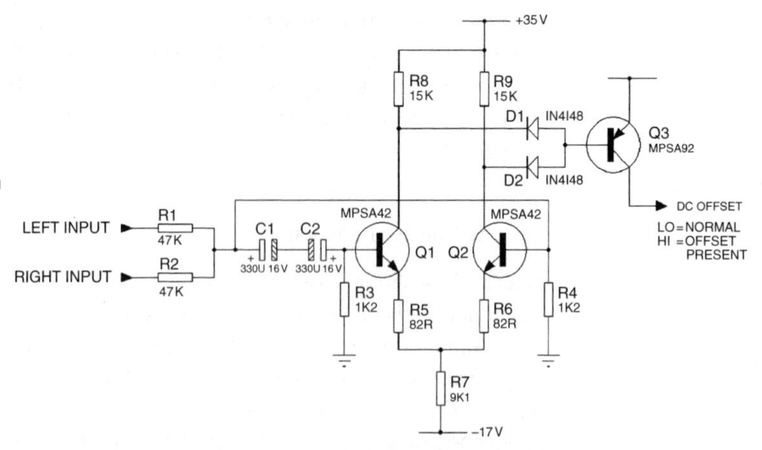

Figure 15.15
The differential detector, which can have very low thresholds. It uses a high-pass rather than a low-pass filter

VLF information. The result is then subtracted from the unfiltered signal by the differential pair, so only the DC and low frequency signals remain.

It works like this; for positive inputs Q2 turns on more and Q1 less, so the voltage on Q2 collector falls and Q3 is turned on via D2, and passes an offset signal to the rest of the system. For negative inputs Q1 turns on more and Q2 less, so the voltage on Q1 collector falls and Q3 is turned on via D1. The thresholds depend on the gain of the pair, set by the ratio of R5,R6 to R8,R9, and whatever voltage is set up on Q3 emitter. The circuit gives excellent threshold symmetry.

The self detector

Figure 15.16 shows my own version of a bidirectional detector. This has two advantages; it is symmetrical in its thresholds, and can be handily converted to a stereo or multichannel form without any loss of sensitivity. The only downside is that the thresholds are relatively high at about ±2.1 V with the component values shown. This is actually quite low enough to protect loudspeakers, and in any case, your typical serious amplifier fault smacks the output hard against the supply rails, and detecting this is not very hard. The exactness of the threshold symmetry depends on the properties of the transistors used, but is more than good enough to eliminate any problems. It works well with transistors such as MPSA42/MPS92 which are designed for high-voltage applications and therefore have low beta.

For positive inputs, D1, Q1, and Q2 conduct, with D4 supplying the base current for Q2. With a negative input, D2, Q2, and Q1 conduct, D3 now supplying base current for Q1. In each case there are two diode drops and two Vbe drops in series, which if each one was a nominal 0.6 V would

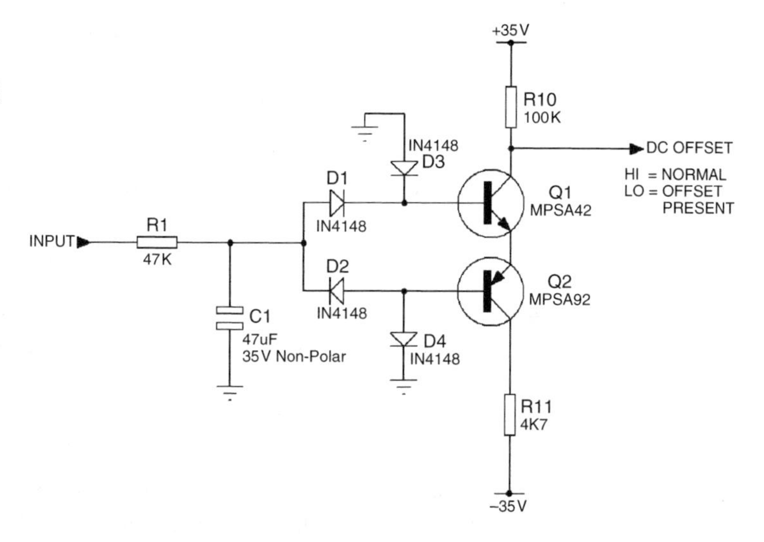

Figure 15.16
The Self detector. Good symmetry and easily expandable for more channels

give thresholds of ±2.4 V; in practice the diode supplying the much smaller base current has a lesser voltage across it, and the real thresholds come to ±2.1 V. Note that R11 is very definitely required to limit the current flowing through Q1, and Q2 when the input goes negative; R10 inherently limits it for positive inputs.

Figure 15.17 shows the stereo version, which uses separate filters for each channel, and two more diodes. The operation is exactly as before for each channel, and so the thresholds are unchanged. Equal-value positive and negative offsets on the two inputs do not cancel, and an offset is always clearly signalled.

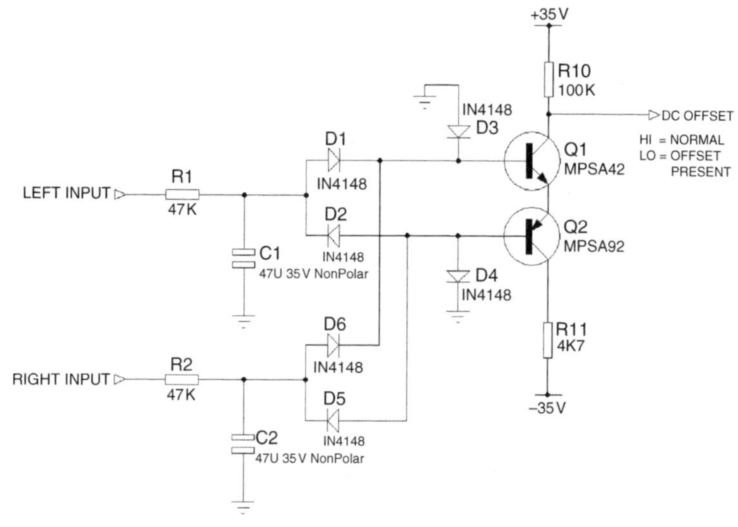

Figure 15.17
The stereo version of Figure 15.16

Having paid for a DC protection relay, it seems only sensible to use it for system muting as well, to prevent thuds and bangs from the upstream parts of the audio system from reaching the speakers at power-up and power down. Most power amplifiers, being dual-rail (i.e., DC-coupled) do not generate enormous thumps themselves, but they cannot be guaranteed to be completely silent, and will probably produce an audible turn-on thud.

An amplifier relay-control system should:

- Leave the relay de-energised when muted. At power-up, there should be a delay of at least 1 sec before the relay closes. This can be increased if required.
- Drop out the relay as fast as is possible at power-down, to stop the dying moans of the pre-amp, etc. from reaching the outside world.

  My preferred technique is a 2 msec (or thereabouts) timer which is held reset by the AC on the mains transformer secondary, except for a brief period around the AC zero-crossing, which is not long enough for the timer to trigger. When the incoming AC disappears, the near-continuous reset is removed, the timer fires, and the relay is dropped out within 10 msec. This will be long before the various reservoir capacitors in the system can begin to discharge. However, if the mains switch contacts are generating RF that is in turn reproduced as a click by the pre-amp, then even this method may not be fast enough to mute it.

- Drop out the relay as fast as is possible when a DC offset of more than 1–2 V, in either direction, is detected at the output of either power amp channel; the exact threshold is not critical. This is normally done by low-pass filtering the output (47 k and 47 μF works OK) and applying it to some sort of absolute-value circuit to detect offsets in either direction. The resulting signal is then OR-ed in some way with the muting signal mentioned above.
- Do not forget that the contacts of a relay have a much lower current rating for breaking DC rather than AC. This is an issue that does not seem to have attracted the attention it deserves.

A block diagram of a relay control system meeting the above requirements is shown in Figure 15.18, which includes over-temperature protection. Any of the three *inhibit* signals can override the turn-on delay and pull out the relay.

Distortion in output relays

Relays remain the only simple and effective method of disconnecting an amplifier from its load. The contacts can carry substantial currents, and it has been questioned whether they can introduce non-linearities.

Figure 15.18
Output relay control combining DC offset protection and power-on/off muting

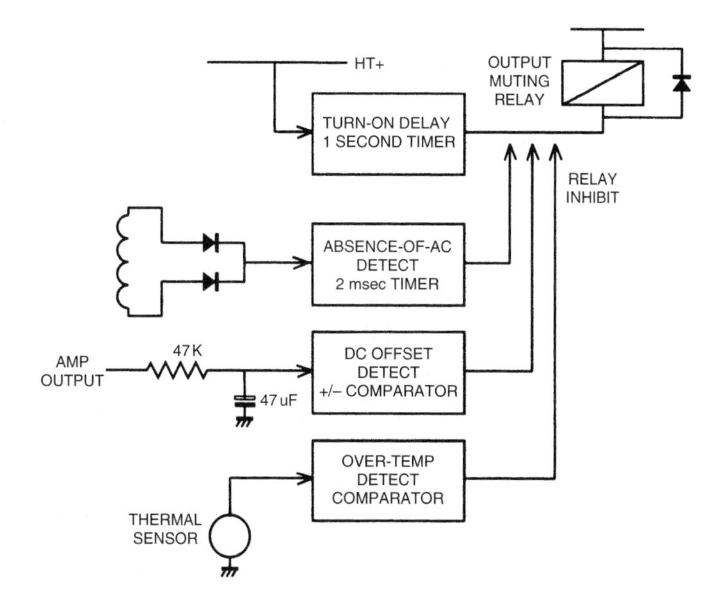

My experience is that silver-based contacts in good condition show effectively perfect linearity. Take a typical relay intended by its manufacturer for output-switching applications, with 'silver alloy' contacts – whatever that means – rated at 10 A. Figure 15.19 shows THD before and after the relay contacts while driving an 8 Ω load to 91 W, giving a current of 3.4 A rms. There is no significant difference; the only reason that the lines do not fall exactly on top of each other is because of the minor bias changes that

Figure 15.19
Demonstrating that relay contacts in themselves are completely distortion-free. Current through contacts was 3.4 A rms

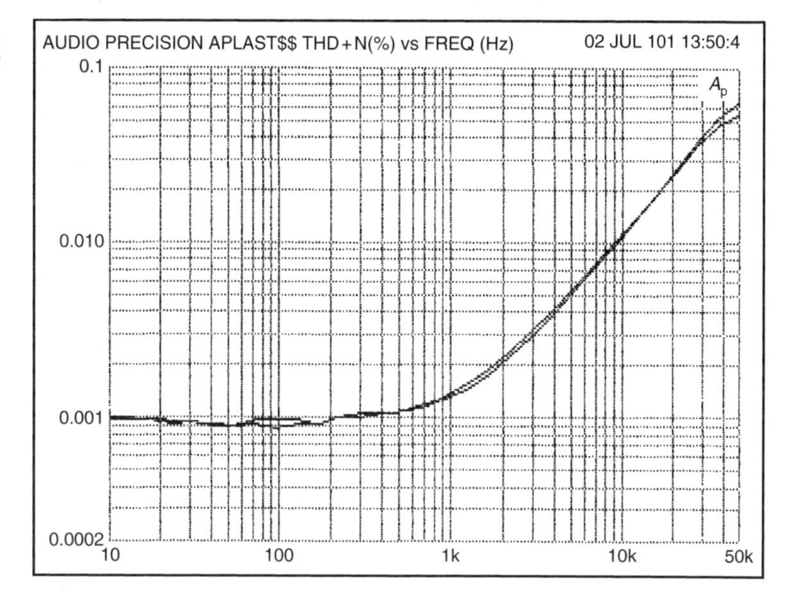

Class B is heir to. This apparently perfect linearity can be badly degraded if the contacts have been maltreated by allowing severe arcing – typically while trying and failing to break a severe DC fault.

Not everyone is convinced of this. If the contacts were non-linear for whatever reason, an effective way of dealing with it would be to include them in the amplifier feedback loop, as shown in Figure 15.20. R1 is the main feedback resistor, and R2 is a subsidiary feedback path that remains closed when the relay contacts open, and hopefully prevents the amplifier from going completely berserk. With the values shown the normal gain is 15.4 times, and with the contacts open it is 151 times. There is a feedback factor of about ten to linearise any relay problems.

The problem of course is that if there is to be a healthy amount of NFB wrapped around the relay contacts, R2 must be fairly high and so the closed-loop gain shoots up. If there is still an input signal, then the amplifier will be driven heavily into clipping. Some designs object to this, but even if the amplifier does not fail it is likely to accumulate various DC offsets on its internal time-constants as a result of heavy clipping, and these could cause unwanted noises when the relay contacts close again. One solution to this is a muting circuit at the amplifier input that removes the signal entirely and prevents clipping. This need not be a sophisticated circuit, as huge amounts of muting are not required; –40 dB should be enough. It must, however, pass the signal cleanly when not muting.

A much more insidious – unexpected – form of non-linearity can occur if the relay is constructed so that its frame makes up part of the switched electrical circuit as well as the magnetic circuit. (This is not the case with the audio application relay discussed above.) A relay frame is made of soft iron, to prevent it becoming permanently magnetised, and this appears to present a non-linear resistance to a loudspeaker level signal, presumably due to magnetisation and saturation of the material. (It should be said at once that this is described by the manufacturer as a 'power relay'

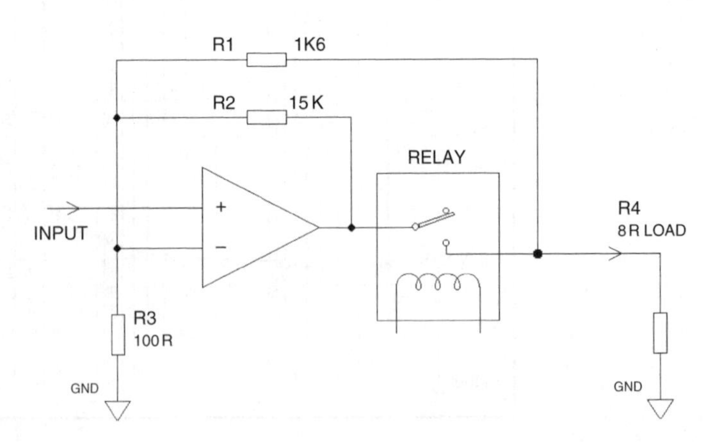

Figure 15.20 How to enclose relay contacts in the feedback loop. The gain shoots up when the relay contacts open, so muting the input signal is desirable

and is apparently not intended for audio use.) A typical example of this construction has massive contacts of silver/cadmium oxide, rated at 30 A AC, which in themselves are linear. However, used as an amplifier output relay, this component generates more – much more – distortion than the power amplifier it is associated with.

The effect increases with increasing current; 4.0 A rms passing through the relay gives 0.0033% THD and 10 A rms gives 0.018%. The distortion level appears to increase with the square of the current. Experiment showed that the distortion was worst where the frame width was narrowest, and hence the current density greatest.

Figure 15.21 shows the effect at 200 W rms/2 Ω (i.e., with 10 A rms through the load) before and after the relay. Trace A is the amplifier alone. This is a Blameless amplifier and so THD is undetectable below 3 kHz, being submerged in the noise floor which sets a measurement limit of 0.0007%.

Trace B adds in the extra distortion from the relay. It seems to be frequency-dependent, but rises more slowly than the usual slope of 6 dB/octave. Trace C shows the effect of closing the relay in the NFB loop using the circuit and component values of Figure 15.20; the THD drops to about a tenth, which is what simple NFB theory would predict. Note that from 10 kHz to 35 kHz the distortion is now lower than before the relay was added; this is due to cancellation of amplifier and relay distortion.

Figure 15.22 was obtained by sawing a 3 mm by 15 mm piece from a relay frame and wiring it in series with the amplifier output, by means of copper wires soldered at each end. As before the level was 200 W rms/2 Ω,

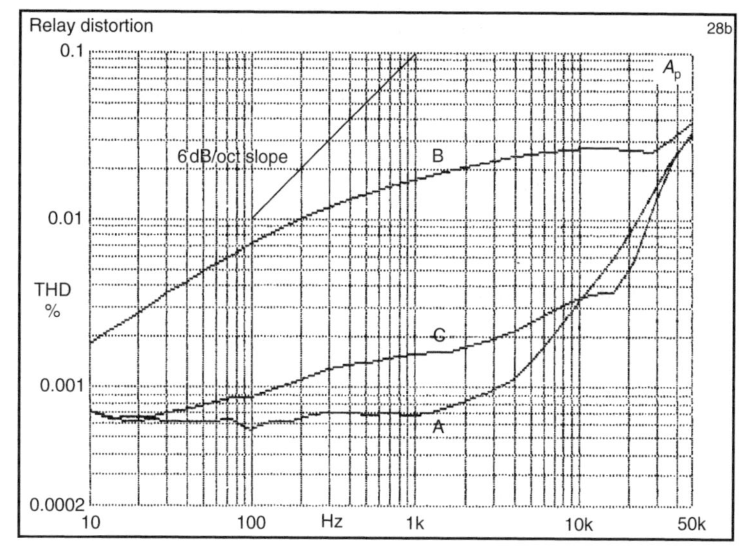

Figure 15.21
A is amplifier distortion alone, B total distortion with power relay in circuit. C shows that enclosing the relay in the feedback loop is not a complete cure

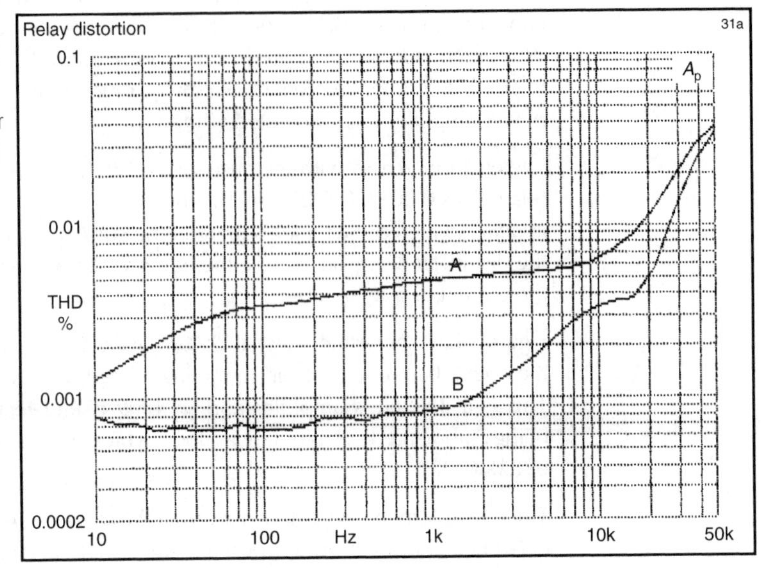

Figure 15.22
Trace A here is total distortion with a sample of the power relay frame material wired in circuit. B is the same, enclosed in the feedback loop as before

i.e., 10 A rms. Trace A is the raw extra distortion; this is lower than shown in Figure 15.21 because the same current is passing through less of the frame material. Trace B is the result of enclosing the frame fragment in the NFB loop exactly as before. This removes all suspicion of interaction with coil or contacts and proves it is the actual frame material itself that is non-linear.

Wrapping feedback around the relay helps but, as usual, is not a complete cure. Soldering on extra wires to the frame to bypass as much frame material as possible is also useful, but it is awkward and there is the danger of interfering with proper relay operation. No doubt any warranties would be invalidated. Clearly it is best to avoid this sort of relay construction if you possibly can, but if high-current switching is required, more than an audio-intended relay can handle, the problem may have to be faced.

Output crowbar DC protection

Since relays are expensive and require control circuitry, and fuse protection is very doubtful, there has for at least two decades been interest in simpler and wholly solid-state solutions to the DC-protection problem. The circuit of Figure 15.23 places a triac across the output, the output signal being low-pass filtered by R and C. If sufficient DC voltage develops on C to fire the diac, it triggers the triac, shorting the amplifier output to ground.

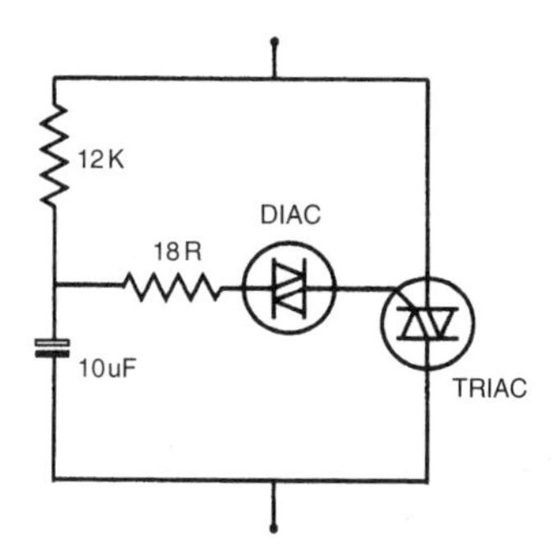

Figure 15.23
Output crowbar DC protection

While this approach has the merit of simplicity, in my (wholly unhappy) experience, it has proved unsatisfactory. The triac needs to be very big indeed if it is to work more than once, because it must pass enough current to blow the HT rail fuses. If these fuses were omitted the triac would have to dump the entire contents of a power-supply reservoir capacitor to ground through a low total resistance, and the demands on it become quite unreasonable.

An output crowbar is also likely to destroy the output devices; the assumption behind this kamikaze crowbar system is that the DC offset is due to blown output devices, and a short across the output can do no more harm. This is quite wrong, because any fault in the small-signal part of the amplifier will also cause the output to saturate positive or negative, with the output devices in perfect working order. The operation of the crowbar under these circumstances may destroy the output devices, for the overload protection may not be adequate to cope with such a very direct short-circuit.

Protection by power-supply shutdown

If your amplifier is powered by a switch-mode supply, it may well have a logic input that gives the option of near-instant shutdown. This can be connected to a DC-detect low-pass filter, and the occurrence of a DC error then gives an apparently foolproof shutdown of everything.

There are (as usual) snags to this. First, the high relative cost of switch-mode supplies means that one will be shared between two or more amplifier channels, and so both channels are lost if one fails. Second, and more worryingly, this provides very dubious protection against a fault in the supply itself. If such a fault causes one of the HT rails to collapse, then it may well also disable the shutdown facility, and all protection is lost.

Conventional transformer power supplies can also be shut down quickly by firing crowbar SCRs across the supply-rails; this overcomes one of the objections to output crowbars, as collateral damage to other parts of the circuit is unlikely, assuming of course you are correctly trying to blow the DC rail fuses, and not the transformer secondary fuses. The latter option would severely endanger the bridge rectifier, and the crowbar circuitry would have to handle enormous amounts of energy as it emptied the reservoir capacitors. Even blowing the DC fuses will require SCRs with a massive peak-current capability.

Thermal protection

This section deals only with protecting the output semiconductors against excessive junction temperature; the thermal safeguarding of the mains transformer is dealt with in Chapter 8.

Output devices that are fully protected against excess current, voltage and power are by no means fully safeguarded. Most electronic overload protection systems allow the devices to dissipate much more power than in normal operation; this can and should be well inside the rated capabilities of the component itself, but this gives no assurance that the increased dissipation will not cause the heatsink to eventually reach such temperatures that the crucial junction temperatures are exceeded and the device fails. If no temperature protection is provided this can occur after only a few minutes drive into a short. Heatsink over-temperature may also occur if ventilation slots, etc. are blocked, or heatsink fins covered up.

The solution is a system that senses the heatsink temperature and intervenes when it reaches a preset maximum. This intervention may be in the form of:

1 Causing an existing muting/DC-protection relay to drop out, breaking the output path to the load. If such a relay is fitted, then it makes sense to use it.
2 Muting or attenuating the input signal so the amplifier is no longer dissipating significant power.
3 Removing the power-supply to the amplifier sections. This normally implies using a bimetallic thermal switch to break the mains supply to the transformer primary, as anywhere downstream of here requires two lines to be broken simultaneously, e.g., the positive and negative HT rails.

Each of these actions may be either self-resetting or latching, requiring the user to initiate a reset. The possibility that a self-resetting system will cycle on and off for long periods, subjecting the output semiconductors to severe temperature changes, must be borne in mind. Such thermal cycling can severely shorten the life of semiconductors.

The two essential parts of a thermal protection system are the temperature sensing element and whatever arrangement performs the intervention. While temperature can be approximately sensed in many ways, e.g., by thermistors, silicon diodes, transistor junctions, etc. these all require some sort of setup or calibration procedure, due to manufacturing tolerances. This is impractical in production, for it requires the heatsink (which normally has substantial thermal inertia) to be brought up to the critical temperature before the circuit is adjusted. This not only takes considerable time, but also requires the output devices to reach a temperature at which they are somewhat endangered.

A much better method is the use of integrated temperature sensors that do not require any calibration. A good example is the National Semiconductor LM35DZ, a three-terminal TO92 device which outputs 10 mV for each degree Centigrade above Zero. Without any calibration procedure, the output voltage may be compared against a fixed reference, usually by an op-amp used as a comparator, and the resulting output used to pull out the muting relay. This approach gives the most trouble-free temperature protection in my experience. IC temperature sensors are more expensive than thermistors, etc. but this is counterbalanced by their accurate and trouble-free operation.

Another pre-calibrated temperature sensor is the thermal switch, which usually operates on the principle of a bistable bimetallic element. These should not be confused with *thermal fuses* which are once-only components that open the circuit by melting an internal fusible alloy; the trouble with these is that they are relatively uncommon, and the chance of a blown thermal fuse being replaced with the correct component in the field is not high.

The physical positioning of the temperature sensor requires some thought. In an ideal world we would judge the danger to the output devices by assessing the actual junction temperature; since this is impractical the sensor must get as close as it can. It is shown elsewhere that the top of a TO3 transistor can gets hotter than the flange, and as for quiescent biasing sensors, the top is the best place for the protection sensor. This does however present some mechanical problems in mounting. This approach may not be equally effective with plastic flat-pack devices such as TO3P, for the outer surface is an insulator; however it still gets hotter than the immediately adjacent heatsink.

Alternatively, the protection sensor can be mounted on the main heatsink, which is mechanically much simpler but imposes a considerable delay between the onset of device heating and the sensor reacting. For this reason a heatsink-mounted sensor will normally need to be set to a lower trip temperature, usually in the region of 80°C, than if it is device-mounted. The more closely the sensor is mounted to the devices, the better they are protected. If two amplifiers share the same heatsink, the sensor should be placed between them; if it was placed at one end the remote amplifier would suffer a long delay between the onset of excess heating and the sensor acting.

One well-known make of PA amplifiers implements temperature protection by mounting a thermal switch in the live mains line on top of one of the TO3 cans in the output stage. This gains the advantage of fast response to dangerous temperatures, but there is the obvious objection that lethal voltages are brought right into the centre of the amplifier circuitry, where they are not normally expected, and this represents a real hazard to service personnel.

Powering auxiliary circuitry

Whenever it is necessary to power auxiliary circuitry, such as the relay control system described above, there is an obvious incentive to use the main HT rails. A separate PSU requires a bridge rectifier, reservoir capacitor, fusing and an extra transformer winding, all of which will cost a significant amount of money.

The main disadvantage is that the HT rails are at an inconveniently high voltage for powering control circuitry. For low-current sections of this circuitry, such as relay timing, the problem is not serious as the same high-voltage small-signal transistors can be used as in the amplifier small-signal sections, and the power dissipation in collector loads, etc. can be controlled simply by making them higher in value. The biggest problem is the relay energising current; many relay types are not available with coil voltages higher than 24 V, and this is not easy to power from a 50 V HT rail without wasting power in a big dropper resistor. This causes unwanted heating of the amplifier internals, and provides a place for service engineers to burn themselves.

One solution in a stereo amplifier is to run the two relays in series; the snag (and for sound reinforcement work it may be a serious one) is that both relays must switch together, so if one channel fails with a DC offset, both are muted. In live work independent relay control is much to be preferred, even though most of the relay control circuitry must be duplicated for each channel.

If the control circuitry is powered from the main HT rails, then its power should be taken off before the amplifier HT fuses. The control circuitry should then be able to mute the relays when appropriate, no matter what faults have occurred in the amplifiers themselves.

If there is additional signal circuitry in the complete amplifier it is not advisable to power it in this way, especially if it has high gain, e.g., a microphone preamplifier. When such signal circuits are powered in this way, it is usually by ±15 V regulators from the HT rails, with series dropper resistors to spread out some of the dissipation. However, bass transients in the power amplifiers can pull down the HT rails alarmingly, and if the regulators drop out large disturbances will appear on the nominally regulated low-voltage rails, leading to very low frequency oscillations which will be extremely destructive to loudspeakers. In this case the use of wholly separate clean rails run from an extra transformer winding is strongly recommended. There will be no significant coupling through the use of a single transformer.

References

1. Bailey, A *Output Transistor Protection in AF Amplifiers* Wireless World, June 1968, p. 154.
2. Becker, R *High-Power Audio Amplifier Design* Wireless World, Feb 1972, p. 79.
3. Motorola *High Power Audio Amplifiers With Sort Circuit Protection* Motorola Application Note AN-485 (1972).
4. Otala, M *Peak Current Requirement of Commerical Loudspeaker Systems* JAES Vol 35, June 1987, p. 455.
5. Baxandall, P *Technique for Displaying Current and Voltage Capability of Amplifiers* JAES Vol 36, Jan/Feb 1988, p. 3.
6. Greiner, R *Amplifier-Loudspeaker Interfacing* JAES. Loudspeakers JAES Vol 28(5), May 1980, pp. 310–315.

16

Grounding and practical matters

Audio amplifier PCB design

This section addresses the special PCB design problems presented by power amplifiers, particularly those operating in Class-B. All power amplifier systems contain the power-amp stages themselves, and usually associated control and protection circuitry; most also contain small-signal audio sections such as balanced input amplifiers, subsonic filters, output meters, and so on.

Other topics that are related to PCB design, such as grounding, safety, reliability, etc., are also dealt with.

The performance of an audio power amplifier depends on many factors, but in all cases the detailed design of the PCB is critical, because of the risk of inductive distortion due to crosstalk between the supply-rails and the signal circuitry; this can very easily be the ultimate limitation on amplifier linearity, and it is hard to over-emphasise its importance. The PCB design will to a great extent define both the distortion and crosstalk performance of the amplifier.

Apart from these performance considerations, the PCB design can have considerable influence on ease of manufacture, ease of testing and repair, and reliability. All of these issues are addressed below.

Successful audio PCB layout requires enough electronic knowledge to fully appreciate the points set out below, so that layout can proceed smoothly and effectively. It is common in many electronic fields for PCB design to be handed over to draughtspersons, who, while very skilled in the use of CAD, have little or no understanding of the details of circuit operation. In some fields this works fine; in power amplifier design it will not, because basic parameters such as crosstalk and distortion are so strongly layout-dependent. At the very least the PCB designer should understand the points set out below.

Crosstalk

All crosstalk has a transmitting end (which can be at any impedance) and a receiving end, usually either at high impedance or virtual-earth. Either way, it is sensitive to the injection of small currents. When interchannel crosstalk is being discussed, the transmitting and receiving channels are usually called the speaking and non-speaking channels, respectively.

Crosstalk comes in various forms:

- Capacitative crosstalk is due to the physical proximity of different circuits, and may be represented by a small notional capacitor joining the two circuits. It usually increases at the rate of 6 dB/octave, though higher dB/octave rates are possible. Screening with any conductive material is a complete cure, but physical distance is usually cheaper.
- Resistive crosstalk usually occurs simply because ground tracks have a non-zero resistance. Copper is not a room-temperature superconductor. Resistive crosstalk is constant with frequency.
- Inductive crosstalk is rarely a problem in general audio design; it might occur if you have to mount two uncanned audio transformers close together, but otherwise you can usually forget it. The notable exception to this rule is the Class-B audio power amplifier, where the rail currents are halfwave sines that seriously degrade the distortion performance if they are allowed to couple into the input, feedback or output circuitry.

In most line-level audio circuitry the primary cause of crosstalk is unwanted capacitative coupling between different parts of a circuit, and in most cases this is defined solely by the PCB layout. Class-B power amplifiers, in contrast, should suffer very low or negligible levels of crosstalk from capacitative effects, as the circuit impedances tend to be low, and the physical separation large; a much greater problem is inductive coupling between the supply-rail currents and the signal circuitry. If coupling occurs to the same channel it manifests itself as distortion, and can dominate amplifier non-linearity. If it occurs to the other (non-speaking) channel it will appear as crosstalk of a distorted signal. In either case it is thoroughly undesirable, and precautions must be taken to prevent it.

The PCB layout is only one component of this, as crosstalk must be both emitted and received. In general the emission is greatest from internal wiring, due to its length and extent; wiring layout will probably be critical for best performance, and needs to be fixed by cable ties, etc. The receiving end is probably the input and feedback circuitry of the amplifier, which will be fixed on the PCB. Designing these sections for maximum immunity is critical to good performance.

Rail induction distortion

The supply-rails of a Class-B power-amp carry large and very distorted currents. As previously outlined, if these are allowed to crosstalk into the audio path by induction the distortion performance will be severely degraded. This applies to PCB conductors just as much as cabling, and it is sadly true that it is easy to produce an amplifier PCB that is absolutely satisfactory in every respect but this one, and the only solution is another board iteration. The effect can be completely prevented but in the present state of knowledge I cannot give detailed guidelines to suit every constructional topology. The best approach is:

Minimise radiation from the supply rails by running the V+ and V− rails as close together as possible. Keep them away from the input stages of the amplifier, and the output connections; the best method is to bring the rails up to the output stage from one side, with the rest of the amplifier on the other side. Then run tracks from the output to power the rest of the amp; these carry no halfwave currents and should cause no problems.

Minimise pickup of rail radiation by keeping the area of the input and feedback circuits to a minimum. These form loops with the audio ground and these loops must be as small in area as possible. This can often best be done by straddling the feedback and input networks across the audio ground track, which is taken across the centre of the PCB from input ground to output ground.

Induction of distortion can also occur into the output and output-ground cabling, and even the output inductor. The latter presents a problem as it is usually difficult to change its orientation without a PCB update.

The mounting of output devices

The most important decision is whether or not to mount the power output devices directly on the main amplifier PCB. There are strong arguments for doing so, but it is not always the best choice.

Advantages

- The amplifier PCB can be constructed so as to form a complete operational unit that can be thoroughly tested before being fixed into the chassis. This makes testing much easier, as there is access from all sides; it also minimises the possibility of cosmetic damage (scratches, etc.) to the metalwork during testing.

- It is impossible to connect the power devices wrongly, providing you get the right devices in the right positions. This is important for such errors usually destroy both output devices and cause other domino-effect faults that are very time-consuming to correct.
- The output device connections can be very short. This seems to help stability of the output stage against HF parasitic oscillations.

Disadvantages

- If the output devices require frequent changing (which obviously indicates something very wrong somewhere) then repeated resoldering will damage the PCB tracks. However, if the worst happens the damaged track can usually be bridged out with short sections of wire, so the PCB need not be scrapped; make sure this is possible.
- The output devices will probably get fairly hot, even if run well within their ratings; a case temperature of 90°C is not unusual for a TO3 device. If the mounting method does not have a degree of resilience, then thermal expansion may set up stresses that push the pads off the PCB.
- The heatsink will be heavy, and so there must be a solid structural fixing between this and the PCB. Otherwise the assembly will flex when handled, putting stress on soldered connections.

Single and double-sided PCBs

Single-sided PCBs are the usual choice for power amplifiers, because of their lower cost; however the price differential between single and double-sided plated-through-hole (PTH) is much less than it used to be. It is not usually necessary to go double-sided for reason of space or convoluted connectivity, because power amplifier components tend to be physically large, determining the PCB size, and in typical circuitry there are a large number of discrete resistors, etc., that can be used for jumping tracks.

Bear in mind that single-sided boards need thicker tracks to ensure adhesion in case desoldering is necessary. Adding one or more ears to pads with only one track leading to them gives much better adhesion, and is highly recommended for pads that may need resoldering during maintenance; unfortunately it is a very tedious task with most CAD systems.

The advantages of double-sided PTH for power amplifiers are as follows:

- No links are required.
- Double-sided PCBs may allow one side to be used primarily as a ground plane, minimising crosstalk and EMC problems.
- Much better pad adhesion on resoldering as the pads are retained by the through-hole plating.

- There is more total room for tracks, and so they can be wider, giving less volt-drop and PCB heating.
- The extra cost is small.

Power supply PCB layout

Power supply subsystems have special requirements due to the very high capacitor-charging currents involved:

- Tracks carrying the full supply-rail current must have generous widths. The board material used should have not less than 2-oz copper. 4-oz copper can be obtained but it is expensive and has long lead-times; not really recommended.
- Reservoir capacitors must have the incoming tracks going directly to the capacitor terminals; likewise the outgoing tracks to the regulator must leave from these terminals. In other words, do not run a tee off to the cap. Failure to observe this puts sharp pulses on the DC and tends to worsen the hum level.
- The tracks to and from the rectifiers carry charging pulses that have a considerably higher peak value than the DC output current. Conductor heating is therefore much greater due to the higher value of I^2R. Heating is likely to be especially severe at PC-mount fuseholders. Wire links may also heat up and consideration should be given to two links in parallel; this sounds crude but actually works very effectively.

  Track heating can usually be detected simply by examining the state of the solder mask after several hours of full-load operation; the green mask materials currently in use discolour to brown on heating. If this occurs then as a very rough rule the track is too hot. If the discoloration tends to dark brown or black then the heating is serious and must definitely be reduced.
- If there are PCB tracks on the primary side of the mains transformer, and this has multiple taps for multi-country operation, then remember that some of these tracks will carry much greater currents at low voltage tappings; mains current drawn on 90 V input will be nearly 3 times that at 240 V.

  Be sure to observe the standard safety spacing of 60 thou between mains tracks and other conductors, for creepage and clearance.

(This applies to all track-track, track-PCB edge, and track-metal-fixings spacings.)

In general PCB tracks carrying mains voltages should be avoided, as presenting an unacceptable safety risk to service personnel. If it must be done, then warnings must be displayed very clearly on both sides of the PCB. Mains-carrying tracks are unacceptable in equipment intended to meet UL regulations in the USA, unless they are fully covered with insulating material that is non-flammable and can withstand at least 120°C (e.g., polycarbonate).

Power amplifier PCB layout details

A simple unregulated supply is assumed.

- Power amplifiers have heavy currents flowing through the circuitry, and all of the requirements for power supply design also apply here. Thick tracks are essential, and 2-oz copper is highly desirable, especially if the layout is cramped.

  If attempting to thicken tracks by laying solder on top, remember that ordinary 60:40 solder has a resistivity of about 6 times that of copper, so even a thick layer may not be very effective.
- The positive and negative rail reservoir caps will be joined together by a thick earth connection; this is called Reservoir Ground (RG). *Do not* attempt to use any point on this track as the audio-ground star-point, as it carries heavy charging pulses and will induce ripple into the signal. Instead take a thick tee from the centre of this track (through which the charging pulses will not flow) and use the end of this as the starpoint.
- Low-value resistors in the output stage are likely to get very hot in operation – possibly up to 200°C. They must be spaced out as much as possible and kept from contact with components such as electrolytic capacitors. Keep them away from sensitive devices such as the driver transistors and the bias-generator transistor.
- Vertical power resistors. The use of these in power amplifiers appears at first attractive, due to the small amount of PCB area they take up. However the vertical construction means that any impact on the component, such as might be received in normal handling, puts a very great strain on the PCB pads, which are likely to be forced off the board. This may result in it being scrapped. Single-sided boards are particularly vulnerable, having much lower pad adhesion due to the absence of vias.
- Solderable metal clips to strengthen the vertical resistors are available in some ranges (e.g., Vitrohm) but this is not a complete solution, and the conclusion must be that horizontal-format power resistors are preferable.
- Rail decoupler capacitors must have a separate ground return to the Reservoir Ground. This ground must *not* share any part of the audio ground system, and must *not* be returned to the Starpoint. See Figure 16.1.
- The exact layout of the feedback takeoff point is criticial for proper operation. Usually the output stage has an *output rail* that connects the emitter power resistors together. This carries the full output current and must be substantial. Take a tee from this track for the output connection, and attach the feedback takeoff point to somewhere along this tee. *Do not* attach it to the track joining the emitter resistors.

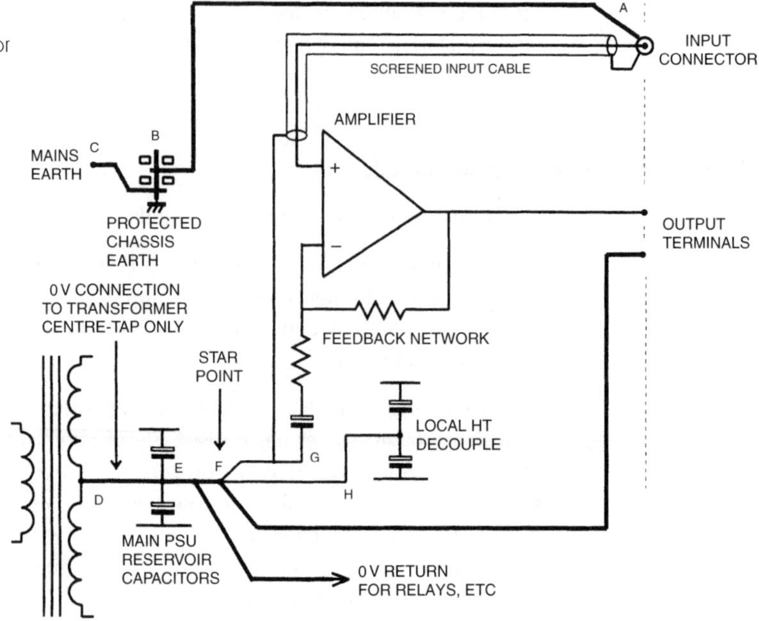

Figure 16.1
Grounding system for a typical power amplifier

- The input stages (usually a differential pair) should be at the other end of the circuitry from the output stage. Never run input tracks close to the output stage. Input stage ground, and the ground at the bottom of the feedback network must be the same track running back to Starpoint. No decoupling capacitors, etc. may be connected to this track, but it seems to be permissible to connect input bias resistors, etc. that pass only very small DC currents.
- Put the input transistors close together. The closer the temperature-match, the less the amplifier output DC offset due to Vbe mismatching. If they can both be hidden from *seeing* the infra-red radiation from the heatsink (for example by hiding them behind a large electrolytic) then DC drift is reduced.
- Most power amplifiers will have additional control circuitry for muting relays, thermal protection, etc. Grounds from this must take a separate path back to Reservoir Ground, and *not* the audio Star point.
- Unlike most audio boards, power amps will contain a mixture of sensitive circuitry and a high-current power-supply. Be careful to keep bridge-rectifier connections, etc., away from input circuitry.
- Mains/chassis ground will need to be connected to the power amplifier at some point. Do not do this at the transformer centre-tap as this is spaced away from the input ground voltage by the return charging pulses, and will create severe groundloop hum when the input ground is connected to mains ground through another piece of equipment.

Connecting mains ground to starpoint is better, as the charging pulses are excluded, but the track resistance between input ground and star will carry any ground-loop currents and induce a buzz.

Connecting mains ground to the input ground gives maximal immunity against groundloops.

- If capacitors are installed the wrong way round the results are likely to be explosive. Make every possible effort to put all capacitors in the same orientation to allow efficient visual checking. Mark polarity clearly on the PCB, positioned so it is still visible when the component is fitted.
- Drivers and the bias generator are likely to be fitted to small vertical heatsinks. Try to position them so that the transistor numbers are visible.
- All transistor positions should have emitter, base and collector or whatever marked on the top-print to aid fault-finding. TO3 devices need also to be identified on the copper side, as any screen-printing is covered up when the devices are installed.
- Any wire links should be numbered to make it easier to check they have all been fitted.

The audio PCB layout sequence

PCB layout must be considered from an early stage of amplifier design. For example, if a front-facial layout shows the volume control immediately adjacent to a loudspeaker routing switch, then a satisfactory crosstalk performance will be difficult to obtain because of the relatively high impedance of the volume control wipers. Shielding metalwork may be required for satisfactory performance and this adds cost. In many cases the detailed electronic design has an effect on crosstalk, quite independently from physical layout.

(a) Consider implications of facia layout for PCB layout.
(b) Circuitry designed to minimise crosstalk. At this stage try to look ahead to see how op-amp halves, switch sections, etc. should be allocated to keep signals away from sensitive areas. Consider crosstalk at above-PCB level; for example, when designing a module made up of two parallel double-sided PCBs, it is desirable to place signal circuitry on the inside faces of the boards, and power and grounds on the outside, to minimise crosstalk and maximise RF immunity.
(c) Facia components (pots, switches, etc.) placed to partly define available board area.
(d) Other fixed components such as power devices, driver heatsinks, input and output connectors, and mounting holes placed. The area left remains for the purely electronic parts of the circuitry that do not have to align with metalwork, etc. and so may be moved about fairly freely.
(e) Detailed layout of components in each circuit block, with consideration towards manufacturability.

(f) Make efficient use of any spare PCB area to fatten grounds and high-current tracks as much as possible. It is not wise to fill in every spare corner of a prototype board with copper as this can be time consuming (depending on the facilities of your PCB CAD system) and some of it will probably have to be undone to allow modifications.

Ground tracks should always be as thick as practicable. Copper is free.

Miscellaneous points

- On double-sided PCBs, copper areas should be solid on the component side, for minimum resistance and maximum screening, but will need to be cross-hatched on the solder side to prevent distortion of the PCB is flow-soldered. A common standard is 10 thou wide non-copper areas; i.e., mostly copper with small square holes; this is determined in the CAD package. If in doubt consult those doing the flow-soldering.
- Do not bury component pads in large areas of copper, as this causes soldering difficulties.
- There is often a choice between running two tracks into a pad, or taking off a tee so that only one track reaches it. The former is better because it holds the pad more firmly to the board if desoldering is necessary. This is *particularly important* for components like transistors that are relatively likely to be replaced; for single-sided PCBs it is absolutely vital.
- If two parallel tracks are likely to crosstalk, then it is beneficial to run a grounded screening track between them. However, the improvement is likely to be disappointing, as electrostatic lines of force will curve over the top of the screen track.
- Jumper options must always be clearly labelled. Assume everyone loses the manual the moment they get it.
- Label pots and switches with their function on the screen-print layer, as this is a great help when testing. If possible, also label circuit blocks, e.g., *DC offset detect*. The labels must be bigger than component ident text to be clearly readable.

Amplifier grounding

The grounding system of an amplifier must fulfil several requirements, amongst which are:

- The definition of a *Star Point* as the reference for all signal voltages.
- In a stereo amplifier, grounds must be suitably segregated for good crosstalk performance. A few inches of wire as a shared ground to the output terminals will probably dominate the crosstalk behaviour.
- Unwanted AC currents entering the amplifier on the signal ground, due to external ground loops, must be diverted away from the critical signal grounds, i.e., the input ground and the ground for the feedback arm.

Any voltage difference between these last two grounds appears directly in the output.

- Charging currents for the PSU reservoir capacitors must be kept out of all other grounds.

Ground is the point of reference for all signals, and it is vital that it is made solid and kept clean; every ground track and wire must be treated as a resistance across which signal currents will cause unwanted voltage-drops. The best method is to keep ground currents apart by means of a suitable connection topology, such as a separate ground return to the Star Point for the local HT decoupling, but when this is not practical it is necessary to make every ground track as thick as possible, and fattened up with copper at every possible point. It is vital that the ground path has no necks or narrow sections, as it is no stronger than the weakest part. If the ground path changes board side then a single via-hole may be insufficient, and several should be connected in parallel. Some CAD systems make this difficult, but there is usually a way to fool them.

Power amplifiers rarely use double-insulated construction and so the chassis and all metalwork must be permanently and solidly grounded for safety; this aspect of grounding is covered in Chapter 15. One result of permanent chassis grounding is that an amplifier with unbalanced inputs may appear susceptible to ground loops. One solution is to connect audio ground to chassis only through a 10 Ω resistor, which is large enough to prevent loop currents becoming significant. This is not very satisfactory as:

- The audio system as a whole may thus not be solidly grounded.
- If the resistor is burnt out due to misconnected speaker outputs, the audio circuitry is floating and could become a safety hazard.
- The RF rejection of the power amplifier is likely to be degraded. A 100 nF capacitor across the resistor may help.

A better approach is to put the audio-chassis ground connection at the input connector, so in Figure 16.1, ground-loop currents must flow through A–B to the Protected Earth at B, and then to mains ground via B–C. They cannot flow through the audio path E–F. This topology is very resistant to ground-loops, even with an unbalanced input; the limitation on system performance in the presence of a ground-loop is now determined by the voltage-drop in the input cable ground, which is outside the control of the amplifier designer. A balanced input could in theory cancel out this voltage drop completely.

Figure 16.1 also shows how the other grounding requirements are met. The reservoir charging pulses are confined to the connection D–E, and do not flow E–F, as there is no other circuit path. E–F–H carries ripple, etc., from the local HT decouplers, but likewise cannot contaminate the crucial audio ground A–G.

Ground loops: how they work and how to deal with them

A ground loop is created whenever two or more pieces of mains-powered equipment are connected together, so that mains-derived AC flows through shields and ground conductors, degrading the noise floor of the system. The effect is worst when two or more units are connected through mains ground as well as audio cabling, and this situation is what is normally meant by the term 'ground loop'. However, ground currents can also flow in systems that are not galvanically grounded; they are of lower magnitude but can still degrade the noise floor, so this scenario is also considered here.

The ground currents may either be inherent in the mains supply wiring (see 'Hum injection by mains grounding currents' below) or generated by one or more of the pieces of equipment that make up the audio system (see sections 'Hum injection by transformer stray magnetic fields' and 'Hum injection' by 'transformer stray capacitance' below).

Once flowing in the ground wiring, these currents will give rise to voltage drops that introduce hum and buzzing noises. This may occur either in the audio interconnects, or inside the equipment itself if it is not well designed. See section 'Ground currents inside equipment', on p. 410.

Here I have used the word 'ground' for conductors and so on, while 'earth' is reserved for the damp crumbly stuff into which copper rods are thrust.

Hum injection by mains grounding currents

Figure 16.2 shows what happens when a so-called 'technical ground' such as a buried copper rod is attached to a grounding system which is already connected to 'mains ground' at the power distribution board. The latter is mandatory both legally and technically, so one might as well accept this and denote as the reference ground. In many cases this 'mains ground' is actually the neutral conductor, which is only grounded at the remote transformer substation. AB is the cable from substation to consumer, which serves many houses from connections tapped off along its length. There is substantial current flowing down the N + E conductor, so point B is often 1 V rms or more above earth. From B onwards, in the internal house wiring, neutral and ground are always separate (in the UK, anyway).

Two pieces of audio equipment are connected to this mains wiring at C and D, and joined to each other through an unbalanced cable F–G. Then an ill-advised connection is made to earth at D; the 1 V rms is now impressed on the path B–C–D, and substantial current is likely to flow through it, depending on the total resistance of this path. There will be a voltage drop from C to D, its magnitude depending on what fraction of the total BCDE resistance is made up by the section C–D. The earth wire C–D will be of

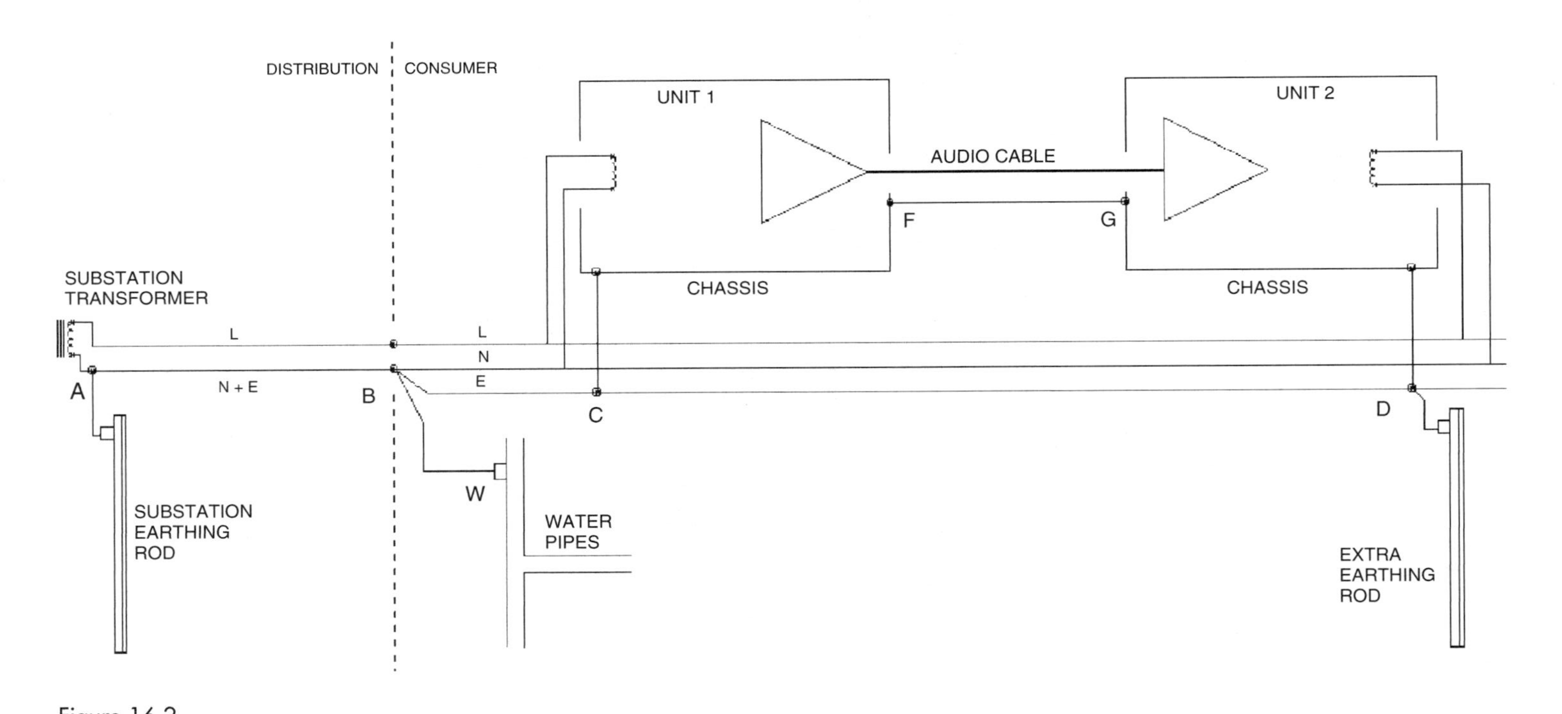

Figure 16.2
The pitfalls of adding a 'technical ground' to a system which is already grounded via the mains

at least 1.5 mm^2 cross-section, and so the extra connection FG down the audio cable is unlikely to reduce the interfering voltage much.

To get a feel for the magnitudes involved, take a plausible ground current of 1 A. The 1.5 mm^2 ground conductor will have a resistance of 0.012 Ω/m, so if the mains sockets at C and D are 1 m apart, the voltage C–D will be 12 mV rms. Almost all of this will appear between F and G, and will be indistinguishable from wanted signal to the input stage of Unit 2, so the hum will be severe, probably only 30 dB below the nominal signal level.

The best way to solve this problem is not to create it in the first place. If some ground current is unavoidable then the use of balanced inputs (or ground-cancel outputs – it is not necessary to use both) should give at least 40 dB of rejection at audio frequencies.

Figure 16.2 also shows a third earthing point, which fortunately does not complicate the situation. Metal water pipes are bonded to the incoming mains ground for safety reasons, and since they are usually electrically connected to an incoming water supply current flows through B–W in the same way as it does through the copper rod link D–E. This water-pipe current does not, however, flow through C–D and cannot cause a ground-loop problem. It may, however, cause the pipes to generate an AC magnetic field which is picked up by other wiring.

Hum injection by transformer stray magnetic fields

Figure 16.3 shows a thoroughly bad piece of physical layout which will cause ground currents to flow even if the system is correctly grounded to just one point.

Here Unit 1 has an external DC power supply; this makes it possible to use an inexpensive frame-type transformer which will have a large stray field. But note that the wire in the PSU which connects mains ground to the outgoing 0 V takes a half-turn around the transformer, and significant current will be induced into it, which will flow round the loop C–F–G–D, and give an unwanted voltage drop between F and G. In this case reinforcing the ground of the audio interconnection is likely to be of some help, as

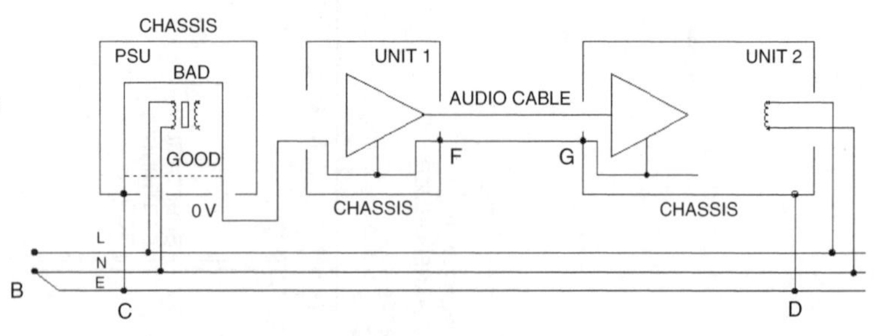

Figure 16.3
Poor cable layout in the PSU at left wraps a loop around the transformer and induces ground currents

it directly reduces the fraction of the total loop voltage which is dropped between F and G.

It is difficult to put any magnitudes to this effect because it depends on many imponderables such as the build quality of the transformer and the exact physical arrangement of the ground cable in the PSU. If this cable is rerouted to the dotted position in the diagram, the transformer is no longer enclosed in a half-turn, and the effect will be much smaller.

Hum injection by transformer stray capacitance

It seems at first sight that the adoption of Class II (double-insulated) equipment throughout an audio system will give inherent immunity to ground-loop problems. Life is not so simple, though it has to be said that when such problems do occur they are likely to be much less severe. This problem afflicts all Class II equipment to a certain extent.

Figure 16.4 shows two Class II units connected together by an unbalanced audio cable. The two mains transformers in the units have stray capacitance from both live and neutral to the secondary. If these capacitances were all identical no current would flow, but in practice they are not, so 50 Hz currents are injected into the internal 0 V rail and flow through the resistance of F–G, adding hum to the signal. A balanced input or ground-cancelling output will remove or render negligible the ill-effects.

Reducing the resistance of the interconnect ground path is also useful – more so than with other types of ground loop, because the ground current is essentially fixed by the small stray capacitances, and so halving the resistance F–G will dependably halve the interfering voltage. There are limits to how far you can take this – while a simple balanced input will give 40 dB of rejection at low cost, increasing the cross-sectional area of copper in the ground of an audio cable by a factor of 100 times is not going to be either easy or cheap. Figure 16.4 shows equipment with metal chassis connected to the 0 V (this is quite acceptable for safety approvals – what counts is the isolation between mains and everything else, not between low-voltage circuitry and touchable metalwork); note the chassis connection, however, has no relevance to the basic effect, which would still occur even if the equipment enclosure was completely non-conducting.

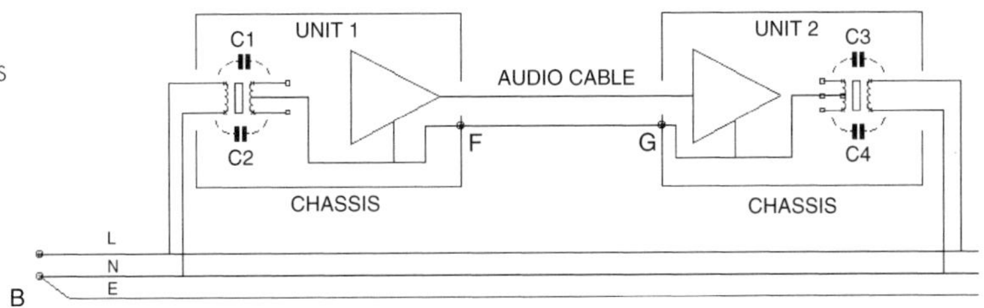

Figure 16.4
The injection of mains current into the ground wiring via transformer inter-winding capacitance

The magnitude of ground current varies with the details of transformer construction, and increases as the size of the transformer grows. Therefore the more power a unit draws, the larger the ground current it can sustain. This is why many systems are subjectively hum-free until the connection of a powered subwoofer, which is likely to have a larger transformer than other components of the system.

Equipment type	Power consumption	Ground current
Turntable, CD, cassette deck	20 W or less	5 μA
Tuners, amplifiers, small TVs	20–100 W	100 μA
Big amplifiers, subwoofers, large TVs	More than 100 W	1 mA

Ground currents inside equipment

Once ground currents have been set flowing, they can degrade system performance in two locations: outside the system units, by flowing in the interconnect grounds, or inside the units, by flowing through internal PCB tracks, etc. The first problem can be dealt with effectively by the use of balanced inputs, but the internal effects of ground currents can be much more severe if the equipment is poorly designed.

Figure 16.5 shows the situation. There is, for whatever reason, ground current flowing through the ground conductor CD, causing an interfering current to flow round the loop CFGD as before. Now, however, the internal design of Unit 2 is such that the ground current flowing through FG also flows through G-G′ before it encounters the ground wire going to point D. G-G′ is almost certain to be a PCB track with higher resistance than any of the cabling, and so the voltage drop across it can be relatively large, and the hum performance correspondingly poor. Exactly similar effects can occur at signal outputs; in this case the ground current is flowing through F-F′.

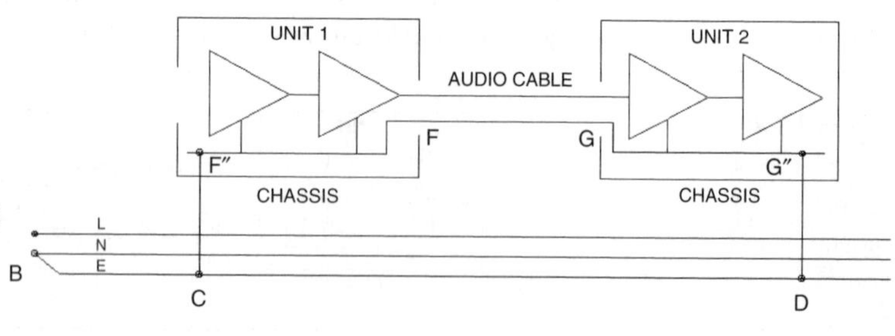

Figure 16.5
If ground current flows through the path F′FGG′ then the relatively high resistance of the PCB tracks produces voltage drops between the internal circuit blocks

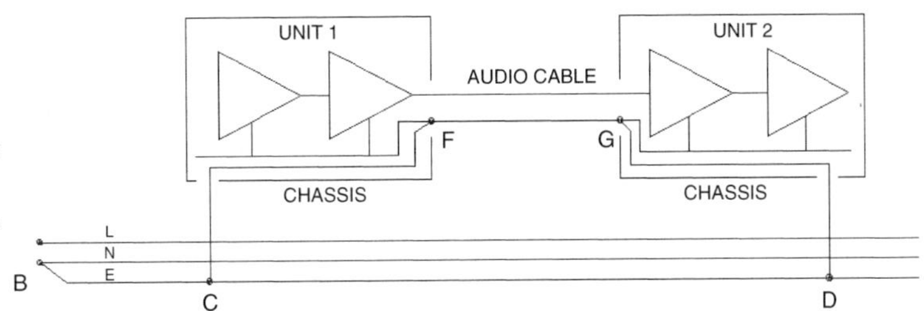

Figure 16.6
The correct method of dealing with ground currents; they are diverted away from internal circuitry

Balanced inputs will have no effect on this; they can cancel out the voltage drop along F–G, but if internal hum is introduced further down the internal signal path, there is nothing they can do about it.

The correct method of handling this is shown in Figure 16.6. The connection to mains ground is made right where the signal grounds leave and enter the units, and are made as solidly as possible. The ground current no longer flows through the internal circuitry. It does, however, still flow through the interconnection at FG, so either a balanced input or a ground-cancelling output will be required to deal with this.

Balanced mains power

There has been speculation in recent times as to whether a balanced mains supply is a good idea. This means that instead of live and neutral (230 V and 0 V) you have live and the other live (115 V–0–115 V) created by a centre-tapped transformer with the tap connected to Neutral (See Figure 16.7).

It has been suggested that balanced mains has miraculous effects on sound quality, makes the sound stage ten dimensional, etc. This is obviously nonsense. If a piece of gear is that fussy about its mains (and I do not believe any such gear exists) then dispose of it.

If there is severe RFI on the mains, an extra transformer in the path may tend to filter it out. However, a proper mains RFI filter will almost certainly

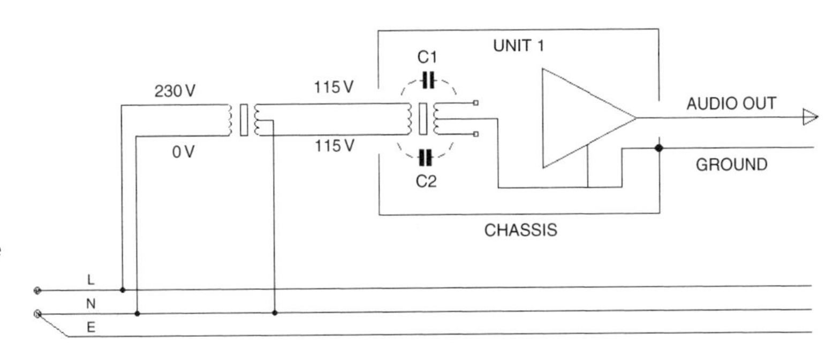

Figure 16.7
Using a balanced mains supply to cancel ground currents stemming from interwinding capacitance in the mains transformer. An expensive solution

be more effective – it is designed for the job, after all – and will definitely be much cheaper.

Where you might gain a real benefit is in a Class II (i.e., double-insulated) system with very feeble ground connections. Balanced mains would tend to cancel out the ground currents caused by transformer capacitance (see Figure 16.4 and above for more details on this) and so reduce hum. The effectiveness of this will depend on C1 being equal to C2 in Figure 16.7, which is determined by the details of transformer construction in the unit being powered. I think that the effect would be small with well-designed equipment and reasonably heavy ground conductors in interconnects. Balanced audio connections are a much cheaper and better way of handling this problem, but if none of the equipment has them then beefing up the ground conductors should give an improvement. If the results are not good enough then as a last resort, balanced mains may be worth considering.

Finally, bear in mind that any transformer you add must be able to handle the maximum power drawn by the audio system at full throttle. This can mean a large and expensive component.

I would not be certain about the whole of Europe, but to the best of my knowledge it is the same as the UK, i.e., not balanced. The neutral line is at earth potential, give or take a volt, and the live is 230 V above this. The 3-phase 11 kV distribution to substations is often described as 'balanced' but this just means that power delivered by each phase is kept as near equal as possible for the most efficient use of the cables.

It has often occurred to me that balanced mains 115 V–0–115 V would be a lot safer. Since I am one of those people that put their hands inside live equipment a lot, I do have a kind of personal interest here.

Class I and Class II

Mains-powered equipment comes in two types: grounded and double insulated. These are officially called Class I and Class II, respectively.

Class I equipment has its external metalwork grounded. Safety against electric shock is provided by limiting the current the live connection can supply with a fuse. Therefore, if a fault causes a short-circuit between live and metalwork, the fuse blows and the metalwork remains at ground potential. A reasonably low resistance in the ground connection is essential to guarantee the fuse blows. A three-core mains lead is mandatory. Two-core IEC mains leads are designed so they cannot be plugged into three-pin Class I equipment. Class I mains transformers are tested to 1.5 kV rms.

Class II equipment is not grounded. Safety is maintained not by interrupting the supply in case of a fault, but by preventing the fault happening in the first place. Regulations require double insulation and a generally high standard of construction to prevent any possible connection between live and the chassis. A two-core IEC mains lead is mandatory; it is not permitted to sell a three-core lead with a Class II product. This would present no hazard in itself, but is presumably intended to prevent confusion as to what kind of product is in use. Class II mains transformers are tested to 3 kV rms, to give greater confidence against insulation breakdown.

Class II is often adopted in an attempt to avoid ground loops. Doing so eliminates the possibility of major problems, at the expense of throwing away all hope of fixing minor ones. There is no way to prevent capacitance currents from the mains transformer flowing through the ground connections. (See section 'Ground loops: how they work and how to deal with, them'). It is also no longer possible to put a grounded electrostatic screen between the primary and secondary windings. This is serious as it deprives you of your best weapon against mains noise coming in and circuit RF emissions getting out. In Class II the external chassis may be metallic, and connected to signal 0 V as often as you like.

If a Class II system is not connected to ground at any point, then the capacitance between primaries and secondaries in the various mains transformers can cause its potential to rise well above ground. If it is touched by a grounded human, then current will flow, and this can sometimes be perceptible, though not directly, as a painful shock like static electricity. The usual complaint is that the front panel of equipment is 'vibrating', or that it feels 'furry'. The maximum permitted touch current (flowing to ground through the human body) permitted by current regulations is 700 μA, but currents well below this are perceptible. It is recommended, though not required, that this limit be halved in the tropics where fingers are more likely to be damp. The current is measured through a 50 k resistance to ground.

When planning new equipment, remember that the larger the mains transformer, the greater the capacitance between primary and secondary, and the more likely this is to be a problem. To put the magnitudes into perspective, I measured a 500 VA toroid (intended for Class II usage and with no interwinding screen) and found 847 pF between the windings. At 50 Hz and 230 V this implies a maximum current of 63 μA flowing into the signal circuitry, the actual figure depending on precisely how the windings are arranged. A much larger 1500 VA toroidal transformer had 1.3 nF between the windings, but this was meant for Class I use and had a screen, which was left floating to get the figure above.

Warning

Please note that the legal requirements for electrical safety are always liable to change. This book does not attempt to give a complete guide to what is required for compliance. The information given here is correct at the time of writing, but it is the designer's responsibility to check for changes to compliance requirements. The information is given here in good faith but the author accepts no responsibility for loss or damage under any circumstances.

Mechanical layout and design considerations

The mechanical design adopted depends very much on the intended market, and production and tooling resources, but I offer a few purely technical points that need to be taken into account.

Cooling

All power amplifiers will have a heatsink that needs cooling, usually by free convection, and the mechanical design is often arranged around this requirement. There are three main approaches to the problem:

(a) The heatsink is entirely internal, and relies on convected air entering the bottom of the enclosure, and leaving near the top (passive cooling).

Advantages

The heatsink may be connected to any voltage, and this may eliminate the need for thermal washers between power device and sink. On the other hand, some sort of conformal material is still needed between transistor and heatsink. A thermal washer is much easier to handle than the traditional white oxide-filled silicone compound, so you will be using them anyway. There are no safety issues as to the heatsink temperatures.

Disadvantages

This system is not suitable for large dissipations, due to the limited fin area possible inside a normal-sized box, and the relatively restricted convection path.

(b) The heatsink is partly internal and partly external, as it forms one or more sides of the enclosure. Advantages and disadvantages are much as above; if any part of the heatsink can be touched then the restrictions on temperature and voltage apply. Greater heat dissipation is possible.
(c) The heatsink is primarily internal, but is fan-cooled (active cooling). Fans always create some noise, and this increases with the amount of air they are asked to move. Fan noise is most unwelcome in a domestic hi-fi environment, but is of little importance in PA applications.

This allows maximal heat dissipation, but requires an inlet filter to prevent the build-up of dust and fluff internally. Persuading people to regularly clean such filters is near-impossible.

Efficient passive heat removal requires extensive heatsinking with a free convective air flow, and this indicates putting the sinks on the side of the amplifier; the front will carry at least the mains switch and power indicator light, while the back carries the in/out and mains connectors, so only the sides are completely free.

The internal space in the enclosure will require some ventilation to prevent heat build-up; slots or small holes are desirable to keep foreign bodies out. Avoid openings on the top surface as these will allow the entry of spilled liquids, and increase dust entry. BS415 is a good starting point for this sort of safety consideration, and this specifies that slots should be no more than 3 mm wide.

Reservoir electrolytics, unlike most capacitors, suffer significant internal heating due to ripple current. Electrolytic capacitor life is very sensitive to temperature, so mount them in the coolest position available, and if possible leave room for air to circulate between them to minimise the temperature rise.

Convection cooling

It is important to realise that the buoyancy forces that drive natural convection are very small, and even small obstructions to flow can seriously reduce the rate of flow, and hence the cooling. If ventilation is by slots in the top and bottom of an amplifier case, then the air must be drawn under the unit, and then execute a sharp right-angle turn to go up through the bottom slots. This change of direction is a major impediment to air flow, and if you are planning to lose a lot of heat then it feeds into the design of something so humble as the feet the unit stands on; the higher the better, for air flow. In one instance the amplifier feet were made 13 mm taller and all the internal amplifier temperatures dropped by 5°C. Standing such a unit on a thick-pile carpet can be a really bad idea, but someone is bound to do it (and then drop their coat on top of it); hence the need for overtemperature cutouts if amplifiers are to be fully protected.

Mains transformers

A toroidal transformer is useful because of its low external field. It must be mounted so that it can be rotated to minimise the effect of what stray fields it does emit. Most suitable toroids have single-strand secondary lead-outs, which are too stiff to allow rotation; these can be cut short and connected to suitably large flexible wire such as 32/02, with carefully sleeved and insulated joints. One prototype amplifier I have built had a sizeable toroid mounted immediately adjacent to the TO3 end of the amplifier PCB; however complete cancellation of magnetic hum (hum and ripple output level below −90 dBu) was possible on rotation of the transformer.

A more difficult problem is magnetic radiation caused by the reservoir charging pulses (as opposed to the ordinary magnetisation of the core, which would be essentially the same if the load current was sinusoidal) which can be picked up by either the output connections or cabling to the power transistors if these are mounted off-board. For this reason the transformer should be kept physically as far away as possible from even the high-current section of the amplifier PCB.

As usual with toroids, ensure the bolt through the middle cannot form a shorted turn by contacting the chassis in two places.

Wiring layout

There are several important points about the wiring for any power amplifier:

- Keep the + and − HT supply wires to the amplifiers close together. This minimises the generation of distorted magnetic fields which may otherwise couple into the signal wiring and degrade linearity. Sometimes it seems more effective to include the 0 V line in this cable run; if so it should be tightly braided to keep the wires in close proximity. For the same reason, if the power transistors are mounted off the PCB, the cabling to each device should be configured to minimise loop formation.
- The rectifier connections should go direct to the reservoir capacitor terminals, and then away again to the amplifiers. Common impedance in these connections superimposes charging pulses on the rail ripple waveform, which may degrade amplifier PSRR.
- Do not use the actual connection between the two reservoir capacitors as any form of star point. It carries heavy capacitor-charging pulses that generate a significant voltage drop even if thick wire is used. As Figure 16.1 shows, the *star-point* is tee-ed off from this connection. This is a star-point only insofar as the amplifier ground connections split off from here, so do not connect the input grounds to it, as distortion performance will suffer.

Semiconductor installation

- Driver transistor installation. These are usually mounted onto separate heatsinks that are light enough to be soldered into the PCB without further fixing. Silicone thermal washers ensure good thermal contact, and spring clips are used to hold the package firmly against the sink. Electrical isolation between device and heatsink is not normally essential, as the PCB need not make any connection to the heatsink fixing pads.

- TO3P power transistor installation. These large flat plastic devices are usually mounted on to the main heatsink with spring clips, which are not only are rapid to install, but also generate less mechanical stress in the package than bolting the device down by its mounting hole. They also give a more uniform pressure onto the thermal washer material.
- TO3 power transistor installation. The TO3 package is extremely efficient at heat transfer, but notably more awkward to mount.

My preference is for TO3s to be mounted on an aluminium thermal-coupler which is bolted against the component side of the PCB. The TO3 pins may then be soldered directly on the PCB solder side. The thermal-coupler is drilled with suitable holes to allow M3.5 fixing bolts to pass through the TO3 flange holes, through the flange, and then be secured on the other side of the PCB by nuts and crinkle washers which will ensure good contact with the PCB mounting pads. For reliability the crinkle washers must cut through the solder-tinning into the underlying copper; a solder contact alone will creep under pressure and the contact force decay over time.

Insulating sleeves are essential around the fixing bolts where they pass through the thermal-coupler; nylon is a good material for these as it has a good high-temperature capability. Depending on the size of the holes drilled in the thermal-coupler for the two TO3 package pins (and this should be as small as practicable to maximise the area for heat transfer), these are also likely to require insulation; silicone rubber sleeving carefully cut to length is very suitable.

An insulating thermal washer must be used between TO3 and flange; these tend to be delicate and the bolts must not be over-tightened. If you have a torque-wrench, then 10 Nw/m is an approximate upper limit for M3.5 fixing bolts. *Do not* solder the two transistor pins to the PCB until the TO3 is firmly and correctly mounted, fully bolted down, and checked for electrical isolation from the heatsink. Soldering these pins and *then* tightening the fixing bolts is likely to force the pads from the PCB. If this should happen then it is quite in order to repair the relevant track or pad with a small length of stranded wire to the pin; 7/02 size is suitable for a very short run.

Alternatively, TO3s can be mounted off-PCB (e.g., if you already have a large heatsink with TO3 drillings) with wires taken from the TO3 pads on the PCB to the remote devices. These wires should be fastened together (two bunches of three is fine) to prevent loop formation; see above. I cannot give a maximum safe length for such cabling, but certainly 8 in. causes no HF stability problems is my experience. The emitter and collector wires should be substantial, e.g., 32/02, but the base connections can be as thin as 7/02.

17

Testing and safety

Testing and fault-finding

Testing power amplifiers for correct operation is relatively easy; faultfinding them when something is wrong is not. I have been professionally engaged with power amplifiers for a long time, and I must admit I still sometimes find it to be a difficult and frustrating business.

There are several reasons for this. First, almost all small-signal audio stages are IC-based, so the only part of the circuit likely to fail can be swiftly replaced, so long as the IC is socketed. A power amplifier is the only place where you are likely to encounter a large number of components all in one big negative feedback loop. The failure of any components may (if you are lucky) simply jam the amplifier output hard against one of the rails, or (if you are not) cause simultaneous failure of all the output devices, possibly with a domino-theory trail of destruction winding through the small-signal section. A certain make of high-power amplifier in the mid-70s was a notorious example of the domino-effect, and when it failed (which was often) the standard procedure was to replace *all* of the semiconductors, back to and including the bridge rectifier.

Component numbers here refer to Figure 6.13.

By far the most important step to successful operation is a careful visual inspection before switch-on. As in all power amplifier designs, a wrongly installed component may easily cause the immediate failure of several others, making fault-finding difficult, and the whole experience generally less than satisfactory. It is therefore most advisable to meticulously check.

- That the supply and ground wiring is correct.
- That all transistors are installed in the correct positions.
- That the drivers and TO3 output devices are not shorted to their respective heatsinks through faulty insulating washers.

- That the circuitry around the bias generator TR13 in particular is correctly built. An error here that leaves TR13 turned off will cause large currents to flow through the output devices and may damage them before the rail fuses can act.

For the Trimodal amplifier in Chapter 9, I recommend that the initial testing is done in Class-B mode. There is the minimum amount of circuitry to debug (the Class-A current-controller can be left disconnected, or not built at all until later) and at the same time the Class-B bias generator can be checked for its operation as a safety-circuit on Class-A/AB mode.

The second stage is to obtain a good sinewave output with no load connected. A fault may cause the output to sit hard up against either rail; this should not in itself cause any damage to components. Since a power-amp consists of one big feedback loop, localising a problem can be difficult. The best approach is to take a copy of the circuit diagram and mark on it the DC voltage present at every major point. It should then be straightforward to find the place where two voltages fail to agree; e.g., a transistor installed backwards usually turns fully on, so the feedback loop will try to correct the output voltage by removing all drive from the base. The clash between *full-on* and *no base-drive* signals the error.

When checking voltages in circuit, bear in mind that C2 is protected against reverse voltage in both directions by diodes which will conduct if the amplifier saturates in either direction.

This DC-based approach can fail if the amplifier is subject to high-frequency oscillation, as this tends to cause apparently anomalous DC voltages. In this situation the use of an oscilloscope is really essential. An expensive oscilloscope is not necessary; a digital scope is actually at a disadvantage here, because HF oscillation is likely to be aliased into nonsense and be hard to interpret.

The third step is to obtain a good sinewave into a suitable high-wattage load resistor. It is possible for faults to become evident under load that are not shown up in Step 2 above.

Setting the quiescent conditions for any Class-B amplifier can only be done accurately by using a distortion analyser. If you do not have access to one, the best compromise is to set the quiescent voltage-drop across both emitter resistors (R16, 17) to 10 mV when the amplifier is at working temperature; disconnect the output load to prevent DC offsets causing misleading current flow. This should be close to the correct value, and the inherent distortion of this design is so low that minor deviations are not likely to be very significant. This implies a quiescent current of approximately 50 mA.

It may simplify faultfinding if D7, D8 are not installed until the basic amplifier is working correctly, as errors in the SOAR protection cannot then confuse the issue. This demands some care in testing, as there is then no short-circuit protection.

Safety

The overall safety record of audio equipment is very good, but no cause for complacency. The price of safety, like that of liberty, is eternal vigilance. Safety regulations are not in general hard to meet so long as they are taken into account at the start of the mechanical design phase. This section considers not only the safety of the user, but also of the service technician.

Many low-powered amplifier designs are inherently safe because all the DC voltages are too low to present any kind of electric-shock hazard. However, high-powered models will have correspondingly high supply rails which are a hazard in themselves, as a DC shock is normally considered more dangerous than the equivalent AC voltage.

Unless the equipment is double-insulated, an essential safety requirement is a solid connection between mains ground and chassis, to ensure that the mains fuse blows if Live contacts the metalwork. British Standards on safety require the mains earth to chassis connection to be a *Protected Earth*, clearly labelled and with its own separate fixing. A typical implementation has a welded ground stud onto which the mains-earth ring-terminal is held by a nut and locking washer; all other internal grounds are installed on top of this and secured with a second nut/washer combination. This discourages service personnel from removing the chassis ground in the unlikely event of other grounds requiring disconnection for servicing. A label warning against *lifting the ground* should be clearly displayed.

There are some specific points that should be considered:

1. An amplifier may have supply-rails of relatively low voltage, but the reservoir capacitors will still store a significant amount of energy. If they are shorted out by a metal finger-ring then a nasty burn is likely. If your bodily adornment is metallic then it should be removed before diving into an amplifier.
2. Any amplifier containing a mains power supply is potentially lethal. The risks involved in working for some time on the powered-up chassis must be considered. The metal chassis *must* be securely earthed to prevent it becoming live if a mains connection falls off, but this presents the snag that if one of your hands touches live, there is a good chance that the other is leaning on chassis ground, so your well-insulated training shoes will not save you. All mains connections (neutral as well as live, in case of mis-wired mains) must therefore be properly insulated so they cannot be accidentally touched by finger or screwdriver. My own preference is for double insulation; for example, the mains inlet connector not only

has its terminals sleeved, but there is also an overall plastic boot fitted over the rear of the connector, and secured with a tie-wrap.

Note that this is a more severe requirement than BS415 which only requires that mains should be inaccessible until you remove the cover. This assumes a tool is required to remove the cover, rather than it being instantly removable. In this context a coin counts as a tool if it is used to undo giant screwheads.

3. A Class-A amplifier runs *hot* and the heatsinks may well rise above 70°C. This is not likely to cause serious burns, but it *is* painful to touch. You might consider this point when arranging the mechanical design. Safety standards on permissible temperature rise of external parts will be the dominant factor.
4. Note the comments on slots and louvres in the section on *Mechanical Design* above.
5. Readers of hi-fi magazines are frequently advised to leave amplifiers permanently powered for optimal performance. Unless your equipment is afflicted with truly doubtful control over its own internal workings, this is quite unnecessary. (And if it *is* so afflicted, personally I would turn it off right now.) While there should be no real safety risk in leaving a soundly constructed power amplifier powered permanently, I see no point and some potential risk in leaving unattended equipment powered; in Class-A mode there may of course be an impact on your electricity bill.

Safety regulations

This section of the book is intended to provide a starting-point in considering safety issues. Its main purpose is to alert you to the various areas that must be considered. For reasons of space it cannot be a comprehensive manual that guarantees equipment compliance; it certainly does not attempt to give a full and complete account of the various safety requirements that a piece of electronic equipment must meet before it can be legally sold. If you plan to manufacture amplifiers and sell them, then it is your responsibility to inform yourself of the regulations. All the information here is given in good faith and is correct at the time of writing, but I accept no responsibility for its use.

European safety standards are defined in a document known as BS EN 60065:2002 'Audio, video and similar electronic apparatus – Safety requirements'. The BS EN classification means it is a European standard (EN) having the force of a British Standard (BS) The latest edition was published in May 2002 It is produced by CENELEC, the European Committee for Electrotechnical Standardisation.

In USA, the safety requirements are set by the Underwriter's Laboratories, commonly known simply as 'UL'. The relevant standards document is UL6500 'Audio/Video and Musical Instrument Apparatus for Household, Commercial, and Similar General Use', ISBN 0-7629-0412-7. The name 'Underwriter's Laboratories' indicates that this institution had its start in the insurance business, allegedly because American houses tend to be wood-framed and are therefore more combustible than their brick counterparts.

The requirements for Asian countries are essentially the same, but it is essential to decide at the start which countries your product will be sold in, so that all the necessary approvals can be obtained at the same time. Changing your mind on this, so things have to be re-tested, is very expensive.

Electrical Safety

This is safety against elecrical shocks. There must be no 'hazard live' parts accessible on the outside of the unit, and precautions must be taken in the internal construction so that parts do not become live due to a fault.

A part is defined as 'hazardous live' if under normal operating conditions it is at 35 V ac peak or 60 V DC with respect to earth. Under fault conditions 70 V AC peak or 120 V DC is permitted. Professional equipment, defined as that not sold to the general public, is permitted 120 V rms; there are also special provision for audio signals. You are strongly advised to consult p. 50 of BS EN 60065:2002 for more detailed information.

Mains connections are always well insulated and protected where they enter the unit, normally by IEC socket or a captive lead, so the likeliest place where such voltages may appear is on the loudspeaker terminals of a power amplifier. An amplifier capable of 80 W into 8 Ω will have 35 V AC peak on the output terminals when at full power.

This seems like a tricky situation, but the current interpretation seems to be that the contacts of loudspeaker terminals, if they are inaccessible when they are fastened down, may be 'hazardous live', provided they are marked with the lightning symbol on the adjacent panel. Strictly speaking one should consider the operation of connecting speaker cables to be 'by hand' and therefore the contacts should be inaccessible at all times i.e., closed or open. However, the general view seems to be that the connection of speaker terminals is a rare event and that adequate user instructions will be sufficient for the 'by hand' clause to be disregarded. The instructions would be of the form: 'Hazardous live voltages may be present on the contacts of the loudspeaker terminals . . . before connecting speaker cables disconnect the amplifier from the mains supply . . . if in doubt consult a qualified electrician' I would remind readers at this point that such an interpretation appears to be the current status quo, but things can change and it is their responsibility to ensure that their equipment complies with the regulations.

Loudspeaker terminals that can accept a 4 mm banana socket from the front have been outlawed for some time. Existing parts can be legally used if an insulating bung is used to block the hole.

In the internal construction, two of the most important requirements to be observed are known as 'Creepage & Clearance'.

Creepage is the distance between two conductors along the surface of an insulating material. This is set to provide protection against surface contamination which might be sufficiently conductive to create a hazard. While the provisions of BS EN 60065:2002 are complex, taking into account the degree of atmospheric pollution and the insulating material involved, the usual distances used are as follows:

Creepage distances between conductors

Conductors	Creepage Distance (mm)
Live to Earth	3
Neutral to Earth	3
Live to Neutral	6
Live to low-voltage circuitry	6

More information can be found on p. 74 of BS EN 60065:2002.

Clearance is the air gap between two conductors, set to prevent any possibility of arcing; obviously the spacing between live conductors and earthed metalwork is the most important. The minimum air spacing is 2 mm More information can be found on p. 70 of BS EN 60065:2002.

Live cables must be fixed so that they cannot become disconnected, and then move about creating a hazard. This is important where a cable is connected directly into a PCB. If the solder joint to the PCB breaks, they must still be restrained. The two most common ways are:

1 Fixing the cable to an adajacent cable with a cable tie or similar restraint. The tie must be close enough to the PCB to prevent the detached cable moving far enough to cause a hazard. Obviously there is an assumption here that two solder joints will not fail at the same time. See Figure 17.1.
2 Passing the cable through a plain hole in the PCB, and then bending it round through 180° to meet the pad and solder joint, as shown in Figure 17.2. This is often called 'hooking' or 'looping'.

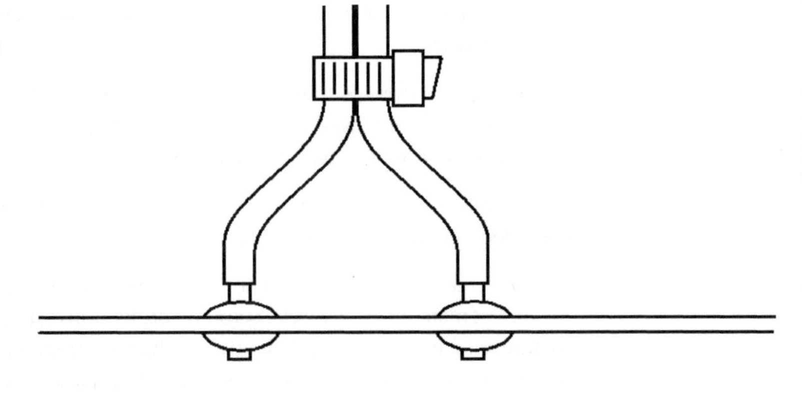

Figure 17.1
Cable restraint by fixing it to an adjacent cable

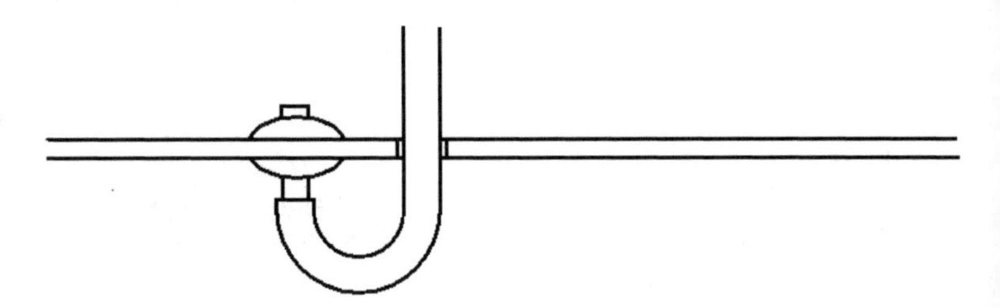

Figure 17.2
Cable restraint by hooking the cable through the PCB

Touch current

As mentioned in the above section on Class I and Class II equipment, the amount of current that can flow to ground via a human being when they touch the casework is an important issue. A Class I (grounded) piece of equipment in normal use should have no touch-current at all, as even a tenuous metallic connection to ground (and hopefully it is not tenuous) will have a negligible resistance compared with the body and no current will flow. For this reason Class I equipment is tested for touch current with the protective earthing connection disconnected.

Class II equipment has no ground connection, and the primary-to-secondary capacitance of the mains transformer can allow enough current to flow through to the casework for it to be perceptible in normal use. Clearly, if the current was big enough it would be hazardous.

Touch current is measured using a special network that connects the equipment to ground via resistors and capacitors, and expressed in terms of the voltage that results; this is then compared with the voltages that make a part 'hazardous live'. The special network is defined in Annexe C of BS EN 60065:2002.

Here are a few more miscellaneous safety requirements, not necessarily enshrined in BS EN 60065:2002.

Mains fuse ratings must be permanently marked, and a legend of the form 'WARNING: replace with rated fuse only' must be marked on the PCB.

Internal wiring does not have to be colour-coded (e.g., brown for live, blue for neutral) except for ground wiring, which must be green with a yellow trace.

Crimp terminals on mains switches do not require color-coding of their plastic shrouds.

It is essential to keep an eye on mains transformer construction. With increasing globalisation, transformers are now being made in parts of the world which do not have a long history of technological manufacturing, and mistakes are sometimes made, for example, not using adequate insulation between primary and secondary.

Case openings

As remarked elsewhere, in the section on mechanical design, case openings are subject to strict dimensional limits. The old 'gold-chain' test has been removed from the latest edition of the standard, and is replaced by a narrow rigid test probe.

Equipment temperature and safety

There are limits on the permissible temperature rise of electronic apparatus, with the simple motivation of preventing people from burning themselves on their cherished hi-fi equipment. The temperature allowed is quoted as a rise above ambient temperature under specified test conditions. These conditions are detailed below. There are two regimes of ambient considered; 'Moderate Climate' where the maximum ambient temperature does not exceed 35°C and 'Tropical Climate' where the maximum ambient temperature does not exceed 45°C. In the Tropical regime, the permitted temperature rises are reduced by 10°C.

The permitted temperature rise also depends on the material of which the relevant part is made. This is because metal at a high temperature causes much more severe burns than non-metallic or insulating material, as its higher thermal conductivity allows more heat to flow into the tissue of the questing finger.

The external parts of a piece of equipment are divided into three categories:

1 Accessible, and likely to be touched often.

   This includes parts which are specifically intended to be touched, such as control knobs and lifting handles.

Metallic, normal operation:	Temperature rise 30°C above ambient
Metallic, fault condition:	Temperature rise 65°C above ambient
Non-metallic, normal operation:	Temperature rise 50°C above ambient
Non-metallic, fault condition:	Temperature rise 65°C above ambient

This is usually an easy condition to meet, as knobs and switches are only connected to the internals of the amplifier via a shaft and a component such as a potentiometer or rotary switch that does not have good heat-conducting paths. Handles can be more difficult as they are likely to be secured to the front panel through a substantial area of metal, in order to have the requisite strength.

2 Accessible, and unlikely to be touched often.

This embraces the front, top and sides of the equipment enclosure.

Metallic, normal operation:	Temperature rise 40°C above ambient
Metallic, fault condition:	Temperature rise 65°C above ambient
Non-metallic, normal operation:	Temperature rise 60°C above ambient
Non-metallic, fault condition:	Temperature rise 65°C above ambient

This is the part of the temperature regulations that usually causes the most grief. To work effectively internal heatsinks have vents in the top panel above them, allowing convective heatflow. The escaping air heats the top panel and this can get very hot. Some amplifier designs have a plastic grille over the heatsink. This has several advantages. Since plastic is more economical to form than metal, the grille can have a structure that is more open and gives a larger exit area, while stil complying with the 3 mm width limit for apertures. The grille itself is also allowed to get 20°C hotter because it is non-metallic, and for the same reason it conducts less heat to the surrounding metal top panel.

3 Not likely to be touched.

This includes rear and bottom panels, unless they carry switches or other controls which are likely to be touched in normal use, external heatsinks and heatsink covers, and any parts of the top enclosure surface that are more than 30 mm below the general level.

Normal operating conditions: Temperature rise 65°C above ambient

The permitted temperature under fault conditions is not specified, but it is probably safe to assume that a rise of 65°C is applicable.

The bottom panel is not likely to get very hot unless heatsinks are directly mounted on it, as it gets the full benefit of the incoming cool air. The rear panel can be a problem as its upper section will be heated by convection, and is typically at much the same temperature as the top of the unit; it also often carries a mains switch, which takes it out of this category.

The test conditions under which these temperatures are measured are as follows:

One-eighth of the rated output power into the rated load
All channels driven and with rated load attached.
The signal source is pink noise which is passed through an IEC filter to define the bandwidth to about 30 Hz–20 kHz. The details of the filter are given in Annexe C of BS EN 60065:2002.
The mains voltage applied is 10% above the nominal mains voltage, so in Europe it is 230 V + 23 V = 253 V.
More information on the test conditions can be found on p. 24–27 of BS EN 60065:2002.

The introduction of temperature rise regulations caused external heatsinks to become a rarity, despite the recognition that heatsinks are rarely going to be touched in normal operation. It is usually much more cost-effective to have the heatsinks completely enclosed by the casework, with suitable vents at top and bottom to allow convection. The heatsinks can then be run much hotter, so they can be smaller, cheaper and lighter, obviously assuming that the semiconductor temperature limits are observed; the limit for power transistors is usually 150°C, and for rectifiers 200°C. This usually allows the heatsinks to be safely run at 90°C or more, depending on the details of transistor mounting and the amount of power dissipated by each device. Hot heat sinks are more effective at dissipating heat by convection, but on the downside the restriction caused by the top and bottom vents, which must be of limited width, impairs the rate of airflow.

An exception to this is the use of massive heatsinks to form part of the case, to make an aesthetic statement. In this case the heatsinks are likely to be much larger for structural reasons than required for heat dissipation, and meeting the temperature-rise requirements is easy. Since aluminium extrusions are relatively expensive, this approach is restricted to 'high-end' equipment.

Instruction manuals

The instruction manual is very often written in a hurry at the end of a design project. However, it must not be overlooked that it is part of the product package, and must be submitted for examination when the equipment itself is submitted for safety testing. There are rules about its contents; certain safety instructions are compulsory, such as warnings about keeping water away from the equipment.

Index